Physiology in Health and Disease

Published on behalf of The American Physiological Society by Springer

Physiology in Health and Disease

This book series is published on behalf of the American Physiological Society (APS) by Springer. Access to APS books published with Springer is free to APS members.
APS publishes three book series in partnership with Springer: *Physiology in Health and Disease* (formerly *Clinical Physiology*), *Methods in Physiology*, and *Perspectives in Physiology* (formerly *People and Ideas*), as well as general titles.

More information about this series at http://www.springer.com/series/11780

Jun Sun • Pradeep K. Dudeja
Editors

Mechanisms Underlying Host-Microbiome Interactions in Pathophysiology of Human Diseases

Springer

aps
American Physiological Society

Editors
Jun Sun
Division of Gastroenterology
and Hepatology, Department of Medicine
University of Illinois at Chicago
Chicago, IL, USA

Pradeep K. Dudeja
Division of Gastroenterology
and Hepatology, Department of Medicine
University of Illinois at Chicago
Chicago, IL, USA

Jesse Brown VA Medical Center
Medical Research Service (600/151)
Chicago, IL, USA

Physiology in Health and Disease
ISBN 978-1-4939-7533-4 ISBN 978-1-4939-7534-1 (eBook)
https://doi.org/10.1007/978-1-4939-7534-1

Library of Congress Control Number: 2017963345

Printed on acid-free paper

This Springer imprint is published by Springer Nature
The registered company is Springer Science+Business Media, LLC
The registered company address is: 233 Spring Street, New York, NY 10013, U.S.A.

I dedicate this book to my family: my father Zong-Xiang Sun, mother Xiao-Yun Fu, husband Yinglin Xia, and sons Yuxuan and Jason. I want to thank them for their love, understanding, and support. —Jun Sun
谨以此书献给我的父亲孙宗祥母亲付小云。—孙**俊**

I dedicate this book to my late parents, Mr. Lachman Dass Dudeja and Mrs. Ved Rani Dudeja, for their constant love and support throughout my life. I also dedicate this book to my wife Renu and my son Amish and daughter Akanksha for their constant love and support. —Pradeep Dudeja

Preface

Only recently has the biomedical community begun to appreciate the roles of microbiome in health and diseases. Some scientists are still skeptical about the link between the gut microbiome and various diseases pertaining to other organs beyond the intestine. In April 2016, we organized a symposium entitled "Mechanisms Underlying Host–Microbial Interactions in the Pathophysiology of Diseases" for the Experimental Biology meeting. The symposium was well attended, even when it was scheduled to start at 8:00 in the morning on the last day of the meeting. We were very encouraged by the scientific content presented by speakers, the active Q&A section, and the enthusiasm of the audience standing at the back of the conference room when we ran out of seats. The American Physiological Society (APS) noted this enthusiasm of the audience with great interest in this symposium on the gut microbiome. Dr. Sun was contacted by Dr. Dee Silverthorn, Chair of the APS Book Committee. She thought that expanding our topic into an APS e-book would be an effective way of reaching more scientists around the world than just those who attended the meeting. Right after the EB, we submitted a book proposal to the APS and started to consider the possibility of creating an e-book of our symposium. We were so glad that the book proposal was supported by the committee members and well-received by the peer review. They were pleased to see something on the emerging subject, and believed that "the microbiome book is very timely, important and of wide interest and the table of contents appeared to be well thought out and should attract a broader community of readers."

In the summer of 2016, we started to invite authors to contribute to the book. The original theme of the EB Symposium focused on the gut microbiome and intestinal diseases. Over the past year, we were able to further develop the chosen topics in the book. In the current book, we have not only included chapters on the role of intestinal bacterial communities in various diseases, but have also included the microbiome from some other organs, such as the oral and lung microbiomes. As the concept of the microbiome includes viruses and fungi, we have, therefore, included chapters covering progress on commensal fungi and virus.

Chicago, IL Jun Sun
Chicago, IL Pradeep K. Dudeja

Acknowledgement

We would like to thank the American Physiology Society for supporting our book proposal and the research supported by the NIH, DOD, and the Department of Veterans Affairs. We thank the dedicated efforts of all our authors, whose excitement about this project is contagious. We also want to thank the diligent and generous support of the reviewers who have contributed to the peer-review process of the book chapters. Thanks to all involved in putting this book together. We hope that readers enjoy the book and send us feedback.

Introduction

The microbiome is the collection of microbes or microorganisms that inhabit an environment, creating a sort of "mini-ecosystem." Our human microbiome is made up of communities of commensal, symbiotic, and pathogenic bacteria, fungi, and viruses. We can consider the human microbiome a newly discovered organ that interacts with other organs and influences the development of diseases. This so-called "microbiome organ" weighs over 1 kg, equivalent to the weight of the human heart or liver. Although it has no distinct structure, the organized system of cells is more akin to the immune system than the liver. The human gut microbiome is dominated by four large groups of bacteria or phyla: Firmicutes, Bacteroidetes, Proteobacteria, and Actinobacteria. Firmicutes and Bacteroidetes are generally the most abundant in the gut microbiota, followed by Proteobacteria and Actinobacteria. The basic functions of the microbiome, the invisible organ, include (1) gleaning indigestible ingredients and synthesizing nutritional factors (e.g., vitamins); (2) producing anti-microbial products that negatively affect pathogenic bacteria through the development of colonization resistance; (3) developing a systemic and intestinal immune system; (4) providing signals for epithelial renewal and maintaining barrier functions; and (5) detoxifying xenobiotics and affecting the host metabotypes.

The complex microbial communities that inhabit most external human surfaces play a critical role in health and diseases. Perturbations of host–microbe interactions can lead to altered host responses that increase the risk of pathogenic processes and promote disorders. It is only recently that we have begun to appreciate the role of the microbiome in health and diseases. Environmental factors and a change of life style, including diet, significantly shape the human microbiome, which in turn appears to modify gut barrier function, affecting nutrient, electrolyte and fluid absorption and triggering inflammation. The functions of the microbiome are vital, because in the absence of the microbiota or in the event of its ablation with long-term broad-spectrum antibiotics, there can be significant consequences, e.g., improper development of the immune system, barrier integrity, metabolic disturbances, and the development of *C. difficile* antibiotics-associated colitis.

Dysbiosis is an imbalance in the structural and/or functional properties of the gut microbiota. Dysbiosis can disrupt host–microbe homeostasis and be involved in various human diseases beyond the digestive system. Three notable areas are: (1) obesity, diabetes, and metabolic syndrome; (2) cardiovascular and renal diseases; and (3) stress/anxiety (gut–brain axis) including irritable bowel syndrome (IBS), autism, and Parkinson's disease. Approaches that can reverse the dysbiosis are represented as reasonable and novel strategies for restoring the balance between host and microbes.

In the current book, we offer a summary and discussion of the advances in our understanding of the pathophysiological mechanisms of microbial–host interactions in human diseases, including necrotizing enterocolitis (NEC), viral infectious diseases, diarrheal diseases, obesity, inflammatory bowel diseases (IBDs), Irritable bowel syndrome (IBS), allergic disorders, and cancers. We discuss not only bacterial community, but also viruses and fungi. In addition to the intestinal microbiome, we have chapters on the microbiome in other organs. For example, a review of the oral microbiome and its potential link to systemic diseases and cancer is included, in addition to a chapter on the lung microbiome.

Microbial colonization plays a significant role in the normal postnatal development of the intestine and other organs. Early-life exposure to microbes decreases the risk of developing allergic disease. Also, exposure to a protected modern lifestyle environment may lead to decreased allergen exposure, potentially creating an immune system that is intolerant to allergens. In particular, Humphrey and Claud focus on the topic of the role of microbiome in intestinal development and outline ways in which poor clinical outcomes in the preterm infant, such as NEC, are related to gut dysbiosis. The benefits of the microbiome are not seen in preterm infants, who experience delayed and altered microbial community colonization after birth. In combination with the reduced intestinal functions in the preterm, dysbiosis can further damage existing intestinal functions and exacerbate the hyperreactive inflammatory state. Perkins and Finn summarize the roles of microbiome from the intestine, skin, and lung in the development of allergic diseases of childhood. They review four of these: food allergy, atopic dermatitis, asthma, and allergic rhinitis. The allergic diseases are related to each other in that having one of these diseases early in life increases the risk of acquiring another allergic disease at a later age.

A growing number of scientists are investigating the role of the microbiome in the development of and protection from disease. One area of particular interest is recovery from infection and injury. Lei et al. outline the pathogenesis, immunity, and role of microbiome/probiotics in enteric virus infections. Liu and Sun update the current understanding of pathogenic *Salmonella* infection, inflammatory response of the host, and anti-inflammatory and apoptotic death mechanisms in infection and cancer. The established experimental models (e.g., organoids, the chronic infected mouse model, and the infected colon cancer model) can be applied to the investigation of other bacteria and their interactions with hosts. Kumar et al. present an overview of the evidence-based effects of probiotics in diarrheal diseases, in addition to a detailed overview of the mechanisms of action of probiotics.

Multiple anti-inflammatory activities can be mediated via various pathways in mammalian cells. This is exemplified by the probiotics story. Thus, probiotics may serve as the paradigm for the multiplicity of the sometimes seemingly contradictory activities of this group of anti-inflammatory agents. Taken together, insights into the anti-inflammatory mechanisms of the bacterial proteins and probiotics should provide promising opportunities for therapeutic intervention.

Healthy microbial–host interactions enhance motility, digestion, and absorption. They also strengthen barrier function and immune homeostasis. The chapter by Raja et al. summarizes the critical roles played by the microbiota in gastrointestinal (GI) motility. They describe the influence of the microbiota in shaping the enteric nervous system. Next, they discuss how microbial metabolites can regulate intestinal motility. Finally, they demonstrate how dysbiosis can lead to motility disorders (e.g., IBS and colonic pseudo-obstruction). Yeoh and Vijay-Kumar discuss altered microbiotas and their metabolism in host metabolic diseases. This chapter examines several key concepts and potential mechanisms that underscore the link between the gut microbiome and metabolic diseases, and provide examples of the extent to which specific bacteria and/or their metabolites affect host metabolism.

It is clear that microbes in the colon, and perhaps in the small intestine, are significant players in the development of colon cancer. Kordahi and DePaolo review the influence of the microbiota on the etiology of colorectal cancer (CRC). They explore the conceptual frameworks through which certain members of the microbiota are believed to cause CRC, and toll-like receptors (TLRs). They discuss the various strategies aimed at manipulating the microbiota and targeting the TLRs in developing new treatment approaches.

Dysbiosis can disrupt host–microbe homeostasis and be involved in various human diseases beyond the digestive system. Vasquez et al. discuss the roles of the oral microbiome, especially the potential link to systemic diseases including cancer. Perkins and Finn focus on the microbiome at different body sites (gut, skin, and lung) that promote resilience or susceptibility to allergic diseases and describe the potential in the inflammatory process of allergic disorders.

Because the human microbiome is made up of communities of bacteria, fungi and viruses, Chen and Huang outline the research progress of fungi *Candida albicans* commensalism and human diseases. They evaluate the roles of *Candida albicans* in specific host niches, including the oral cavity, reproductive tract, and GI tract.

Microbiome studies are likely to facilitate diagnosis, functional studies, drug development, and personalized medicine. It requires a multi-disciplinary team effort, involving basic, translational, and clinical investigators. Further, we discuss the current knowledge and future directions of probiotics and fetal microbiome transplantation (FMT) in various diseases. The chapter by Chis et al. takes us through the key aspects of FMT, including methodology, physician and patient attitudes, safety and regulation, and its therapeutic potential for the treatment of *Clostridium difficile* infection and other GI conditions, including IBD, obesity, IBS, and CRC.

The next phase of research investigation of the gut microbiome should be guided by specific biological questions relevant to the clinical aspects and natural history of the disease, utilizing the full spectrum of "omic" technologies, bioinformatic analysis, and experimental models. To emphasize the significant roles of bioinformatic and biostatistical methods in gut microbiome studies, we also include a chapter by Xia and Sun focusing on statistical models and analysis of microbiome data.

Taken together, our book highlights the microbiome in the context of health and disease, focusing on mechanistic concepts that underlie the complex relationships between host and microbes.

Chicago, IL Jun Sun
Chicago, IL Pradeep Dudeja

Contents

1 **Impact of Microbes on the Intestinal Development of the Preterm Infant** 1
Elizabeth Humphrey and Erika Claud

2 **Microbiome: Allergic Diseases of Childhood** 35
Ozge Nur Aktas, Benjamin Turturice, David L. Perkins, and Patricia W. Finn

3 **Pathogenesis, Immunity and the Role of Microbiome/Probiotics in Enteric Virus Infections in Humans and Animal Models** 55
Shaohua Lei, Erica Twitchell, and Lijuan Yuan

4 **Enteric Bacterial Regulation of the Wnt/β-Catenin Signaling** 79
Xingyin Liu and Jun Sun

5 **Mechanisms Underlying the Beneficial Role of Probiotics in Diarrheal Diseases: Host–Microbe Interactions** 89
Anoop Kumar, Pradhan Bhat, Alip Borthakur, Waddah A. Alrefai, and Pradeep K. Dudeja

6 **The Influence of Microbiota on Gastrointestinal Motility** 113
Shreya Raja, Vivek Batra, and Shanthi Srinivasan

7 **Altered Microbiota and Their Metabolism in Host Metabolic Diseases** 129
Beng San Yeoh and Matam Vijay-Kumar

8 **The Influence of the Microbiota on the Etiology of Colorectal Cancer** 167
Melissa C. Kordahi and R. William DePaolo

9 Oral Microbiome: Potential Link to Systemic Diseases and Oral Cancer 195
Adrian A. Vasquez, Jeffrey L. Ram, Milad S. Qazazi, Jun Sun, and Ikuko Kato

10 *Candida albicans* Commensalism and Human Diseases 247
Changbin Chen and Xinhua Huang

11 Fecal Microbiota Transplants: Current Knowledge and Future Directions 279
Roxana Chis, Prameet M. Sheth, and Elaine O. Petrof

12 Statistical Models and Analysis of Microbiome Data from Mice and Humans 303
Yinglin Xia and Jun Sun

Index 373

About the Editors

Jun Sun is a tenured Associate Professor at the University of Illinois at Chicago, Chicago, USA. She is a Fellow of the American Gastroenterological Association. Her research interests are host–microbiome interactions in inflammation and cancer, and her key achievements include: (1) characterization of the role of vitamin D receptor in the regulation of the gut microbiome in intestinal homeostasis and inflammation; (2) identification of dysbiosis and gut dysfunction in amyotrophic lateral sclerosis (ALS); (3) identification of the role of gut bacteria in regulating intestinal stem cells, and (4) identification and characterization of the *Salmonella* effector protein AvrA in host–bacterial interactions in infection and cancer. Dr. Sun has published over 140 scientific articles in peer-reviewed journals, including *Cell Stem Cells*, *Nature Genetics*, *Gut*, *JBC*, *American Journal of Pathology*, and *American Journal of Physiology-GI*. She is on the editorial board of more than 10 peer-reviewed, international scientific journals. She serves on the National Institutes of Health (NIH), the American Cancer Society, and other national and international research foundations. Her research is supported by the NIH, the Department of Defense (DOD), and other industry-sponsored awards. Using several models, including transgenic mice, germ-free animals, and human samples, her laboratory is currently pursuing the following research topics: (1) novel roles of the gut microbiome in colon cancer, inflammatory bowel diseases (IBDs), infectious diseases, obesity, ALS, and other human diseases; (2) bacterial regulation of vitamin D/vitamin D receptor in inflammation and cancer; and (3) bacterial regulation of intestinal stem cells.

Pradeep K. Dudeja is a Professor of Physiology in Medicine at the University of Illinois at Chicago and a Senior Research Career Scientist at the Jesse Brown VA Medical Center, Chicago, IL, USA. His research primarily focuses on an understanding of the pathophysiology of diarrheal diseases and on developing better therapeutic interventions. His focus has been on host–microbe interactions with regard to the mechanisms underlying infectious diarrhea as it relates to infections by a food-borne pathogen enteropathogenic *E. coli* and diarrhea elicited by *C. difficile* infection. He is also investigating the mechanisms underlying the antidiarrheal role of probiotics. Another focus of his group has been to understand the mechanisms of the absorption of key bacterial metabolites: short chain fatty acids and their role in intestinal health in general and the implications for gut fluid absorption and gut inflammation. Dr. Dudeja has published over 200 original peer-reviewed articles in journals including *Gastroenterology*, *Journal of Clinical Investigation*, *Journal of Inflammatory Bowel Diseases* and *American Journal of Physiology*. He serves as an Editor for *Comprehensive Physiology* and on the editorial board of many journals including *Gastroenterology*, *AJP-GI-Liver*, *Digestive Diseases & Sciences*, and *Physiological Reports*. He has been serving on many grant review committees, including the NIH and the Department of Veterans Affairs.

Chapter 1
Impact of Microbes on the Intestinal Development of the Preterm Infant

Elizabeth Humphrey and Erika Claud

Abstract The preterm intestine is not ready for life outside the womb because of its impaired digestive, absorptive, and motility capabilities. Intestinal barrier function is inadequate and enterocyte contributions to innate immunity are hyper-responsive, predisposing the infant to inflammatory disease and sepsis. Microbial colonization plays a significant role in normal postnatal development of the intestine. Microbial–host interactions can enhance motility, digestion, and absorption, in addition to strengthening barrier function and encouraging immune homeostasis. These benefits are not seen in preterm infants who experience delayed and altered microbial community colonization after birth, termed dysbiosis. In combination with the reduced gut functions in the preterm infant, dysbiosis can further damage existing gut functions and exacerbate the hyper-reactive inflammatory state, which increases the risk for inflammatory diseases such as necrotizing enterocolitis (NEC). This chapter details the role of the microbiome in intestinal maturation and outlines ways in which poor clinical outcomes in the preterm infant, such as NEC, could be circumvented through clinical interventions that optimize the microbiome community.

List of Abbreviations

BB	Brush border
BL	Basolateral
BM	Basement membrane
CI	Confidence interval
EEC	Enteroendocrine cell
EN	Enteric nutrition
ENS	Enteric nervous system

E. Humphrey • E. Claud (✉)
University of Chicago, Chicago, IL, USA
e-mail: elizabeth.humphrey@uchospitals.edu; eclaud@peds.bsd.uchicago.edu

J. Sun, P.K. Dudeja (eds.), *Mechanisms Underlying Host-Microbiome Interactions in Pathophysiology of Human Diseases*, Physiology in Health and Disease,
https://doi.org/10.1007/978-1-4939-7534-1_1

GALT	Gut-associated lymphoid tissue
GC	Goblet cell
GF	Germ-free
HMOs	Human milk oligosaccharides
IEC	Intestinal epithelial cell
LPS	Lipopolysaccharide
MAMPs	Microbial-associated molecular patterns
NEC	Necrotizing enterocolitis
NICU	Neonatal intensive care unit
NLR	NOD-like receptor
PC	Paneth cell
PN	Parenteral nutrition
RR	Relative risk
TJs	Tight junctions
TLR	Toll-like receptor

1.1 Intestinal Development and the Preterm Infant

Human intestinal development occurs in three overlapping stages that begin in utero and continue after birth: morphogenesis and cell proliferation, cell differentiation, and functional maturation (Colony 1983). Highlights of morphogenesis during gestation are summarized in Table 1.1.

1.1.1 Morphogenesis and Differentiation

Formation of the primitive gut tube begins during gastrulation at week 3 of gestation, with the gut tube largely closed by week 4 (Montgomery et al. 1999). At this point, the intestine consists of endoderm surrounded by a layer of mesenchyme. As early as 8 weeks, villi and microvilli begin to form in a cranial–caudal direction as the subendodermal mesenchyme forms fingerlike projections into the central lumen (Lebenthal and Lebenthal 1999). Simultaneously, the endoderm covering the villi transitions to a columnar epithelium. The resulting polarized enterocytes have an apical surface with a brush border (BB) membrane and a basolateral (BL) surface with an underlying basement membrane (BM).

Following villus development, intestinal crypts form and are initially lined by undifferentiated columnar cells. Proliferation of stem cells populating the crypts gives rise to the four major epithelial cell lines of the intestine (Montgomery et al. 1999). Most small intestinal epithelial cells (IECs) are columnar absorptive enterocytes. Goblet cells (GCs) produce the mucin coating of the intestine and trefoil factors that strengthen the mucus coating. Enteroendocrine cells (EECs) export

Table 1.1 General landmarks during gestation (Montgomery et al. 1999)

Developmental landmark	Gestational week
Gastrulation begins	Week 3
Gut tube closes	Week 4
Villus formation begins	Week 8
Crypts develop	Week 10
Morphogenesis complete	Week 20
Functional maturation	Week 20 and beyond

peptide hormones. Paneth cells (PCs) produce antimicrobial substances called cryptdins, in addition to various growth factors and digestive enzymes (Lebenthal and Lebenthal 1999). Between 17 and 20 weeks, the muscularis mucosae starts to develop near the base of the crypts, and M cells, another specialized cell population, are also detectable overlaying Peyer's patches at 17 weeks (Moxey and Trier 1978).

By 20 weeks, the fetal intestine has undergone both morphogenesis and cell differentiation and resembles the adult intestine with the exception of the fetal colon, which retains absorptive enterocytes, digestive enzymes, and villi until later in gestation (Lacroix et al. 1984). Although the fetal intestine is grossly similar to the adult intestine at this point, functional maturation and growth are still incomplete.

1.1.2 Functional Maturation

The human intestine carries out a wide range of functions. Although some functions reach adult levels in utero, many continue to develop through childhood to reach adult levels. The maturation reached by term is, however, sufficient for normal infant needs. Figure 1.1 shows a timeline of major developmental events. A preterm infant delivered as early as 22 weeks' gestation must accomplish much of the functional maturation ex utero. Many possible deficiencies at preterm may contribute to inflammatory conditions such as necrotizing enterocolitis (NEC).

1.1.2.1 Motility

The nervous system in the gut is collectively called the enteric nervous system (ENS) and is located in two plexuses: the myenteric plexus, between the inner and outer smooth muscle layer, and the submucosal plexus. The ENS generates intrinsic motility patterns that churn luminal contents and propel them through the entire gastrointestinal tract. As early as 15 weeks, the swallowing reflex is initiated in the esophagus and the GI tract begins to contain a small amount of amniotic fluid. After this point, intestinal motility patterns appear. There are four developmental stages of motility: disorganized motility from 25 to 30 weeks, the fetal complex from 30 to

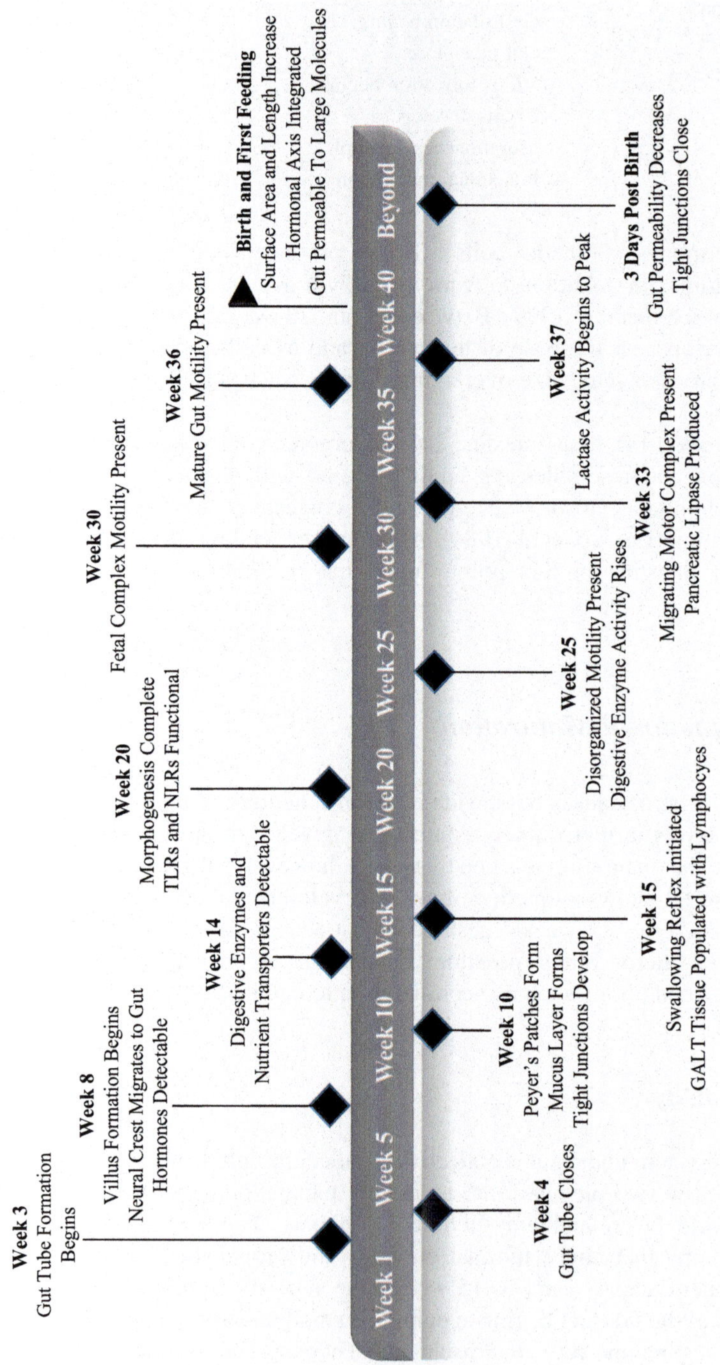

Fig. 1.1 Functional maturation landmarks during gestation (for sources, see the text)

33 weeks, propagation of the migrating motor complex from 33 to 36 weeks, and mature motility from 36 weeks onward (Dumont and Rudolph 1994). As the fetal intestine progresses through these stages, the waves of coordinated contraction increase in amplitude and number while becoming more coordinated and propulsive. Both the duration of the quiescent period between bursts of activity and the amount of cluster activity increases.

In healthy and normal term infants, intestinal motility is mature, although the ENS continues to develop as the infant grows (Burns et al. 2009). In contrast, a preterm infant may be able to swallow, but display immature motility patterns. This severely impairs its ability to tolerate enteral feeding. Without propulsive motility, nutrients remain in the intestine longer. These nutrients sitting in the lumen can attract neutrophils, cause bacterial overgrowth, and ultimately trigger inflammatory reactions (Clark and Miller 1990) such as NEC.

1.1.2.2 Normal Digestion and Absorption

Intestinal digestive and absorptive capabilities are sufficient for the normal infant's needs at 40 weeks, if fewer than full adult capacity. This allows for the digestion and absorption of key nutrients in colostrum and breast milk.

Proteins

Proteins are digested in the stomach by pepsin and further cleaved in the small intestine by BB peptidases and pancreatic enzymes. Absorption of the resulting small peptides occurs by co-transport with hydrogen ions into the cytoplasm, where they are hydrolyzed and exported to the blood stream (Adibi et al. 1975). Individual amino acids are transported via Na^+ dependent transporters in the apical membrane and contribute to the overall osmotic gradient driving H_2O absorption in the intestine.

Overall, the ability to breakdown proteins is still developing at term. Gastric acid, pepsin, and intestinal and pancreatic proteolytic enzymes do not peak until at least 3 months after birth (Agunod et al. 1969; Antonowicz and Lebenthal 1977; Lebenthal and Lee 1980; Kelly et al. 1993). Amino acid and peptide transporters are also present at low levels at birth, but drastically increase in number after birth (Malo 1991). However, newborn intestinal permeability is significantly increased compared with that of the adult (Lebenthal et al. 1981). Thus, the infant, though still developing the ability to break down and transport proteins, can absorb intact proteins passing through the intestine during the first days after birth. This higher gut permeability allows the absorption of intact antibodies from colostrum, but declines after the first few days of life, when peptide digestion and transport are still developing.

Carbohydrates

Carbohydrate digestion begins with the breakdown of starches by salivary and pancreatic amylases. The disaccharides created by amylase action are further digested by BB hydrolases, including sucrase-isomaltase and lactase. Once broken down, the resulting glucose monomers are transported across the BB and BL membranes by active transport.

At term, the intestine has the enzymatic capability to digest the important carbohydrates in the infant diet, largely breast milk (Raul et al. 1988; Triadou and Zweibaum 1985; Dewit et al. 1990). It is also capable of absorbing the monosaccharides generated by digestion. As soon as the epithelium begins to differentiate around 8 weeks, sugar transport proteins, including sodium-dependent glucose transporter 1 (glucose), are expressed in the BB membranes of enterocytes (Buddington and Malo 1996). The BL membrane glucose transporter 2 is also present as the epithelium differentiates (Davidson et al. 1992). Both BB and BL transporters greatly increase in number in the last weeks of gestation before birth.

Lipids

Lipid digestion in the intestine requires the action of bile acids and various lipases to yield fatty acids and monoglycerides. As bile acid production does not fully develop until after weaning, lipases do the bulk of the work digesting lipids in the infant intestine. Pancreatic lipase is active at 32 weeks, but remains low at birth, increasing in the 10 weeks after birth (Cleghorn et al. 1988). Lingual and gastric lipases, however, are detected earlier and at birth are able to break down most lipids ingested with the aid of lipases and esterases present in breast milk (Hamosh et al. 1981; Alemi et al. 1981).

Intestinal absorption of lipids occurs in multiple ways. The primary mechanism is the simple diffusion of fatty acids, monoglycerides, and cholesterol across enterocyte membranes (Black et al. 1990). Other pathways include pinocytosis of intact triglycerides and carrier-mediated transport (Berendsen and Blanchette-Mackie 1979; Black et al. 1990). All these systems have been shown to be functional in human infant intestine cells, starting as early as 14 weeks. Lipoproteins, which help carry lipids inside the body, are also produced at that time (Thibault et al. 1992).

Intestinal Size

For the intestine to perform adequate digestion and absorption, it must achieve sufficient size and surface area. Although some of this growth, particularly in length, occurs during gestation, a significant portion of intestinal growth occurs upon the first feeding of colostrum, the protein-rich substance produced by the mother's mammary glands right after birth. Studies of piglets show that after the first feeding of colostrum, the entire GI tract undergoes rapid changes, nearly doubling in weight during the first 3 days of life (Xu et al. 1992). In the small intestine, most of this weight comes from increases in the mucosa. Villus height and width increase to allow a larger surface area for absorption. In the large intestine, both mucosal and nonmucosal structures enlarge. The primitive villi present in the colon during gestation also disappear after 3 days of life. These changes, triggered by the enteral feeding of colostrum, are important for digestion and absorption in early life.

1.1.2.3 Digestion/Absorption in the Preterm Infant

As many of the digestive and absorptive functions do not mature until shortly before or after term birth at 40 weeks, the preterm infant is born ill-prepared to digest enteral nutrition. Low lactase levels are particularly concerning, as the main nutrient in breast milk is lactose. Bile acid levels and lipase levels are decreased as well, which impedes breakdown and absorption of the lipids in breast milk. These deficits, combined with the lack of motility patterns, make enteral feeding of the preterm infant challenging. Thus, parenteral nutrition (PN)—intravenous feeding—is used until the infant can tolerate enteral nutrition (EN). PN, while protective against the effects of static luminal contents, has adverse effects of its own. Short-term complications can include infection, hyperglycemia, electrolyte abnormalities, hypertriglyceridemia, and decreased GI barrier function (Commare and Tappenden 2007). Long-term complications such as cholestasis, osteopenia, and catheter sepsis are also likely. Additionally, the preterm infant receiving PN does not ingest colostrum, which impairs the massive growth of the intestine that normally occurs in the 3 days after birth. This pre-disposes the infant to villous atrophy.

1.1.2.4 Gut–Brain Axis

The ENS communicates with the central nervous system (CNS) via neural and hormonal signals. These signals are released by EECs in the intestine in response to the presence of nutrients in the lumen. These signals communicate with the brain and other parts of the body to affect energy homeostasis, adjusting intake and expenditure. Some important peptides and hormones secreted by the intestine are listed in Table 1.2 (Ter Beek et al. 2008; Bauer et al. 2015).

Most digestive hormones and neuropeptides appear at around the same time that digestive enzymes first appear in the intestine, as early as 8 weeks into gestation. The signaling pathways do not begin working until term, when the infant first encounters large quantities of oral nutrients. For the preterm infant, release of these hormones and full integration of the neural–hormonal axis is delayed until the infant receives EN.

Table 1.2 Intestinal hormones and peptides (Ter Beek et al. 2008; Bauer et al. 2015)

Signal	Location	Signal target site	Effect
Cholecystokinin (CCK)	Proximal small intestine	Hypothalamus	Decreases energy intake
		Pancreas	Secretes pancreatic enzymes
			Secretes bile acids
		Enteric neurons	Decrease gut motility
Glucagon-like peptide (GLP-1)	Distal small intestine/ colon	Hypothalamus	Satiety
			Increases energy expenditure
Peptide YY	Distal small intestine/ colon	Hypothalamus	Decreases energy intake
Motilin	Small intestine	Enteric neurons	Regulate motility patterns

1.1.2.5 Immunity

The adaptive immune system is naïve at birth, as the term infant has not yet developed its own specific immunity. Therefore, passive immunity acquired from the mother and the infant's own innate immune system is the primary defense against pathogens. The mucosal surface of the intestinal tract plays a critical role in acquired passive immunity and innate immune defense. Specialized components of intestinal immunity include the intestinal mucus layer and tight junctions (TJs) between enterocytes. Epithelial enterocytes are also capable of inducing innate immune responses. For the preterm infant, the innate immune system is immature at birth, making it unprepared to encounter life outside the sterile environment of the womb and develop immunity.

Barrier Functions

Intestinal mucus provides a thick, viscous layer of protection against pathogen invasion of the intestinal wall. It has two distinct layers: a dense, adherent inner layer and a loose non-adherent outer layer that is more densely populated by the microbiota than the dense layer (Johansson et al. 2011). It comprises mucins (large glycoproteins), water, ions, secretory IgA (sIgA), and antimicrobial peptides. Mucin-producing GCs and PCs, which secrete a wide variety of antimicrobial peptides, contribute heavily to generating the mucus layer. These cells develop as the intestinal epithelium matures between 9 and 20 weeks' gestation (Poulsen et al. 1996; Kim and Ho 2010). Mucus layers can reach adult levels by 27 weeks' gestation (Buisine et al. 1998).

Tight junctions between enterocytes form a selective barrier between the intestinal lumen and internal tissues. Paracellular movement of water and small molecules is controlled by these structures. They include transmembrane proteins, such as claudins, bound to peripheral scaffolding proteins. These proteins in turn interact with actin and microtubules to allow the passage of water, ions, and molecules between cells in response to various stimuli (Van Itallie and Anderson 2014).

Tight junctions are developed by 10 weeks, but the machinery necessary to regulate the permeability of the junctions is only partially developed at term. Thus, the infant is born with a highly permeable gut. This permeability is crucial for the acquisition of passive immunity from the mother's colostrum, which contains large numbers of sIgA antibodies. The infant is unable to produce sIgA of its own until at least 2 weeks after birth (González-Ariki and Husband 2000). This gut permeability is transient and appears to end after the first 3 days of life in term infants (Vukavic 1983). The gut then seals under the influence of growth factors in colostrum and breast milk, and through interactions with commensal microbes.

Preterm infants born before mucus layer maturation have reduced levels of all components of intestinal mucus compared with their normal term counterparts. Therefore, their intestinal tract is much more vulnerable to pathogen invasion and it is highly likely that their IECs interact with microbes more frequently than would term infant IECs. Additionally, it has been shown that preterm gut permeability is increased for up to 10 days compared with a term infant (Beach et al. 1982; Riezzo et al. 2009). Because preterm infants often do not receive EN, any benefits of increased gut permeability with regard to colostrum and passive immunity are not seen. In fact, this increased gut permeability increases susceptibility to pathogen invasion. Preterm infants also experience alterations in normal, healthy colonization patterns, which affect the maturation of TJs.

Antigen Recognition

Enterocytes participate in innate immunity through recognition of microbial-associated molecular patterns (MAMPs), such as lipopolysaccharide (LPS) and flagellin, via toll-like receptors (TLRs) on the cell surface and NOD-like receptors (NLRs) located in the cytoplasm. These receptors constantly sample the intestinal environment to detect pathogens or other possibly dangerous substances. Binding of MAMPs to either TLRs or NLRs induces the production of cytokines, chemokines, growth factors, and adhesion proteins to generate an immune response. Most of these effector molecules require phosphorylation and activation of NF*K*B, a transcription factor that regulates cell growth, survival, apoptosis, and inflammation (Oeckinghaus and Ghosh 2009).

Although these systems appear functional at 18–21 weeks (Fusunyan et al. 2001), TLR4 binding to LPS in preterm infants results in an inappropriately large immune response, rather than an attenuated response. In fact, studies suggest that the premature neonate intestine might be predisposed to exaggerated innate inflammation in response to antigens. Compared with adult IECs, human fetal IECs produce more

IL-8 in response to both pathogenic and commensal bacteria and inflammatory mediators such as TNFα and IL-1γ (Nanthakumar et al. 2011; Claud et al. 2004). Both in vitro and in vivo studies reveal that immature enterocytes have increased NF-κB activity associated with decreased baseline expression of IκB (Claud et al. 2004). Inflammatory stimuli induce increased degradation of IκBα and prolonged NF-κB activation in immature enterocytes (Claud et al. 2004). Research by Egan et al. showed that increased TLR4 expression and signaling in preterm intestinal cells is specifically linked to increased recruitment and activation of T lymphocytes (2016). Collectively, this evidence suggests that the preterm infant might have an exaggerated, immature immune response compared with term infants.

Adaptive Immunity

In addition to innate immunity, the intestine plays a role in adaptive immunity. Adaptive immunity development is supported by circulating lymphocytes in the lamina propria and the gut-associated lymphoid tissues (GALTs). GALTs secrete sIgA and help the immune system to discriminate between harmful antigens and nonharmful antigens, such as food. The primary GALTs in the intestine are Peyer's patches, lymphoid aggregates beneath the mucosa of the small intestine. Interspersed Peyer's patches can be found at around 10 weeks' gestation (MacDonald and Spencer 1994) and become populated with lymphocytes at around 12–16 weeks (Brugman et al. 2015). By 40 weeks, these tissues are ready to encounter antigens and aid in the acquisition of specific immunity.

After birth, both Peyer's patches and isolated lymphoid follicles continue to develop in the intestine owing to interactions with commensal microbes. Because microbial colonization is delayed in preterm infants receiving PN, GALT development is also significantly delayed, which means that it takes longer for the infant to develop specific immunity and defenses such as sIgA.

1.2 The Healthy Microbiome and Postnatal Development

This chapter has thus far been an overview of intestinal development during gestation and highlighted areas in which preterm infants face unique challenges. (summarized in Table 1.3). Earlier gestational age at birth makes these deficits more likely, particularly infants born before week 25 when motility patterns are absent and enzyme levels are low. They are not ready to receive EN because of impaired digestion, motility, and absorption in the intestine. Their gut–brain axis may be underdeveloped at birth and does not begin working until the infant receives EN. Intestinal barrier function is inadequate, and enterocyte contributions to innate immunity are hyper-responsive, predisposing the infant to inflammatory disease and sepsis. All these problems, specifically immune function, are made worse by

Table 1.3 Possible deficiencies in functional maturation for infants born before term (summarized from the text)

Function	Deficit
Motility	Abnormal or absent motility; stasis
Digestion/absorption	Severe enzyme deficiencies
Gut–brain axis	Delayed neural–hormonal axis integration
Immunity	Decreased mucus production
	Immature tight junctions and increased gut permeability
	Exaggerated immune responses
	Reduced GALT

GALT gut-associated lymphoid tissue

the fact that preterm infants experience delayed and altered microbial colonization after birth. The next section discusses microbial colonization after birth and the role of the microbiome in postnatal development.

1.2.1 Microbial Colonization Patterns in Healthy Term and Preterm Infants

Although current research indicates that the fetus may interact with microbes in the womb (Koleva et al. 2015), most intestinal colonization begins at birth. Both beneficial and potentially harmful microorganisms are capable of colonizing the gut. Healthy communities may include anaerobic species such as *Bifidobacterium* and *Lactobacillus*. *Escherichia coli* and *Bacteroides* species may be beneficial or pathogenic. Microbes such as *Pseudomonas aeruginosa*, *Clostridia* species, and *Staphylococcus* species are opportunistic pathogens (Claud and Walker 2008).

It is important to note that the presence of specific beneficial microbe species does not necessarily indicate a healthy microbiome, but rather, that the diversity and function of the overall microbial community contributes to gut health. Broadly speaking, the healthy, vaginally born, term infant microbiome is dominated by members of the Bacteroidetes phyla (Palmer et al. 2007; Yassour et al. 2016). This community structure appears to have the most beneficial effects on development.

These microbes largely live in the outer, loose aspect of the mucus layer in the intestine. The thickness of intestinal mucus changes throughout the intestinal tract. It is thinner in the proximal small intestine and becomes thicker in the distal small and large intestine. These changes correlate with the local bacterial load in these regions (10^3–10^5 to about 10^{12} organisms per gram content from the duodenum to the colon) (Johannson et al. 2011).

Many factors are known to influence the composition of the infant microbiome (Table 1.4). Initially, the newborn is colonized with microbes of maternal origin. In particular, diet and antibiotic use during pregnancy have been implicated in infant microbial colonization (Neu 2015). The offspring of pregnant, antibiotic-treated

Table 1.4 Factors that affect microbiome composition

Colonization factor	Effect on colonization
Method of delivery	Infants delivered via cesarean section have microbiomes resembling their mothers' skin (Dominguez-Bello et al. 2010)
	Bacteroidetes phyla are more abundant in vaginally born infants (Jakobsson et al. 2014)
Feeding	Breast-fed infants and formula-fed infants have *Bifidobacterium* as a primary organism; formula-fed infants also have Clostridia and *Staphylococcus* (Harmsen et al. 2000)
	IgA from breast milk helps to protect against early pathogen colonization (Rogier et al. 2014)
Maternal factors	Antibiotic use during pregnancy reduces the diversity of the infant microbiome (Tormo-Badia et al. 2014)
	Maternal high fat diet decreases Bacteroidetes organisms in the microbiome (Chu et al. 2016)
Preterm-specific factors	
Mode of nutrition intake	Enteral nutrient deprivation increases Proteobacteria phyla members (Demehri et al. 2013)
Antibiotic use	Antibiotic-treated infants have a microbiome with increased Proteobacteria and decreased Firmicutes members (Claud et al. 2013; Mai et al. 2011)
	The microbiome is rendered less diverse by antibiotic treatment (Greenwood et al. 2014)
Other medications	Infants treated with H2 blockers are at an increased risk for colonization with possibly pathogenic organisms: *Staphylococcus*, *Streptococcus*, and *E. coli* (Terrin et al. 2012)
	Opioids disrupt intestinal motility and allow adherence of pathogenic bacteria (Claud and Walker 2001)

mice show a reduction in the diversity of their gut microbiota (Tormo-Badia et al. 2014). A human longitudinal cohort study of infant–mother pairs showed that a maternal high-fat diet alters the neonatal and infant gut microbiome in early life. Specifically, it resulted in a relative depletion of Bacteroidetes in the neonates exposed to a maternal high-fat gestational diet. This difference was present in samples of fecal meconium and in infant fecal samples up to 6 weeks of age (Chu et al. 2016).

Delivery method greatly influences the microbiome that the infant receives from its mother. A 2010 study of mother–infant pairs found that the primary determinant of a newborn's bacterial community composition was his or her mode of delivery (Dominguez-Bello et al. 2010). The microbiomes of vaginally delivered infants contain bacterial communities that mirror the composition of their mother's vaginal community, whereas babies born via cesarean section lack these vaginal bacterial communities (Dominguez-Bello et al. 2010). Infants delivered via cesarean section instead have bacterial communities that resemble those found on the skin of their mothers (Dominguez-Bello et al. 2010). Differences in microbial composition between vaginally born and cesarean section-delivered infants can be noted as long as 2 years after birth (Jakobsson et al. 2014; Yassour et al. 2016), with Bacteroidetes phyla members more abundant in vaginally born infants.

Other postnatal factors contribute to microbiome colonization. Feeding, in particular, is known to affect colonization patterns. Breast-fed infants and formula-fed infants have been both shown to have *Bifidobacterium* as a primary organism, but formula-fed infants also have minor components of possibly pathogenic species such as *Clostridia* species, and *Staphylococcus* species (Harmsen et al. 2000). It is also possible that breast milk may temper the long-ranging effects of cesarean section on microbial colonization. One study evaluating fecal microbiota from healthy infants at 4 months of age showed infants receiving breast milk had only minor differences whether or not they were born by cesarean section or vaginal delivery (Azad et al. 2013).

These effects may be mediated by both the nutrient composition of breast milk and its immunological components. Secretory IgA from breast milk has been shown to promote barrier function, preventing systemic infection by potential pathogens (Rogier et al. 2014). It may also be crucial to the long-term maintenance of a healthy gut microbiota and regulation of gene expression in IECs (Rogier et al. 2014).

Gestational age at birth is also a major determinant of microbiome colonization. It is known that healthy full-term breast-fed, vaginally delivered infants are colonized by *Bifidobacterium* by day 7 of life, whereas preterm infants are not (Butel et al. 2007). For preterm infants, the precise definition of a healthy microbiome has not yet been established. Generally, preterm infants show increased Proteobacteria phyla members and decreased anaerobes such as *Bifidobacterium* and *Bacteroides* compared with full-term infants (Arboleya et al. 2012). Additionally, there may be certain gestational age thresholds at which microbes colonize the gut. Studies by Butel et al. and LaRosa et al. (2007, 2014) suggest that the preterm infant microbiome might follow a temporal evolution from unstable populations at birth to stable populations of anaerobes such as *Clostridia* or *Bifidobacterium* species at around 33–36 weeks. Other factors that affect preterm colonization are also summarized in Table 1.4 and are discussed later.

1.2.2 Healthy Commensals and Postnatal Development

Once microbes colonize the infant gut, they affect the structural and functional maturation of the gut in various ways. A healthy microbiome tends to enhance the functions of the infant gut and its maturation (Table 1.5).

1.2.2.1 Motility

The gut microbiome is believed to play a role in regulating intestinal motility. Short-chain fatty acids (SCFAs) from microbial fermentation have been shown to affect local and distant motor events in the intestine (Rondeau et al. 2003). A study by

Table 1.5 Microbiome and functional maturation of the intestine (summarized from the text)

Function	Healthy microbiome benefit
Motility	Increased nerve density, frequency of amplitude of muscle contractions, and increased neural excitability
Digestion/ absorption	Increased digestion of HMOs and other indigestible carbohydrates
	Production of short chain fatty acids
	Augmented intestinal growth and surface area
Immunity	Increased mucin thickness
	Strengthened tight junctions
	Increased production of anti-inflammatory cytokines
	Inhibited activation of the NF-κB pathway
	Development of isolated lymphoid follicles
	Maturation of Peyer's patches
	Formation of IgA-producing plasma cells and mature T lymphocytes
	Specific immunity to pathogens

HMOs human milk oligosaccharides

Anitha et al. (2012) demonstrates that murine intestines depleted of microbes show delays in intestinal motility and reduced numbers of neurons. Other studies confirm this finding: germ-free (GF) mice display decreases in nerve density, decreased frequency of amplitude of muscle contractions, and decreased neural excitability (McVey Neufeld et al. 2013; Collins et al. 2014). Although the precise mechanism of the microbiome regulation of motility is unknown, it may involve TLR signaling, specifically TLR4 on enteric neurons. Anitha et al. also found gastrointestinal emptying delays and reduced intestinal motility in TLR-4 −/− mice (2012). This indicates that the LPS–TLR4 interaction may modulate gut motility and postnatal neural development in the intestine.

1.2.2.2 Digestion/Absorption

The third most abundant nutrient in breastmilk is human milk oligosaccharides (HMOs); yet, infants lack the specific enzymes necessary for its absorption (Marcobal and Sonnenburg 2012). *Bifidobacterium* and *Bacteroides* species are beneficial to the infant because they possess enzymes that break HMOs down into components that the infant can absorb (Marcobal and Sonnenburg 2012; Underwood et al. 2015a, b). The gut microbiota is also responsible for the breakdown of indigestible carbohydrates and the production of short-chain fatty acids (SCFAs) (Fernandes et al. 2014). Microbes also aid in the breakdown of toxins and drugs, vitamin synthesis, and ion absorption. Furthermore, healthy commensals may augment intestinal growth and surface area. The small intestines of GF mice are underdeveloped compared with colonized mice with irregular villi, reduced regenerative capacity, and reduced surface area (Smith et al. 2007).

1.2.2.3 Immunity and the Microbiome

Barrier Function

A healthy microbiome living in the outer mucus layer has long been known to enhance mucus synthesis, secretion, and chemical composition (Cornick et al. 2015). Decades ago, studies in GF rodents showed reduced GC counts and a mucus layer that is up to two times thinner than in conventionally raised mice (Enss et al. 1992; Kandori et al. 1996). When stimulated with bacterial products (LPS and peptidoglycan), GF mice develop thicker mucus layers (Petersson et al. 2011).

The mucosal barrier also contains secretory immunoglobulin A (sIgA). As discussed earlier, newborn infants secrete little to no sIgA. However, sIgA increases after 4 weeks of life, and this can be attributed to microbial stimulation of GALT (Nahmias et al. 1991). GF mice demonstrate a complete lack of secretory IgA and plasma cells, but this defect is reversible by colonizing these mice with commensals (Hapfelmeier et al. 2010). This demonstrates that the microbiota can direct the production of intestinal plasma cells and secretory IgA. Faster development of sIgA is found in infants from countries where they are exposed to a heavier microbial load (Mellander et al. 1985). Neonatal mice raised on antibiotics exhibit impaired barrier function and decreased claudin 3 expression (Patel et al. 2012). This suggests that commensal bacteria might induce the maturation of TJs in the infant intestine.

Antigen Detection and Immune Tolerance

Many benefits of commensal bacteria are mediated by the interactions of MAMPs with TLRs located on IECs and mucosal-dwelling immune cells (Rakoff-Nahoum et al. 2006). TLR–MAMP interactions support the development of immune tolerance via anti-inflammatory processes. Under homeostatic conditions, commensals stimulate IEC secretion of cytokines, including TGFβ, to encourage maturation of tolerogenic macrophages and dendritic cells (DCs) (Mazmanian et al. 2008; Atarashi et al. 2011). These macrophages secrete IL-10, an anti-inflammatory cytokine, whereas the DCs influence maturation of the host-generated immune responses. The cytokines produced by these macrophages and DCs maintain the anti-inflammatory balance of the intestine by inhibiting potential responses.

Commensal bacteria also inhibit the TLR-activated NF-κB pathway by encouraging the stabilization of the NF-κB inhibitor IκBα (Neish and Naumann 2011). Recent in vivo experiments comparing GF mice, specific pathogen-free (SPF) mice, and mice colonized with a healthy preterm infant's microbiome show this NF-κB inhibition in both SPF mice and preterm-colonized mice, but not in GF mice (Lu et al. 2015).

Adaptive Immunity

The MAMP–TLR interactions stimulate normal adaptive immune system development and influence the nature of host-generated immune responses (Hooper et al. 2012). MAMPs are sensed by receptors on IECs and DCs in the mucosa. These cells release cytokines to recruit B cells and T cells, causing the maturation of isolated lymphoid follicles. These isolated lymphoid follicles are a significant source of IgA-producing plasma cells (Maynard et al. 2012). Intestinal commensals can also help to generate specific immunity to pathogens. Non-pathogenic strains of *Escherichia coli* are often among the first organisms evident in the intestine of newborn infants, and in animal models, these commensal *E. coli* strains have been shown to inhibit invasive *E. coli* O157:H7 growth in the intestine (Leatham et al. 2009). This same effect has been demonstrated using toxigenic and non-toxigenic strains of *Bacteroides fragilis* (Hecht et al. 2016).

1.3 Preterm Dysbiosis and Its Effects on the Developing Intestine

The section above details the developmental benefits of the microbiome for the infant. A dysbiotic microbiome, however, is harmful, particularly for preterm infants, and correlates specifically with sepsis, high mortality rates, and NEC.

1.3.1 Necrotizing Enterocolitis and Dysbiosis

Necrotizing enterocolitis is a form of inflammatory bowel necrosis that primarily affects low birth weight, preterm infants (Neu and Walker 2011). The pathogenesis of this disease is not well understood, and specific cellular markers have not been identified to aid diagnosis in the early stages. Symptoms of NEC include feeding intolerance, abdominal distention, and bloody stools after 8–10 days of age (Neu and Walker 2011). Within hours, these symptoms can progress from subtle signs to abdominal discoloration, intestinal perforation, and peritonitis, culminating in systemic hypotension that requires intensive medical support and often surgery (Neu and Walker 2011). NEC has been shown to be influenced by the preterm infant immune response under conditions of intestinal immaturity and dysbiosis. Mortality rates are high, ranging between 30 and 50%, with the highest rates among infants requiring surgical treatment, and those who do survive are at risk of long-term neurodevelopmental and intestinal complications (Neu and Walker 2011).

Although no specific microbe has been linked to NEC development, the following alterations in community structure have been seen in preterm infants up to 3 weeks before the clinical diagnosis with NEC: a decrease in Firmicutes phyla,

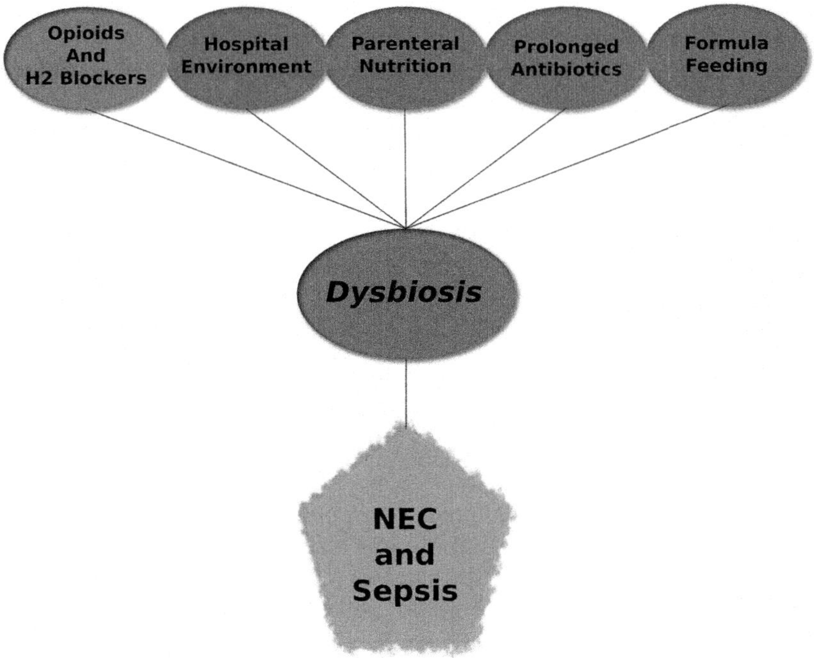

Fig. 1.2 Causes of dysbiosis in the preterm infant after birth

an increase in Proteobacteria phyla, and a decrease in species richness (Claud et al. 2013). These alterations can be collectively termed dysbiosis. Healthy preterm infant microbiomes do not display these characteristics, and instead show a slow progression toward the microbiome known to be typical of healthy term infants (Claud et al. 2013). These dysbiotic changes in microbiome community structure are believed to precede NEC, and they occur in the preterm population for several reasons (summarized in Fig. 1.2 and Table 1.4).

1.3.2 Causes of Dysbiosis in Preterm Infants

Parenteral nutrition can cause dysbiosis in preterm populations. Preterm infants in the neonatal intensive care unit (NICU) may be fed with PN, which deprives the initial microbial colonists of nutrients. Mouse models of PN dependence suggest that enteral nutrient deprivation might shift the intestinal microbiota to predominantly gram-negative Proteobacteria phyla members instead of Firmicutes (Demehri et al. 2013). Proteobacteria members are believed to increase the intestinal inflammatory basal state in developing infants, as their presence is associated with an increase in pro-inflammatory cytokines in intestinal mucosa and a decrease in growth factors that regulate cell proliferation and apoptosis (Demehri et al. 2013). PN dependence has also been shown to reduce the presence of antimicrobial peptides in the preterm gut, which may also have an effect on pathogen colonization (Barrett et al. 2015).

Formula feeding may possibly lead to dysbiosis and pathogen colonization. As mentioned previously, feeding can affect community colonization patterns, and formula-fed infants are more likely to have pathogenic microbes in their gut than age-matched breast-fed infants (Harmsen et al. 2000). Azad et al. (2013) reports lower bacterial richness and diversity in the stools of 4-month-old breast-fed infants compared with formula-fed infants, and this effect may be seen in preterm infants as well. Studies of preterm infants born before 32 weeks' gestation show lower bacterial diversity in infants fed with formula (Song et al. 2013; Gregory et al. 2016). Furthermore, formula may negatively affect the immature gut epithelial border by increasing permeability, gut epithelial cell toxicity, and inflammatory responses (Taylor et al. 2009; Penn et al. 2012).

Dysbiosis in preterm infants can also be explained by the common use of broad spectrum antibiotics. Ampicillin and gentamicin are the two most frequently used drugs in the NICU (Clark et al. 2006). Studies show that the use of antibiotics and increased duration of antibiotic treatment correlates with NEC, sepsis, and death (Cotten et al. 2009; Alexander et al. 2011). This is explained by the altered intestinal microbial ecology and the loss of diversity that occurs with antibiotics treatment (Greenwood et al. 2014). Antibiotics-treated infant microbiomes show a decrease in Firmicutes and an increase in Proteobacteria phyla members (Mai et al. 2011; Claud et al. 2013). Furthermore, the hospital environment is rife with opportunistic, antibiotics-resistant pathogens. Antibiotics-induced microbiome loss increases the likelihood of normal, healthy community members being depleted and pathogens thriving in their place, colonizing the neonate via hospital equipment such as nasogastric tubes or catheters.

Other medications in the NICU cause dysbiosis besides antibiotics. Opioids delay intestinal transit time and affect bacterial adherence and colonization (Claud and Walker 2008). Gastric acidity, a major defense mechanism against infection, can be depleted by the H2 blockers often used to treat acid gastroesophageal reflux in preterm infants, who already have low gastric acid production in the first weeks after birth (Hyman et al. 1985). Preterm infants treated with H2 blockers have been shown to have a microbiome with reduced diversity (Gupta et al. 2013) in addition to being at an increased risk for NEC (Terrin et al. 2012).

1.3.3 Dysbiosis, NEC, and Preterm Development

Dysbiosis is concerning in the preterm infant because it is associated with NEC, sepsis, and high mortality. It also interferes with preterm intestinal development. The preterm intestine is functionally impaired; thus, it does not have the necessary digestive, motor, and absorptive capabilities, and its immune system is hyper-responsive. Dysbiosis and NEC tend to exacerbate these impairments.

1.3.3.1 Dysbiosis, NEC, and Motility

As mentioned in our discussion of beneficial microbiome effects, GF mice and mice treated with antibiotics display decreases in nerve density, a decreased frequency of muscle contractions, and decreased neural excitability (McVey Neufeld et al. 2013; Collins et al. 2014). This indicates that a lack of healthy commensal populations has deleterious effects on development of the ENS and intrinsic motility networks. Additionally, intrinsic ENS immaturity may contribute to the development of NEC under dysbiotic conditions (Berseth 1996).

Post-NEC infants demonstrate a loss of enteric neurons in the submucosa and alterations in the myenteric plexus (Wedel et al. 1998; Sigge et al. 1998). Other reported post-NEC complications include intestinal dysmotility, stricture, and recurrent abdominal distention (Beardmore et al. 1978). Neurons of the ENS are similar to those of the central nervous system, in that they have limited potential for post-injury regeneration. Failure to reverse NEC neuronal cell loss may lead to compromised intestinal function long after the infant's recovery. Overall, dysbiosis and NEC reduce normal intestinal motility and may even irreversibly damage the ENS.

1.3.3.2 Dysbiosis, NEC, and Digestion/Absorption

The lack of EN for the preterm infant prevents the immediate effects of dysbiosis on digestion and the absorption of nutrients. However, it has been shown that the small intestines of GF mice are underdeveloped compared with colonized mice with irregular villi, reduced regenerative capacity, and reduced surface area (Smith et al. 2007). This indicates that the lack of a healthy microbiome early on may affect the digestive capacity of the intestine later in life. Additionally, NEC survivors often experience long-term consequences for digestive health due to inflammatory damage to their GI tract, including intestinal strictures and short bowel syndrome (Hintz et al. 2005).

1.3.3.3 Dysbiosis, NEC, and Immunity

A recent paper by Lu et al. (2015) analyzed the effect of preterm dysbiotic communities on inflammation in the intestine. Pregnant mice were colonized with the microbiomes from a preterm infant with poor growth. This microbiome was characterized by increased Proteobacteria, decreased Firmicutes, and decreased diversity. When these pregnant mice gave birth, their pups also acquired this microbiome. Analysis of the pups' intestines showed an increased intestinal inflammatory profile compared with specific pathogen-free and healthy preterm infant microbiome controls. Immunohistochemistry of distal ileum segments revealed increased activation and translocation of NF-κB. Genes and pathways involving the following innate immune and inflammatory responses were upregulated in the intestinal samples:

chemotaxis, chemokine/cytokine signaling, chemokine receptor expression and phagocyte-, monocyte-, and leukocyte-rolling and adhesion.

These findings indicate that dysbiotic communities may induce decreased immune-regulation and increased inflammation, both of which may contribute to NEC. It is known that deregulated interactions between commensals and TLRs promote the kind of chronic inflammation seen in inflammatory bowel diseases (Barbara et al. 2005). In the face of dysbiosis caused by antibiotic use or other factors implicated in prematurity, MAMPs direct the production and secretion of pro-inflammatory cytokines by IECs (IL-6, IL-1, and IL-18) (Maynard et al. 2012). They also stimulate the maturation of intestinal DCs and macrophages, which secrete cytokines that induce development of CD4+ T cells T_H1 and T_H17. NEC is also associated with increases in TLR4, NF-κB, TNF-alpha, and IL-6, IL-8, and IL-1β (Nanthakumar et al. 2011). This massive release of cytokines disrupts the $T_H1/T_H2/T_H17/T_{reg}$ immune balance. These disruptions may contribute to NEC exacerbation and development of sepsis. Long-term effects may include the development of immune disorders such as asthma (Murk et al. 2011), Crohn's disease (Hviid et al. 2011), and celiac disease (Eggesbø et al. 2003).

1.4 Optimizing the Microbiome of the Preterm Infant

1.4.1 Research Supporting the Existence of Optimal Communities in Preterm Infants

Given the negative short- and long-term effects of dysbiosis, including NEC, clinicians and researchers have worked to find ways to optimize the microbiome for preterm infants in the NICU to avoid dysbiosis and encourage eubiosis—healthy microbial communities—and healthy growth. As mentioned before, there is currently no specific definition of a healthy microbiome composition in preterm infants. Recent studies indicate, however, that there is indeed an optimal microbial colonization community that is both protective and developmentally beneficial for preterm infants.

1.4.1.1 Specific Community Alterations Lead to NEC

In a study by Claud et al. (2013), weekly fecal samples from 10 preterm infants, 5 of whom went on to develop NEC, and 5 healthy controls, were obtained and sequenced. Subtle differences in community structure could be seen up to 3 weeks before NEC development, with a gradual progression leading to an increased Proteobacteria phyla presence, decreased Firmicutes phyla members, and decreased diversity in patients with NEC. No specific microbe was identified to be protective in controls or harmful in NEC patients (Claud et al. 2013), which may indicate that all the actions of the microbial community contributed to protection.

The study also compared twins, one of whom later developed NEC and one of whom did not. Differentially abundant genes in the NEC twin were associated with carbohydrate metabolism and mapped to members of the family *Enterobacteriaceae* (Proteobacteria phylum), although not to a specific species (Claud et al. 2013). Furthermore, the functional gene profile of the pre-NEC sample reflected a narrowing of the overall community diversity.

These findings, especially the results from the twin study, indicate that there may be a preterm microbial community that is protective, whereas the dysbiotic alterations seen weeks before NEC development promote disease.

1.4.1.2 Community Composition Influences Development

Two recent studies using humanized mouse models indicate that healthy preterm microbial communities have positive effects on overall growth and development. In both these studies, pregnant mice were gavaged with fecal samples from two preterm infants; one exhibiting a pattern of poor weight gain and one with normal weight gain. When the pregnant mice gave birth, their pups acquired the microbiota of interest. Lu et al. (2015) showed that mice with the microbiome of normal weight preterm infants exhibited the normal growth phenotype of their human counterparts. These mice also showed decreased basal intestinal inflammation (Lu et al. 2015) in comparison with both GF mice and low-birth-weight mice.

Using the same model, Yu et al. (2016) showed that a microbiome from a normal weight preterm infant can benefit intestinal growth and overall growth. Mice with this microbiome showed increased villus height and crypt depth, increased cell proliferation, increased numbers of GCs and PCs, and enhanced TJs (Yu et al. 2016). In contrast, the microbiota from the preterm infant with poor weight gain failed to induce these positive changes. Instead, mice with this microbiome had a decreased intestinal surface area and reduced crypt depth (Yu et al. 2016). These findings are comparable with what is found in studies on GF animals.

For both studies, the composition of the different mouse microbiomes (obtained from the preterm infants) differed in significant ways. Consistent with other analyses of preterm infant fecal microbiota, the microbiota of these two preterm infants were both dominated by Proteobacteria and Firmicutes, with a small portion of Bacteroidetes and Actinobacteria. The microbiota of the normal-weight preterm infant had a greater contribution of Bacteroidetes (8.30 vs 3.42%) and Actinobacteria (8.53 vs 0.57%). Also, the microbiome of the normal-weight infant showed fewer Proteobacteria and more Firmicutes than its low-weight counterpart (Lu et al. 2015). These community differences may be part of the reason why the microbiota of the normal-weight infant induced such distinct changes in inflammation and development in murine intestines.

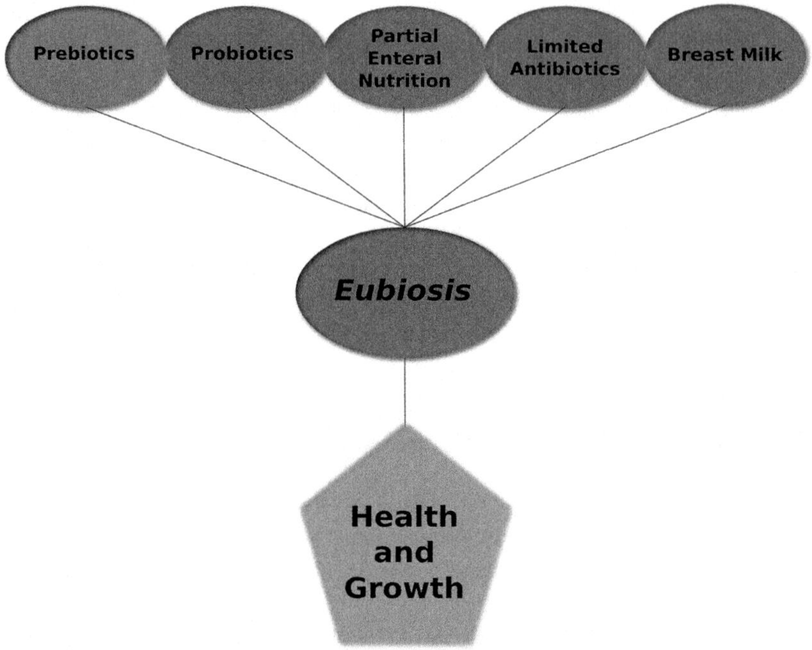

Fig. 1.3 Potential clinical interventions to promote eubiosis and preterm health and growth

Although these studies only compared the microbiomes of two infants, these studies begin to demonstrate that a healthy preterm microbial community—a eubiotic community—does exist, with both protective effects against disease and positive effects on intestinal development. Current clinical research is focused on optimizing the preterm microbial community to achieve these two effects and secure all the previously mentioned benefits of the microbiome for long-term development. Possible therapeutic options and the research that supports their use are outlined below and summarized in Fig. 1.3.

1.4.2 Probiotics

The World Health Organization (WHO) defines probiotics as "live microorganisms, which when administered in adequate amounts confer a health benefit on the host" (http://www.who.int/foodsafety/fs_management/en/probiotic_guidelines.pdf). It has been shown that specific probiotic bacteria may benefit IECs, the microbiome, and the immune system via several mechanisms of action. Probiotic organisms may displace potential pathogenic organisms, stimulate the production of antimicrobial compounds, and improve barrier function. They may also modulate the immune response by inducing tolerogenic lymphocytes and downregulating TLR response and NF-κB pathways (Collado et al. 2009; Ganguli et al. 2013).

Systematic review and meta-analyses of both randomized, placebo-controlled trials and observational studies seem to validate the hope that routine use of probiotics in the NICU could prevent NEC. Out of 26 clinical trials included in one analysis, probiotics, including Bifidobacteria, had an overall preventive effect on NEC in preterm infants (relative risk [RR] 0.47 [95% CI 0.36–0.60], $p < 0.00001$) (Aceti et al. 2015). A recent meta-analysis of 12 cohort studies, including 10,800 premature neonates (5144 receiving prophylactic probiotics and 5656 controls) showed a significantly decreased incidence of NEC (RR = 0.55, 95% CI, 0.39–0.78; $p = 0.0006$) (Olsen et al. 2016).

There are some unanswered questions that need to be addressed before probiotics can be widely used in NICUs. The 26 studies in the 2015 meta-analysis were heterogeneous, differing in terms of probiotic strain, dosage, duration of supplementation, and target population (Aceti et al. 2015). Additionally, few studies definitively documented effective colonization of the infants' gut with the desired probiotic strain and none of the studies sufficiently addressed the effects of probiotics on high-risk groups, such as extremely-low-birth-weight infants (Aceti et al. 2015). Probiotics may be a way forward for NEC prevention, but these questions remain unanswered.

Furthermore, probiotics generally contain a single bacteria species, and although that organism may be beneficial, it is likely that the overall functioning of the microbiome community contributes most to host health. In fact, there is clinical evidence to suggest that single species supplementation provides no benefit. The results of one randomized controlled trial on a single species *Bifidobacterium* probiotic showed no evidence of benefit for NEC prevention (RR 0.93, 95% CI 0.68–1.27) (Costeloe et al. 2016). A review of three trials using one *Lactobacillus* strain failed to show statistically significant prevention of either NEC (RR 0.69; 95% CI, 0.47–1.01) or mortality (RR 0.79; 95% CI, 0.57–1.09) (Athalye-Jape et al. 2016). Reduced microbial diversity appears to be a hallmark in pre-NEC patients, and supplementing the gut with a single species of bacteria may not be the best way to encourage species richness. Also, artificially introducing a single microbe into a community structure may have unpredictable effects. By narrowing its focus to specific species to find single protective agents, current probiotic research cannot account for the contribution of overall community function or species richness to host health. Thus, further studies should clarify the specific effect of probiotics and species combinations on microbial communities in the short term and long term before probiotics can be widely used. These studies must be subjected to rigorous quality controls and identify the most effective probiotic, proper dosage, and duration of use. Potential long-term effects on immunity, host gene expression, and gut function must be explored as well.

1.4.3 Prebiotics

Prebiotics are nondigestible food ingredients such as HMOs that have been shown to promote the growth of nonpathogenic commensals and increase species richness, thus helping preterm infants to develop a healthy microbiome (Yu et al. 2013). One of the benefits of prebiotics is that they do not simply increase the levels of one healthy microbial species such as probiotics. Instead, they affect the whole microbial community to promote diversity and healthy species growth. Prebiotic stimulation of bacterial community growth has wide-ranging effects on the infant intestine. For example, the prebiotic oligofructose strengthens barrier function, decreases endotoxemia, and improves glucose tolerance (Cani et al. 2009; Reinhardt et al. 2009). These benefits arise because a healthy bacterial community can trigger the release of gastrointestinal hormones, such as GLP-1, that decrease serum glucose and decrease energy intake. Prebiotics have also demonstrated anti-inflammatory effects, because they foster a healthy microbiome that has anti-inflammatory properties (Ganguli et al. 2013). Prebiotics have been used as an effective treatment for ulcerative colitis (Furrie et al. 2005); thus, it is possible that they could additionally be used to treat NEC, which has a similar pathogenesis.

Prebiotics may be a viable, safe therapeutic option for the treatment of dysbiosis in preterm infants, as they only have mild adverse effects, such as flatulence and increased stool output (Teitelbaum and Walker 2002). Although only a few studies have been performed in NICUs, they have had promising results. A 2013 meta-analysis of prebiotic clinical trials indicated that supplementation with prebiotic oligosaccharides is safe and results in significantly higher growth of beneficial microbes and improved species diversity. More research is warranted in this area, however, as these trials did not find an overall decreased incidence of NEC in prebiotics-treated infants (Srinivasjois et al. 2013). Prebiotic supplementation may not be beneficial for preterm infants receiving their mother's milk. Underwood et al. show no significant increase in beneficial Bifidobacteria for prebiotics-supplemented infants already receiving maternal breast milk (2014). Probiotics combined with prebiotics may also be more effective than prebiotics alone. Dilli et al. showed that prebiotics alone have no benefit for the prevention of NEC or mortality, but prebiotics plus probiotics may be beneficial in reducing the incidence of NEC and mortality (2015).

1.4.4 Breast Milk Feeding and Minimal Enteric Nutrition

The Section on Breastfeeding of the American Academy of Pediatrics recommends that all preterm infants should receive human milk, with pasteurized donor milk rather than premature infant formula the preferred alternative if a mother is unable to provide enough for her infant (AAP 2012). Breast milk has been shown to provide many benefits; in particular, it decreases rates of NEC (Sisk et al. 2007). Receiving

>50 ml/kg/day of maternal milk decreases the risk of late-onset sepsis and NEC compared with <50 ml/kg/day (Meinzen-Derr et al. 2009) and there is a 5% reduction in hospital readmission rate for each 10 ml/kg/day increase in human milk (Vohr et al. 2007). In extremely preterm infants given exclusive diets of preterm formula vs human milk, there was a significantly higher rate of NEC, requiring surgical treatment in infants receiving preterm formula (Cristofalo et al. 2013).

The protection provided by mother's milk against NEC stems from many factors. Breast milk contains sIgA, lactoferrin, lysozyme, bile salt-stimulating lipase, growth factors, and HMOs, which may all provide protective benefits (Rogier et al. 2014; Underwood et al. 2015a, b). With regard to the microbiome and dysbiosis, breast milk is thought to have both prebiotic and probiotic effects (Underwood et al. 2015a, b). HMOs are known to influence the composition of the microbiome, encouraging colonization with a healthy community of Bacteroides phyla members. Breast milk also contains many polyamines, which are known to protect beneficial microbiota (Plaza-Zamora et al. 2013). Additionally, breast milk contains micro-organisms, which may contribute to colonization of the preterm gut with a beneficial microbial community (Nakamura et al. 2009; Gregory et al. 2016).

Both mother's milk and donor pasteurized breast milk have been shown to protect against NEC compared with formula (Chang et al. 2013; Quigley and Maguire 2014). The standard for breast milk, however, is fresh milk directly from the breast of the infant's mother, because this preserves all the biological benefits of the milk. Storage and processing can counteract many of the benefits of donor milk, including macronutrient, hormone, and bacterial content (Meier et al. 2017). Donor milk has also been associated with slower growth in preterm infants (Schanler et al. 2005; Quigley and Maguire 2014). This suggests that although donor milk may protect against NEC, its benefits may stem from avoiding formula rather than using donor milk itself, as formula has a strong association with dysbiosis (Meier et al. 2017). Thus, donor milk may still be beneficial in cases where there is no option for mother's milk.

The beneficial effects of mother's milk, and the beneficial growth spurt provided by the first milk colostrum, cannot be seen if the preterm infant is fed entirely PN. Partial EN—giving small amounts of enteral solution to the infant receiving PN—may allow PN-fed infants to receive some of these benefits before transitioning to full EN. Intake for partial EN often ranges between 10 and 20 ml/kg/day (Commare and Tappenden 2007). The goal of partial EN is to enhance gut function, encourage development of a healthy microbiome, and increase bile flow to decrease the likelihood of developing cholestasis, a major adverse side effect of PN (Commare and Tappenden 2007). Partial EN has been shown to increase overall functional maturation of the intestine: mucosal mass (Berseth et al. 1983), lactase activity (Park et al. 1999), intestinal motility (Berseth et al. 1992), and hormone/peptide release (Meetze et al. 1992).

1.4.5 Limiting Prolonged Antibiotic Use

A 2009 study by Cotton et al. revealed that each empirical treatment day of antibiotic treatment in the NICU—that is, treatment when cultures are clear—was associated with increased odds of death, NEC, and the composite measure of NEC or death (Cotten et al. 2009). This led the American Academy of Pediatrics to release a recommendation to discontinue antimicrobial therapy for early onset sepsis in preterm infants at 48 h in clinical situations in which the probability of sepsis is low. Antibiotic duration affects both the timeline and type colonization of the preterm intestine, especially in the NICU environment where opportunistic, antibiotic-resistant pathogens are abundant. Limiting use of antibiotics when cultures are clear, allowing earlier colonization of the gut by healthy commensals, which is important for avoiding NEC and dysbiosis. Consequences of prolonged early antibiotic use and microbiome disruption are suspected to be even more far-reaching than the immediate effects of reducing the diversity of bacterial communities, including predisposing the infant to allergy (Madan et al. 2012) and childhood obesity (Ajslev et al. 2011).

1.5 Conclusion

In this chapter, the timeline of infant intestinal development is laid out and areas are highlighted in which preterm infants are ill-prepared for life ex utero and vulnerable to developing inflammatory diseases such as NEC. The process of intestinal microbiome colonization and factors that influence both normal colonization in term infants and the development of dysbiosis in preterm infants are described. Further, the various ways in which a healthy microbiome benefits intestinal development are enumerated and the ways in which dysbiosis and NEC can harm intestinal development are detailed. Finally, the latest research on clinical interventions is discussed that may optimize the microbiome composition of the preterm infant to avoid the deleterious effects of dysbiosis and NEC and restore a microbiome that provides developmental benefits. More research is needed in these areas, but it is clear that the microbiome does play a crucial role in preterm intestinal development, and that optimizing the microbiome in the preterm population is a worthy goal for the scientific community.

References

Aceti A, Gori D, Barone G, Callegari ML, Di Mauro A, Fantini MP, Indrio F, Maggio L, Meneghin F, Morelli L, Zuccotti G, Corvaglia L (2015) Probiotics for prevention of necrotizing enterocolitis in preterm infants: systematic review and meta-analysis. Ital J Pediatr 41:89

Adibi SA et al (1975) Evidence for two different modes of tripeptide disappearance in human intestine. Uptake by peptide carrier systems and hydrolysis by peptide hydrolases. J Clin Invest 56:1355–1363

Agunod M, Yamaguchi N, Lopez R, Luhby AL, Glass GB (1969) Correlative study of hydrochloric acid, pepsin, and intrinsic factor secretion in newborns and infants. Am J Dig Dis 14:400–414

Ajslev TA, Andersen CS, Gamborg M, Sørensen TIA, Jess T (2011) Childhood overweight after establishment of the gut microbiota: the role of delivery mode, pre-pregnancy weight and early administration of antibiotics. Int J Obes (Lond) 35:522–529

Alemi B, Hamosh M, Scanlon JW, Salzman-Mann C, Hamosh P (1981) Fat digestion in very low-birth-weight infants: effect of addition of human milk to low-birth-weight formula. Pediatrics 68:484–489

Alexander VN, Northrup V, Bizzarro MJ (2011) Antibiotic exposure in the newborn intensive care unit and the risk of necrotizing enterocolitis. J Pediatr 159:392–397

Anitha M, Vijay-Kumar M, Sitaraman SV, Gewirtz AT, Srinivasan S (2012) Gut microbial products regulate murine gastrointestinal motility via Toll-like receptor 4 signaling. Gastroenterology 143:1006–1016.e4

Antonowicz I, Lebenthal E (1977) Developmental pattern of small intestinal enterokinase and disaccharidase activities in the human fetus. Gastroenterology 72:1299–1303

Arboleya S, Binetti A, Salazar N, Fernández N, Solís G, Hernández-Barranco A, Margolles A, de los Reyes-Gavilán CG, Gueimonde M (2012) Establishment and development of intestinal microbiota in preterm neonates. FEMS Microbiol Ecol 79:763–772

Atarashi K, Tanoue T, Shima T, Imaoka A, Kuwahara T, Momose Y et al (2011) Induction of colonic regulatory T cells by indigenous Clostridium species. Science 331:337–341

Athalye-Jape G, Rao S, Patole S (2016) Lactobacillus reuteri DSM 17938 as a probiotic for preterm neonates: a strain-specific systematic review. J Parenter Enteral Nutr 40:783–794

Azad MB, Konya T, Maughan H, Guttman DS, Field CJ, Chari RS, Sears MR, Becker AB, Scott JA, Kozyrskyj AL (2013) Gut microbiota of healthy Canadian infants: profiles by mode of delivery and infant diet at 4 months. CMAJ 185:385–394

Barbara G, Stanghellini V, Brandi G, Cremon C, Di Nardo G, De Giorgio R, Corinaldesi R (2005) Interactions between commensal bacteria and gut sensorimotor function in health and disease. Am J Gastroenterol 100:2560–2568

Barrett M, Demehri FR, Teitelbaum DH (2015) Intestine, immunity and parenteral nutrition in an era of preferred enteral feeding. Curr Opin Clin Nutr Metab Care 18:496–500

Bauer PV, Hamr SC, Duca FA (2015) Regulation of energy balance by a gut–brain axis and involvement of the gut microbiota. Cell Mol Life Sci 73:737–755

Beach RC, Menzies IS, Clayden GS, Scopes JW (1982) Gastrointestinal permeability changes in the preterm neonate. Arch Dis Child 57:141–145

Beardmore HE, Rodgers BM, Outerbridge E (1978) Necrotizing enterocolitis (ischemic enteropathy) with the sequel of colonic atresia. Gastroenterology 74:914–917

Berendsen PB, Blanchette-Mackie EJ (1979) Milk lipid absorption and chylomicron production in the suckling rat. Anat Rec 195:397–414

Berseth CL (1996) Gastrointestinal motility in the neonate. Clin Perinatol 23:179–190

Berseth CL, Lichtenberger LM, Morriss FH (1983) Comparison of the gastrointestinal growth-promoting effects of rat colostrum and mature milk in newborn rats in vivo. Am J Clin Nutr 37:52–60

Berseth CL, Nordyke CK, Valdes MG, Furlow BL, Go VL (1992) Responses of gastrointestinal peptides and motor activity to milk and water feedings in preterm and term infants. Pediatr Res 31:587–590

Black DD, Rohwer-Nutter PL, Davidson NO (1990) Intestinal apolipoprotein A-IV gene expression in the piglet. J Lipid Res 31:497–505

Brugman S, Perdijk O, van Neerven RJJ, Savelkoul HFJ (2015) Mucosal immune development in early life: setting the stage. Arch Immunol Ther Exp (Warsz) 63:251–268

Buddington RK, Malo C (1996) Intestinal brush-border membrane enzyme activities and transport functions during prenatal development of pigs. J Pediatr Gastroenterol Nutr 23:51–64
Buisine MP, Devisme L, Savidge TC, Gespach C, Gosselin B, Porchet N, Aubert JP (1998) Mucin gene expression in human embryonic and fetal intestine. Gut 43:519–524
Burns AJ, Roberts RR, Bornstein JC, Young HM (2009) Development of the enteric nervous system and its role in intestinal motility during fetal and early postnatal stages. Semin Pediatr Surg 18:196–205
Butel M-J, Suau A, Campeotto F, Magne F, Aires J, Ferraris L, Kalach N, Leroux B, Dupont C (2007) Conditions of bifidobacterial colonization in preterm infants: a prospective analysis. J Pediatr Gastroenterol Nutr 44:577–582
Cani PD, Possemiers S, Van de Wiele T, Guiot Y, Everard A, Rottier O, Geurts L, Naslain D, Neyrinck A, Lambert DM, Muccioli GG, Delzenne NM (2009) Changes in gut microbiota control inflammation in obese mice through a mechanism involving GLP-2-driven improvement of gut permeability. Gut 58:1091–1103
Chang J-C, Chen C-H, Fang L-J, Tsai C-R, Chang Y-C, Wang T-M (2013) Influence of prolonged storage process, pasteurization, and heat treatment on biologically-active human milk proteins. Pediatr Neonatol 54:360–366
Chu DM, Antony KM, Ma J, Prince AL, Showalter L, Moller M, Aagaard KM (2016) The early infant gut microbiome varies in association with a maternal high-fat diet. Genome Med 8:77
Clark DA, Miller MJ (1990) Intraluminal pathogenesis of necrotizing enterocolitis. J Pediatr 117 (1 Pt 2):S64–S67
Clark RH, Bloom BT, Spitzer AR, Gerstmann DR (2006) Reported medication use in the neonatal intensive care unit: data from a large national data set. Pediatrics 117:1979–1987
Claud EC, Walker WA (2001) Hypothesis: inappropriate colonization of the premature intestine can cause neonatal necrotizing enterocolitis. FASEB J 15:1398–1403
Claud EC, Walker WA (2008) Bacterial colonization, probiotics, and necrotizing enterocolitis. J Clin Gastroenterol 42(Suppl 2):S46–S52
Claud EC, Lu L, Anton PM, Savidge T, Walker WA, Cherayil BJ (2004) Developmentally regulated IκB expression in intestinal epithelium and susceptibility to flagellin-induced inflammation. PNAS 101:7404–7408
Claud EC, Keegan KP, Brulc JM, Lu L, Bartels D, Glass E, Chang EB, Meyer F, Antonopoulos DA (2013) Bacterial community structure and functional contributions to emergence of health or necrotizing enterocolitis in preterm infants. Microbiome 1:20
Cleghorn G, Durie P, Benjamin L, Dati F (1988) The ontogeny of serum immunoreactive pancreatic lipase and cationic trypsinogen in the premature human infant. Biol Neonate 53:10–16
Collado MC, Isolauri E, Salminen S, Sanz Y (2009) The impact of probiotic on gut health. Curr Drug Metab 10:68–78
Collins J, Borojevic R, Verdu EF, Huizinga JD, Ratcliffe EM (2014) Intestinal microbiota influence the early postnatal development of the enteric nervous system. Neurogastroenterol Motil 26:98–107
Colony PC (1983) Successive phases of human fetal intestinal development. In: Kretchmer N, Miinkowski A (eds) Nutritional adaptation of the gastrointestinal tract of the newborn. Vevey/Raven, New York, pp 3–28
Commare CE, Tappenden KA (2007) Development of the infant intestine: implications for nutrition support. Nutr Clin Pract 22:159–173
Cornick S, Tawiah A, Chadee K (2015) Roles and regulation of the mucus barrier in the gut. Tissue Barriers 3:e982426. https://doi.org/10.4161/21688370.2014.982426
Costeloe K, Bowler U, Brocklehurst P, Hardy P, Heal P, Juszczak E et al (2016) A randomised controlled trial of the probiotic Bifidobacterium breve BBG-001 in preterm babies to prevent sepsis, necrotising enterocolitis and death: the Probiotics in Preterm Infants (PiPS) trial. Health Technol Assess 20:1–194
Cotten CM, Taylor S, Stoll B, Goldberg RN, Hansen NI, Sanchez PJ, Ambalavanan N, Benjamin DK (2009) Prolonged duration of initial empirical antibiotic treatment is associated with

increased rates of necrotizing enterocolitis and death for extremely low birth weight infants. Pediatrics 123:58–66
Cristofalo EA, Schanler RJ, Blanco CL, Sullivan S, Trawoeger R, Kiechl-Kohlendorfer U, Dudell G, Rechtman DJ, Lee ML, Lucas A, Abrams S (2013) Randomized trial of exclusive human milk versus preterm formula diets in extremely premature infants. J Pediatr 163:1592–1595
Davidson NO, Hausman AM, Ifkovits CA et al (1992) Human intestinal glucose transporter expression and localization of GLUT5. Am J Physiol 262:C795–C800
Demehri FR, Barrett M, Ralls MW, Miyasaka EA, Feng Y, Teitelbaum DH (2013) Intestinal epithelial cell apoptosis and loss of barrier function in the setting of altered microbiota with enteral nutrient deprivation. Front Cell Infect Microbiol 3:105
Dewit O, Dibba B, Prentice A (1990) Breast-milk amylase activity in English and Gambian mothers: effects of prolonged lactation, maternal parity, and individual variations. Pediatr Res 28:502–506
Dilli D, Aydin B, Fettah ND, Özyazıcı E, Beken S, Zenciroğlu A, Okumuş N, Özyurt BM, İpek MŞ, Akdağ A, Turan Ö, Bozdağ Ş (2015) The propre-save study: effects of probiotics and prebiotics alone or combined on necrotizing enterocolitis in very low birth weight infants. J Pediatr 166:545–551.e1
Dominguez-Bello MG, Costello EK, Contreras M, Magris M, Hidalgo G, Fierer N, Knight R (2010) Delivery mode shapes the acquisition and structure of the initial microbiota across multiple body habitats in newborns. PNAS 107:11971–11975
Dumont RC, Rudolph CD (1994) Development of gastrointestinal motility in the infant and chilD. Gastroenterol Clin North Am 23:655–671
Egan CE, Sodhi CP, Good M, Lin J, Jia H, Yamaguchi Y et al (2016) Toll-like receptor 4–mediated lymphocyte influx induces neonatal necrotizing enterocolitis. J Clin Invest 126:495–508
Eggesbø M, Botten G, Stigum H, Nafstad P, Magnus P (2003) Is delivery by cesarean section a risk factor for food allergy? J Allergy Clin Immunol 112:420–426
Enss ML, Grosse-Siestrup H, Schmidt-Wittig U, Gärtner K (1992) Changes in colonic mucins of germfree rats in response to the introduction of a "normal" rat microbial flora. J Exp Anim Sci 35:110–119
Fernandes J, Su W, Rahat-Rozenbloom S, Wolever TMS, Comelli EM (2014) Adiposity, gut microbiota and faecal short chain fatty acids are linked in adult humans. Nutr Diab 4:e121. https://doi.org/10.1038/nutD2014.23
Furrie E, Macfarlane S, Kennedy A, Cummings JH, Walsh SV, O'Neil DA, Macfarlane GT (2005) Synbiotic therapy (Bifidobacterium longum/Synergy 1) initiates resolution of inflammation in patients with active ulcerative colitis: a randomised controlled pilot trial. Gut 54:242–249
Fusunyan RD, Nanthakumar NN, Baldeon ME, Walker WA (2001) Evidence for an innate immune response in the immature human intestine: toll-like receptors on fetal enterocytes. Pediatr Res 49:589–593
Ganguli K, Meng D, Rautava S, Lu L, Walker WA, Nanthakumar N (2013) Probiotics prevent necrotizing enterocolitis by modulating enterocyte genes that regulate innate immune-mediated inflammation. Am J Physiol Gastrointest Liver Physiol 304:G132–G141
González-Ariki S, Husband AJ (2000) Ontogeny of IgA+ cells in lamina propria: effects of sympathectomy. Dev Comp Immunol 24:61–69
Greenwood C, Morrow AL, Lagomarcino AJ, Altaye M, Taft DH, Yu Z, Newburg DS, Ward DV, Schibler KR (2014) Early empiric antibiotic use in preterm infants is associated with lower bacterial diversity and higher relative abundance of Enterobacter. J Pediatr 165:23–29
Gregory KE, Samuel BS, Houghteling P, Shan G, Ausubel FM, Sadreyev RI, Walker WA (2016) Influence of maternal breast milk ingestion on acquisition of the intestinal microbiome in preterm infants. Microbiome 4:68
Gupta RW, Tran L, Norori J, Ferris MJ, Eren AM, Taylor CM, Dowd SE, Penn D (2013) Histamine-2 receptor blockers alter the fecal microbiota in premature infants. J Pediatr Gastroenterol Nutr 56:397–400

Hamosh M, Scanlon JW, Ganot D, Likel M, Scanlon KB, Hamosh P (1981) Fat digestion in the newborn. Characterization of lipase in gastric aspirates of premature and term infants. J Clin Invest 67:838–846
Hapfelmeier S, Lawson MAE, Slack E, Kirundi JK, Stoel M et al (2010) Reversible microbial colonization of germ-free mice reveals the dynamics of IgA immune responses. Science 328:1705–1709
Harmsen HJ, Wildeboer-Veloo AC, Raangs GC et al (2000) Analysis of intestinal flora development in breast-fed and formula-fed infants by using molecular identification and detection methods. J Pediatr Gastroenterol Nutr 30:61–67
Hecht AL, Casterline BW, Earley ZM, Goo YA, Goodlett DR, Bubeck Wardenburg J (2016) Strain competition restricts colonization of an enteric pathogen and prevents colitis. EMBO Rep 17:1281–1291
Hintz SR, Kendrick DE, Stoll BJ, Vohr BR, Fanaroff AA, Donovan EF, Poole WK, Blakely ML, Wright L, Higgins R (2005) Neurodevelopmental and growth outcomes of extremely low birth weight infants after necrotizing enterocolitis. Pediatrics 115:696–703
Hooper LV, Littman DR, Macpherson AJ (2012) Interactions between the microbiota and the immune system. Science 336:1268–1273
Hviid A, Svanström H, Frisch M (2011) Antibiotic use and inflammatory bowel diseases in childhood. Gut 60:49–54
Hyman PE, Clarke DD, Everett SL, Sonne B, Stewart D, Harada T, Walsh JH, Taylor IL (1985) Gastric acid secretory function in preterm infants. J Pediatr 106:467–471
Jakobsson HE, Abrahamsson TR, Jenmalm MC, Harris K, Quince C, Jernberg C, Björkstén B, Engstrand L, Andersson AF (2014) Decreased gut microbiota diversity, delayed Bacteroidetes colonisation and reduced Th1 responses in infants delivered by Caesarean section. Gut 63:559–566
Johansson MEV, Ambort D, Pelaseyed T, Schütte A, Gustafsson JK, Ermund A et al (2011) Composition and functional role of the mucus layers in the intestine. Cell Mol Life Sci 68:3635–3641
Kandori H, Hirayama K, Takeda M, Doi K (1996) Histochemical, lectin-histochemical and morphometrical characteristics of intestinal goblet cells of germfree and conventional mice. Exp Anim 45:155–160
Kelly EJ, Newell SJ, Brownlee KG, Primrose JN, Dear PR (1993) Gastric acid secretion in preterm infants. Early Hum Dev 35:215–220
Kim YS, Ho SB (2010) Intestinal goblet cells and mucins in health and disease: recent insights and progress. Curr Gastroenterol Rep 12:319–330
Koleva PT, Kim J-S, Scott JA, Kozyrskyj AL (2015) Microbial programming of health and disease starts during fetal life. Birth Defects Res C Embryo Today 105:265–277
La Rosa PS, Warner BB, Zhou Y, Weinstock GM, Sodergren E, Hall-Moore CM, Stevens HJ et al (2014) Patterned progression of bacterial populations in the premature infant gut. Proc Natl Acad Sci U S A 111:12522–12527
Lacroix B et al (1984) Developmental pattern of brush border enzymes in the human fetal colon. Correlation with some morphogenetic events. Early Hum Dev 9:95–103
Leatham MP, Banerjee S, Autieri SM, Mercado-Lubo R, Conway T, Cohen PS (2009) Precolonized human commensal Escherichia coli strains serve as a barrier to E. coli O157:H7 growth in the streptomycin-treated mouse intestine. Infect Immun 77:2876–2886
Lebenthal A, Lebenthal E (1999) The ontogeny of the small intestinal epithelium. J Parenter Enteral Nutr 23:S3–S6
Lebenthal E, Lee PC (1980) Development of functional responses in human exocrine pancreas. Pediatrics 66:556–560
Lebenthal E, Rossi TM, Nord KS, Branski D (1981) Recurrent abdominal pain and lactose absorption in children. Pediatrics 67:828–832

Lu L, Yu Y, Guo Y, Wang Y, Chang EB, Claud EC (2015) Transcriptional modulation of intestinal innate defense/inflammation genes by preterm infant microbiota in a humanized gnotobiotic mouse model. PLoS One 10:e0124504

MacDonald TT, Spencer J (1994) Ontogeny of the gut-associated lymphoid system in man. Acta Paediatr Suppl 83:3–5

Madan JC, Farzan SF, Hibberd PL, Karagas MR (2012) Normal neonatal microbiome variation in relation to environmental factors, infection and allergy. Curr Opin Pediatr 24:753–759

Mai V, Young CM, Ukhanova M, Wang X, Sun Y, Casella G, Theriaque D, Li N, Sharma R, Hudak M, Neu J (2011) Fecal microbiota in premature infants prior to necrotizing enterocolitis. PLoS One 6:e20647

Malo C (1991) Multiple pathways for amino acid transport in brush border membrane vesicles isolated from the human fetal small intestine. Gastroenterology 100:1644–1652

Marcobal A, Sonnenburg JL (2012) Human milk oligosaccharide consumption by intestinal microbiota. Clin Microbiol Infect 18:12–15

Maynard CL, Elson CO, Hatton RD, Weaver CT (2012) Reciprocal interactions of the intestinal microbiota and immune system. Nature 489:231–241

Mazmanian SK, Round JL, Kasper DL (2008) A microbial symbiosis factor prevents intestinal inflammatory disease. Nature 453:620–625. https://doi.org/10.1038/nature07008

McVey Neufeld KA, Mao YK, Bienenstock J, Foster JA, Kunze WA (2013) The microbiome is essential for normal gut intrinsic primary afferent neuron excitability in the mouse. Neurogastroenterol Motil 25:183–e88

Meetze WH, Valentine C, McGuigan JE, Conlon M, Sacks N, Neu J (1992) Gastrointestinal priming prior to full enteral nutrition in very low birth weight infants. J Pediatr Gastroenterol Nutr 15:163–170

Meier P, Patel A, Esquerra-Zwiers A (2017) Donor human milk update: evidence, mechanisms, and priorities for research and practice. J Pediatr 180:15–21

Meinzen-Derr J, Poindexter B, Wrage L, Morrow AL, Stoll B, Donovan EF (2009) Role of human milk in extremely low birth weight infants' risk of necrotizing enterocolitis or death. J Perinatol 29:57–62

Mellander L, Carlsson B, Jalil F, Söderström T, Hanson LA (1985) Secretory IgA antibody response against Escherichia coli antigens in infants in relation to exposure. J Pediatr 107:430–433

Montgomery RK et al (1999) Development of the human gastrointestinal tract: twenty years of progress. Gastroenterology 116:702–731

Moxey PC, Trier JS (1978) Specialized cell types in the human fetal small intestine. Anat Rec 191:269–285

Murk W, Risnes KR, Bracken MB (2011) Prenatal or early-life exposure to antibiotics and risk of childhood asthma: a systematic review. Pediatrics 127:1125–1138

Nahmias A, Stoll B, Hale E, Ibegbu C, Keyserling H, Innis-Whitehouse W, Holmes R, Spira T, Czerkinsky C, Lee F (1991) IgA-secreting cells in the blood of premature and term infants: normal development and effect of intrauterine infections. Adv Exp Med Biol 310:59–69

Nakamura N, Gaskins HR, Collier CT, Nava GM, Rai D, Petschow B et al (2009) Molecular ecological analysis of fecal bacterial populations from term infants fed formula supplemented with selected blends of prebiotics. Appl Environ Microbiol 75:1121–1128

Nanthakumar N, Meng D, Goldstein AM, Zhu W, Lu L, Uauy R, Llanos A, Claud EC, Walker WA (2011) The mechanism of excessive intestinal inflammation in necrotizing enterocolitis: an immature innate immune response. PLoS One 6:e17776

Neish AS, Naumann M (2011) Microbial-induced immunomodulation by targeting the NF-κB system. Trends Microbiol 19:596–605

Neu J (2015) Developmental aspects of maternal-fetal, and infant gut microbiota and implications for long-term health. Matern Health Neonatol Perinatol 1:6

Neu J, Walker WA (2011) Necrotizing enterocolitis. N Engl J Med 364:255–264

Oeckinghaus A, Ghosh S (2009) The NF-κB family of transcription factors and its regulation. Cold Spring Harb Perspect Biol 1:a000034

Olsen R, Greisen G, Schrøder M, Brok J (2016) Prophylactic probiotics for preterm infants: a systematic review and meta-analysis of observational studies. Neonatology 109:105–112

Palmer C, Bik EM, DiGiulio DB, Relman DA, Brown PO (2007) Development of the human infant intestinal microbiota. PLoS Biol 5:e177

Park YK, Monaco MH, Donovan SM (1999) Enteral insulin-like growth factor-I augments intestinal disaccharidase activity in piglets receiving total parenteral nutrition. J Pediatr Gastroenterol Nutr 29:198–206

Patel RM, Myers LS, Kurundkar AR, Maheshwari A, Nusrat A, Lin PW (2012) Probiotic bacteria induce maturation of intestinal claudin 3 expression and barrier function. Am J Pathol 180:626–635

Penn AH, Altshuler AE, Small JW, Taylor SF, Dobkins KR, Schmid-Schönbein GW (2012) Digested formula but not digested fresh human milk causes death of intestinal cells in vitro: implications for necrotizing enterocolitis. Pediatr Res 72:560–567

Petersson J, Schreiber O, Hansson GC, Gendler SJ, Velcich A, Lundberg JO, Roos S, Holm L, Phillipson M (2011) Importance and regulation of the colonic mucus barrier in a mouse model of colitis. Am J Physiol Gastrointest Liver Physiol 300:G327–G333

Plaza-Zamora J, Sabater-Molina M, Rodríguez-Palmero M et al (2013) Polyamines in human breast milk for preterm and term infants. Br J Nutr 10:524–528

Poulsen SS, Kryger-Baggesen N, Nexø E (1996) Immunohistochemical localization of epidermal growth factor in the second-trimester human fetus. Histochem Cell Biol 105:111–117

Quigley M, McGuire W (2014) Formula versus donor breast milk for feeding preterm or low birth weight infants. Cochrane Database Syst Rev 22:CD002971

Rakoff-Nahoum S, Hao L, Medzhitov R (2006) Role of toll-like receptors in spontaneous commensal-dependent colitis. Immunity 25:319–329

Raul F, Gosse F, Doffoel M, Darmenton P, Wessely JY (1988) Age related increase of brush border enzyme activities along the small intestine. Gut 29:1557–1563

Reinhardt C, Reigstad CS, Bäckhed F (2009) Intestinal microbiota during infancy and its implications for obesity. J Pediatr Gastroenterol Nutr 48:249–256

Riezzo G, Indrio F, Raimondi F, Montagna O, Salvia G, Massimo B, Polimeno L, Cavallo L, Francavilla R (2009) Maturation of gastric electrical activity, gastric emptying and intestinal permeability in preterm newborns during the first month of life. Ital J Pediatr 35:6. https://doi.org/10.1186/1824-7288-35-6

Rogier EW, Frantz AL, Bruno MEC, Wedlund L, Cohen DA, Stromberg AJ, Kaetzel CS (2014) Secretory antibodies in breast milk promote long-term intestinal homeostasis by regulating the gut microbiota and host gene expression. Proc Natl Acad Sci U S A 111:3074–3079. https://doi.org/10.1073/pnas.1315792111

Rondeau MP, Meltzer K, Michel KE, McManus CM, Washabau RJ (2003) Short chain fatty acids stimulate feline colonic smooth muscle contraction. J Feline Med Surg 5:167–173

Schanler RJ, Lau C, Hurst NM, Smith EO (2005) Randomized trial of donor human milk versus preterm formula as substitutes for mothers' own milk in the feeding of extremely premature infants. Pediatrics 116:400–406

Sigge W, Wedel T, Kühnel W, Krammer HJ (1998) Morphologic alterations of the enteric nervous system and deficiency of non-adrenergic non-cholinergic inhibitory innervation in neonatal necrotizing enterocolitis. Eur J Pediatr Surg 8:87–94

Sisk PM, Lovelady CA, Dillard RG, Gruber KJ, O'Shea TM (2007) Early human milk feeding is associated with a lower risk of necrotizing enterocolitis in very low birth weight infants. J Perinatol 27:428–433

Smith K, McCoy KD, Macpherson AJ (2007) Use of axenic animals in studying the adaptation of mammals to their commensal intestinal microbiota. Semin Immunol 19:59–69

Song SJ, Dominguez-Bello MG, Knight R (2013) How delivery mode and feeding can shape the bacterial community in the infant gut. Can Med Assoc J 185:373–374

Srinivasjois R, Rao S, Patole S (2013) Prebiotic supplementation in preterm neonates: updated systematic review and meta-analysis of randomised controlled trials. Clin Nutr 32:958–965
Taylor SN, Basile LA, Ebeling M, Wagner CL (2009) Intestinal permeability in preterm infants by feeding type: mother's milk versus formula. Breastfeed Med 4:11–15
Teitelbaum JE, Walker WA (2002) Nutritional impact of pre- and probiotics as protective gastro-intestinal organisms. Annu Rev Nutr 22:107–138
Ter Beek WP, Muller ESM, van den Berg M, Meijer MJ, Biemond I, Lamers CB (2008) Motilin receptor expression in smooth muscle, myenteric plexus, and mucosa of human inflamed and noninflamed intestine. Inflamm Bowel Dis 14:612–619
Terrin G, Passariello A, Curtis MD, Manguso F, Salvia G, Lega L, Messina F, Paludetto R, Canani RB (2012) Ranitidine is associated with infections, necrotizing enterocolitis, and fatal outcome in newborns. Pediatrics 129:e40–e45
Thibault L, Ménard D, Loirdighi N, Levy E (1992) Ontogeny of intestinal lipid and lipoprotein synthesis. Biol Neonate 62:100–107
Tormo-Badia N, Håkansson Å, Vasudevan K, Molin G, Ahrné S, Cilio CM (2014) Antibiotic treatment of pregnant non-obese diabetic mice leads to altered gut microbiota and intestinal immunological changes in the offspring. Scand J Immunol 80:250–260
Triadou N, Zweibaum A (1985) Maturation of sucrase-isomaltase complex in human fetal small and large intestine during gestation. Pediatr Res 19:136–138
Underwood MA, Kalanetra KM, Bokulich NA, Mirmiran M, Barile D, Tancredi DJ, German JB, Lebrilla CB, Mills DA (2014) Prebiotic oligosaccharides in premature infants. J Pediatr Gastroenterol Nutr 58:352–360
Underwood MA, German JB, Lebrilla CB, Mills DA (2015a) Bifidobacterium longum subspecies infantis: champion colonizer of the infant gut. Pediatr Res 77:229–235
Underwood MA, Gaerlan S, De Leoz MLA, Dimapasoc L, Kalanetra KM, Lemay DG, German JB, Mills DA, Lebrilla CB (2015b) Human milk oligosaccharides in premature infants: absorption, excretion, and influence on the intestinal microbiota. Pediatr Res 78:670–677
Van Itallie CM, Anderson JM (2014) Architecture of tight junctions and principles of molecular composition. Semin Cell Dev Biol 36:157–165
Vohr BR, Poindexter BB, Dusick AM, McKinley LT, Higgins RD, Langer JC, Poole WK, National Institute of Child Health and Human Development National Research Network (2007) Persistent beneficial effects of breast milk ingested in the neonatal intensive care unit on outcomes of extremely low birth weight infants at 30 months of age. Pediatrics 120:e953–e959
Vukavic T (1983) Intestinal absorption of IgA in the newborn. J Pediatr Gastroenterol Nutr 2:248–251
Wedel T, Krammer HJ, Kühnel W, Sigge W (1998) Alterations of the enteric nervous system in neonatal necrotizing enterocolitis revealed by whole-mount immunohistochemistry. Pediatr Pathol Lab Med 18:57–70
Xu RJ, Mellor DJ, Tungthanathanich P, Birtles MJ, Reynolds GW, Simpson HV (1992) Growth and morphological changes in the small and the large intestine in piglets during the first three days after birth. J Dev Physiol 18:161–172
Yassour M, Vatanen T, Siljander H, Hämäläinen A-M, Härkönen T, Ryhänen SJ, Xavier RJ, DIABIMMUNE Study Group et al (2016) Natural history of the infant gut microbiome and impact of antibiotic treatment on bacterial strain diversity and stability. Sci Transl Med 8:343–381
Yu Z-T, Chen C, Kling DE, Liu B, McCoy JM, Merighi M, Heidtman M, Newburg DS (2013) The principal fucosylated oligosaccharides of human milk exhibit prebiotic properties on cultured infant microbiota. Glycobiology 23:169–177
Yu Y, Lu L, Sun J, Petrof EO, Claud EC (2016) Preterm infant gut microbiota affects intestinal epithelial development in a humanized microbiome gnotobiotic mouse model. Am J Physiol Gastrointest Liver Physiol 311:G521–G532

Chapter 2
Microbiome: Allergic Diseases of Childhood

Ozge Nur Aktas, Benjamin Turturice, David L. Perkins, and Patricia W. Finn

Abstract Recent interest has focused on the microbiome modulating immune responses, and thus playing a significant role in the development of many diseases, including allergic responses. "Dysbiosis," alteration in the normal microbiome, does have an effect on inflammation and may also influence the course of the disease. In this chapter, we discuss the influence of the microbiome on common pediatric allergic diseases: food allergy, atopic dermatitis, asthma, and allergic rhinitis. These diseases stem from common immune mechanisms, and are part of a progressive "Atopic March" phenomenon, which will be introduced in this chapter. We explain how the microbiome is related to allergic diseases in both human and murine studies. Studying the microbiome in the context of allergic diseases has the potential to elucidate ways of manipulating the microbiome in disease in general. With further studies in the field, we may be able to modulate the immune response and the disease course by understanding the relationship between the microbiome and the immune response.

O.N. Aktas
Koc University School of Medicine, Istanbul, Turkey

B. Turturice
Department of Microbiology and Immunology, University of Illinois at Chicago, Chicago, IL, USA

D.L. Perkins (✉)
Division of Nephrology, Department of Medicine, University of Illinois at Chicago, Chicago, IL, USA

Department of Surgery, University of Illinois at Chicago, Chicago, IL, USA

Department of Bioengineering, University of Illinois at Chicago, Chicago, IL, USA
e-mail: perkinsd@uic.edu

P.W. Finn
Department of Microbiology and Immunology, University of Illinois at Chicago, Chicago, IL, USA

Division of Pulmonary, Critical Care, Sleep, and Allergy, Department of Medicine, University of Illinois at Chicago, Chicago, IL, USA
e-mail: pwfinn@uic.edu

J. Sun, P.K. Dudeja (eds.), *Mechanisms Underlying Host-Microbiome Interactions in Pathophysiology of Human Diseases*, Physiology in Health and Disease,
https://doi.org/10.1007/978-1-4939-7534-1_2

2.1 Introduction

The prevalence of allergic diseases affecting the pediatric population has been increasing during the last decade. Early-life exposure to microbes decreases the risk of developing allergic disease, as articulated in the hygiene hypothesis (Ball et al. 2000; Kramer et al. 1999; Strachan 1989, 1997). Also, exposure to a protected modern life-style environment may lead to decreased allergen exposure, potentially creating an immune system that is intolerant to allergens. Taken together, allergens may include food, topical exposures, or aero-allergens. In this chapter, we review four common types of allergic diseases of childhood: food allergy, atopic dermatitis (AD), asthma, and allergic rhinitis (AR). The allergic diseases are related to each other in that having one of these diseases early in life increases the risk of acquiring another allergic disease when older. This temporal development of allergic diseases is termed the "Atopic March" (Bantz et al. 2014). Specifically, progressive allergic disease history typically begins with food allergy and/or AD and leads to an increased risk of developing asthma and AR later in life. This "progressive" clinical course illustrates the potential of common immunological pathways mediating allergic responses. This putative common pathway suggests that intervention or treatment in earlier atopic diseases might prevent the development of later allergic phenotypes.

The mechanisms that determine sensitivity or tolerance to a specific allergen are complex. Numerous types of immune cells reside in the tissues (e.g., the lymphoid system) or in the bloodstream. In this chapter, we focus on a specific cell type termed T helper 2 (Th2) cells, which play a leading role in the development of allergic diseases, and describe other cell types and receptors as well. Th2 cells produce "signaling molecules" called interleukins (ILs) including IL-4, IL-5, and IL-13, and immunoglobulin E (IgE) antibodies. When Th2 cells encounter an allergen, they trigger allergic symptoms such as itching, erythema, sneezing, or an aberrant response to a food (e.g., diarrhea). Another type of T cell, the regulatory T cell (Treg), may also regulate allergic responses. Tregs are suppressor cells that can inhibit the response of other cell types, including Th2 cells and mast cells; thus, the aberrant or deficient function of Tregs can also lead to allergic responses. Mast cells and eosinophils also play roles in developing allergic responses by producing histamine and other inflammatory cytokines. Being a common mechanism for allergic diseases mentioned, Th2-related immune responses should be considered a systemic response, rather than local inflammation existing in a single organ. For instance, even diseases not directly affecting the intestinal tissue itself (e.g., asthma) are demonstrated to influence the gut microbiome (Abrahamsson et al. 2014; Arrieta et al. 2015).

Recent interest has also focused on the microbiome modulating immune responses, and thus playing a significant role in the development of allergic responses. Changes in the microbiome have been demonstrated in all four of the allergic diseases previously mentioned. In this chapter, we discuss how the microbiome at different body sites could promote resilience or susceptibility to allergic diseases, and describe its potential in the inflammatory process of allergic disorders.

2.2 Food Allergy and the Microbiome

Food allergy is defined as an adverse reaction to specific types of food allergens and usually starts in the first 2 years of life with a prevalence of 5–8% (Gupta et al. 2011; Jackson et al. 2013). The diagnosis is highly dependent on patient history together with the help of the diagnostic work-up, including serum-specific Ig analyses, skin prick tests, or oral food challenges. The disease is usually characterized by gastrointestinal symptoms such as diarrhea, vomiting, mucus or blood in the stool or abdominal discomfort. There can be skin manifestations such as eczema or urticaria (hives), which can accompany the clinical picture. In general, food allergies are categorized into three groups: IgE-mediated, non-IgE-mediated food allergy, and mixed reactions. We focus primarily on the IgE-mediated food allergy phenotype as a part of the Atopic March and the ways in which it interacts with the microbiome and the body's immune system.

Why do some people develop adverse reactions to specific types of food, whereas others do not? A healthy response to what we eat depends on the immunological process that the food antigen (Ag) encounters inside our body. A healthy response is actually an unresponsiveness to various food Ags, which we come into contact with every day, referred to as oral tolerance (Faria and Weiner 1999). When this required tolerance is not accomplished, food Ags can induce Th2 cells, producing inflammatory ILs (e.g., IL-4 and IL-13) that induce B cells to produce Ag-specific IgE (Iweala and Burks 2016). Large amounts of IgE bind to the mast cells, which evokes mast cell degranulation and release of inflammatory cytokines such as histamine in the case of re-exposure to the food Ag. This reaction chain leads to symptoms such as diarrhea, gastrointestinal upset, or anaphylaxis, a multisystemic, life-threatening reaction against a food allergen. Oral tolerance was originally considered to be mediated by food Ag-specific Tregs. Ag encountered in the gut lamina propria is endocytosed and transported by CD103+ intestinal dendritic cells (DCs) to mesenteric lymph nodes (mLNs), inducing differentiation of Ag-specific naïve T cells into Tregs through retinoic acid and TGF-β signaling (Fig. 2.1) (Benson et al. 2007; Coombes et al. 2007; Sun et al. 2007). Additionally, retinoic acid and TGF-β upregulate the gut homing receptors CCR-9 and α4 β7 on these differentiated Tregs to recruit them back into the intestinal lamina propria. The Tregs are enriched via IL-10 secreted by residing CX3CR1 macrophages in the gut (Hadis et al. 2011). Some Tregs circulate via the bloodstream to develop

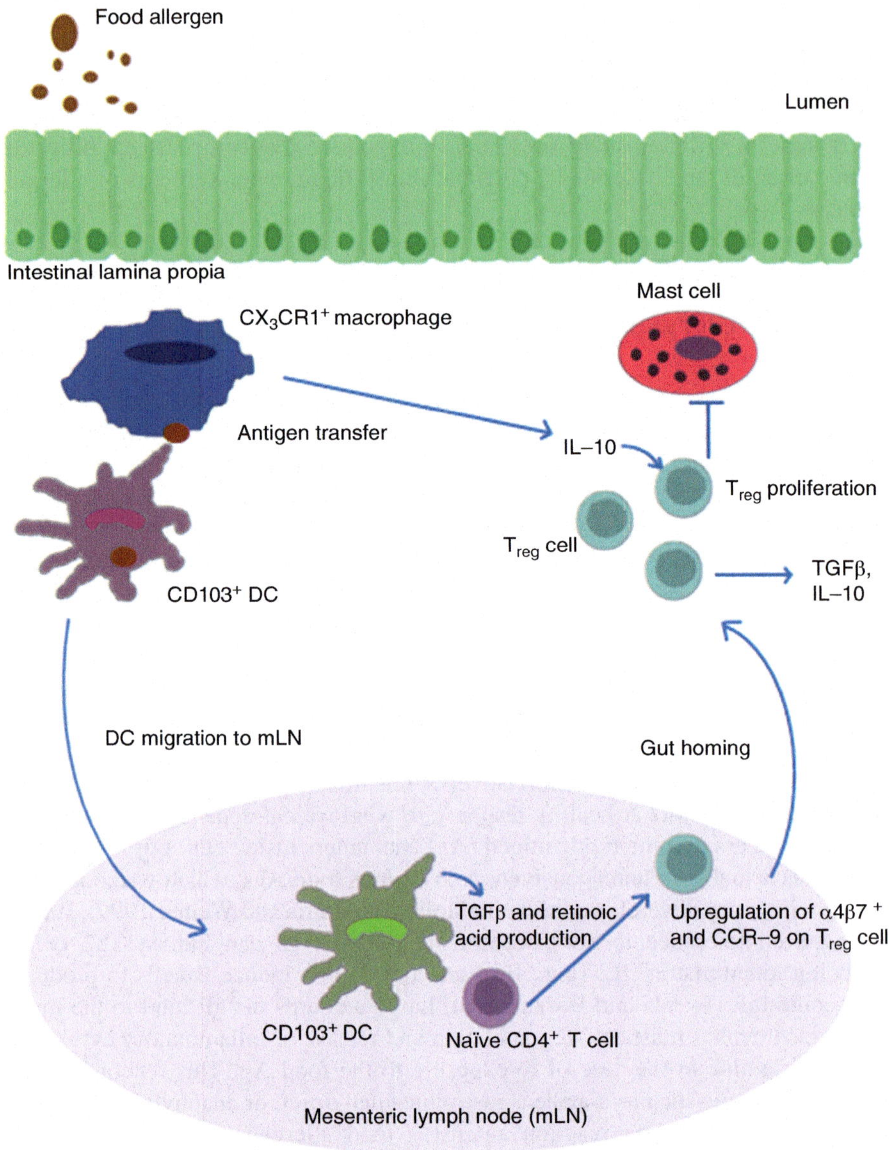

Fig. 2.1 Development of oral tolerance to food allergens. CX3CR1+ macrophages transfer antigens from the intestinal lumen to dendritic cells (DCs) in the intestinal lamina propria. A subset of DCs migrates to the mesenteric lymph nodes, where the DCs express transforming growth factor-β (TGF-β) and retinoic acid, thereby inducing the differentiation of naïve T cells and upregulating the gut homing receptors CCR-9 and α4β7 on differentiated Tregs to recruit them back into the intestinal lamina propria. Macrophages also secrete IL-10, which leads to Treg proliferation in the lamina propria

systemic tolerance to oral Ags (Husby et al. 1985). Any error in these pathways impairs the development of oral tolerance. As an example, oral Ag exposure induces murine allergic responses when vitamin A is deficient (Yokota-Nakatsuma et al. 2014). Retinoic acid-deficient mLN DCs induce abnormal Th2 type rather than Treg responses, precipitating aberrant responses to food specific Ag (Yokota-Nakatsuma et al. 2014).

The interplay between the intestinal microbiome and the immune system has an impact on oral tolerance. Spore-forming Clostridia species (Firmicutes) are said to protect against an inflammatory response in the gut by induction of colonic Tregs (Atarashi et al. 2011, 2013). Clostridia-induced Tregs are associated with decreased food allergen sensitization (Atarashi et al. 2011, 2013; Stefka et al. 2014). Additionally, food Ag exposure can affect the quantity of Tregs, suggesting that food Ags and bacteria work together to mediate immune balance and oral tolerance. Interestingly, different types of stimulation (either from bacteria or food Ag) favor a different subgroup of Tregs. Administration of antibiotics to mice reduced retinoic acid receptor-related orphan receptor gamma t positive (RORγt^{+}) Tregs in colonic lamina propria, which interact with the microbes, whereas an Ag-free diet resulted in decreased RORγt^{-} Tregs in the small intestine (Kim et al. 2016). Not only the type, but also the location of the Tregs differed. One potential explanation for these differences may be based on the frequency of encountering a food Ag or bacteria in different parts of the intestine. The small intestine has a large surface area to welcome all food allergens and handle most digestive function; yet, the colon is known to be enriched with the microbiota. Thus, both food Ag and bacteria induced Tregs in different parts of the gut may either influence each other or work together to establish an appropriate immune response in the intestine.

Gut colonization in early infancy has an impact on the risk of a child developing food sensitization, or a food allergy later on. Low levels of Bacteroidaceae and high levels of Enterobacteriaceae (a member of the Proteobacteria phylum) are observed in the gut microbiome of food-sensitized infants at 3 months and 1 year of age respectively (Azad et al. 2015). Lower microbiota richness was demonstrated in 3-month-old subjects in the same study (Azad et al. 2015), signifying that early gut colonization can be a determinant of developing food sensitization or a robust clinical picture of food allergy.

The intestinal microbiota also varies depending on multiple factors, such as mode of delivery (Biasucci et al. 2010; Dominguez-Bello et al. 2010; Fallani et al. 2010), postnatal age (Backhed et al. 2015; Fujimura et al. 2016), diet (Azad et al. 2016; David et al. 2014; Jost et al. 2014), and antibiotic use (Choo et al. 2017; Guo et al. 2017). In children, the gut microbiome is under the influence of other unique factors, such as breastfeeding. A recent study exploring the gut microbiome in the first year of life demonstrated that an important shift in the gut microbiome occurs after the cessation of breastfeeding (Backhed et al. 2015). Breastfed infants have a high abundance of *Bifidobacterium*, *Lactobacillus*, and some members of the Clostridia class (Collinsella and Veillonella), which were also found to be members of breast milk microbiota in previous studies (Backhed et al. 2015; Jost et al. 2014).

Thus, breastfeeding is an important factor that could shape the intestinal microbiome.

The microbiome may modulate the immune system, interfering with the risk of developing an allergy, as demonstrated in previous studies (Stefka et al. 2014; Sudo et al. 1997). Reconstitution of intestinal flora of germ-free mice with *Bifidobacterium infantis* restores the oral tolerance response by controlling Th2 skewing and increased IL-4 production with oral allergen ovalbumin (OVA) challenge (Sudo et al. 1997). Interestingly, this effect was only observed in young progeny during the neonatal period, but not in adult mice. These data suggest that there might be a window for allergic sensitization and intervention to recover a healthy immune response in the intestine. In a similar gnotobiotic (germ-free) mouse model, mice treated with antibiotics showed increased susceptibility to peanut allergen, characterized by an increase in peanut-specific IgE and anaphylaxis symptoms in the absence of microbiota (Stefka et al. 2014).

Toll-like receptor (TLR) is a pattern recognition receptor in the innate immune system that can detect microbial DNA, lipoteichoic acid (LTA), or lipopolysaccharides (LPS). TLR4, a specific subtype detecting LPS, plays an important role in limiting inflammation in the gut (Rakoff-Nahoum et al. 2004). Introduction of LPS to germ-free mice establishes oral tolerance (Bashir et al. 2004). TLR4-deficient mice exposed to intragastric administration of peanut allergen, together with cholera toxin, generate increased allergen-specific IgE response compared with TLR4-sufficient mice (Bashir et al. 2004). In addition to TLR receptors leading to host responses, metabolites of commensal bacteria in the gut, termed short chain fatty acids (SCFAs), also have an effect on inducing immune responses. SCFAs are released via metabolism of indigestible fibers by commensal bacteria in the gut (Sonnenburg and Backhed 2016). Acetate, propionate, and butyrate are examples of SCFAs produced by intestinal commensal inhabitants that induce a local and systemic immunomodulatory response (Arpaia et al. 2013; Brown et al. 2003; Smith et al. 2013; Tan et al. 2016; Thangaraju et al. 2009). These SCFAs signal through G protein-coupled receptors (Thangaraju et al. 2009; Singh et al. 2014), and inhibit histone deacetylases, affecting transcriptional regulation (Furusawa et al. 2013). Furthermore, low levels of SCFAs are associated with an increased inflammatory response and allergic phenotypes, as increasing levels of SCFAs can prevent the development of allergic diseases (Tan et al. 2014; Thorburn et al. 2015; Trompette et al. 2014). The proposed mechanism of the prevention of allergic response by SCFAs may stem from the induction of colonic Tregs. In previous murine studies, a high fiber diet increased SCFA production, and protected against colonic inflammation and food allergy (Smith et al. 2013; Tan et al. 2016). Furthermore, in a recent study, mice fed a high-fiber diet showed significantly reduced symptoms of anaphylaxis and lower IgE levels compared with those on a non-fiber diet. In the same study, protection from food allergy also correlated with decreased levels of Th2-specific cytokines, such as IL-4, IL-5, and IL-13 from stimulated lymphocytes in MLNs by using peanut extract (Tan et al. 2016). These examples support the hypothesis that the intestinal microbiome has an influence on intestinal homeostasis and the prevention of food sensitization by means of their metabolites.

Previous studies of food allergy influencing alteration of the gut microbiome mainly focused on IgE-mediated food allergies. The relative abundance of Bacteroidetes, Proteobacteria, and Actinobacteria phyla in food allergy subjects are significantly reduced compared with the healthy gut microbiome (Ling et al. 2014). In the same study, infants with IgE-mediated food allergy had increased levels of *Clostridium sensu stricto* and *Anaerobacter*, and decreased levels of *Bacteroides* and *Clostridium* cluster XVIII (Ling et al. 2014). Even though Clostridia are thought to play a protective role in the gut, different species belonging to the *Clostridium* genus, as seen in the example above, could either be influenced by inflammation or affect the inflammation process itself in a different way. One explanation for this observation could be that environmental change inside the gut lumen may alter the balance between different microbial communities, and how they interact with each other. Another explanation could be that microbial communities are not only influenced by the disease process, but are active decision-makers according to varying conditions. Different species may increase or decrease their abundance to maintain functional balance inside a niche to fight against an unusual state (e.g., a food allergy) and gut inflammation.

Human breast milk is the first food introduced to the infant gut for most newborns. Breast milk has many bioactive components and unique nutritional factors that help the development of the innate immune system of the infants (Cacho and Lawrence 2017). Previous work suggests a protective effect of breast milk on the development of atopic diseases, including food allergy (Gdalevich et al. 2001a, b; Midodzi et al. 2010; Mimouni Bloch et al. 2002; Muraro et al. 2004; Sears et al. 2002; Snijders et al. 2007). Breastfeeding was proposed to be prophylactic against atopic diseases in a previous study with a 17-year follow-up (Saarinen and Kajosaari 1995). According to that study, the prevalence of food allergy was highest at 1–3 years in infants who had been breastfed for less than 1 month ($p = 0.02$, ANOVA) (Saarinen and Kajosaari 1995). Other studies also support the protective effect of breastfeeding on atopic diseases (Gdalevich et al. 2001a, b; Midodzi et al. 2010; Mimouni Bloch et al. 2002; Muraro et al. 2004; Snijders et al. 2007; Oddy et al. 1999). Not only the bioactive components but also the microbiota of breast milk are under investigation with the advance of nonculture-dependent sequencing technology. Breast milk is shown to shape the gut microbiome during infancy (Jost et al. 2014, 2015; Benito et al. 2015; Martin et al. 2012; Pannaraj et al. 2017), and the effect of the breast milk microbiome on food allergies is yet to be discovered.

Overall, the immune system and the microbiome could interact and alter the course of the allergic process. An initial inflammatory response may be induced by altered quantities, functions or metabolites of the intestinal microbiota. In addition, although murine studies provide valuable information, one limitation is the exposure to only specific types of model food (e.g., OVA). Thus, the effect of a complex diet on intestinal Tregs and the intestinal environment has not yet been clarified

(Plunkett and Nagler 2017). More studies are needed to further the understanding of food allergies, and the role of interventions utilizing the microbiome.

2.3 Atopic Dermatitis and the Microbiome

Atopic dermatitis is the most common chronic skin inflammatory disease, with a prevalence of 10–20% among the pediatric population (Flohr and Mann 2014; Silverberg and Simpson 2014; Weidinger and Novak 2016). AD is characterized by intense itching and dry scaly skin, together with recurrent eczematous skin lesions. In about 60% of cases, disease symptoms start in the first year of life, but AD can affect children at any age (Garmhausen et al. 2013; Illi et al. 2004). The first manifestation of the disease is dry, scaly skin in infants. Eczematous lesions are not evident before the second month of life. Despite appropriate treatment, the clinical course of AD can be relapsing-remitting (Garmhausen et al. 2013; Illi et al. 2004). Birth cohort studies demonstrate that in up to 70% of cases, the disease greatly improves or resolves by late childhood (Illi et al. 2004; Peters et al. 2010).

The strongest risk factor for the development of AD is family history (Apfelbacher et al. 2011). Genetic factors can affect the development of AD. The most common and well-known genetic determinant is the filaggrin (FLG) mutation, which, under normal conditions, encodes for a key structural protein in the epidermis (Irvine et al. 2011). Thus, it is an important element to maintain the skin barrier and its integrity. Although the FLG mutation can play a role in some affected individuals, most AD patients do not have the FLG mutation (Irvine et al. 2011); thus, it is neither sufficient nor necessary for development of the disease.

In addition to genetic factors, cutaneous inflammation is a hallmark of AD characterized by CD4+ cell infiltration. Even unaffected skin areas in patients with AD have signs of subclinical inflammation with Th2 cells, Th22 cells, and to a lesser extent, Th17 cells (Suarez-Farinas et al. 2011). Furthermore, an increased number of type 2 innate lymphoid cells, which produce Th2-type ILs, is found in the skin lesions, and contributes to local inflammation (Imai et al. 2013; Salimi et al. 2013). Th2-type cytokines such as IL-4, IL-5, and IL-13 have been shown to play a role in AD pathogenesis. In murine studies, IL-4 and IL-13 induce eczema-like skin features (Chan et al. 2001; Chen et al. 2004; Zheng et al. 2009). In human studies, IL-4 and IL-5 are evident in higher amounts, together with IL-13 in either acute or chronic cases of AD (Hamid et al. 1994, 1996). One example of the relationship of IL-4 with the skin barrier is that IL-4 alters the expression of multiple genes, which, in turn, leads to the aberrant regulation of epidermal barrier function (Sehra et al. 2010). A compromised barrier can increase the vulnerability of skin, allowing penetration of allergens and bacteria, which could then contribute to the development of AD.

The skin, known to be the largest organ, harbors a diverse microbiota. A healthy skin microbiome consists of the genera *Propionibacterium*, *Corynebacterium*, *Staphylococcus*, and *Streptococcus* (Shi et al. 2016). Microbiome composition is dependent on many factors, including age. The skin microbiome was previously demonstrated to vary in healthy adults and children (Shi et al. 2016). The diversity of the healthy skin microbiome is distinct, depending on age. Younger children have a greater diversity of skin microbiota than adults (Shi et al. 2016). However, age is not the only factor that can affect the skin microbiota. Inflammatory diseases of the skin, in particular AD, can alter the composition and diversity of the skin microbiome. Overall, bacterial diversity of the skin in AD patients is lower than in healthy subjects (Kong et al. 2012). Even among AD patients, the sampling site and phase of the disease, which is individual and disease course-specific, are important determinants of the skin microbiome. Lesion and nonlesion sites can differ tremendously in bacterial content, and warrant further analyses. *Staphylococcus* was significantly more abundant in nonlesional skin of AD patients compared with healthy subjects (Shi et al. 2016). *Staphylococcus aureus* also is more abundant in the flare period of AD, with an increase in *S. epidermidis*, whereas the abundance of *Streptococcus*, *Propionibacterium*, and *Corynebacterium* are decreased in the patients' skin microbiota (Kong et al. 2012).

Atopic dermatitis is one of the first manifestations of atopy at early ages. Early changes in the skin microbiome, cutaneous dysbiosis, may lead to AD in later life. Two-month-old infants who developed AD in follow-ups had significantly lower numbers of commensal *Staphylococcus* species in their antecubital fossae, which is a very common site for AD presentation in that age group (Kennedy et al. 2017). These data suggest that dysbiosis in the skin microbiome might be associated with the process of developing AD at very early ages.

Fungi are less abundant members of the skin microbiome that share the same niche with bacteria on the skin. Although less frequently studied, fungal members of the skin microbiome are altered in AD patients. Nearly all *Malassezia* members are reported to be depleted in AD patients. Interestingly, enrichment of *Malassezia dermatis* and the increased diversity of non-*Malassezia* species (*Aspergillus*, *Candida albicans*, and *Cryptococcus diffluens*) are shown in AD (Chng et al. 2016; Oh et al. 2013; Zhang et al. 2011). There is an alteration of the skin microbiome affecting various members in AD, but further studies are needed to understand the significance of this change, and how it could correlate with the immune response in the skin.

The relationship among the immune system, the host cells, and the skin microbiome is a subject of intense research. Th2-mediated inflammatory mechanisms can affect microbial availability and interaction with the skin barrier. An increase in Th2-type responses promotes *S. aureus* binding and colonization in the skin (Brauweiler et al. 2014). Moreover, Th2-mediated inflammation and cytokine production can indirectly interfere with natural habitants in the skin. As an example, IL-4 and IL-13 inhibit the production of antimicrobial peptides in the skin, predisposing the skin to *S. aureus* infection, leading to inflammation and disease exacerbation in AD patients (Eyerich et al. 2009; Kasraie et al. 2010; Lehmann et al. 2004; Nakamura et al. 2013; Niebuhr et al. 2010; Ong et al. 2002).

In addition to Th2-mediated mechanisms, mast cells (MCs) also interact with microbial communities. MCs are derived from bone marrow and are not released to the bloodstream as mature cells. They settle in peripheral sites of the body, such as the gastrointestinal system or skin, and differentiate maturity induced by cytokines in the surrounding microenvironment (Halova et al. 2012; Liu et al. 2010; Metcalfe et al. 1997). MCs produce IL-22, which promotes epidermal hyperplasia. They also induce antimicrobial peptides in both human and murine studies, and elicit a protective function at lesion sites (Mashiko et al. 2015; Wolk et al. 2006; Zheng et al. 2007). The way in which MCs communicate with the skin microbiome is not well defined. Notably, murine MCs mature with the help of skin microbiota. Germ-free mice contain largely undifferentiated MCs, and administration of the Gram-positive cell wall component, LTA, to germ-free mice induces MC maturity, underscoring the importance of the microbiome in the differentiation of immune cells (Wang et al. 2017).

Thus, there is increasing evidence that the skin microbiome is an essential part of the pathogenesis of AD. Additional studies are needed to explore the changes in the skin microbiome, the ways in which the members of the microbiome interact with each other, and the inflammatory response created by AD.

2.4 Allergic Diseases of the Airways and the Microbiome

Asthma and AR are additional allergic diseases that are subsets of the "Atopic March," and are highly associated with AD and food allergies seen at early ages. Asthma is a chronic inflammatory disease of the airways. Airway inflammation subsequently causes airway hyperresponsiveness, limitation of airflow, and respiratory symptoms such as wheezing. Mast cells, eosinophils, T lymphocytes, macrophages, and neutrophils play a role in the pathophysiology of asthma, creating a hyperreactive airway that leads to airway obstruction. In this section, we focus on atopic asthma in the early years of life. Notably, demonstration of aero-allergen-specific Th2 cells in cord blood underscores that the allergic inflammatory response is initiated in utero, earlier than previously thought (Piccinni et al. 1993; Prescott et al. 1998; Yabuhara et al. 1997).

Allergic asthma is described primarily as a Th2-mediated disorder involving a complex immunological response. Th2-related cytokines (IL-4, IL-5, and IL-13) lead to IgE overproduction, an increase in the eosinophil count, and airway hyperresponsiveness. Tregs, defined as a control for Th2-mediated response, are decreased in asthmatics. In contrast, natural killer (NK) cells, which produce more Th1- and Th2-related cytokines, worsen the inflammatory response (Akbari et al. 2006; Larche et al. 2003). The decision of either tolerance or sensitivity to an aeroallergen does not solely depend on the previously described mechanisms above. Airway mucosal dendritic cells (AMDCs) are responsible for local immunity in the respiratory epithelium (Holt et al. 1990). AMDCs have the ability to communicate with the immune system via transmitting signals from airway mucosa. AMDCs can help to create a balance between Th2 (inflammatory) and Treg (anti-inflammatory)-mediated immune responses to a specific type of aero-allergen. The immunological response created on the mucosal airway surface is not restricted to a local area. There is cross-talk between mucosal immunity and the central immune system through AMDCs. The immunological processes in asthma are not limited to the cell types or ILs described in this chapter. Other environmental and genetic influences play a role in asthma pathogenesis. For this section, we focus on the microbiome and how microbial communities could affect the development and prognosis of asthma.

The analysis of the airway microbiome has been a developing topic over the last decade. Until recently, airways were considered to be sterile (Dickson et al. 2016), partly stemming from the challenge that bacteria found in airways are not culturable. Owing to advances in sequencing technology, several studies indicate that the healthy lung harbors a microbial community (Dickson et al. 2017; Hilty et al. 2010; Morris et al. 2013). Even considering differences in the sampling methods, e.g., bronchoalveolar lavage (BAL) fluid or protected specimen brushings (PSBs), the microbiome of the airways detected by bronchoscopy appears to more similar to the oropharynx than the nasopharynx (Morris et al. 2013). In a study of healthy adults, Enterobacteriaceae, *Haemophilus*, *Methylobacterium*, and *Ralstonia* species were demonstrated in the lung and in the oral flora (Morris et al. 2013), although *Tropheryma* was only found in the lung and not inside the mouth (Morris et al. 2013), indicating that oral flora is not the only source of the lung microbial communities. One speculation is that inhalation of the allergen, or dispersion along the bronchial mucosa, could be the other entrance routes of microorganisms to the lower airways (Dickson et al. 2017).

Analyses of the airway microbiota of healthy adults revealed that the most abundant phyla are Bacteroidetes and Firmicutes, and to a lesser extent Actinobacteria and Proteobacteria. This microbial community mainly consists of *Pseudomonas*, *Streptococcus*, *Prevotella*, *Fusobacterium*, *Veillonella*, *Haemophilus*, *Neisseria* and *Porphyromonas* at the genus level (Charlson et al. 2011; Erb-Downward et al. 2011). Although less intensely studied than the bacterial microbiome, fungal and viral components of the murine microbiome, especially fungi, possess a significant relationship with the health or disease status (Noverr et al. 2005).

Even on the first day of life, infants' lungs harbor microbial communities. Healthy airway microbiota on the first day of life consists of Firmicutes and Proteobacteria predominantly, together with Actinobacteria, Bacteroidetes, Tenericutes, Fusobacteria, Cyanobacteria, and Verrucomicrobia (Lal et al. 2016). As seen in healthy subjects, a microbial population is present in the airways. Could an aberrant change in the airway microbiome promote a risk for the development of pulmonary allergic inflammation?

As demonstrated in murine studies, a low diversity and low load of microorganisms are associated with an aberrant immune response and susceptibility to allergic inflammation. Studies using germ-free mice demonstrated that the absence of microbial colonization at early ages increases the risk of an exaggerated response to allergens (Herbst et al. 2011).

The lung microbiome is not the only microbial habitat that could affect the risk of developing asthma. Recent studies examine the relationship of other parts of the airways with asthma. A low diversity nasal microbiome is found to be a key feature in children with asthma compared with healthy children. Additionally, a greater abundance of *Moraxella* in the nasal microbiota is noted in asthmatic children (Depner et al. 2017). The development of a recurrent wheeze and asthma had also been predicted as a result of early-life colonization of the upper respiratory tract with specific members and possible pathogens of the bacterial community, such as *Haemophilus influenzae*, *Moraxella catarrhalis*, and *Streptococcus pneumoniae* (Bisgaard et al. 2007). In children with asthma, the nasal microbiota was found to a have greater abundance of Proteobacteria. In the Childhood Asthma Study (CAS), the nasopharyngeal aspirate (NPA) samples and the clinical information on 234 high-risk infants were examined for the development of asthma/allergy. In this study focusing on the relationship between the NPA microbiome and the dynamics of respiratory health and illnesses, one finding was significant for the risk of developing asthma. Early colonization with *Streptococcus* species in NPA was strongly associated with the risk of the subsequent development of a chronic wheeze, a strong predictor of future asthma, supporting previous studies (Teo et al. 2015). Although there is a debate regarding which species are primarily affected by airway dysbiosis of asthmatic children, the microbiome appears to have a relationship with the pulmonary allergic inflammatory process.

Not only the airway microbiome, but also the microbiome at other body sites could contribute to an increased risk of asthma. The gut microbiome has been proposed to have an influence on asthma development. Stool samples of babies who are sensitized to an allergen in skin prick tests have fewer Lactobacilli, Bacteroidetes, Bifidobacteria, and a greater abundance of Clostridia and Enterococci. Additionally, low gut microbiota diversity in the first month of life is associated with asthma development at 7 years of age (Abrahamsson et al. 2014). In a comprehensive study, the Canadian Healthy Infant Longitudinal Development study, 319 infants were examined to assess the relationship between the gut microbiota and the risk of asthma development. This study highlighted that the risk of asthma development correlated with the gut microbial dysbiosis during the first 100 days of life. Furthermore, the relative abundance of the bacterial genera *Lachnospira*, *Veillonella*, *Faecalibacterium*, and *Rothia* was significantly decreased in children at risk of asthma. In the same study, inoculation of germ-free mice with those four taxa ameliorated airway inflammation in adult mice, supporting the causal role of the microbiome in asthma development (Arrieta et al. 2015). These examples signify that early changes in the intestinal microbiome could be of significance. Also, future therapeutic interventions may target young ages for those at a high risk of asthma development.

In addition to human data, murine studies show the effect of the intestinal microbiome on asthma. Allergic airway inflammation induced by OVA can be limited by feeding a mix of *Clostridium* strains to pathogen-free experimental mice models. Food rich with *Clostridium* mixed strains induces Treg expansion in the circulation, and reduces the inflammatory response to OVA challenge (Atarashi et al. 2011). Another study demonstrated that feeding with bacteria strains is not always necessary for protection from allergic response. A high-fiber diet could also alter the ratio of bacteria in the gut, and increase the level of SCFAs produced as the by-product of bacterial metabolism, which could be protective against allergic inflammation in the lung (Trompette et al. 2014).

Compared with other sites of the body, the lung microbiome is challenging because of the difficulties in sampling the lower airways, but showing the similarity of the microbiome of the upper airways or oropharynx to the lower airways provides insights. Multiple studies illustrated the significant change in the asthmatic airway microbiome compared with healthy airways. Additional studies regarding the microbiome and host interactions are warranted to understand how microbiota change could affect the immunological disease process.

Allergic rhinitis, another component of the "Atopic March," is initiated when exposed to an aero-allergen, which creates an IgE-mediated inflammatory response in the nasal mucosa. AR presents with rhinorrhea, sneezing, pruritis, and nasal congestion. Although it is typically self-limited, AR is a global problem affecting 500 million patients worldwide, and has a major impact on the quality of life (Brozek et al. 2010; Meltzer et al. 2009). AR is the least intensively studied of the allergic diseases in terms of the microbiome. There are a few studies that analyzed the nasal microbiota of AR patients (Choi et al. 2014; Lal et al. 2017). One study proposed that seasonal AR subjects have increased bacterial diversity in the middle meatus compared with non-allergic subjects during the allergy season (Choi et al. 2014). Interestingly, the same study did not demonstrate any difference in microbiota between AR subjects and healthy controls in the nasal vestibule, which emphasizes, once again, the importance of the sampling site in microbiome studies. Furthermore, the age of the subjects recruited to the study could have an effect on the composition of the microbiome. A study comparing the microbiome of nares in healthy children and adults showed that young children have a predominance of Streptococcaceae, other Firmicutes, Bacteroidetes, and alpha/beta Proteobacteria, whereas adult nares are dominated by Corynebacterineae and Propionibacteriaceae (Oh et al. 2012).

To date, information is still limited regarding the ways in which an alteration in the nasal microbiome could affect AR development, prognosis, and immunopathogenesis. Further studies are essential to reveal the relation between AR and microorganisms inhabiting the airways to increase understanding of this disease.

2.5 Future Directions

With improvements in the sequencing technology, whole genome sequencing (WGS) methods have been becoming more accessible and convenient to use. Previous microbiome studies predominantly utilize 16S approaches. Compared with 16S-based approaches, WGS affords greater depth of coverage, even for bacteria with low abundance (Ranjan et al. 2016), and illustrates other members of the microbial community including viruses, protozoa, and fungi. WGS may present challenges depending on body sites. Considering all the technical advantages and disadvantages of those methods, it is convenient to choose a method that is applicable and most suitable for the study. In the future, 16S-based approaches used in most microbiome studies may be complemented by WGS. Examination of other microbiome members offers us an analysis of the relationship among all community members, in addition to the host cells and the immune system. Changes in the microbiome have been shown in different disease processes (Ling et al. 2014; Gupta et al. 2010; Viljoen et al. 2015), including allergic diseases. It is also significant to consider the clinical picture, and the ways in which to use this combined information to intervene in the process of allergic diseases from the beginning, at a very early age. Further studies are necessary to be able to control, or even prevent the allergic phenotype to progress to other types of Th2-mediated allergic disease.

References

Abrahamsson TR et al (2014) Low gut microbiota diversity in early infancy precedes asthma at school age. Clin Exp Allergy 44(6):842–850

Akbari O et al (2006) CD4+ invariant T-cell-receptor+ natural killer T cells in bronchial asthma. N Engl J Med 354(11):1117–1129

Apfelbacher CJ, Diepgen TL, Schmitt J (2011) Determinants of eczema: population-based cross-sectional study in Germany. Allergy 66(2):206–213

Arpaia N et al (2013) Metabolites produced by commensal bacteria promote peripheral regulatory T-cell generation. Nature 504(7480):451–455

Arrieta MC et al (2015) Early infancy microbial and metabolic alterations affect risk of childhood asthma. Sci Transl Med 7(307):307ra152

Atarashi K et al (2011) Induction of colonic regulatory T cells by indigenous Clostridium species. Science 331(6015):337–341

Atarashi K et al (2013) Treg induction by a rationally selected mixture of Clostridia strains from the human microbiota. Nature 500(7461):232–236

Azad MB et al (2015) Infant gut microbiota and food sensitization: associations in the first year of life. Clin Exp Allergy 45(3):632–643

Azad MB et al (2016) Impact of maternal intrapartum antibiotics, method of birth and breastfeeding on gut microbiota during the first year of life: a prospective cohort study. BJOG 123(6):983–993

Backhed F et al (2015) Dynamics and stabilization of the human gut microbiome during the first year of life. Cell Host Microbe 17(5):690–703

Ball TM et al (2000) Siblings, day-care attendance, and the risk of asthma and wheezing during childhood. N Engl J Med 343(8):538–543

Bantz SK, Zhu Z, Zheng T (2014) The atopic march: progression from atopic dermatitis to allergic rhinitis and asthma. J Clin Cell Immunol 5(2). https://doi.org/10.4172/2155-9899.1000202

Bashir ME et al (2004) Toll-like receptor 4 signaling by intestinal microbes influences susceptibility to food allergy. J Immunol 172(11):6978–6987

Benito D et al (2015) Characterization of Staphylococcus aureus strains isolated from faeces of healthy neonates and potential mother-to-infant microbial transmission through breastfeeding. FEMS Microbiol Ecol 91(3). https://doi.org/10.1093/femsec/fiv007

Benson MJ et al (2007) All-trans retinoic acid mediates enhanced T reg cell growth, differentiation, and gut homing in the face of high levels of co-stimulation. J Exp Med 204(8):1765–1774
Biasucci G et al (2010) Mode of delivery affects the bacterial community in the newborn gut. Early Hum Dev 86(Suppl 1):13–15
Bisgaard H et al (2007) Childhood asthma after bacterial colonization of the airway in neonates. N Engl J Med 357(15):1487–1495
Brauweiler AM, Goleva E, Leung DY (2014) Th2 cytokines increase Staphylococcus aureus alpha toxin-induced keratinocyte death through the signal transducer and activator of transcription 6 (STAT6). J Invest Dermatol 134(8):2114–2121
Brown AJ et al (2003) The Orphan G protein-coupled receptors GPR41 and GPR43 are activated by propionate and other short chain carboxylic acids. J Biol Chem 278(13):11312–11319
Brozek JL et al (2010) Allergic Rhinitis and its Impact on Asthma (ARIA) guidelines: 2010 revision. J Allergy Clin Immunol 126(3):466–476
Cacho NT, Lawrence RM (2017) Innate immunity and breast milk. Front Immunol 8:584
Chan LS, Robinson N, Xu L (2001) Expression of interleukin-4 in the epidermis of transgenic mice results in a pruritic inflammatory skin disease: an experimental animal model to study atopic dermatitis. J Invest Dermatol 117(4):977–983
Charlson ES et al (2011) Topographical continuity of bacterial populations in the healthy human respiratory tract. Am J Respir Crit Care Med 184(8):957–963
Chen L et al (2004) Early up-regulation of Th2 cytokines and late surge of Th1 cytokines in an atopic dermatitis model. Clin Exp Immunol 138(3):375–387
Chng KR et al (2016) Whole metagenome profiling reveals skin microbiome-dependent susceptibility to atopic dermatitis flare. Nat Microbiol 1(9):16106
Choi CH et al (2014) Seasonal allergic rhinitis affects sinonasal microbiota. Am J Rhinol Allergy 28(4):281–286
Choo JM et al (2017) Divergent relationships between fecal microbiota and metabolome following distinct antibiotic-induced disruptions. mSphere 2(1). https://doi.org/10.1128/mSphere.00005-17
Coombes JL et al (2007) A functionally specialized population of mucosal CD103+ DCs induces Foxp3+ regulatory T cells via a TGF-beta and retinoic acid-dependent mechanism. J Exp Med 204(8):1757–1764
David LA et al (2014) Diet rapidly and reproducibly alters the human gut microbiome. Nature 505(7484):559–563
Depner M et al (2017) Bacterial microbiota of the upper respiratory tract and childhood asthma. J Allergy Clin Immunol 139(3):826–34.e13
Dickson RP et al (2016) The microbiome and the respiratory tract. Annu Rev Physiol 78:481–504
Dickson RP et al (2017) Bacterial topography of the healthy human lower respiratory tract. MBio 8(1). https://doi.org/10.1128/mBio.02287-16
Dominguez-Bello MG et al (2010) Delivery mode shapes the acquisition and structure of the initial microbiota across multiple body habitats in newborns. Proc Natl Acad Sci USA 107 (26):11971–11975
Erb-Downward JR et al (2011) Analysis of the lung microbiome in the "healthy" smoker and in COPD. PLoS One 6(2):e16384
Eyerich K et al (2009) IL-17 in atopic eczema: linking allergen-specific adaptive and microbial-triggered innate immune response. J Allergy Clin Immunol 123(1):59–66.e4
Fallani M et al (2010) Intestinal microbiota of 6-week-old infants across Europe: geographic influence beyond delivery mode, breast-feeding, and antibiotics. J Pediatr Gastroenterol Nutr 51(1):77–84
Faria AM, Weiner HL (1999) Oral tolerance: mechanisms and therapeutic applications. Adv Immunol 73:153–264
Flohr C, Mann J (2014) New insights into the epidemiology of childhood atopic dermatitis. Allergy 69(1):3–16
Fujimura KE et al (2016) Neonatal gut microbiota associates with childhood multisensitized atopy and T cell differentiation. Nat Med 22(10):1187–1191

Furusawa Y et al (2013) Commensal microbe-derived butyrate induces the differentiation of colonic regulatory T cells. Nature 504(7480):446–450

Garmhausen D et al (2013) Characterization of different courses of atopic dermatitis in adolescent and adult patients. Allergy 68(4):498–506

Gdalevich M et al (2001a) Breast-feeding and the onset of atopic dermatitis in childhood: a systematic review and meta-analysis of prospective studies. J Am Acad Dermatol 45(4):520–527

Gdalevich M, Mimouni D, Mimouni M (2001b) Breast-feeding and the risk of bronchial asthma in childhood: a systematic review with meta-analysis of prospective studies. J Pediatr 139(2):261–266

Guo Y et al (2017) Long-term use of ceftriaxone sodium induced changes in gut microbiota and immune system. Sci Rep 7:43035

Gupta A, Madani R, Mukhtar H (2010) Streptococcus bovis endocarditis, a silent sign for colonic tumour. Color Dis 12(3):164–171

Gupta RS et al (2011) The prevalence, severity, and distribution of childhood food allergy in the United States. Pediatrics 128(1):e9–17

Hadis U et al (2011) Intestinal tolerance requires gut homing and expansion of FoxP3+ regulatory T cells in the lamina propria. Immunity 34(2):237–246

Halova I, Draberova L, Draber P (2012) Mast cell chemotaxis – chemoattractants and signaling pathways. Front Immunol 3:119

Hamid Q, Boguniewicz M, Leung DY (1994) Differential in situ cytokine gene expression in acute versus chronic atopic dermatitis. J Clin Invest 94(2):870–876

Hamid Q et al (1996) In vivo expression of IL-12 and IL-13 in atopic dermatitis. J Allergy Clin Immunol 98(1):225–231

Herbst T et al (2011) Dysregulation of allergic airway inflammation in the absence of microbial colonization. Am J Respir Crit Care Med 184(2):198–205

Hilty M et al (2010) Disordered microbial communities in asthmatic airways. PLoS One 5(1):e8578

Holt PG et al (1990) A contiguous network of dendritic antigen-presenting cells within the respiratory epithelium. Int Arch Allergy Appl Immunol 91(2):155–159

Husby S, Jensenius JC, Svehag SE (1985) Passage of undegraded dietary antigen into the blood of healthy adults. Quantification, estimation of size distribution, and relation of uptake to levels of specific antibodies. Scand J Immunol 22(1):83–92

Illi S et al (2004) The natural course of atopic dermatitis from birth to age 7 years and the association with asthma. J Allergy Clin Immunol 113(5):925–931

Imai Y et al (2013) Skin-specific expression of IL-33 activates group 2 innate lymphoid cells and elicits atopic dermatitis-like inflammation in mice. Proc Natl Acad Sci USA 110 (34):13921–13926

Irvine AD, McLean WH, Leung DY (2011) Filaggrin mutations associated with skin and allergic diseases. N Engl J Med 365(14):1315–1327

Iweala OI, Burks AW (2016) Food allergy: our evolving understanding of its pathogenesis, prevention, and treatment. Curr Allergy Asthma Rep 16(5):37

Jackson KD, Howie LD, Akinbami LJ (2013) Trends in allergic conditions among children: United States, 1997–2011. NCHS Data Brief (121):1–8

Jost T et al (2014) Vertical mother-neonate transfer of maternal gut bacteria via breastfeeding. Environ Microbiol 16(9):2891–2904

Jost T et al (2015) Impact of human milk bacteria and oligosaccharides on neonatal gut microbiota establishment and gut health. Nutr Rev 73(7):426–437

Kasraie S, Niebuhr M, Werfel T (2010) Interleukin (IL)-31 induces pro-inflammatory cytokines in human monocytes and macrophages following stimulation with staphylococcal exotoxins. Allergy 65(6):712–721

Kennedy EA et al (2017) Skin microbiome before development of atopic dermatitis: early colonization with commensal staphylococci at 2 months is associated with a lower risk of atopic dermatitis at 1 year. J Allergy Clin Immunol 139(1):166–172

Kim KS et al (2016) Dietary antigens limit mucosal immunity by inducing regulatory T cells in the small intestine. Science 351(6275):858–863

Kong HH et al (2012) Temporal shifts in the skin microbiome associated with disease flares and treatment in children with atopic dermatitis. Genome Res 22(5):850–859
Kramer U et al (1999) Age of entry to day nursery and allergy in later childhood. Lancet 353(9151):450–454
Lal CV et al (2016) The airway microbiome at birth. Sci Rep 6:31023
Lal D et al (2017) Mapping and comparing bacterial microbiota in the sinonasal cavity of healthy, allergic rhinitis, and chronic rhinosinusitis subjects. Int Forum Allergy Rhinol 7:561–569
Larche M, Robinson DS, Kay AB (2003) The role of T lymphocytes in the pathogenesis of asthma. J Allergy Clin Immunol 111(3):450–463. quiz 464
Lehmann HS et al (2004) Staphylococcal enterotoxin-B-mediated stimulation of interleukin-13 production as a potential aetiologic factor in eczema in infants. Int Arch Allergy Immunol 135(4):306–312
Ling Z et al (2014) Altered fecal microbiota composition associated with food allergy in infants. Appl Environ Microbiol 80(8):2546–2554
Liu C et al (2010) Molecular regulation of mast cell development and maturation. Mol Biol Rep 37(4):1993–2001
Martin V et al (2012) Sharing of bacterial strains between breast milk and infant feces. J Hum Lact 28(1):36–44
Mashiko S et al (2015) Human mast cells are major IL-22 producers in patients with psoriasis and atopic dermatitis. J Allergy Clin Immunol 136(2):351–9.e1
Meltzer EO et al (2009) Sleep, quality of life, and productivity impact of nasal symptoms in the United States: findings from the Burden of Rhinitis in America survey. Allergy Asthma Proc 30(3):244–254
Metcalfe DD, Baram D, Mekori YA (1997) Mast cells. Physiol Rev 77(4):1033–1079
Midodzi WK et al (2010) Early life factors associated with incidence of physician-diagnosed asthma in preschool children: results from the Canadian Early Childhood Development cohort study. J Asthma 47(1):7–13
Mimouni Bloch A et al (2002) Does breastfeeding protect against allergic rhinitis during childhood? A meta-analysis of prospective studies. Acta Paediatr 91(3):275–279
Morris A et al (2013) Comparison of the respiratory microbiome in healthy nonsmokers and smokers. Am J Respir Crit Care Med 187(10):1067–1075
Muraro A et al (2004) Dietary prevention of allergic diseases in infants and small children. Part III: critical review of published peer-reviewed observational and interventional studies and final recommendations. Pediatr Allergy Immunol 15(4):291–307
Nakamura Y et al (2013) Staphylococcus delta-toxin induces allergic skin disease by activating mast cells. Nature 503(7476):397–401
Niebuhr M et al (2010) Staphylococcal exotoxins are strong inducers of IL-22: a potential role in atopic dermatitis. J Allergy Clin Immunol 126(6):1176–83.e4
Noverr MC et al (2005) Development of allergic airway disease in mice following antibiotic therapy and fungal microbiota increase: role of host genetics, antigen, and interleukin-13. Infect Immun 73(1):30–38
Oddy WH et al (1999) Association between breast feeding and asthma in 6 year old children: findings of a prospective birth cohort study. BMJ 319(7213):815–819
Oh J et al (2012) Shifts in human skin and nares microbiota of healthy children and adults. Genome Med 4(10):77
Oh J et al (2013) The altered landscape of the human skin microbiome in patients with primary immunodeficiencies. Genome Res 23(12):2103–2114
Ong PY et al (2002) Endogenous antimicrobial peptides and skin infections in atopic dermatitis. N Engl J Med 347(15):1151–1160
Pannaraj PS et al (2017) Association between breast milk bacterial communities and establishment and development of the infant gut microbiome. JAMA Pediatr 171(7):647–654
Peters AS et al (2010) Prediction of the incidence, recurrence, and persistence of atopic dermatitis in adolescence: a prospective cohort study. J Allergy Clin Immunol 126(3):590–5.e1–3

Piccinni MP et al (1993) Aeroallergen sensitization can occur during fetal life. Int Arch Allergy Immunol 102(3):301–303

Plunkett CH, Nagler CR (2017) The influence of the microbiome on allergic sensitization to food. J Immunol 198(2):581–589

Prescott SL et al (1998) *Transplacental priming of the human immune system to environmental allergens: universal skewing of initial T cell responses toward the Th2 cytokine profile.* J Immunol 160(10):4730–4737

Rakoff-Nahoum S et al (2004) Recognition of commensal microflora by toll-like receptors is required for intestinal homeostasis. Cell 118(2):229–241

Ranjan R et al (2016) Analysis of the microbiome: Advantages of whole genome shotgun versus 16S amplicon sequencing. Biochem Biophys Res Commun 469(4):967–977

Saarinen UM, Kajosaari M (1995) Breastfeeding as prophylaxis against atopic disease: prospective follow-up study until 17 years old. Lancet 346(8982):1065–1069

Salimi M et al (2013) A role for IL-25 and IL-33-driven type-2 innate lymphoid cells in atopic dermatitis. J Exp Med 210(13):2939–2950

Sears MR et al (2002) Long-term relation between breastfeeding and development of atopy and asthma in children and young adults: a longitudinal study. Lancet 360(9337):901–907

Sehra S et al (2010) IL-4 regulates skin homeostasis and the predisposition toward allergic skin inflammation. J Immunol 184(6):3186–3190

Shi B et al (2016) The skin microbiome is different in pediatric versus adult atopic dermatitis. J Allergy Clin Immunol 138(4):1233–1236

Silverberg JI, Simpson EL (2014) Associations of childhood eczema severity: a US population-based study. Dermatitis 25(3):107–114

Singh N et al (2014) Activation of Gpr109a, receptor for niacin and the commensal metabolite butyrate, suppresses colonic inflammation and carcinogenesis. Immunity 40(1):128–139

Smith PM et al (2013) The microbial metabolites, short-chain fatty acids, regulate colonic Treg cell homeostasis. Science 341(6145):569–573

Snijders BE et al (2007) Breastfeeding and infant eczema in the first year of life in the KOALA birth cohort study: a risk period-specific analysis. Pediatrics 119(1):e137–e141

Sonnenburg JL, Backhed F (2016) Diet-microbiota interactions as moderators of human metabolism. Nature 535(7610):56–64

Stefka AT et al (2014) Commensal bacteria protect against food allergen sensitization. Proc Natl Acad Sci USA 111(36):13145–13150

Strachan DP (1989) Hay fever, hygiene, and household size. BMJ 299(6710):1259–1260

Strachan DP (1997) Allergy and family size: a riddle worth solving. Clin Exp Allergy 27(3):235–236

Suarez-Farinas M et al (2011) Nonlesional atopic dermatitis skin is characterized by broad terminal differentiation defects and variable immune abnormalities. J Allergy Clin Immunol 127(4):954–64.e1–4

Sudo N et al (1997) The requirement of intestinal bacterial flora for the development of an IgE production system fully susceptible to oral tolerance induction. J Immunol 159(4):1739–1745

Sun CM et al (2007) Small intestine lamina propria dendritic cells promote de novo generation of Foxp3 T reg cells via retinoic acid. J Exp Med 204(8):1775–1785

Tan J et al (2014) The role of short-chain fatty acids in health and disease. Adv Immunol 121:91–119

Tan J et al (2016) Dietary fiber and bacterial SCFA enhance oral tolerance and protect against food allergy through diverse cellular pathways. Cell Rep 15(12):2809–2824

Teo SM et al (2015) The infant nasopharyngeal microbiome impacts severity of lower respiratory infection and risk of asthma development. Cell Host Microbe 17(5):704–715

Thangaraju M et al (2009) GPR109A is a G-protein-coupled receptor for the bacterial fermentation product butyrate and functions as a tumor suppressor in colon. Cancer Res 69(7):2826–2832

Thorburn AN et al (2015) Evidence that asthma is a developmental origin disease influenced by maternal diet and bacterial metabolites. Nat Commun 6:7320

Trompette A et al (2014) Gut microbiota metabolism of dietary fiber influences allergic airway disease and hematopoiesis. Nat Med 20(2):159–166

Viljoen KS et al (2015) Quantitative profiling of colorectal cancer-associated bacteria reveals associations between fusobacterium spp., enterotoxigenic Bacteroides fragilis (ETBF) and clinicopathological features of colorectal cancer. PLoS One 10(3):e0119462

Wang Z et al (2017) Skin microbiome promotes mast cell maturation by triggering stem cell factor production in keratinocytes. J Allergy Clin Immunol 139(4):1205–16.e6

Weidinger S, Novak N (2016) Atopic dermatitis. Lancet 387(10023):1109–1122

Wolk K et al (2006) IL-22 regulates the expression of genes responsible for antimicrobial defense, cellular differentiation, and mobility in keratinocytes: a potential role in psoriasis. Eur J Immunol 36(5):1309–1323

Yabuhara A et al (1997) TH2-polarized immunological memory to inhalant allergens in atopics is established during infancy and early childhood. Clin Exp Allergy 27(11):1261–1269

Yokota-Nakatsuma A et al (2014) Retinoic acid prevents mesenteric lymph node dendritic cells from inducing IL-13-producing inflammatory Th2 cells. Mucosal Immunol 7(4):786–801

Zhang E et al (2011) Characterization of the skin fungal microbiota in patients with atopic dermatitis and in healthy subjects. Microbiol Immunol 55(9):625–632

Zheng Y et al (2007) Interleukin-22, a T(H)17 cytokine, mediates IL-23-induced dermal inflammation and acanthosis. Nature 445(7128):648–651

Zheng T et al (2009) Transgenic expression of interleukin-13 in the skin induces a pruritic dermatitis and skin remodeling. J Invest Dermatol 129(3):742–751

Chapter 3
Pathogenesis, Immunity and the Role of Microbiome/Probiotics in Enteric Virus Infections in Humans and Animal Models

Shaohua Lei, Erica Twitchell, and Lijuan Yuan

Abstract The gut microbiota has a profound impact on the resistance, pathogenesis, and immunity of enteric viral pathogens. Commensal microbes may prevent the host from infection or enhance infection by altering virus stability, attachment or cellular entry. Additionally, microbiota members can stimulate or suppress host immune responses to the viral infection. In most cases, the gut microbiota plays a role in host resistance against invading enteric viral pathogens; hence, germ-free animals are more susceptible to infection of various enteric pathogens. However, increasing evidence has demonstrated that certain commensal bacteria can enhance enteric viral infection. Exact mechanisms by which specific bacteria carry out these effects are not clearly understood in most instances. In this chapter, human norovirus (HuNoV) and human rotavirus (HRV), the two most important viral pathogens causing gastroenteritis, are chosen for the discussion of the impacts and mechanisms of microbiome–host interactions on viral gastroenteritis. The pathogenesis and immunity of HuNoV and HRV in humans and in germ-free animal models, particularly gnotobiotic (Gn) mice and pigs, and Gn pigs transplanted with human gut microbiota are reviewed. Findings from studies on host–microbiome interactions on the pathogenesis and immunity of the two viruses, and mechanisms of probiotics/prebiotics in ameliorating their infection and diseases, are summarized. Unraveling the role of microbiome and specific probiotics in the infectivity, pathogenesis, and immunity of HuNoV and HRV facilitates the development of strategies for manipulating the microbiome against viral infections. Further studies are needed to improve our understanding of mechanisms underlying host–microbiome interactions in the pathophysiology of enteric viral diseases.

S. Lei • E. Twitchell • L. Yuan (✉)
Department of Biomedical Sciences and Pathobiology, Virginia-Maryland College of Veterinary Medicine, Virginia Polytechnic Institute and State University, Blacksburg, VA, USA
e-mail: lyuan@vt.edu

J. Sun, P.K. Dudeja (eds.), *Mechanisms Underlying Host-Microbiome Interactions in Pathophysiology of Human Diseases*, Physiology in Health and Disease,
https://doi.org/10.1007/978-1-4939-7534-1_3

List of Abbreviations

ASC	Antibody-secreting cells
AttHRV	Attenuated human rotavirus
Dpi	Days post-inoculation
EcN	*Escherichia coli* Nissle 1917
Gn	Gnotobiotic
HBGA	Histo-blood group antigen
HHGM	Healthy human gut microbiota
HRV	Human rotavirus
HuNoV	Human norovirus
LGG	*Lactobacillus rhamnosus* GG
MNCs	Mononuclear cells
NHPs	Nonhuman primates
UHGM	Unhealthy human gut microbiota
VirHRV	Virulent human rotavirus

3.1 HuNoV Pathogenesis and Immunity in Humans and in Animal Models

3.1.1 HuNoV Gastroenteritis, Pathogenesis, and Cell Tropism in Humans

Human noroviruses (HuNoVs) are positive-sense, single-stranded, non-enveloped RNA viruses that belong to the genus *Norovirus* in the family Caliciviridae (Zheng et al. 2006). Since the introduction of rotavirus vaccines (RotaTeq in 2006 and Rotarix in 2008), HuNoVs have become the predominant cause of viral epidemic acute gastroenteritis across the globe (Pringle et al. 2015; Hemming et al. 2013; Payne et al. 2013). Viral transmission occurs via the fecal–oral route by contaminated food or water and person-to-person spread (Patel et al. 2009). HuNoV gastroenteritis is generally self-limiting, with a duration of 2–3 days and consists of moderate to severe acute diarrhea episodes, sudden onset of vomiting, and mild or no fever (O'Ryan et al. 2010), but the diseases can become more severe and prolonged in infants, the elderly, and individuals with impaired immunity (Karst 2010). Despite its importance in public health, no virus-specific therapeutics or vaccines are currently available to treat or prevent HuNoV gastroenteritis (Kocher and Yuan 2015), mainly because HuNoV research has been hampered by the long absence of a robust cell culture system and small-animal model. HuNoV biology has been explored most frequently by viral challenge studies in human volunteers (Karst 2010), chimpanzees (Bok et al. 2011), gnotobiotic (Gn) calves and pigs (Souza et al. 2008; Bui et al. 2013; Cheetham et al. 2006), and immunodeficient mice (Taube et al. 2013) (Table 3.1).

Table 3.1 Features of animal models for human norovirus (HuNoV) infection

Host	Gem-free condition	Viral strain	Route of infection	Viral antigen location	Fecal virus shedding	Viremia	Diseases	Pathological changes
Immunocompetent humans	–	Multiple	Oral	N/A	+	±	Vomiting and severe diarrhea	Yes (Agus et al. 1973; Schreiber et al. 1973, 1974; Dolin et al. 1975; Karst et al. 2014)
Immunocompromised humans	–	Multiple	Oral	IECs	+	N/A	Vomiting and severe diarrhea	Yes (Karandikar et al. 2016; Bok and Green 2012; Green 2014)
Chimpanzees	–	GI.1	Oral (Wyatt et al. 1978) intravenous (Bok et al. 2011)	Intestinal DC and B cells	+	–	Asymptomatic	No (Bok et al. 2011)
Balb/c RAG/$\gamma c^{-/-}$ mice	–	GII mix	Intraperitoneally	Mφ-like cells in liver and spleen	–	N/A	Asymptomatic	No (Taube et al. 2013)
Gnotobiotic pigs	+	GII.4	Oral	IECs	+	+	Mild diarrhea	Yes (Bui et al. 2013; Cheetham et al. 2006)
Gnotobiotic calves	+	GII.4	Oral	IECs and intestinal Mφ-like cells	+	+	Mild diarrhea	Yes (Souza et al. 2008)

IECs intestinal epithelial cells, *Mφ* macrophage, *N/A* not available

Challenging HuNoV in immunocompetent volunteers resulted in acute gastroenteritis, and biopsy specimens from the individuals who acquired clinical gastroenteritis displayed histological changes in the small intestine, including mucosal inflammation, villus blunting, microvillus shortening, and abnormal organelles such as endoplasmic reticulum and mitochondria (Agus et al. 1973; Schreiber et al. 1973, 1974; Dolin et al. 1975). Although intestinal epithelial cells (IECs) are the target for most enteric pathogens, the presence of HuNoV virions or antigen have not been reported in these biopsies from immunocompetent humans, and the cellular tropism of HuNoV has long been elusive (Agus et al. 1973; Schreiber et al. 1973, 1974; Dolin et al. 1975; Karst et al. 2014). Chronic HuNoV infection occurs in immunocompromised transplant patients. A recent study using intestinal biopsies from a patient cohort showed that HuNoV infection was observed in duodenal and jejunal enterocytes, and HuNoV-associated histopathological changes were present as the flattening of epithelial cells and the severe loss of villin in enterocytes (Karandikar et al. 2016). In addition, stem cell-derived and nontransformed human intestinal enteroids have been recently established as a reproducible cultivation system for multiple HuNoV strains, confirming enterocytes as target cell types for HuNoV infection in vitro and in vivo (Ettayebi et al. 2016). B cells were suggested to be a permissive cell type for HuNoV replication in vitro, which is a novel HuNoV cultivation system in the BJAB cell line supplemented with free histo-blood group antigen (HBGA) or HBGA-expressing inactivated enteric bacteria (Jones et al. 2014). However, this cell culture system produced inconsistent results in other laboratories (Jones et al. 2015; Lei et al. 2016c), and HuNoV infection was observed in B cell-deficient patients and Gn pigs (Brown et al. 2016; Lei et al. 2016b), along with the low virus yields in such an in vitro cell system compared with high-level virus shedding in patients (Bok and Green 2012), suggesting that B cells might not be the primary target cell of HuNoV.

3.1.2 HuNoV Infection and Pathophysiology in Conventional Animal Models

Nonhuman primates (NHPs), particularly chimpanzee (99%) (Kehrer-Sawatzki and Cooper 2007), share the greatest genome similarities with humans, which makes them desirable models for studies on several fastidious viral pathogens, such as human immunodeficiency virus and hepatitis viruses (O'Neil et al. 2000; Pfaender et al. 2014; Purcell and Emerson 2001). The chimpanzee was presented as a viable animal model for subclinical GI.1 HuNoV infection, characterized by intravenous inoculation, asymptomatic fecal virus shedding, and viral associated serum antibody responses (Bok et al. 2011). Biopsies from the jejunum and duodenum showed no histological changes after HuNoV infection, although the viral genome was detectable up to 21 days post-inoculation. Interestingly, viral capsid antigen was only observed in cells of the duodenal and jejunal lamina propria, and further investigations indicated that viral antigen-positive cells were dendritic cells and B lymphocytes (Bok et al. 2011). However, the chimpanzee is not available for biomedical research any longer owing to ethical concerns.

Another animal model of subclinical HuNoV infection is the Balb/c mouse deficient in recombination activation gene (RAG) and common gamma chain (γc or IL2RG), which lacks T cells, B cells, and natural killer cells. In this mouse model, a HuNoV GII mix was inoculated intraperitoneally (Taube et al. 2013). Although virus shedding and gastrointestinal diseases were not observed in those Balb/c RAG/$\gamma c^{-/-}$ mice, viral genome was detected in the intestinal and systemic sites, with increased levels over the input virus 1–2 days post-inoculation. Viral structural and nonstructural proteins were observed in cells morphologically resembling macrophages in the liver and spleen, validating HuNoV propagation (Taube et al. 2013). Moreover, Balb/c RAG/$\gamma c^{-/-}$ mice can be used for the evaluation of anti-HuNoV drugs such as the nucleoside analog 2′-C-methylcytidine, which inhibited HuNoV replication in vivo (Kolawole et al. 2016).

3.1.3 HuNoV Infection and Pathophysiology in Gnotobiotic Large Animal Models

The neonatal Gn pig model is well suited for the evaluation of HuNoV pathogenesis and vaccine efficacy, and it reflects HuNoV biology in terms of supporting the natural oral route of infection, resulting in diarrhea, transient viremia, and virus shedding in feces (Cheetham et al. 2006; Bui et al. 2013; Kocher et al. 2014; Souza et al. 2007a, b). Viral structural and nonstructural proteins were detected in enterocytes in wild-type Gn pigs experimentally infected with HuNoV genotype GII.4 (Bui et al. 2013; Cheetham et al. 2006; Lei et al. 2016c), indicating viral infection and replication in Gn pigs. HuNoV-induced diarrhea in Gn pigs was associated with mild villus atrophy and cytopathological changes in the small intestine, manifested as blunting and shortening of microvilli and necrosis and apoptosis of enterocytes (Bui et al. 2013; Cheetham et al. 2006), which recapitulate the hallmark pathological features in humans.

Twenty-four units of the P domain of HuNoV capsid protein can form P particle, which efficiently induces innate, humoral, and cellular immune responses in mice (Fang et al. 2013). Together with its easy and economical preparation in *E. coli*, P particle has gained recognition as a promising vaccine candidate against HuNoV infection (Kocher and Yuan 2015). In the study of P particle vaccination in Gn pigs, P particle exhibited 47% cross-variant protection against HuNoV diarrhea, and the protection correlated positively with T cell expansion in the ileum and spleen, while correlating inversely with T cell expansion in the duodenum (Kocher et al. 2014). Persistent HuNoV infection in immunocompromised patients can lead to increasingly debilitating and life-threatening gastroenteritis with prolonged virus shedding (Bok and Green 2012; Green 2014). Similarly, in RAG2/IL2RG-deficient Gn pigs,

HuNoV infection was severe and prolonged owing to the severe combined immunodeficiency of the host, and enterocytes of the duodenum and jejunum were sites of HuNoV infection (Lei et al. 2016b).

Neonatal Gn calves serve as another large animal model that supports GII.4 HuNoV infection; viral capsid protein was detected in enterocytes of the jejunum and ileum, and in cells morphologically resembling macrophages in the lamina propria (Souza et al. 2008). Similar to the findings in Gn pigs, HuNoV challenge in Gn calves resulted in diarrhea along with intestinal lesions and mild villous atrophy, fecal virus shedding, transient viremia, and intestinal and systemic immune responses (Souza et al. 2008).

Notably, pigs are natural hosts of noroviruses GII (genotypes 11, 18, and 19); however, all porcine noroviruses were detected from conventional pigs without clinical signs (Knowles and Reuter 2012). Porcine norovirus has been detected in many countries and geographical distribution indicates the worldwide occurrence of porcine noroviruses among pigs on farms. The QW101/2003/US (GII.18) isolate from a healthy adult pig was genetically and antigenically related to HuNoVs and replicated in Gn pigs with fecal shedding coincident with mild diarrhea (Wang et al. 2005). Seroprevalence of norovirus GII in pigs was reported to be 97% in the USA. Attempts have been made, but failed to infect conventional Göttingen miniature pigs (Marshall BioResources, North Rose, NY, USA) with HuNoV (Tin et al. 2017). The miniature pigs shed neither virus nor seroconvert after oral and intravenous HuNoV inoculation. The difference in the susceptibility to norovirus infection and lack of disease in conventional pigs suggest that the gut microbiota or maternal antibodies might be protective. Effects of the gut microbiota on the resistance and immunity to norovirus infection are currently under investigation.

3.2 Effects of the Microbiome on Norovirus Infection, Immunity, and Disease

The notion that commensal bacteria can enhance enteric viral infection was demonstrated by two landmark studies published in 2011 using poliovirus, reovirus, and mouse mammary tumor virus (Kuss et al. 2011; Kane et al. 2011). When intestinal bacteria were depleted by administering a cocktail of antibiotics to mice, poliovirus infection was dramatically attenuated in comparison with normal mice with gut microbiota, as characterized by the reduced fecal virus shedding and mortality (Kuss et al. 2011). In addition, the reduced poliovirus infection was reversed by fecal transplantation to reconstitute intestinal microbes, and the status of the intestinal microbiota did not affect viral infectivity when poliovirus was inoculated intraperitoneally (Kuss et al. 2011), indicating the role of intestinal bacteria in enhancing enteric viral infection. Poliovirus was shown to directly bind to the bacterial outer-membrane component lipopolysaccharide, resulting in virion thermo-stabilization and attachment to host cells (Kuss et al. 2011; Wilks et al. 2015). As a result, the interactions between host-microbiome and enteric viruses have been gaining intense attention. However, the understanding of effects of the intestinal microbiota on HuNoV has been impeded by the absence of a suitable cell culture system and animal model. Limited studies analyzing stool samples from human patients showed that HuNoV infection could alter microbial composition (Nelson et al. 2012).

Murine norovirus (MNV) was first identified in 2002 from the brain of an immunocompromised mouse, RAG/STAT1$^{-/-}$ strain, because of its lethal infection (Karst et al. 2003). Since then, MNV has been used widely as a surrogate to explore HuNoV biology regarding viral pathogenesis, host immunity, and interplays with gut microbiota. Antibiotic treatment reduced the acute MNV infection, and lower virus titers in the distal ileum, mesenteric lymph nodes, and colon were observed compared with control mice (Jones et al. 2014). Antibiotics also prevented persistent MNV infection in mice, but persistent infection could be restored by microbial colonization (Baldridge et al. 2015), indicating the stimulatory role of microbiota in MNV infectivity. However, major disruptions of the microbiome were not observed following acute or persistent MNV infection in mice (Nelson et al. 2013). MNV infection is asymptomatic in wild-type mice, but mucosal inflammation was observed in IL-10$^{-/-}$ mice maintained in a specific pathogen-free environment, and MNV-induced pathological changes such as reduced tight junction proteins and inflammatory lesions were lacking in germ-free IL-10$^{-/-}$ mice, suggesting that MNV-triggered intestinal diseases might be induced via bacterial microbiota (Basic et al. 2014).

3.3 Mechanisms of Probiotics/Prebiotics in Ameliorating Norovirus Infection and Disease

Probiotics have been increasingly recognized as vaccine adjuvants and therapeutic agents to ameliorate pediatric acute gastroenteritis (Schnadower et al. 2015; Licciardi and Tang 2011). The underlying mechanisms of their beneficial health effects include modulating gut microbiota composition, enhancing intestinal barrier function, and promoting mucosal immunity (Wen et al. 2009). Notably, *Lactobacillus* spp. exhibit promising properties against viral infection and diseases in human clinical trials (Guandalini et al. 2000; Sindhu et al. 2014; Szajewska et al. 2014), and their binding capacity with viral P particles holds great promise for reducing HuNoV infectivity and disease in vivo (Rubio-del-Campo et al. 2014). Evaluation of the effects of consuming *Lactobacillus casei* strain Shirota fermented milk on HuNoV gastroenteritis during an outbreak in Japan demonstrated that the elderly HuNoV-infected patients (about 84 years old) who continuously consumed the milk experienced a shorter duration of fever than the nontreated patients

(1.5 vs. 2.9 days), although the incidence of HuNoV gastroenteritis did not differ between the two groups (Nagata et al. 2011).

Probiotic bacteria can bind HuNoV P particles on their surface in vitro, and the presence of *L. casei* BL23 and *Escherichia coli* Nissle 1917 (EcN) may inhibit P particle attachment to epithelial cells (Rubio-del-Campo et al. 2014). *Lactobacillus rhamnosus* GG (LGG) is another probiotic strain with HuNoV-binding capacity, and a recent study showed that the binding between HuNoV and LGG/EcN was associated with their inhibitory role of HuNoV shedding in Gn pigs (Lei et al. 2016a). In addition, daily supplement of prebiotic rice bran in LGG/EcN colonized Gn pigs was highly protective against HuNoV diarrhea and shedding. The mechanism involves enhancement of IFN-γ-producing T cell responses, increased production of total intestinal IgA and IgG antibodies, and maintenance of longer villi compared with the non-rice bran-fed and non-probiotic-colonized control group (Lei et al. 2016a). Norovirus infection leads to epithelial barrier dysfunction and an increase in epithelial apoptosis, which results in a reduction in villus height (Troeger et al. 2009). The antiviral effects of IFN-γ and mucosal antibodies induced by rice bran can attenuate the damage to the intestinal epithelia by HuNoV infection to reduce diarrhea and maintain longer villi. In another study, the presence of *Bifidobacterium adolescentis* inhibited the attachment of HuNoV GI.1 virus-like particle to epithelial cells in vitro (Li et al. 2016), indicating the inhibitory role of probiotics on the initial viral infection stage. However, instead of affecting the viral attachment, *B. adolescentis* decreased the replication of MNV in RAW 264.7 cells (Li et al. 2016). Vitamin A was shown to inhibit MNV replication in mice by upregulating lactobacilli in gut microbiota, and anti-MNV effects of lactobacilli were confirmed in RAW264.7 cells (Lee and Ko 2016). Given the natural source and commercial accessibility, probiotic and prebiotic treatments may constitute a novel, safe, and effective measure in clinical practice against HuNoV infection and disease.

3.4 Rotavirus Pathogenesis and Immunity in Humans and in Animal Models

3.4.1 Rotavirus Pathogenesis and Immunity

Rotaviruses are double-stranded, segmented, non-enveloped, RNA viruses belonging to the Reoviridae family. Worldwide, rotaviruses are a major cause of acute gastroenteritis in infants and young children, which is characterized by diarrhea, vomiting, and dehydration. They were responsible for approximately 500,000 deaths a year, mostly in low-middle income countries before the two commercial vaccines (Rotarix and RotaTaq) became available (Desselberger 2014; Desselberger and Huppertz 2011). Diarrhea is caused by viral damage to enterocytes, villus ischemia, action of the enterotoxin NSP4, and activation of the enteric nervous system (Desselberger 2014; Desselberger and Huppertz 2011).

Rotaviruses replicate in mature, nondividing enterocytes near the tips of the villi. The pathological changes due to rotavirus infection are mostly limited to the small intestine (Ramig 2004). Systemic rotavirus infections have been documented in humans and animals and systemic disease does occur in rare cases (Ramig 2007). In humans, after primary symptomatic or asymptomatic rotavirus infection, the patient is typically protected from subsequent severe disease (Desselberger and Huppertz 2011). Correlates of protection include rotavirus-specific serum IgA and fecal IgA (Desselberger 2014; Desselberger and Huppertz 2011). In some studies, there is a lack of correlation between neutralizing antibody titers and protection (Desselberger 2014). Rotavirus-specific T cells help to eliminate virus after infection and memory B cells provide long-term protection (Desselberger 2014). In humans, newborns are provided with additional protection through transplacental and breast milk antibodies (Desselberger and Huppertz 2011).

3.4.2 Animal Models of Rotavirus Infection and Disease

In addition to humans, many animals are susceptible to rotavirus infection and disease, and can be used as models (i.e., pigs, calves, lambs, rats, rabbits, mice, and NHPs) to study rotavirus pathogenesis and immunity. These models have been reviewed in extensive detail elsewhere (Yuan and Wen 2017). The Gn pig model has many benefits over other animal models. Pigs and humans share high genomic and protein sequence homologies, omnivorous diet, similar gastrointestinal physiology, and similar immune systems (Wang and Donovan 2015; Saif et al. 1996). Additionally, there is no transfer of maternal antibodies across the porcine placenta and Gn pigs are deprived of sow colostrum/milk, which prevents maternal antibodies from interfering with studies. Gn pigs are susceptible to genotype 1 (G1) and genotype 3 (G3) human rotavirus (HRV) infections. Inoculation of Gn pigs, up to at least 8 weeks of age with Wa strain (G1P1A[8]) HRV results in diarrhea (Yuan et al. 1998). Based on duodenal biopsies from children with acute rotavirus infection, the histopathological changes are similar to those found in piglets (Barnes and Townley 1973; Davidson and Barnes 1979; Ward et al. 1996a). Extensive work has been done with Gn pigs and Wa strain HRV. The virulent Wa human rotavirus strain (VirHRV) allows assessment of host response to natural infection, whereas the attenuated human rotavirus (AttHRV) can be used to study vaccination (Yuan and Saif 2002; Saif et al. 1997).

After oral inoculation with VirHRV, Gn pigs develop diarrhea, shed virus, and develop nearly complete protection against subsequent clinical disease and viral shedding when challenged with VirHRV after recovery (Yuan et al. 1996; Ward et al. 1996b; Iosef et al. 2002). Diarrhea develops approximately 13 h after inoculation and is associated with viral antigen in enterocytes at villus tips; villus atrophy develops 24 h post-infection and correlates with peak fecal viral titers

(Ward et al. 1996a). Gn pigs orally inoculated with AttHRV seroconvert, but have little to no virus shedding and no clinical signs, and protection from diarrhea and viral shedding after challenge is less efficacious than what is seen in pigs receiving primary VirHRV oral inoculation (Yuan et al. 1996; Ward et al. 1996b; Iosef et al. 2002).

Gnotobiotic calves have also been used to study rotavirus; however, reports are not as numerous as those in Gn pigs. Gn calves can be infected with some HRV strains, but clinical illness does not always develop (Tzipori et al. 1980). In a study in which calves successfully developed diarrhea after administration of an HRV strain, they had histological lesions consistent with rotavirus infection (Mebus et al. 1977). In addition to the fact that Gn calves are not as consistent as Gn pigs as a model of HRV infection and disease, ruminant microbiota is very different from that of humans; therefore, calves are not a proper model for the study of the role of microbiota in HRV infection and immunity.

Despite the close genetic relationship between NHPs and humans, they are not a superior rotavirus animal model compared with Gn pigs. Often, HRV strains are naturally attenuated in NHPs (McNeal et al. 2005). There have been reports of oral inoculation of simian (SA11) or human (Wa) rotavirus into different NHPs with development of diarrhea; however, it is usually during the first week of life, after which disease is not observed, and older animals may not shed virus or even seroconvert (McNeal et al. 2005). Even in a study evaluating a wild-type macaque rotavirus in 14- to 42-day-old macaques, they remained clinically normal, despite shedding large amounts of virus (McNeal et al. 2005).

Mice are attractive animal models because of their size, ease of maintenance compared with Gn pigs, and availability of numerous strains and genetic knockouts. The major downside of the murine rotavirus model is that mice are only susceptible to disease for approximately 15 days after birth (Ward et al. 1990). Adult mice can be used to study rotavirus infection; however, infections are subclinical and often do not predict protective efficacy of interventions against clinical infection (Ward et al. 1990; Yuan and Saif 2002).

3.5 Effects of Microbiome on Rotavirus Infection, Immunity, and Disease

3.5.1 Studies Comparing Conventional and Germ-Free Mice

A French research group pioneered the study on the impact of the microbiota on rotavirus pathogenesis nearly 30 years ago (Heyman et al. 1987). They compared intestinal absorption of macromolecules during murine rotavirus infection in conventional versus germ-free newborn mice derived from seronegative dams. The study showed that rotavirus infection caused a transient increase in gut permeability to undegraded proteins; this increase occurred earlier after infection in conventional pups and later in germ-free pups. Furthermore, the length of virus excretion was different in conventional and germ-free mice; rotavirus titers in intestinal homogenates were still high at 8 days post-inoculation (dpi) in conventional mice, whereas they become very low in germ-free mice. However, there was no correlation between virus excretion and diarrhea in mice, as diarrhea was observed from 2 to 8 dpi in both conventional and germ-free mice, and no differences were detected on diarrhea kinetics. When the intestinal microbiota was absent, clinical and physiological disturbance were more severe, i.e., greater weight loss after rotavirus infection, and a more marked and long-lasting augmentation in intestinal permeability to intact proteins. This study indicates that intestinal microbiota has a significant impact on both rotavirus replication and pathogenesis, as supported by the timing, magnitude, and duration of increased epithelial permeability and virus excretion (Heyman et al. 1987).

A recent study showed that rotavirus infection was reduced by 42% and diarrhea was decreased in incidence and duration in germ-free mice (via ablation of microbiota by antibiotics) compared with mice with conventional microbiota (Uchiyama et al. 2014). Based on the non-altered ratio of positive to negative sense rotavirus RNA strands, the authors suggested that the antibiotics used to ablate the microbiota affected entry of the virus into host cells (Uchiyama et al. 2014). These antibiotic-treated mice had more durable mucosal and systemic humoral immune response and the durability correlated with small intestinal rotavirus-specific IgA antibody-secreting cells (ASCs) (Uchiyama et al. 2014). Mice treated with low levels of dextran sodium sulfate to increase exposure of immune cells to the microbiota had decreased rotavirus-specific antibodies. Further studies are needed to understand how the microbiota and antibiotics interact to induce the immunological differences between the mouse groups. The contradictory findings between the two studies on the role of microbiota in rotavirus infection and diarrhea are most likely due to the difference between using true germ-free newborn mice (Heyman et al. 1987) versus using mice ablated of the microbiota with antibiotics (Uchiyama et al. 2014). In addition to killing bacteria, antibiotics have many effects on the physiology and immune cell development of the host, which need to be taken into consideration and should be properly controlled in these types of studies.

3.5.2 Studies in Gn Pigs and Human Gut Microbiota-Transplanted Gn Pigs

To identify the influence of microbiota in the response of the Gn pig to HRV and to more closely mimic human infants with the model, Gn pigs transplanted with newborn human gut microbiota (HGM) and infected with HRV have been evaluated (Zhang et al. 2014). HGM successfully colonized the Gn pig intestine after three oral inoculations. Sequencing of the V4 region of 16S rRNA genes demonstrated that the pigs carried a microbiome similar to that of the cesarean-section-delivered human infant donor (Zhang et al. 2014). This model was used to test the effects of probiotics on the gut microbiome structure during a VirHRV infection and the development of AttHRV vaccine-induced immune responses were compared between the HGM- and non-HGM-transplanted Gn pigs (Wen et al. 2014). The AttHRV vaccine conferred overall similar protection against rotavirus diarrhea and virus shedding in Gn pigs and HGM-transplanted Gn pigs. HGM promoted the development of the neonatal immune system, significantly enhancing IFN-γ-producing T cell responses and reducing Treg cell responses in the AttHRV-vaccinated pigs (Wen et al. 2014).

Furthermore, a Gn pig model of enteric dysbiosis and rotavirus immunity has been developed (Twitchell et al. 2016). Using this model, it has been shown that after vaccination with AttHRV, pigs colonized by gut microbiota from children who had a good immune response to oral rotavirus vaccination and low enteropathy scores (healthy human gut microbiota, HHGM) had more rotavirus-specific IFN-γ T cells in the ileum, spleen, and blood than pigs colonized by microbiota from children who did not respond well to the oral rotavirus vaccine and showed evidence for enteropathy (unhealthy human gut microbiota, UHGM) (Twitchell et al. 2016). UHGM pigs had higher viral shedding titers and more severe clinical signs than HHGM pigs after challenge with VirHRV (Twitchell et al. 2016). There was a significantly positive correlation between *Collinsella* and significantly negative correlations between *Clostridium* spp. or *Anaerococcus* and frequencies of IFN-γ T cells at the time of challenge. HHGM pigs had an increased mean relative abundance of *Bacteroides* after VirHRV challenge (Twitchell et al. 2016). As the only variable that differed between these groups was microbiota composition, this study clearly demonstrated that the differences in immune responses and clinical disease are due to the influence of the different microbiota.

It has been shown that human intestinal cells incubated with soluble factors from *Bacteroides thetaiotaomicron* and *L. casei* were protected from rotavirus infection (Varyukhina et al. 2012). The protection was attributed to increased cell surface galactose induced by the bacterial factors, which blocked rotavirus infection. This mechanism is significant in rotavirus infection because these viruses use glycan recognition to attach to enterocytes (Varyukhina et al. 2012). Perhaps a similar mechanism was at play in the Gn pig enteric dysbiosis study and may partially explain why HHGM pigs had decreased viral shedding compared with UHGM pigs (Twitchell et al. 2016).

3.5.3 Studies of the Microbiome in Rotavirus Infection and Vaccination in Humans

The abundance of *Bacteroides* species in rotavirus infected and uninfected children was different. *B. fragilis* was increased whereas *B. vulgatus* and *B. stercoris* were decreased in the intestines of infected children (Zhang et al. 2009). A rotavirus vaccine study in rural Ghana evaluated pre-vaccination fecal microbiome of vaccine responders and non-responders and then compared them with age-matched healthy Dutch infants (Harris et al. 2016). The Ghanaian vaccine responder microbiome was more like the healthy Dutch infant microbiome than the Ghanaian nonresponders. Vaccine response correlated with an increased abundance of *Streptococcus bovis* and decreased Bacteroidetes phylum (Harris et al. 2016). The significance of these findings needs to be further elucidated. These studies suggest that certain bacterial components of microbiome might play a modulatory role in the development of rotavirus infection and immunity. Although the underlying mechanisms of specific host–bacteria and virus–bacteria interactions that lead to the different outcomes in enteric viral diseases and immunity have not been identified, studies of probiotics shed some light on the mechanisms.

3.6 Mechanisms of Probiotics/Prebiotics in Ameliorating Rotavirus Infection and Disease

Prebiotics and probiotics are being developed as a nonpharmacological means of preventing or ameliorating gastroenteritis caused by enteropathogens, and as vaccine adjuvants. Mechanisms by which prebiotics and probiotics affect infection, disease, and immunity are an active area of study. Effects vary based on strain, dose, and frequency of administration.

3.6.1 Mechanisms for Reducing Rotavirus Diarrhea Using Probiotics

Among all commercially available probiotics, LGG has been most extensively studied in rotavirus infection, disease, and immunity. LGG has been shown to protect the intestinal barrier in studies using conventional pigs and Gn pigs. When supplemented with LGG and then challenged with rotavirus, conventional pigs had increased villus-to-crypt ratios, villus height, sIgA, IL-4, mucin1 and mucin2 concentrations, and ZO-1, occludin, and Bcl-2 mRNA in jejunal mucosa, and decreased Bax mRNA and NSP4 in the jejunum compared with controls (Mao et al. 2016). Gn pigs supplemented with LGG were partially protected from HRV-induced increases in adherens junction proteins α-catenin and β-catenin, tight junction proteins occludin, claudin-3 and claudin-4, and leakage of protein claudin-2 compared with controls (Liu et al. 2013). In both studies, LGG-supplemented pigs had reduced diarrhea compared with controls after rotavirus challenge (Mao et al. 2016; Liu et al. 2013). One mechanism by which LGG may reduce diarrhea is by protecting small intestinal barrier function. A recent study showed that metabolites of *L. casei* and *B. adolescentis* significantly reduced NSP4 production and $Ca2^{++}$ liberation in vitro, suggesting activity against rotavirus infection (Olaya Galan et al. 2016).

LGG can improve innate immunity. It has been shown that LGG increases mRNA levels of TLR3 when incubated with intestinal organoids (Aoki-Yoshida et al. 2016). In vivo, single and a 7-day course of LGG increased TLR3 mRNA levels in the small intestine of C57/BL6N mice (Aoki-Yoshida et al. 2016). Co-colonization of Gn pigs with LGG and *Bifidobacterium lactis* Bb12 (Bb12) induced upregulation of TLR3 after VirHRV challenge and downregulation of TLR 2 and TLR4 expressing mononuclear cells (MNCs) after AttHRV vaccination (Vlasova et al. 2013). *L. ruminis* SPM02111, *Bifidobacterium longum* SPM1205 and SPM1206 were able to inhibit rotavirus replication in neonatal mice, inhibit Wa HRV infection of Caco-2 cells, increase IFN-α and IFN-β, and increase gene expression of IFN-inducible antiviral effectors when compared to controls (Kang et al. 2015). *Lactobacillus reuteri* and *Lactobacillus acidophilus* with HRV infection have an additive effect on TLR2 and TLR9 expressing antigen presenting cell responses in Gn pigs (Wen et al. 2009). This same study demonstrated increased IFN-γ and IL-4 responses in serum of the probiotic colonized pigs while serum IFN-α response to HRV were reduced (Wen et al. 2009).

Lactobacillus rhamnosus GG, in combination with other probiotics, has been shown to influence T cell and humoral responses. Nonvaccinated Gn pigs colonized with LGG and Bb12 challenged with VirHRV had less diarrhea and viral shedding than nonvaccinated, noncolonized pigs, and the increased protection was associated with higher T regulatory cells before and after challenge; higher serum TGF-β; and lower proinflammatory cytokines after viral challenge (Chattha et al. 2013). AttHRV-vaccinated pigs colonized with these two probiotics had enhanced serum IFN-α, splenic and blood IFN-γ-producing T cells, and serum Th1 cytokines compared with noncolonized vaccinated pigs (Chattha et al. 2013). Gn pigs colonized with LGG and *Bifidobacterium animalis lactis* Bb12 had lower diarrhea scores and viral shedding after AttHRV vaccination and VirHRV challenge than noncolonized vaccinated pigs (Kandasamy et al. 2014). The decreased clinical signs in the colonized pigs correlated with higher intestinal rotavirus-specific IgA titers, and rotavirus-specific IgA ASC (Kandasamy et al. 2014).

Modulation of microbiome by probiotics has been studied in Gn pig models. AttHRV-vaccinated Gn pigs colonized with infant gut microbiota showed that LGG prevented changes in the microbiome structure caused by VirHRV challenge that were seen in non-LGG-supplemented groups (Zhang et al. 2014).

Bifidobacterium spp. are commonly studied probiotics. *B. thermophilum* RBL67 is thought to inhibit rotavirus infection by competing for adherence on cells, as demonstrated in vitro with Caco-2 and HT-29 cells (Gagnon et al. 2016). When incubated before rotavirus infection of cells to assess exclusion and incubated with rotavirus to assess competition, there was decreased viral attachment in the *B. thermophilum*-treated cells; however, the probiotic did not appear to displace virus already adhered (Gagnon et al. 2016). *B. longum* subsp. infantis CECT 7210 can inhibit rotavirus replication in vitro via an 11-aminoacid peptide (11-mer peptide) released into supernatant along with a protease that releases the 11-mer peptide (Chenoll et al. 2016). In vivo studies have shown the effectiveness of *B. thermophilum* during rotavirus infection. Administration of *B. thermophilum* RBL67 to CD-1 suckling mice before challenge with simian rotavirus SA-11 decreased the duration of diarrhea, viral replication, recovery time, and histological lesions, and stimulated rotavirus-specific IgG and IgM (Gagnon et al. 2016).

The combination of EcN and LGG have been evaluated in vivo and in vitro. Gn pigs colonized by EcN had lower mean peak viral shedding titers and mean cumulative fecal scores compared with LGG or noncolonized pigs (Kandasamy et al. 2016). Total IgA levels after challenge in the intestine and before challenge in serum were higher in EcN than LGG-colonized pigs (Kandasamy et al. 2016). EcN but not LGG induced IL-6, IL-10, and IgA in MNCs treated with EcN or LGG in vitro (Kandasamy et al. 2016). EcN colonization was associated with better protection against HRV and induced higher frequencies of plasmacytoid dendritic cells (pDCs), increased NK-cell function, and decreased frequencies of apoptotic and TLR4+MNC compared with LGG-colonized pigs (Vlasova et al. 2016). EcN-treated splenic or intestinal MNC produced higher levels of IFN-α, IL-12, and IL-10, compared with MNC treated with LGG (Vlasova et al. 2016). These studies demonstrate that different probiotic strains do not have the same immunomodulatory functions and that strain selection should be based on the effect desired.

Bacteria are not the only microorganisms used as probiotics. When the yeast *Saccharomyces boulardii* was given to children with acute rotavirus diarrhea, the mean duration of diarrhea and hospitalization were shorter than in controls; however, there was no difference between the groups in the number of children requiring parenteral rehydration or who had diarrhea lasting beyond 7 days (Das et al. 2016). It is believed that *S. boulardii* decreases diarrhea by preventing rotavirus-induced oxidative stress and thus activation of NSP4 and subsequent chloride secretion based on results obtained in Caco-2 cells and human intestinal organ culture (Buccigrossi et al. 2014).

3.6.2 Dose Effects of Probiotics in Modulating Rotavirus Vaccine-Induced Immune Responses

Differences in the dosing schedule of the probiotics influence host immune response. A Gn pig study looking at the influence of LGG on protection provided by AttHRV vaccination showed that rotavirus-specific intestinal memory B cell responses and virus-specific intestinal IgA ASCs were enhanced by a five-dose regimen of LGG, but not nine-dose regimen, although both doses enhanced the rotavirus-specific serum IgA response and rotavirus-specific IFN-γ producing effector/memory T cell responses, with the nine-dose regimen having a stronger effect (Wen et al. 2015). This study demonstrated how the dosing regimen can affect the immune response; in this case, the five-dose regimen favored a mucosal

IgA response, whereas the higher dosing schedule favored the T cell response (Wen et al. 2015). Another AttHRV vaccine study in Gn pigs showed that pigs receiving 14 doses of LGG had increased large intestinal LGG titers and a shifted microbiota structure, which correlated with increased rotavirus-specific IFN-γ-producing T cells, suggesting a Th1 adjuvant effect (Wang et al. 2016). However, pigs in the same study receiving nine doses of LGG had enhanced TLR9 signaling, which may suggest that this dosing regimen might have enhanced innate immunity (Wang et al. 2016). A third study also demonstrated a differential effect from LGG dosing schedules. In this study, using HGM=transplanted Gn pigs, it was shown that a 14-dose regimen of LGG enhanced rotavirus-specific, IFN-γ-producing T cell response to AttHRV vaccination, whereas a nine-dose regimen was ineffective (Wen et al. 2014). The effects of dosing schedules are seen with other probiotics in addition to LGG. Gn pigs colonized with the *L. acidophilus* NCFM, vaccinated with AttHRV, and challenged with VirHRV demonstrated that a nine-dose regimen of *L. acidophilus* but not a 14- or five-dose regimen improved protection provided by the vaccine and this was associated with enhanced rotavirus-specific antibody, ASC, and memory B cell responses to the vaccination (Liu et al. 2014). Neither the high-dose (14) nor the low-dose (5) regimen enhanced antibody or ASC responses, and thus did not improve vaccine efficacy (Liu et al. 2014). The differential modulating effects of different doses of probiotics are intriguing. The underlying mechanisms require further investigation. It has been reported that the effect of low-dose microbe-associated molecular patterns (MAMPs), such as lipopolysaccharide, was strikingly different than that of high-dose MAMPs on macrophage cell functions: low-dose lipopolysaccharide induced a strong inflammatory response in macrophages (Maitra et al. 2011). It is plausible that a similar interaction occurs between the MAMPs from probiotics and immune cells in the gut. Future studies are needed to identify the molecular mechanisms of the dose responses of different MAMPs.

3.6.3 Mechanisms for Reducing Rotavirus Diarrhea Using Prebiotics

Prebiotics are another nonpharmacological category of agents being investigated for treatment or prevention of diarrhea with or without concurrent probiotic administration. Rice bran contains phytochemicals that can promote intestinal mucosal immunity to enteropathogens (Yang et al. 2014). Gn pigs fed rice bran were protected from diarrhea after VirHRV challenge and AttHRV was more protective in these pigs than in nonrice bran-fed pigs (Yang et al. 2014). IFN-γ-producing CD4 + and CD8+ T cells were increased in intestinal and systemic lymphoid tissues, IgM ASC, IgA ASC, total serum IgM, IgG, IgA, and rotavirus-specific IgA intestinal titers were increased in rice bran-fed pigs compared with nonrice bran-fed pigs (Yang et al. 2014). Results support rice bran as a stimulator of nonspecific and rotavirus-specific immune responses (Yang et al. 2014). Gn pigs colonized with LGG and EcN were fed a diet supplemented with rice brain daily (Yang et al. 2015). Rice bran completely prevented rotavirus diarrhea in the colonized pigs after VirHRV challenge and promoted growth of both probiotic strains LGG and EcN compared with nonrice bran-fed pigs (Yang et al. 2015). In addition, after VirHRV challenge, the rice bran-fed pigs had increased intestinal IFN-γ and total IgA levels, and fewer histological changes in the ileum, compared with the nonrice bran-fed group (Yang et al. 2015).

Prebiotics are often evaluated with probiotics as they can have synergistic effects on each other. A study evaluating *B. lactis* B94 and inulin in children with acute diarrhea showed that the duration and amount of diarrhea was reduced in the group receiving the prebiotics and probiotics (Islek et al. 2014). The clinical effects were most pronounced in cases of rotavirus diarrhea (Islek et al. 2014).

3.7 Concluding Remarks

Germ-free animal models provide an indispensable tool for the study of the consequences of bacterial colonization and mechanisms underlying host–microbiome interactions in enteric virus infection and gastroenteritis. The Gn pig model, with its distinct advantages, has greatly contributed to studies on the effects and mechanisms of gut microbiota and probiotics on enteric virus infections and vaccines. However, the drawback of using pig models is the decreased availability of species-specific molecular reagents and gene knockout pigs compared with mouse models, which hinders in-depth mechanistic studies. Further optimization of the pig models, including genetic modification using CRISPR/Cas9 technology, humanization of the immune system through stem cell transfer, and transplantation with HGM from donors representing different health, disease, and immune statuses will further improve the usefulness and reliability of pig models for mimicking HuNoV and HRV infection and disease in humans. Unraveling the role of the microbiome and specific probiotics in the infectivity, pathogenesis, and immunity of HuNoV and HRV will facilitate the development of strategies for manipulating the microbiome against viral infections.

References

Agus SG, Dolin R, Wyatt RG, Tousimis AJ, Northrup RS (1973) Acute infectious nonbacterial gastroenteritis: intestinal histopathology. Histologic and enzymatic alterations during illness produced by the Norwalk agent in man. Ann Intern Med 79(1):18–25

Aoki-Yoshida A, Saito S, Fukiya S, Aoki R, Takayama Y, Suzuki C, Sonoyama K (2016) Lactobacillus rhamnosus GG increases Toll-like receptor 3 gene expression in murine small intestine ex vivo and in vivo. Benefic Microbes 7(3):421–429. https://doi.org/10.3920/BM2015.0169

Baldridge MT, Nice TJ, McCune BT, Yokoyama CC, Kambal A, Wheadon M, Diamond MS, Ivanova Y, Artyomov M, Virgin HW (2015) Commensal microbes and interferon-lambda determine persistence of enteric murine norovirus infection. Science 347(6219):266–269. https://doi.org/10.1126/science.1258025

Barnes GL, Townley RR (1973) Duodenal mucosal damage in 31 infants with gastroenteritis. Arch Dis Child 48(5):343–349

Basic M, Keubler LM, Buettner M, Achard M, Breves G, Schroder B, Smoczek A, Jorns A, Wedekind D, Zschemisch NH, Gunther C, Neumann D, Lienenklaus S, Weiss S, Hornef MW, Mahler M, Bleich A (2014) Norovirus triggered microbiota-driven mucosal inflammation in interleukin 10-deficient mice. Inflamm Bowel Dis 20(3):431–443. https://doi.org/10.1097/01.MIB.0000441346.86827.ed

Bok K, Green KY (2012) Norovirus gastroenteritis in immunocompromised patients. N Engl J Med 367(22):2126–2132. https://doi.org/10.1056/NEJMra1207742

Bok K, Parra GI, Mitra T, Abente E, Shaver CK, Boon D, Engle R, Yu C, Kapikian AZ, Sosnovtsev SV, Purcell RH, Green KY (2011) Chimpanzees as an animal model for human norovirus infection and vaccine development. Proc Natl Acad Sci USA 108(1):325–330. https://doi.org/10.1073/pnas.1014577107

Brown JR, Gilmour K, Breuer J (2016) Norovirus infections occur in B-cell-deficient patients. Clin Infect Dis 62(9):1136–1138. https://doi.org/10.1093/cid/ciw060

Buccigrossi V, Laudiero G, Russo C, Miele E, Sofia M, Monini M, Ruggeri FM, Guarino A (2014) Chloride secretion induced by rotavirus is oxidative stress-dependent and inhibited by Saccharomyces boulardii in human enterocytes. PLoS One 9(6):e99830. https://doi.org/10.1371/journal.pone.0099830

Bui T, Kocher J, Li Y, Wen K, Li G, Liu F, Yang X, LeRoith T, Tan M, Xia M, Zhong W, Jiang X, Yuan L (2013) Median infectious dose of human norovirus GII.4 in gnotobiotic pigs is decreased by simvastatin treatment and increased by age. J Gen Virol 94(Pt 9):2005–2016. https://doi.org/10.1099/vir.0.054080-0

Chattha KS, Vlasova AN, Kandasamy S, Rajashekara G, Saif LJ (2013) Divergent immunomodulating effects of probiotics on T cell responses to oral attenuated human rotavirus vaccine and virulent human rotavirus infection in a neonatal gnotobiotic piglet disease model. J Immunol 191(5):2446–2456. https://doi.org/10.4049/jimmunol.1300678

Cheetham S, Souza M, Meulia T, Grimes S, Han MG, Saif LJ (2006) Pathogenesis of a genogroup II human norovirus in gnotobiotic pigs. J Virol 80(21):10372–10381. https://doi.org/10.1128/JVI.00809-06

Chenoll E, Casinos B, Bataller E, Buesa J, Ramon D, Genoves S, Fabrega J, Rivero Urgell M, Moreno Munoz JA (2016) Identification of a peptide produced by Bifidobacterium longum CECT 7210 with antirotaviral activity. Front Microbiol 7:655. https://doi.org/10.3389/fmicb.2016.00655

Das S, Gupta PK, Das RR (2016) Efficacy and safety of Saccharomyces boulardii in acute rotavirus diarrhea: double blind randomized controlled trial from a developing country. J Trop Pediatr 62:464–470. https://doi.org/10.1093/tropej/fmw032

Davidson GP, Barnes GL (1979) Structural and functional abnormalities of the small intestine in infants and young children with rotavirus enteritis. Acta Paediatr Scand 68(2):181–186

Desselberger U (2014) Rotaviruses. Virus Res 190:75–96. https://doi.org/10.1016/j.virusres.2014.06.016

Desselberger U, Huppertz HI (2011) Immune responses to rotavirus infection and vaccination and associated correlates of protection. J Infect Dis 203(2):188–195. https://doi.org/10.1093/infdis/jiq031

Dolin R, Levy AG, Wyatt RG, Thornhill TS, Gardner JD (1975) Viral gastroenteritis induced by the Hawaii agent. Jejunal histopathology and serologic response. Am J Med 59(6):761–768

Ettayebi K, Crawford SE, Murakami K, Broughman JR, Karandikar U, Tenge VR, Neill FH, Blutt SE, Zeng XL, Qu L, Kou B, Opekun AR, Burrin D, Graham DY, Ramani S, Atmar RL, Estes

MK (2016) Replication of human noroviruses in stem cell-derived human enteroids. Science 353(6306):1387–1393. https://doi.org/10.1126/science.aaf5211

Fang H, Tan M, Xia M, Wang L, Jiang X (2013) Norovirus P particle efficiently elicits innate, humoral and cellular immunity. PLoS One 8(4):e63269. https://doi.org/10.1371/journal.pone.0063269

Gagnon M, Vimont A, Darveau A, Fliss I, Jean J (2016) Study of the ability of bifidobacteria of human origin to prevent and treat rotavirus infection using colonic cell and mouse models. PLoS One 11(10):e0164512. https://doi.org/10.1371/journal.pone.0164512

Green KY (2014) Norovirus infection in immunocompromised hosts. Clin Microbiol Infect 20 (8):717–723. https://doi.org/10.1111/1469-0691.12761

Guandalini S, Pensabene L, Zikri MA, Dias JA, Casali LG, Hoekstra H, Kolacek S, Massar K, Micetic-Turk D, Papadopoulou A, de Sousa JS, Sandhu B, Szajewska H, Weizman Z (2000) Lactobacillus GG administered in oral rehydration solution to children with acute diarrhea: a multicenter European trial. J Pediatr Gastroenterol Nutr 30(1):54–60

Harris VC, Armah G, Fuentes S, Korpela KE, Parashar U, Victor JC, Tate J, de Weerth C, Giaquinto C, Wiersinga WJ, Lewis KD, de Vos WM (2016) The infant gut microbiome correlates significantly with rotavirus vaccine response in rural Ghana. J Infect Dis 215:34–41. https://doi.org/10.1093/infdis/jiw518

Hemming M, Rasanen S, Huhti L, Paloniemi M, Salminen M, Vesikari T (2013) Major reduction of rotavirus, but not norovirus, gastroenteritis in children seen in hospital after the introduction of RotaTeq vaccine into the National Immunization Programme in Finland. Eur J Pediatr 172 (6):739–746. https://doi.org/10.1007/s00431-013-1945-3

Heyman M, Corthier G, Petit A, Meslin JC, Moreau C, Desjeux JF (1987) Intestinal absorption of macromolecules during viral enteritis: an experimental study on rotavirus-infected conventional and germ-free mice. Pediatr Res 22(1):72–78

Iosef C, Chang KO, Azevedo MS, Saif LJ (2002) Systemic and intestinal antibody responses to NSP4 enterotoxin of Wa human rotavirus in a gnotobiotic pig model of human rotavirus disease. J Med Virol 68(1):119–128. https://doi.org/10.1002/jmv.10178

Islek A, Sayar E, Yilmaz A, Baysan BO, Mutlu D, Artan R (2014) The role of Bifidobacterium lactis B94 plus inulin in the treatment of acute infectious diarrhea in children. Turk J Gastroenterol 25(6):628–633. https://doi.org/10.5152/tjg.2014.14022

Jones MK, Watanabe M, Zhu S, Graves CL, Keyes LR, Grau KR, Gonzalez-Hernandez MB, Iovine NM, Wobus CE, Vinje J, Tibbetts SA, Wallet SM, Karst SM (2014) Enteric bacteria promote human and mouse norovirus infection of B cells. Science 346(6210):755–759. https://doi.org/10.1126/science.1257147

Jones MK, Grau KR, Costantini V, Kolawole AO, de Graaf M, Freiden P, Graves CL, Koopmans M, Wallet SM, Tibbetts SA, Schultz-Cherry S, Wobus CE, Vinje J, Karst SM (2015) Human norovirus culture in B cells. Nat Protoc 10(12):1939–1947. https://doi.org/10.1038/nprot.2015.121

Kandasamy S, Chattha KS, Vlasova AN, Rajashekara G, Saif LJ (2014) Lactobacilli and bifidobacteria enhance mucosal B cell responses and differentially modulate systemic antibody responses to an oral human rotavirus vaccine in a neonatal gnotobiotic pig disease model. Gut Microbes 5(5):639–651. https://doi.org/10.4161/19490976.2014.969972

Kandasamy S, Vlasova AN, Fischer D, Kumar A, Chattha KS, Rauf A, Shao L, Langel SN, Rajashekara G, Saif LJ (2016) Differential effects of Escherichia coli Nissle and Lactobacillus rhamnosus strain GG on human rotavirus binding, infection, and B cell immunity. J Immunol 196(4):1780–1789. https://doi.org/10.4049/jimmunol.1501705

Kane M, Case LK, Kopaskie K, Kozlova A, MacDearmid C, Chervonsky AV, Golovkina TV (2011) Successful transmission of a retrovirus depends on the commensal microbiota. Science 334(6053):245–249. https://doi.org/10.1126/science.1210718

Kang JY, Lee DK, Ha NJ, Shin HS (2015) Antiviral effects of Lactobacillus ruminis SPM0211 and Bifidobacterium longum SPM1205 and SPM1206 on rotavirus-infected Caco-2 cells and a

neonatal mouse model. J Microbiol 53(11):796–803. https://doi.org/10.1007/s12275-015-5302-2
Karandikar UC, Crawford SE, Ajami NJ, Murakami K, Kou B, Ettayebi K, Papanicolaou GA, Jongwutiwes U, Perales MA, Shia J, Mercer D, Finegold MJ, Vinje J, Atmar RL, Estes MK (2016) Detection of human norovirus in intestinal biopsies from immunocompromised transplant patients. J Gen Virol 97(9):2291–2300. https://doi.org/10.1099/jgv.0.000545
Karst SM (2010) Pathogenesis of noroviruses, emerging RNA viruses. Viruses 2(3):748–781. https://doi.org/10.3390/v2030748
Karst SM, Wobus CE, Lay M, Davidson J, HWt V (2003) STAT1-dependent innate immunity to a Norwalk-like virus. Science 299(5612):1575–1578. https://doi.org/10.1126/science.1077905
Karst SM, Wobus CE, Goodfellow IG, Green KY, Virgin HW (2014) Advances in norovirus biology. Cell Host Microbe 15(6):668–680. https://doi.org/10.1016/j.chom.2014.05.015
Kehrer-Sawatzki H, Cooper DN (2007) Understanding the recent evolution of the human genome: insights from human-chimpanzee genome comparisons. Hum Mutat 28(2):99–130. https://doi.org/10.1002/humu.20420
Knowles NJ, Reuter G (2012) Porcine caliciviruses. In: Zimmerman JJ, Karriker LA, Ramirez A, Schwartz KJ, Stevenson GW (eds) Diseases of swine, 10th edn. Wiley, New York, pp 493–500
Kocher J, Yuan L (2015) Norovirus vaccines and potential antinorovirus drugs: recent advances and future perspectives. Futur Virol 10(7):899–913. https://doi.org/10.2217/fvl.15.57
Kocher J, Bui T, Giri-Rachman E, Wen K, Li G, Yang X, Liu F, Tan M, Xia M, Zhong W, Jiang X, Yuan L (2014) Intranasal P particle vaccine provided partial cross-variant protection against human GII.4 norovirus diarrhea in gnotobiotic pigs. J Virol 88(17):9728–9743. https://doi.org/10.1128/JVI.01249-14
Kolawole AO, Rocha-Pereira J, Elftman MD, Neyts J, Wobus CE (2016) Inhibition of human norovirus by a viral polymerase inhibitor in the B cell culture system and in the mouse model. Antiviral Res 132:46–49. https://doi.org/10.1016/j.antiviral.2016.05.011
Kuss SK, Best GT, Etheredge CA, Pruijssers AJ, Frierson JM, Hooper LV, Dermody TS, Pfeiffer JK (2011) Intestinal microbiota promote enteric virus replication and systemic pathogenesis. Science 334(6053):249–252. https://doi.org/10.1126/science.1211057
Lee H, Ko G (2016) Antiviral effect of vitamin A on norovirus infection via modulation of the gut microbiome. Sci Rep 6:25835. https://doi.org/10.1038/srep25835
Lei S, Ramesh A, Twitchell E, Wen K, Bui T, Weiss M, Yang X, Kocher J, Li G, Giri-Rachman E, Trang NV, Jiang X, Ryan EP, Yuan L (2016a) High protective efficacy of probiotics and rice bran against human norovirus infection and diarrhea in gnotobiotic pigs. Front Microbiol 7:1699. https://doi.org/10.3389/fmicb.2016.01699
Lei S, Ryu J, Wen K, Twitchell E, Bui T, Ramesh A, Weiss M, Li G, Samuel H, Clark-Deener S, Jiang X, Lee K, Yuan L (2016b) Increased and prolonged human norovirus infection in RAG2/IL2RG deficient gnotobiotic pigs with severe combined immunodeficiency. Sci Rep 6:25222. https://doi.org/10.1038/srep25222
Lei S, Samuel H, Twitchell E, Bui T, Ramesh A, Wen K, Weiss M, Li G, Yang X, Jiang X, Yuan L (2016c) Enterobacter cloacae inhibits human norovirus infectivity in gnotobiotic pigs. Sci Rep 6:25017. https://doi.org/10.1038/srep25017
Li D, Breiman A, le Pendu J, Uyttendaele M (2016) Anti-viral Effect of Bifidobacterium adolescentis against noroviruses. Front Microbiol 7:864. https://doi.org/10.3389/fmicb.2016.00864
Licciardi PV, Tang ML (2011) Vaccine adjuvant properties of probiotic bacteria. Discov Med 12(67):525–533
Liu F, Li G, Wen K, Wu S, Zhang Y, Bui T, Yang X, Kocher J, Sun J, Jortner B, Yuan L (2013) Lactobacillus rhamnosus GG on rotavirus-induced injury of ileal epithelium in gnotobiotic pigs. J Pediatr Gastroenterol Nutr 57(6):750–758. https://doi.org/10.1097/MPG.0b013e3182a356e1

Liu F, Wen K, Li G, Yang X, Kocher J, Bui T, Jones D, Pelzer K, Clark-Deener S, Yuan L (2014) Dual functions of Lactobacillus acidophilus NCFM as protection against rotavirus diarrhea. J Pediatr Gastroenterol Nutr 58(2):169–176. https://doi.org/10.1097/MPG.0000000000000197

Maitra U, Gan L, Chang S, Li L (2011) Low-dose endotoxin induces inflammation by selectively removing nuclear receptors and activating CCAAT/enhancer-binding protein delta. J Immunol 186(7):4467–4473. https://doi.org/10.4049/jimmunol.1003300

Mao X, Gu C, Hu H, Tang J, Chen D, Yu B, He J, Yu J, Luo J, Tian G (2016) Dietary Lactobacillus rhamnosus GG supplementation improves the mucosal barrier function in the intestine of weaned piglets challenged by porcine rotavirus. PLoS One 11(1):e0146312. https://doi.org/10.1371/journal.pone.0146312

McNeal MM, Sestak K, Choi AH, Basu M, Cole MJ, Aye PP, Bohm RP, Ward RL (2005) Development of a rotavirus-shedding model in rhesus macaques, using a homologous wild-type rotavirus of a new P genotype. J Virol 79(2):944–954

Mebus CA, Wyatt RG, Kapikian AZ (1977) Intestinal lesions induced in gnotobiotic calves by the virus of human infantile gastroenteritis. Vet Pathol 14(3):273–282

Nagata S, Asahara T, Ohta T, Yamada T, Kondo S, Bian L, Wang C, Yamashiro Y, Nomoto K (2011) Effect of the continuous intake of probiotic-fermented milk containing Lactobacillus casei strain Shirota on fever in a mass outbreak of norovirus gastroenteritis and the faecal microflora in a health service facility for the aged. Br J Nutr 106(4):549–556. https://doi.org/10.1017/S000711451100064X

Nelson AM, Walk ST, Taube S, Taniuchi M, Houpt ER, Wobus CE, Young VB (2012) Disruption of the human gut microbiota following norovirus infection. PLoS One 7(10):e48224. https://doi.org/10.1371/journal.pone.0048224

Nelson AM, Elftman MD, Pinto AK, Baldridge M, Hooper P, Kuczynski J, Petrosino JF, Young VB, Wobus CE (2013) Murine norovirus infection does not cause major disruptions in the murine intestinal microbiota. Microbiome 1(1):7. https://doi.org/10.1186/2049-2618-1-7

O'Neil SP, Novembre FJ, Hill AB, Suwyn C, Hart CE, Evans-Strickfaden T, Anderson DC, deRosayro J, Herndon JG, Saucier M, McClure HM (2000) Progressive infection in a subset of HIV-1-positive chimpanzees. J Infect Dis 182(4):1051–1062. https://doi.org/10.1086/315823

O'Ryan ML, Pena A, Vergara R, Diaz J, Mamani N, Cortes H, Lucero Y, Vidal R, Osorio G, Santolaya ME, Hermosilla G, Prado VJ (2010) Prospective characterization of norovirus compared with rotavirus acute diarrhea episodes in chilean children. Pediatr Infect Dis J 29 (9):855–859. https://doi.org/10.1097/INF.0b013e3181e8b346

Olaya Galan NN, Ulloa Rubiano JC, Velez Reyes FA, Fernandez Duarte KP, Salas Cardenas SP, Gutierrez Fernandez MF (2016) In vitro antiviral activity of Lactobacillus casei and Bifidobacterium adolescentis against rotavirus infection monitored by NSP4 protein production. J Appl Microbiol 120(4):1041–1051. https://doi.org/10.1111/jam.13069

Patel MM, Hall AJ, Vinje J, Parashar UD (2009) Noroviruses: a comprehensive review. J Clin Virol: the official publication of the Pan American Society for Clinical Virology 44(1):1–8. https://doi.org/10.1016/j.jcv.2008.10.009

Payne DC, Vinje J, Szilagyi PG, Edwards KM, Staat MA, Weinberg GA, Hall CB, Chappell J, Bernstein DI, Curns AT, Wikswo M, Shirley SH, Hall AJ, Lopman B, Parashar UD (2013) Norovirus and medically attended gastroenteritis in U.S. children. N Engl J Med 368 (12):1121–1130. https://doi.org/10.1056/NEJMsa1206589

Pfaender S, Brown RJ, Pietschmann T, Steinmann E (2014) Natural reservoirs for homologs of hepatitis C virus. Emerg Microbes Infect 3(3):e21. https://doi.org/10.1038/emi.2014.19

Pringle K, Lopman B, Vega E, Vinje J, Parashar UD, Hall AJ (2015) Noroviruses: epidemiology, immunity and prospects for prevention. Futur Microbiol 10(1):53–67. https://doi.org/10.2217/fmb.14.102

Purcell RH, Emerson SU (2001) Animal models of hepatitis A and E. ILAR J 42(2):161–177

Ramig RF (2004) Pathogenesis of intestinal and systemic rotavirus infection. J Virol 78 (19):10213–10220

Ramig RF (2007) Systemic rotavirus infection. Expert Rev Anti-Infect Ther 5(4):591–612
Rubio-del-Campo A, Coll-Marques JM, Yebra MJ, Buesa J, Perez-Martinez G, Monedero V, Rodriguez-Diaz J (2014) Noroviral p-particles as an in vitro model to assess the interactions of noroviruses with probiotics. PLoS One 9(2):e89586. https://doi.org/10.1371/journal.pone.0089586
Saif LJ, Ward LA, Yuan L, Rosen BI, To TL (1996) The gnotobiotic piglet as a model for studies of disease pathogenesis and immunity to human rotaviruses. Arch Virol Suppl 12:153–161
Saif L, Yuan L, Ward L, To T (1997) Comparative studies of the pathogenesis, antibody immune responses, and homologous protection to porcine and human rotaviruses in gnotobiotic piglets. Adv Exp Med Biol 412:397–403
Schnadower D, Finkelstein Y, Freedman SB (2015) Ondansetron and probiotics in the management of pediatric acute gastroenteritis in developed countries. Curr Opin Gastroenterol 31 (1):1–6. https://doi.org/10.1097/MOG.0000000000000132
Schreiber DS, Blacklow NR, Trier JS (1973) The mucosal lesion of the proximal small intestine in acute infectious nonbacterial gastroenteritis. N Engl J Med 288(25):1318–1323. https://doi.org/10.1056/NEJM197306212882503
Schreiber DS, Blacklow NR, Trier JS (1974) The small intestinal lesion induced by Hawaii agent acute infectious nonbacterial gastroenteritis. J Infect Dis 129(6):705–708
Sindhu KN, Sowmyanarayanan TV, Paul A, Babji S, Ajjampur SS, Priyadarshini S, Sarkar R, Balasubramanian KA, Wanke CA, Ward HD, Kang G (2014) Immune response and intestinal permeability in children with acute gastroenteritis treated with Lactobacillus rhamnosus GG: a randomized, double-blind, placebo-controlled trial. Clin Infect Dis: an official publication of the Infectious Diseases Society of America 58(8):1107–1115. https://doi.org/10.1093/cid/ciu065
Souza M, Cheetham SM, Azevedo MS, Costantini V, Saif LJ (2007a) Cytokine and antibody responses in gnotobiotic pigs after infection with human norovirus genogroup II.4 (HS66 strain). J Virol 81(17):9183–9192. https://doi.org/10.1128/JVI.00558-07
Souza M, Costantini V, Azevedo MS, Saif LJ (2007b) A human norovirus-like particle vaccine adjuvanted with ISCOM or mLT induces cytokine and antibody responses and protection to the homologous GII.4 human norovirus in a gnotobiotic pig disease model. Vaccine 25 (50):8448–8459. https://doi.org/10.1016/j.vaccine.2007.09.040
Souza M, Azevedo MS, Jung K, Cheetham S, Saif LJ (2008) Pathogenesis and immune responses in gnotobiotic calves after infection with the genogroup II.4-HS66 strain of human norovirus. J Virol 82(4):1777–1786. https://doi.org/10.1128/JVI.01347-07
Szajewska H, Urbanska M, Chmielewska A, Weizman Z, Shamir R (2014) Meta-analysis: Lactobacillus reuteri strain DSM 17938 (and the original strain ATCC 55730) for treating acute gastroenteritis in children. Benefic Microbes 5(3):285–293. https://doi.org/10.3920/BM2013.0056
Taube S, Kolawole AO, Hohne M, Wilkinson JE, Handley SA, Perry JW, Thackray LB, Akkina R, Wobus CE (2013) A mouse model for human norovirus. MBio 4(4):e00450–e00413. https://doi.org/10.1128/mBio.00450-13
Tin CM, Yuan L, Dexter RJ, Parra GI, Bui T, Green KY, Sosnovtsev SV (2017) A Luciferase Immunoprecipitation System (LIPS) assay for profiling human norovirus antibodies. J Virol Methods 248:116–129
Troeger H, Loddenkemper C, Schneider T, Schreier E, Epple HJ, Zeitz M, Fromm M, Schulzke JD (2009) Structural and functional changes of the duodenum in human norovirus infection. Gut 58(8):1070–1077. https://doi.org/10.1136/gut.2008.160150
Twitchell EL, Tin C, Wen K, Zhang H, Becker-Dreps S, Azcarate-Peril MA, Vilchez S, Li G, Ramesh A, Weiss M, Lei S, Bui T, Yang X, Schultz-Cherry S, Yuan L (2016) Modeling human enteric dysbiosis and rotavirus immunity in gnotobiotic pigs. Gut Pathog 8:51. https://doi.org/10.1186/s13099-016-0136-y

Tzipori SR, Makin TJ, Smith ML (1980) The clinical response of gnotobiotic calves, pigs and lambs to inoculation with human, calf, pig and foal rotavirus isolates. Aust J Exp Biol Med Sci 58(3):309–318

Uchiyama R, Chassaing B, Zhang B, Gewirtz AT (2014) Antibiotic treatment suppresses rotavirus infection and enhances specific humoral immunity. J Infect Dis 210(2):171–182. https://doi.org/10.1093/infdis/jiu037

Varyukhina S, Freitas M, Bardin S, Robillard E, Tavan E, Sapin C, Grill JP, Trugnan G (2012) Glycan-modifying bacteria-derived soluble factors from Bacteroides thetaiotaomicron and Lactobacillus casei inhibit rotavirus infection in human intestinal cells. Microbes Infect 14 (3):273–278. https://doi.org/10.1016/j.micinf.2011.10.007

Vlasova AN, Chattha KS, Kandasamy S, Liu Z, Esseili M, Shao L, Rajashekara G, Saif LJ (2013) Lactobacilli and bifidobacteria promote immune homeostasis by modulating innate immune responses to human rotavirus in neonatal gnotobiotic pigs. PLoS One 8(10):e76962. https://doi.org/10.1371/journal.pone.0076962

Vlasova AN, Shao L, Kandasamy S, Fischer DD, Rauf A, Langel SN, Chattha KS, Kumar A, Huang HC, Rajashekara G, Saif LJ (2016) Escherichia coli Nissle 1917 protects gnotobiotic pigs against human rotavirus by modulating pDC and NK-cell responses. Eur J Immunol 46 (10):2426–2437. https://doi.org/10.1002/eji.201646498

Wang M, Donovan SM (2015) Human microbiota-associated swine: current progress and future opportunities. ILAR J 56(1):63–73. https://doi.org/10.1093/ilar/ilv006

Wang QH, Han MG, Cheetham S, Souza M, Funk JA, Saif LJ (2005) Porcine noroviruses related to human noroviruses. Emerg Infect Dis 11(12):1874–1881

Wang H, Gao K, Wen K, Allen IC, Li G, Zhang W, Kocher J, Yang X, Giri-Rachman E, Li GH, Clark-Deener S, Yuan L (2016) Lactobacillus rhamnosus GG modulates innate signaling pathway and cytokine responses to rotavirus vaccine in intestinal mononuclear cells of gnotobiotic pigs transplanted with human gut microbiota. BMC Microbiol 16(1):109. https://doi.org/10.1186/s12866-016-0727-2

Ward RL, McNeal MM, Sheridan JF (1990) Development of an adult mouse model for studies on protection against rotavirus. J Virol 64(10):5070–5075

Ward LA, Rosen BI, Yuan L, Saif LJ (1996a) Pathogenesis of an attenuated and a virulent strain of group A human rotavirus in neonatal gnotobiotic pigs. J Gen Virol 77(Pt 7):1431–1441

Ward LA, Yuan L, Rosen BI, To TL, Saif LJ (1996b) Development of mucosal and systemic lymphoproliferative responses and protective immunity to human group A rotaviruses in a gnotobiotic pig model. Clin Diagn Lab Immunol 3(3):342–350

Wen K, Azevedo MS, Gonzalez A, Zhang W, Saif LJ, Li G, Yousef A, Yuan L (2009) Toll-like receptor and innate cytokine responses induced by lactobacilli colonization and human rotavirus infection in gnotobiotic pigs. Vet Immunol Immunopathol 127(3–4):304–315. https://doi.org/10.1016/j.vetimm.2008.10.322

Wen K, Tin C, Wang H, Yang X, Li G, Giri-Rachman E, Kocher J, Bui T, Clark-Deener S, Yuan L (2014) Probiotic Lactobacillus rhamnosus GG enhanced Th1 cellular immunity but did not affect antibody responses in a human gut microbiota transplanted neonatal gnotobiotic pig model. PLoS One 9(4):e94504. https://doi.org/10.1371/journal.pone.0094504

Wen K, Liu F, Li G, Bai M, Kocher J, Yang X, Wang H, Clark-Deener S, Yuan L (2015) Lactobacillus rhamnosus GG dosage affects the adjuvanticity and protection against rotavirus diarrhea in gnotobiotic pigs. J Pediatr Gastroenterol Nutr 60(6):834–843. https://doi.org/10.1097/MPG.0000000000000694

Wilks J, Lien E, Jacobson AN, Fischbach MA, Qureshi N, Chervonsky AV, Golovkina TV (2015) Mammalian lipopolysaccharide receptors incorporated into the retroviral envelope augment virus transmission. Cell Host Microbe 18(4):456–462. https://doi.org/10.1016/j.chom.2015.09.005

Wyatt RG, Greenberg HB, Dalgard DW, Allen WP, Sly DL, Thornhill TS, Chanock RM, Kapikian AZ (1978) Experimental infection of chimpanzees with the Norwalk agent of epidemic viral gastroenteritis. J Med Virol 2(2):89–96

Yang X, Wen K, Tin C, Li G, Wang H, Kocher J, Pelzer K, Ryan E, Yuan L (2014) Dietary rice bran protects against rotavirus diarrhea and promotes Th1-type immune responses to human rotavirus vaccine in gnotobiotic pigs. Clin Vaccine Immunol 21(10):1396–1403. https://doi.org/10.1128/CVI.00210-14

Yang X, Twitchell E, Li G, Wen K, Weiss M, Kocher J, Lei S, Ramesh A, Ryan EP, Yuan L (2015) High protective efficacy of rice bran against human rotavirus diarrhea via enhancing probiotic growth, gut barrier function, and innate immunity. Sci Rep 5:15004. https://doi.org/10.1038/srep15004

Yuan L, Saif LJ (2002) Induction of mucosal immune responses and protection against enteric viruses: rotavirus infection of gnotobiotic pigs as a model. Vet Immunol Immunopathol 87 (3–4):147–160

Yuan L, Wen K (2017) Rotavirus. In: Liu D (ed) Laboratory models for foodborne infections, Food microbiology. CRC Press/Taylor & Francis, Boca Raton, pp 95–107

Yuan L, Ward LA, Rosen BI, To TL, Saif LJ (1996) Systematic and intestinal antibody-secreting cell responses and correlates of protective immunity to human rotavirus in a gnotobiotic pig model of disease. J Virol 70(5):3075–3083

Yuan L, Kang SY, Ward LA, To TL, Saif LJ (1998) Antibody-secreting cell responses and protective immunity assessed in gnotobiotic pigs inoculated orally or intramuscularly with inactivated human rotavirus. J Virol 72(1):330–338

Zhang M, Zhang M, Zhang C, Du H, Wei G, Pang X, Zhou H, Liu B, Zhao L (2009) Pattern extraction of structural responses of gut microbiota to rotavirus infection via multivariate statistical analysis of clone library data. FEMS Microbiol Ecol 70(2):21–29. https://doi.org/10.1111/j.1574-6941.2009.00694.x

Zhang H, Wang H, Shepherd M, Wen K, Li G, Yang X, Kocher J, Giri-Rachman E, Dickerman A, Settlage R, Yuan L (2014) Probiotics and virulent human rotavirus modulate the transplanted human gut microbiota in gnotobiotic pigs. Gut Pathog 6:39. https://doi.org/10.1186/s13099-014-0039-8

Zheng DP, Ando T, Fankhauser RL, Beard RS, Glass RI, Monroe SS (2006) Norovirus classification and proposed strain nomenclature. Virology 346(2):312–323. https://doi.org/10.1016/j.virol.2005.11.015

Chapter 4
Enteric Bacterial Regulation of the Wnt/β-Catenin Signaling

Xingyin Liu and Jun Sun

Abstract Enteric bacteria such as *Salmonella* use a type three secretion system to inject bacterial pathogenic proteins, known as effectors, into host cells. These injected virulent effectors mimic the activity of eukaryotic proteins and debilitate host–cell signaling pathways. *Salmonella* infection is a common public health problem that can become chronic and increase the risk of cancer. In this chapter, we summarize the research progress on *Salmonella* regulation of Wnt/β-catenin signaling. The Wnt/β-catenin signaling is critical in intestinal renewal and development, inflammation, and tumorigenesis. We discuss in vitro and in vivo experimental models, especially the recently developed organoid system used for investigating *Salmonella*–host interactions. Finally, we highlight the novel roles of *Salmonella* effector protein AvrA in chronically activating Wnt/β-catenin signaling, impacting intestinal renewal and thus promoting colitis-associated cancer. These findings indicate the importance of understanding the complex interactions between bacteria and hosts in infection, inflammation, and cancer. The established experimental models (e.g., organoids, the chronic infected mouse model, and the infected colon cancer model) can be applied to investigating other bacteria and their interactions with hosts.

X. Liu (✉)
Department of Pathogen Biology, Key Laboratory of Pathogen of Jiangsu Province, Nanjing Medical University, Nanjing, China
e-mail: Xingyinliu@njmu.edu.cn

J. Sun (✉)
Division of Gastroenterology and Hepatology, Department of Medicine, University of Illinois at Chicago, Chicago, IL, USA
e-mail: junsun7@uic.edu

J. Sun, P.K. Dudeja (eds.), *Mechanisms Underlying Host-Microbiome Interactions in Pathophysiology of Human Diseases*, Physiology in Health and Disease,
https://doi.org/10.1007/978-1-4939-7534-1_4

4.1 Introduction

Salmonella infection appears as enteric fever, gastroenteritis, bacteremia, or extraintestinal focal infection. In addition to salmonellosis, the acute enteric infection, *Salmonella* colonization can become chronic (Grassl et al. 2008) and increase the risk for other gastrointestinal diseases, including chronic inflammation and cancer (Gradel et al. 2009; Kato et al. 2013). Type three secretion system (TTSS or T3SS) is a nanomachine that delivers bacterial proteins into the cytosol of eukaryotic cells. It consists of both the cylindrical basal structure spanning the two bacterial membranes and the peptidoglycan, which is connected to a hollow needle that is eventually followed by a filament (animal pathogens) or a long pilus (plant pathogens) (Cornelis 2010). *Salmonella* uses the TTSS system to inject bacterial pathogenic proteins known as effectors into host cells. These bacterial effectors then mimic the activity of eukaryotic proteins and debilitate host–cell signaling pathways (Du and Galan 2009).

The canonical Wnt/β-catenin signaling pathway is a critical mediator of intestinal epithelial and mucosal homeostasis (Clevers 2006). As shown in Fig. 4.1, in the absence of Wnt signaling, β-catenin does not accumulate in the cytoplasm as a destruction

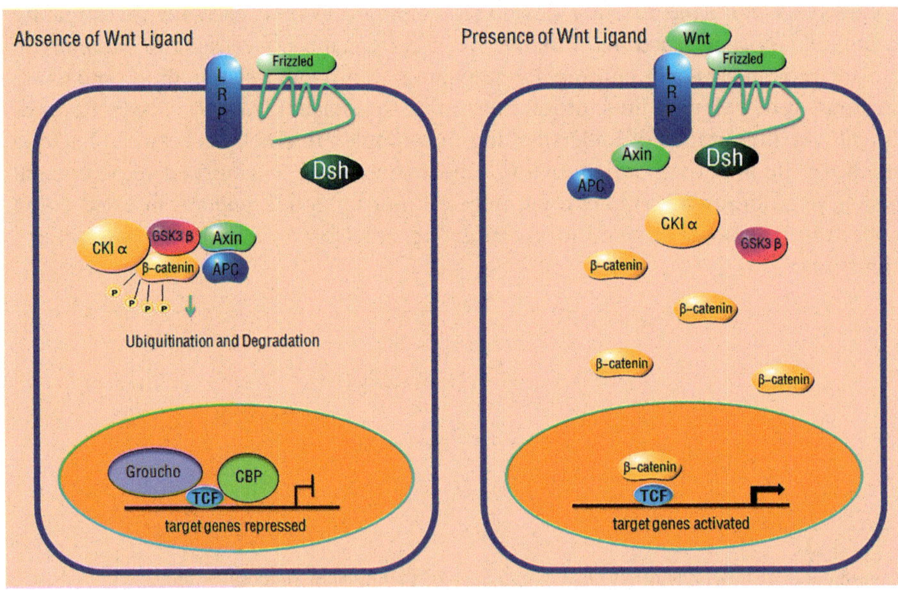

Fig. 4.1 Wnt/β-catenin signaling pathway. (**a**) In the absence of Wnt ligands, β-catenin does not accumulate in the cytoplasm as a destruction complex normally degrades it. This destruction complex includes the following proteins: Axin, adenomatous polyposis coli (*APC*), glycogen synthase kinase-3 (*GSK-3*), and casein kinase 1α (*CK1α*). It degrades β-catenin by targeting it for ubiquitination, which subsequently moves β-catenin to the proteasome for degradation. (**b**) In the presence of Wnt ligands, Wnt binds Fz and LRP5/6, disrupting the function of the destruction complex. It allows β-catenin to accumulate and translocate to the nucleus, where β-catenin acts as a transcription factor, thus leading to regulation of its target genes

complex normally degrades it. This destruction complex includes the following proteins: Axin, adenomatous polyposis coli, glycogen synthase kinase-3 (GSK-3), and casein kinase 1α (CK1α). It degrades β-catenin by targeting it for ubiquitination, which subsequently moves β-catenin to the proteasome for degradation. However, in the presence of a Wnt ligand, Wnt binds Fz and LRP5/6, disrupting the function of the destruction complex. It allows β-catenin to accumulate and translocate to the nucleus, where β-catenin acts as a transcription factor, thus leading to regulation of its target genes (Moon 2005). Although it is known that canonical Wnt/β-catenin signaling promotes proliferation of intestinal stem cells (Espada et al. 2009), knowledge about Wnt signaling activity that is regulated by bacterial infection is limited.

In the current chapter, we review research progress on the role of *Salmonella* in regulating Wnt/β-catenin signaling. Specifically, we discuss the effector protein AvrA in activating Wnt signaling and its impact on inflammation and colon cancer development. These studies not only provide insights into the complex interactions between enteric bacteria and hosts in infection, inflammation, and cancer, but also help to establish experimental models, including organoids, a chronic infected mouse model, and an infected colon cancer model, which can be applied when investigating other bacteria and their interactions with the host.

4.2 *Salmonella* Activation of Wnt/β-Catenin Signaling

4.2.1 Deubiquitinating β-Catenin

Salmonella strains can attenuate the host innate immune response by preventing the ubiquitination of IκBα and thus the activation of the NF-κB proinflammatory pathway (Neish et al. 2000). AvrA is a bacterial effector present in 80% of *Salmonella enterica* serovars and *E. coli*. Previous studies have shown that AvrA-related family members include the *Yersinia* virulence factor, YopJ, and the *Xanthomonas ampestris* pv *vesicatoria* protein, AvrBsT (Orth et al. 2000). The AvrA protein from nonpathogenic *Salmonella* PhoP^c has been shown to inhibit activation of NF-κB in cultured human epithelial cells (Collier-Hyams et al. 2002). β-Catenin is a negative regulator of the proinflammatory NF-κB pathway in epithelial cells (Deng et al. 2002). We found that bacterial effector AvrA enhances β-catenin-mediated signaling and blocks the activation of NF-κB signaling in epithelial cells (Sun et al. 2004). To determine whether expression of AvrA mediates the inhibition of β-catenin ubiquitination, cells were colonized with *Salmonella* PhoP^c with AvrA expressing PhoP^c lacking AvrA (PhoP^c AvrA$^-$), or PhoP^c AvrA$^-$ transcomplemented with a plasmid-encoding WT AvrA (PhoP^c AvrA$^-$/AvrA$^+$) (Sun et al. 2004). There was greater expression of ubiquitinated β-catenin in cells colonized with PhoP^c AvrA$^-$, compared with PhoP^c and PhoP^{c}AvrA$^-$/AvrA$^+$ groups. These data suggest that the expression of *Salmonella* AvrA effector in PhoP^c might decrease the ubiquitination of β-catenin. Furthermore, Ye et al. tested whether AvrA might inhibit the ubiquitination of β-catenin by acting as a deubiquitinating protease. Using the cell-free system, the purified GST-AvrA was mixed with ubiquitinated β-catenin. The ubiquitinated forms of β-catenin disappeared after reacting for 60 min. In contrast, in the mixture of the mutant AvrA C186A proteins and ubiquitinated β-catenin, there was no change in ubiquitinated β-catenin (Ye et al. 2007). These data indicate that AvrA is a deubiquitinase that removes ubiquitin from β-catenin. Consistent with the analysis, colonization with nonvirulent *Salmonella* promoted nuclear translocation of β-catenin and activated the target genes of the Wnt/β-catenin pathway, such as *c-myc and cyclinD1*, which were correspondingly upregulated with AvrA expression (Sun et al. 2004). In mouse models infected with AvrA-deficient and AvrA-sufficient *Salmonella* strains (Sun et al. 2004), Ye et al. (2007) found that increased β-catenin activity in AvrA-sufficient *Salmonella* strains negatively regulates the NF-κB pathway. These findings suggest that AvrA might play an important role in regulating host inflammatory responses through β-catenin signaling.

4.2.2 Enhancing Expression of Wnts

To understand the effects of Wnt/β-catenin signaling in controlling intestinal epithelial homeostasis, we used cultured epithelial cells, a *Salmonella* colitis mouse model, and a gnotobiotic mouse model for host–*Salmonella* interactions. We found that mRNA and protein expression levels of Wnt2 (as a canonical Wnt mediator) and Wnt11 (as a noncanonical Wnt signaling mediator) were elevated after bacterial infection (Liu et al. 2011, 2012). We observed enhanced staining for Wnt2 and Wnt11 in epithelial cells lining the crypt region, in addition to the villus and surface epithelium, after *Salmonella* infection. Recently, using a Wnt polymer chain reaction array, Neumann et al. (2014) also identified that Wnt5a, Wnt8b, and Wnt11 were differentially expressed along the colon length. They found that Wnt signatures were associated with differential epithelial cell proliferation and migration in the proximal versus distal colon, which suggests that physiological heterogeneity of the proximal and distal colon can be explained by differential Wnt signaling. Our study indicated that the elevation of Wnt2 and Wnt11 was a strategy for host defense to inhibit cell apoptosis and inflammatory responses to infection. Using Wnt2 and Wnt11 small interfering ribonucleic acid analysis, we showed that there was enhanced inflammatory cytokine interleukin- (IL-8) expression in epithelial cells. In contrast, cells that overexpressed Wnt2 and Wnt11 had less bacterial-induced IL-8 secretion. In a mono-associated mouse model, *E. coli* F18 expressing AvrA increased Wnt2 expression and changed Wnt2 distribution in the intestine. Moreover, the bacterial protein AvrA from *Salmonella* and *E. coli* stabilized Wnt2 protein expression and induced Wnt11 protein secretion. In addition, Wnt2 had no effects on inhibiting bacterial invasion, whereas Wnt11 significantly decreased *Salmonella* invasion of epithelial cells. The data suggest that although Wnt family members have different functions, they can sometimes complement each other.

Wnt3, Wnt6, and Wnt9A are known to regulate intestinal stem cells (Katoh 2007). Using a *Salmonella* colitis mouse model, we assessed mRNA expression of these three Wnts before and after *Salmonella* infection (Liu et al. 2010). We found that the mRNA levels of Wnt3, Wnt6, and 9a were significantly upregulated in the mouse intestinal epithelial cells by *Salmonella* infection. Because β-catenin is downstream of the Wnt pathway, we reasoned that pathogenic *Salmonella* modulates the intestinal stem cells and that AvrA activates stem cell niches through the Wnt pathway. Therefore, we tested whether AvrA changes phosphorylated β-catenin (Ser552) expression as an intestinal stem cell marker (Liu et al. 2010). In vivo, phosphorylated β-catenin (Ser552) is decreased by infection with an AvrA-deficient bacterial strain (AvrA−), but not by bacterial strains expressing AvrA in vivo. The number of stem cells and proliferative cells increased in the intestine infected with *Salmonella* that expressed AvrA. Thus, these results suggest that AvrA may contribute to stem cell maintenance in *Salmonella*-infected mice. Our study provides new insights into the mechanism by which bacteria effectors can expand stem cells by activating the Wnt/β-catenin pathway in vivo.

4.2.3 Decreasing Axin-1 and Suppressing GSK-3β

Belonging to the Axin family, Axin-1 is a negative regulator of the Wnt signaling pathway and a key player in the developmental processes and pathogenesis of human diseases (Logan and Nusse 2004; Kishida et al. 1998). In normal cells, Axin-1 forms a complex with GSK-3β and β-catenin and promotes GSK-3-β-dependent phosphorylation of β-catenin (Kishida et al. 1998; Itoh et al. 1998). Zhang et al. (2012) investigated the molecular mechanisms and physiological roles of Axin-1–*Salmonella* interactions. They found that pathogenic *Salmonella* colonization decreased Axin-1 protein expression in intestinal epithelial cells at the post-transcriptional level and that Axin-1 inhibits *Salmonella* invasion and bacterial inflammation. The resulting data indicate that intestinal Axin-1 plays a novel role in modulating host defense against pathogen-induced inflammation.

The major function of GSK-3β is its role in the phosphorylation of cytosolic β-catenin (Doble and Woodgett 2003; Zhou et al. 2004). It was reported that *Salmonella* first activates PI3K and consequently Akt, which leads to inhibitory phosphorylation (Ser9) of GSK-3β (Tahoun et al. 2012). Inactivation of GSK-3β leads to an increase in cytosolic β-catenin levels, and further induces the excessive β-catenin translocated into the nucleus, leading to the induction of both the receptor activator of the NF-κB ligand (RANKL) and its receptor, RANK. M cells constitute a small subset of highly specialized follicle-associated epithelium (FAE) enterocytes overlying lymphoid follicles in intestine. The autocrine activation of RelB-expressing FAE enterocytes by upregulated RANKL/RANK further guides the transcription factor Slug, which marks epithelial transdifferentiation into M cells (Tahoun et al. 2012). Contrasting with the role of the AvrA effector in intestinal epithelial cells, *Salmonella* SopB effector transforms primed epithelial cells into M cells to promote host colonization and invasion.

Taken together, Wnt/β-catenin signaling and its upregulators (Wnt2 and Wnt11), negative regulators (Axin-1 and GSK-3β), and downstream transcription factor β-catenin, are all involved in bacteria-induced inflammation in the intestine.

4.3 *Salmonella* Modulation of the Intestinal Stem Cells in the Organoid System

Salmonella infection regulates stem cell proliferation and differentiation through the Wnt signaling pathway in bacterial infected mammalian intestines (Liu et al. 2010; Lu et al. 2012). To further study bacterial infection and intestinal stem cells in the in vitro system, Zhang et al. established a *Salmonella*-infected organoid culture system. Using the stem-cell-derived mouse organoids, they reported that wild-type *Salmonella*-infected organoids significantly regulate the expression of the intestinal stem cell markers, Lgr5 and Bmi1. Using GFP-labeled Lgr5 organoids, they further visualized the changes in the Lgr5-labeled organoids and quantitated the change in fluorescence values affected by wild-type *Salmonella*. This study demonstrated that the *Salmonella*-infected organoid culture system is a new experimental model that is suitable for studying host–bacteria interactions. However, the role of *Salmonella* AvrA has not yet been tested in the organoid system.

Recent studies have shown that the human organoids derived from human-induced pluripotent stem cells and intestinal tissue specimens are similar to the native intestine because of the presence of Paneth-like cells and villus-like protrusions (Zachos et al. 2016; Foulke-Abel et al. 2016; Forbester et al. 2015). The availability of the organoid system allows us to further investigate the bacterial regulation of Wnt signaling mechanisms of intestinal stem cells.

4.4 Activation of Wnt-β-Catenin Signaling in the Colon Chronically Infected with *Salmonella*

Live, mutated, noninvasive *Salmonella* species have been used as vectors to specifically target cancer cells (Zhao et al. 2007). However, to our knowledge, the chronic effects and molecular mechanisms of infection with nonpathogenic or mutated *Salmonella* in the host are largely unexplored. Our previous study (Lu et al. 2010) showed the chronic effects of *Salmonella* in vivo after 27 weeks of infection. We demonstrated that in chronically colon-infected mice there was constant activation of the Wnt/β-catenin pathway by *Salmonella* effector protein AvrA (Lu et al. 2012). These data identified that AvrA has long-term effects in the *Salmonella*-infected intestine, including activating Wnt/β-catenin signaling and increasing cell proliferation, thus leading to the dysregulation of intestinal responses to chronic bacterial infections.

Colorectal cancer is the third most common cancer type worldwide and causes 600,000 deaths every year (Bonnet et al. 2014). Increasing evidence has shown that chronic bacterial infection and ensuing colonic inflammation contribute to tumor initiation and progression (Mantovani 2009). An imbalance of the microbiota in favor of opportunistic pathogens contributes to higher mucosal permeability, bacterial translocation, and the activation of components of both the innate and adaptive immune systems (Shanahan 2013). We have demonstrated that the *Salmonella* $PhoP^c$

strain expressing AvrA activates the Wnt/β-catenin pathway in chronic infection (Lu et al. 2012). We further tested the effect of chronic *Salmonella* infection on tumor development in the AOM/DSS-induced colon cancer model (Lu et al. 2014). We showed that AvrA enhanced proliferation, promoted colonic tumorigenesis and tumor progression, and concomitantly activated proto-oncogene β-catenin and its downstream targets, c-Myc and cyclin D1. In part, upregulated β-catenin signals may be mediated by increased phosphorylation of Ser552 and decreased β-catenin ubiquitination. AvrA expression also increased the abundance of *Salmonella* in tumors, confirming a direct effect of AvrA on *Salmonella* invasiveness. Specifically, our studies identify a novel tumor-promoting role of bacterial AvrA. In addition, this study provides potentially important insights into how bacterial mechanisms promote colonic tumorigenesis and tumor promotion by activating Wnt/β-catenin signaling.

4.5 Conclusion

In summary, as shown in Fig. 4.2a, *Salmonella* uses various strategies in regulating post-translational modification of Axin-1, GSK-3β and β-catenin to activate Wnt signaling. The activation of Wnt/β-catenin by *Salmonella* infection is involved in cell proliferation, inflammation, apoptosis, transdifferentiation, and tumorigenesis. *Salmonella* AvrA also enhances Wnt2 and Wnt11 expression to inhibit cell apoptosis and inflammatory responses to infection (Fig. 4.2b). The current studies have increased our understanding of the relationship between bacteria and hosts (e.g., through Wnt signaling) in infectious diseases and colon cancer. The effects of *Salmonella* on tumorigenesis also urge caution regarding the use of mutant *Salmonella* organisms as vectors for anti-cancer therapy. A recent study has demonstrated the presence of *Salmonella* AvrA in colorectal tumors and its precursor lesions in human specimens (Lu et al. 2017), suggesting that this bacterial protein might play a novel role in colon cancer. Overall, it is of critical importance to study the complex interactions between bacteria and hosts in infection, inflammation, and cancer for a better understanding of the pathophysiology of diseases and to develop better therapeutic interventions.

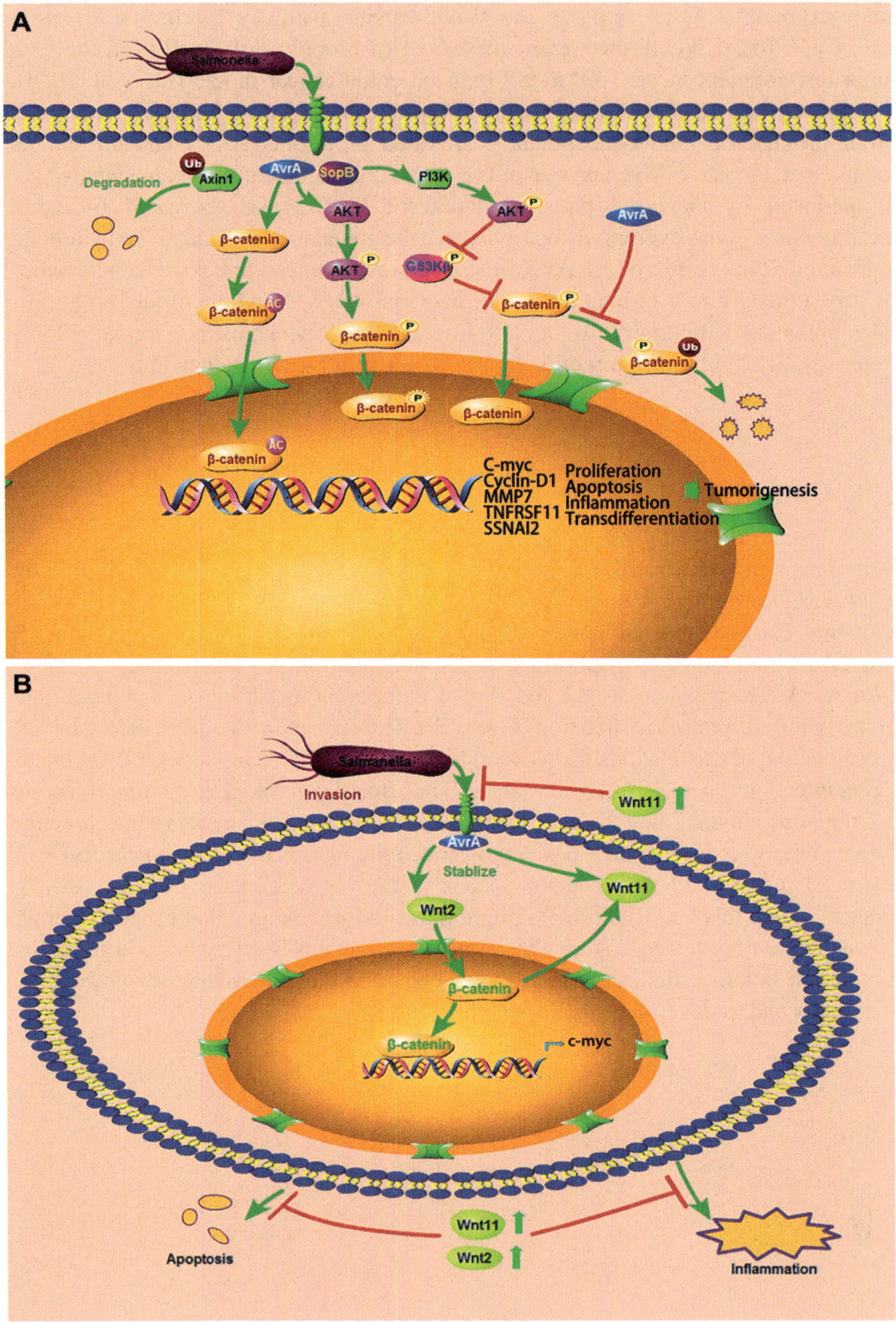

Fig. 4.2 *Salmonella* regulate Wnt signaling. (**a**) *Salmonella* effector AvrA and SopB regulate Axin1, GSK-3β and β-catenin respectively. (**b**) *Salmonella* AvrA enhances Wnt2 and Wnt11 expression to inhibit cell apoptosis and inflammatory responses to infection. *Ub-Axin1* ubiquitination-Axin-1, *Ub-β-catenin* ubiquitinated β-catenin, *Ac-β-catenin* acetylated β-catenin, *P-β-catenin* phosphorylated β-catenin, *P-Akt* phosphorylated Akt, *P-GSK-3B* phosphorylated GSK-3β

Acknowledgements This work was supported by NIDDK R01 DK105118 to Jun Sun and NSFC 8167080961 to Xingyin Liu.

Conflicts of Interest The authors have no conflicts of interest.

References

Bonnet M, Buc E, Sauvanet P, Darcha C, Dubois D, Pereira B, Dechelotte P, Bonnet R, Pezet D, Darfeuille-Michaud A (2014) Colonization of the human gut by E. coli and colorectal cancer risk. Clin Cancer Res 20(4):859–867

Clevers H (2006) Wnt/beta-catenin signaling in development and disease. Cell 127(3):469–480

Collier-Hyams LS, Zeng H, Sun J, Tomlinson AD, Bao ZQ, Chen H, Madara JL, Orth K, Neish AS (2002) Cutting edge: Salmonella AvrA effector inhibits the key proinflammatory, anti-apoptotic NF-kappa B pathway. J Immunol 169(6):2846–2850

Cornelis GR (2010) The type III secretion injectisome, a complex nanomachine for intracellular 'toxin' delivery. Biol Chem 391(7):745–751

Deng J, Miller SA, Wang HY, Xia W, Wen Y, Zhou BP, Li Y, Lin SY, Hung MC (2002) beta-catenin interacts with and inhibits NF-kappa B in human colon and breast cancer. Cancer Cell 2(4):323–334

Doble BW, Woodgett JR (2003) GSK-3: tricks of the trade for a multi-tasking kinase. J Cell Sci 116 (Pt 7):1175–1186

Du F, Galan JE (2009) Selective inhibition of type III secretion activated signaling by the Salmonella effector AvrA. PLoS Pathog 5(9):e1000595

Espada J, Calvo MB, Diaz-Prado S, Medina V (2009) Wnt signalling and cancer stem cells. Clin Transl Oncol 11(7):411–427

Forbester JL, Goulding D, Vallier L, Hannan N, Hale C, Pickard D, Mukhopadhyay S, Dougan G (2015) Interaction of Salmonella enterica serovar typhimurium with intestinal organoids derived from human induced pluripotent stem cells. Infect Immun 83(7):2926–2934

Foulke-Abel J, In J, Yin JY, Zachos NC, Kovbasnjuk O, Estes MK, de Jonge H, Donowitz M (2016) Human enteroids as a model of upper small intestinal ion transport physiology and pathophysiology. Gastroenterology 150(3):638–649

Gradel KO, Nielsen HL, Schonheyder HC, Ejlertsen T, Kristensen B, Nielsen H (2009) Increased short- and long-term risk of inflammatory bowel disease after Salmonella or campylobacter gastroenteritis. Gastroenterology 137(2):495–501

Grassl GA, Valdez Y, Bergstrom KS, Vallance BA, Finlay BB (2008) Chronic enteric Salmonella infection in mice leads to severe and persistent intestinal fibrosis. Gastroenterology 134 (3):768–780

Itoh K, Krupnik VE, Sokol SY (1998) Axis determination in Xenopus involves biochemical interactions of axin, glycogen synthase kinase 3 and beta-catenin. Curr Biol 8(10):591–594

Kato I, Boleij A, Kortman GA, Roelofs R, Djuric Z, Severson RK, Tjalsma H (2013) Partial associations of dietary iron, smoking and intestinal bacteria with colorectal cancer risk. Nutr Cancer 65(2):169–177

Katoh M (2007) Networking of WNT, FGF, Notch, BMP, and Hedgehog signaling pathways during carcinogenesis. Stem Cell Rev 3(1):30–38

Kishida S, Yamamoto H, Ikeda S, Kishida M, Sakamoto I, Koyama S, Kikuchi A (1998) Axin, a negative regulator of the wnt signaling pathway, directly interacts with adenomatous polyposis coli and regulates the stabilization of beta–catenin. J Biol Chem 273(18):10823–10826

Liu X, Lu R, Wu S, Sun J (2010) Salmonella regulation of intestinal stem cells through the Wnt/ beta-catenin pathway. FEBS Lett 584(5):911–916

Liu X, Wu S, Xia Y, Li XE, Xia Y, Zhou ZD, Sun J (2011) Wingless homolog Wnt11 suppresses bacterial invasion and inflammation in intestinal epithelial cells. Am J Physiol Gastrointest Liver Physiol 301(6):G992–G1003

Liu X, Lu R, Wu S, Zhang YG, Xia Y, Sartor RB, Sun J (2012) Wnt2 inhibits enteric bacterial-induced inflammation in intestinal epithelial cells. Inflamm Bowel Dis 18(3):418–429

Logan CY, Nusse R (2004) The Wnt signaling pathway in development and disease. Annu Rev Cell Dev Biol 20:781–810

Lu R, Wu S, Liu X, Xia Y, Zhang YG, Sun J (2010) Chronic effects of a Salmonella type III secretion effector protein AvrA in vivo. PLoS One 5(5):e10505

Lu R, Liu X, Wu S, Xia Y, Zhang YG, Petrof EO, Claud EC, Sun J (2012) Consistent activation of the beta-catenin pathway by Salmonella type-three secretion effector protein AvrA in chronically infected intestine. Am J Physiol Gastrointest Liver Physiol 303(10):G1113–G1125

Lu R, Wu S, Zhang YG, Xia Y, Liu X, Zheng Y, Chen H, Schaefer KL, Zhou Z, Bissonnette M, Li L, Sun J (2014) Enteric bacterial protein AvrA promotes colonic tumorigenesis and activates colonic beta-catenin signaling pathway. Oncogenesis 3:e105

Lu R, Bosland M, Xia Y, Zhang Y, Kato I, Sun J (2017) Presence of *Salmonella* AvrA in colorectal tumor and its precursor lesions in mouse intestine and human specimens. Oncotarget. https://doi.org/10.18632/oncotarget.19052

Mantovani A (2009) Cancer: inflaming metastasis. Nature 457(7225):36–37

Moon RT (2005) Wnt/beta-catenin pathway. Science's STKE: Signal Transduct Knowl Environ 2005(271):cm1

Neish AS, Gewirtz AT, Zeng H, Young AN, Hobert ME, Karmali V, Rao AS, Madara JL (2000) Prokaryotic regulation of epithelial responses by inhibition of IkappaB-alpha ubiquitination. Science 289(5484):1560–1563

Neumann PA, Koch S, Hilgarth RS, Perez-Chanona E, Denning P, Jobin C, Nusrat A (2014) Gut commensal bacteria and regional Wnt gene expression in the proximal versus distal colon. Am J Pathol 184(3):592–599

Orth K, Xu Z, Mudgett MB, Bao ZQ, Palmer LE, Bliska JB, Mangel WF, Staskawicz B, Dixon JE (2000) Disruption of signaling by Yersinia effector YopJ, a ubiquitin-like protein protease. Science 290(5496):1594–1597

Shanahan F (2013) The colonic microbiota in health and disease. Curr Opin Gastroenterol 29(1):49–54

Sun J, Hobert ME, Rao AS, Neish AS, Madara JL (2004) Bacterial activation of beta-catenin signaling in human epithelia. Am J Physiol Gastrointest Liver Physiol 287(1):G220–G227

Tahoun A, Mahajan S, Paxton E, Malterer G, Donaldson DS, Wang D, Tan A, Gillespie TL, O'Shea M, Roe AJ, Shaw DJ, Gally DL, Lengeling A, Mabbott NA, Haas J, Mahajan A (2012) Salmonella transforms follicle-associated epithelial cells into M cells to promote intestinal invasion. Cell Host Microbe 12(5):645–656

Ye Z, Petrof EO, Boone D, Claud EC, Sun J (2007) Salmonella effector AvrA regulation of colonic epithelial cell inflammation by deubiquitination. Am J Pathol 171(3):882–892

Zachos NC, Kovbasnjuk O, Foulke-Abel J, In J, Blutt SE, de Jonge HR, Estes MK, Donowitz M (2016) Human enteroids/colonoids and intestinal organoids functionally recapitulate normal intestinal physiology and pathophysiology. J Biol Chem 291(8):3759–3766

Zhang YG, Wu S, Xia Y, Chen D, Petrof EO, Claud EC, Hsu W, Sun J (2012) Axin1 prevents Salmonella invasiveness and inflammatory response in intestinal epithelial cells. PLoS One 7(4):e34942

Zhao M, Geller J, Ma H, Yang M, Penman S, Hoffman RM (2007) Monotherapy with a tumor-targeting mutant of Salmonella typhimurium cures orthotopic metastatic mouse models of human prostate cancer. Proc Natl Acad Sci USA 104(24):10170–10174

Zhou BP, Deng J, Xia W, Xu J, Li YM, Gunduz M, Hung MC (2004) Dual regulation of Snail by GSK-3beta-mediated phosphorylation in control of epithelial-mesenchymal transition. Nat Cell Biol 6(10):931–940

Chapter 5
Mechanisms Underlying the Beneficial Role of Probiotics in Diarrheal Diseases: Host–Microbe Interactions

Anoop Kumar, Pradhan Bhat, Alip Borthakur, Waddah A. Alrefai, and Pradeep K. Dudeja

Abstract Probiotics are nonpathogenic microorganisms which, when administered in adequate amounts, confer health benefits to hosts. Advances in our understanding of the gut microbiome have spurred the use of probiotics for the treatment of a wide variety of gastrointestinal pathological conditions. Of these conditions, probiotic treatment in diarrheal diseases has shown particular promise. Multiple pre-clinical and clinical studies over the past decade have shown probiotics to significantly attenuate the effects of both acute and chronic diarrheal phenotypes. Only recently, studies have begun to unravel the mechanisms by which probiotics increase electrolyte and nutrient absorption, decrease secretion and counteract diarrheal diseases associated with infection and inflammation. However, the lack of a detailed mechanistic understanding of their beneficial effects in gut limits the development of probiotics-based novel therapeutics. This review provides an overview of the evidence-based analysis of the effects of probiotics, including a detailed description of the knowledge of mechanisms by which probiotics show benefits in diarrheal diseases.

List of Abbreviations

CD	Crohn's disease
CFTR	Cystic fibrosis transmembrane Conductance regulator
DRA	Downregulated in adenoma
EPEC	Enteropathogenic *E-coli*

A. Kumar • P. Bhat • A. Borthakur
Division of Gastroenterology and Hepatology, Department of Medicine, University of Illinois at Chicago, Chicago, IL, USA

W.A. Alrefai • P.K. Dudeja (✉)
Division of Gastroenterology and Hepatology, Department of Medicine, University of Illinois at Chicago, Chicago, IL, USA

Jesse Brown VA Medical Center, Medical Research Service (600/151), Chicago, IL, USA
e-mail: pkdudeja@uic.edu

J. Sun, P.K. Dudeja (eds.), *Mechanisms Underlying Host-Microbiome Interactions in Pathophysiology of Human Diseases*, Physiology in Health and Disease,
https://doi.org/10.1007/978-1-4939-7534-1_5

FSK	Forskolin
IBD	Inflammatory bowel disease
IBS	Irritable bowel syndrome
IECs	Intestinal epithelial cells
IFN-γ	Interferon-γ
MCT1	Monocarboxylate transporter 1
NEC	Necrotizing enterocolitis
NHE	Na^+/H^+ exchanger
NKCC1	Na-K-Cl co-transporter 1
PepT1	Peptide transporter 1
RCT	Randomized controlled trial
SCFA	Short chain fatty acid
SGLT-1	Sodium-dependent glucose cotransporter 1
STp	*Escherichia coli* heat-stable enterotoxin
TcpC	Toll/IL-1 receptor-containing (TIR-containing) protein C
TGF-β	Transforming growth factor beta
TLR	Toll-like receptor
UC	Ulcerative colitis
VRE	Vancomycin-resistant enterococci
ZO	Zonula occludens

5.1 Introduction

Diarrheal diseases are a significant contributor to the global disease burden and a leading cause of death in the developing nations. They affect over 1.7 billion people per year and can be fatal in one in nine children (Priyamvada et al. 2015). Diarrhea is clinically defined as three or more liquid stools per day and its pathogenesis is characterized by dysregulation of electrolyte, solute, water absorption and/or secretion, and gut barrier dysfunction in response to multiple factors. Common etiologies include enteric infections by bacterial or viral pathogens, gut immune dysregulation commonly seen in inflammatory bowel disease (IBD), neurogenic causes seen in irritable bowel syndrome (IBS), uncontrolled antibiotics usage, and necrotizing enterocolitis (NEC) associated with premature birth (Priyamvada et al. 2015). Classically, treatment of this assortment of diseases has been focused on either remission maintenance or symptom control. However, new therapeutic modalities, in particular probiotics, have shown promise in both preventing diarrheal disease and significantly attenuating the severity of diarrhea (Culligan et al. 2009).

Probiotics are defined by the World Health Organization (WHO) as "living organisms which provide benefits to the host when administered in adequate quantities." The human gastrointestinal system has previously been proven to host a multitude of symbiotic microorganisms, but only recently have we begun to understand their role in the prevention of disease and improving health outcomes. Indeed, clinical trial data have supported the promise of probiotics in the treatment of certain gastrointestinal pathological conditions, including infectious diarrhea, ulcerative colitis, and antibiotics-associated diarrhea (Ritchie and Romanuk 2012; Salari et al. 2012; Miele et al. 2009).

As clinical data support the effectiveness of probiotics in certain pathological conditions, studies have sought to better understand how probiotics exert therapeutic effects in diarrheal diseases. Indeed, several mechanisms have been proposed for the beneficial effects of probiotics including: pathogen exclusion; maintenance of intestinal epithelial integrity; secretion of anti-microbial factors and immunomodulation. However, increasing evidence now indicates that these microorganisms may also play a key role in affecting ion, nutrient, short chain fatty acid transport, and ion channel function via distinct host–microbe interactions (Borthakur et al. 2008; Kumar et al. 2012, 2016). This review presents a summary of the evidence-based effects of probiotics on diarrheal disease in addition to a detailed overview of the mechanisms of action of probiotics.

5.2 Clinical Evidence for Probiotic Efficacy in Diarrheal Diseases

Different probiotic regimens have been evaluated for their efficacy in treating infectious diarrhea, including *Helicobacter pylori* infection, *Clostridium difficile* infection, rotavirus infection, and travelers' diarrhea. A systematic review of 23 randomized controlled trials (RCTs) involving 1917 patients demonstrated that probiotics might significantly reduce the duration of acute gastroenteritis compared with placebo (Gallo et al. 2016). Three RCTs (including 1043 children) tested *Lactobacillus rhamnosus* LGG supplementation, yielding a significantly reduced incidence of nosocomial rotavirus-induced diarrhea (Gallo et al. 2016). In addition, meta-analyses by Salari et al. (2012) showed that probiotics decreased the duration of diarrhea and fever in children. Another meta-analysis by Ritchie and Romanuk (2012) showed that a mixture of 11 species of probiotics had a positive and significant effect on infection or IBD-associated diarrhea. Unfortunately, the therapeutic benefits of probiotics did not translate to traveler's diarrhea or NEC.

Clinical models investigating the use of probiotic mixture VSL #3, (*Lactobacillus plantarum*, *L. delbrueckii* subsp. *bulgaricus*, *L. casei*, *L. acidophilus*, *Bifidobacterium breve*, *B. longum*, *B. infantis*, and *Streptococcus salivarius* subsp.) found probiotics to be beneficial in the treatment of inflammatory diarrhea in IBD. For example, 29 children with ulcerative colitis were randomized by Miele et al. to receive either VSL#3 or placebo together with steroid (induction) or mesalamine (remission) (Miele et al. 2009). Results demonstrated disease remission in 13 patients treated with VSL#3 compared with 4 patients treated with placebo. In a separate clinical trial, adults with ulcerative colitis (UC) were randomly treated with VSL#3 ($n = 77$) or placebo ($n = 70$), twice daily for 12 weeks, showing that both at week 6 and week 12, improvement in the clinical score was

Table 5.1 Clinical evidence for probiotic efficacy in diarrheal diseases

Genus/ Formulation	Species	Functional implications	Citation
VSL #3	Mixture	⇑ Remission of pouchitis, AAD, ID, IBS, HP, and CDD	Ritchie et al. (2012) (Meta-analysis)
VSL #3	Mixture	⇑ IBD remission	Miele et al. (2009)
VSL #3	Mixture	⇑ UC remission	Mardini et al. (2014) (Meta-analysis)
VSL #3	Mixture	⇓ UC associated diarrhea	Sang et al. (2010) (Meta-analysis)
Lactobacillus	*Rhamnosus*	⇓ Nosocomial rotavirus-induced diarrhea	Gallo et al. (2016)
	Rhamnosus	⇑ Pouchitis remission	Ritchie and Romanuk (2012) (Meta-analysis)
Bifidobacterium	*Bifidum*	⇓ UC-associated diarrhea	Sang et al. (2010) (Meta-analysis)

VSL #3 is a mixture of *L. Acidophilus*, *L. plantarum*, *L. casei*, *L. delbrueckii*, *B. breve*, *B. longum*, *B. longum*, *B. infantis*, and *S. thermophilus*
AAD antibiotics-associated diarrhea, *ID* infectious diarrhea, *IBS* irritable bowel syndrome, *HP Helicobacter pylori* infection, *CDD Clostridium difficile* disease, *IBD* inflammatory bowel disease, *UC* ulcerative colitis
⇑ indicates an increase and ⇓ indicates a decrease

significantly higher in the probiotics-treated group than in the placebo group (Mardini and Grigorian 2014). Another study showed that probiotics treatment was more effective compared with the placebo group in maintaining remission for UC (Sang et al. 2010). In their meta-analysis, Shen et al. hypothesized that inconclusive probiotic efficacy in Crohn's disease (CD) patients compared with UC patients may be due to different inflammatory patterns between the two diseases. For example, an impaired production of the regulatory cytokines has been found in CD4+ T cells of CD patients, but not those of UC patients (Shen et al. 2014). Further investigation is needed to better understand if probiotics play a role in the treatment of either CD or NEC. However, most of these studies support the use of probiotics as a therapy in diarrheal diseases. The effects of probiotics in clinical studies for diarrheal disorders are outlined in Table 5.1.

5.3 Mechanisms of the Beneficial Effects of Probiotics in Diarrheal Diseases

5.3.1 Pathogen Exclusion

Probiotics may confer protective effects to the host by competitively inhibiting the binding of diarrheal pathogens to the gastrointestinal mucosa. Termed "competitive exclusion," this mechanism is especially relevant in counteracting pathogen infections that involve direct interactions of the pathogen with the intestinal epithelium (Nishiyama et al. 2015; Hecht et al. 2016). Indeed, the attachment of several species of pathogenic bacteria is a key step in the disruption of the commensal microbiota, deterioration of the intestinal epithelium, and development of conditions germane to further bacterial, viral, or parasitic infection. In response, probiotics are uniquely equipped to inhibit both adhesion and survival of such pathogens.

One mechanism by which probiotics ameliorate infectious diarrhea is a successful competition for mucosal adhesion sites via interaction with cell–surface proteins. In vitro studies have thus far associated this mechanism of pathogen expulsion with strains of *Lactobacilli* and *Bifidobacteria*. Investigation by Nishiyama et al. found aggregation-promoting factors to mediate adhesion of probiotic *Lactobacillus gasseri* (LG2055) and subsequent competitive exclusion of diarrheal pathogen *Campylobacter jejuni* in an in vivo chicken model (Nishiyama et al. 2015). Additionally, Acosta et al. demonstrated exposure of epithelial cell surface receptor DC-SIGN (dendritic cell-specific intercellular adhesion molecule-3-grabbing non-integrin; CD209) to the surface layer proteins of four species of *Lactobacilli* (*L. acidophilus, L. brevis, L. helveticus,* and *L. kefiri*) to inhibit foreign bacterial growth (Prado Acosta et al. 2016). More specifically, the investigation revealed significant inhibition of *E. coli*, *S. typhi*, and *Klebsiella pneumoniae* growth by *L. acidophilus*, a key bacterial species found in commercial and clinical probiotic formulations.

Probiotics may also play a key role in the inhibition of intracellular pathogen internalization as a means of pathogen exclusion. Hirano et al. showed that *Lactobacillus rhamnosus* GG was capable of preventing the internalization of enterohemorrhagic *E. coli* (EHEC) in the C2BBe1 human colonic cell line (Hirano et al. 2003). Furthermore, *L. rhamnosus* GG has been shown to ameliorate diarrheal disease driven by highly resistant nosocomial vancomycin-resistant enterococci (VRE). Tytgat et al. (2016) reported that there is significant sequence homology of mucus binding pili found on *L. rhamnosus* GG and VRE. This underlies the mechanism responsible for successful inhibition of VRE colonization by permitting *L. rhamnosus* GG to competitively bind to intestinal epithelial cells (IECs). Furthermore, *Bifidobacteria* have been shown to result in decreased rotavirus adhesion to Caco-2 and HT-29 cells. *B. thermophilum* has also been shown to have prophylactic effects when administered before rotavirus infection in CD-1 suckling mice (Gagnon et al. 2016).

Pathogen exclusion is another possible mechanism underlying the efficacy of probiotics in IBD models. Recently, IBD has been shown to be associated with modifications to the natural intestinal microbiome (Kostic et al. 2014). Patients with CD were found to harbor the pathogen adherent-invasive *E. coli* (AIEC) (Conte et al. 2014). In vitro studies by Ingrassia et al. (2005) showed *Lactobacillus casei* strain DN-114001 to inhibit interaction between AIEC and IECs. These results were further supported by other studies showing *L. helveticus* and *L. rhamnosus* to be effective in the attenuation of infectious colitis by *Citrobacter rodentium* (Johnson-Henry et al. 2005; Collins et al. 2014). The findings indicated that this effect was possibly due to reduced pathogen adhesion to the mouse colonic epithelium. Detailed studies are needed to better understand the in-depth mechanisms underlying the pathogen exclusion effects of probiotics.

5.3.2 Secretion of Anti-Microbial Factors that Prevent Diarrhea

In a corollary to pathogen exclusion, probiotics have also been found to secrete anti-microbial and bioactive factors to enhance their competitive advantage against diarrheal pathogens within the intestinal microbiome (Zanfardino et al. 2017). These anti-microbial compounds may be further classified as either organic acids or small heat-stable peptides (Gibson and Barrett 2010; Lievin et al. 2000; Hassan et al. 2012).

The anti-microbial effects of lactic or acetic acid secretion by probiotic bacteria have been shown to exert a significant inhibitory effect on Gram-negative bacteria (Alakomi et al. 2000). The acid enters the bacteria and dissociates in the cytoplasm, thereby reducing the intracellular pH and activating signaling pathways, leading to death of the pathogen (Russell and Diez-Gonzalez 1997). Indeed, accumulation of lactic acid was found to mediate the anti-microbial activity exhibited by *Lactobacillus rhamnosus* GG against *Salmonella typhimurium* (Castillo et al. 2011). Furthermore, organic acids are also shown to permeabilize the Gram-negative bacterial outer membrane, allowing for diffusion of other antimicrobial compounds from probiotic bacteria, in addition to the host Paneth cells into the bacterial cytoplasm (Helander and Mattila-Sandholm 2000).

Different species of probiotics produce heat-stable peptides, termed bacteriocins, which are associated with anti-microbial effects (Hassan et al. 2012). In general, bacteriocins have been shown to promote pathogen death by inhibiting cell-wall synthesis or initiating pore formation in bacterial cell walls. For example, *L. curvatus* DN317 was shown to produce a bacteriocin termed curvaticin DN317, with anti-microbial effects against the infectious diarrheal agent *Campylobacter jejuni* (Zommiti et al. 2016). Several other studies demonstrated that different strains of *Lactobacilli* produce similar effects mediated by bacteriocins in infectious models. *Bifidobacterium* has also been implicated in the production of bacteriocins (Martinez et al. 2013). Liévin et al. isolated 14 human *Bifidobacteria* strains from infant stools and examined samples for antimicrobial activity (Lievin et al. 2000). Two strains (CA1 and F9) were found to inhibit cellular entry and exhibit cytotoxic effects against intracellular *S. typhimurium* SL1344 in Caco-2 cells (Lievin et al. 2000).

5.3.3 Maintenance of Gut Epithelial Barrier Function

The intestinal epithelium plays a critical role in protecting the human body from the contents of the gut lumen. Disruption of this barrier may either directly disrupt ion and water transport or induce an inflammatory response leading to diarrhea (Krishna Rao and Samak 2013). The human microbiome is intricately involved in maintaining the integrity of the gut epithelium and, increasing evidence has now demonstrated the novel role of probiotics in maintaining epithelial barrier function. The detailed mechanisms by which probiotics enhance intestinal barrier function are currently under investigation by many groups.

Maintenance, upregulation or phosphorylation of tight junction proteins are some of the proposed mechanisms of enhancing intestinal epithelial integrity. Several studies have found different probiotic species to utilize this mechanism to confer therapeutic benefit. For example, strains of *Lactobacilli* in a T84 cell-line model were found to upregulate expression of adherens junction proteins such as E-cadherin and β-catenin (Seth et al. 2008). *Lactobacilli* were also shown to enhance phosphorylation of adherens junction proteins, augmenting intestinal barrier function (Resta-Lenert and Barrett 2003). Seth et al. demonstrated that *Lactobacillus rhamnosus* GG-produced soluble proteins (p40 and p75) attenuated a hydrogen peroxide-induced decrease in transepithelial resistance (TER) and increase in inulin permeability via PKC- and MAP kinase-dependent mechanisms (Seth et al. 2008). Studies have additionally shown an increase in TER upon exposure of HT29 and Caco-2 cells to probiotic species such as *Streptococcus thermophilus* and *Lactobacillus acidophilus* (Resta-Lenert and Barrett 2006). This effect was accompanied by enhanced phosphorylation of tight-junction proteins including zonula occludens 1 (ZO-1) and occludin. In an in vivo murine model of DSS colitis, the probiotic *E. coli* Nissle 1917 (EcN) was shown to inhibit the diarrheal phenotype and to upregulate expression of the tight junction proteins ZO-1, ZO-2, and claudin-14 (Schroeder et al. 2006). A more recent study has attributed this upregulation of tight junction proteins to factors released by EcN outer membrane vesicles and secreted protein Toll/IL-1 receptor-containing (TIR-containing) protein C (Alvarez et al. 2016).

Probiotics have also been shown to strengthen intestinal barrier function via immunomodulation. Pro-inflammatory cytokines produced in IBD are involved in the deterioration of intestinal barrier function (Schmitz et al. 1999). Further studies showed these effects to be due to decreased tight junction protein function and/or expression leading to a diarrheal phenotype (Landy et al. 2016). Two commercially available probiotic mixtures have been associated with downregulation of pro-inflammatory pathway mediators and epithelial defense. Krishnan et al. (2016) found probiotic mixture VSL#3 to reduce interferon gamma (IFN-γ)-induced epithelial barrier disruption in an in vitro HT29 cell model via direct inhibition of the T-cell protein tyrosine phosphatase, a key protein product of IBD-associated gene *PTPN2*. In a separate study, probiotics treatment with a commercial probiotics mixture containing *Lactobacillus acidophilus*, *L. plantarum*, *L. salivarius*, and *Bifidobacterium lactis* (Lactibiane Tolerance®, or LT) significantly prevented the epithelial barrier disruption induced by the intracolonic infusion of fecal supernatant from IBS patients and the LPS-induced increase in paracellular permeability in T-84 cells (Nébot-Vivinus et al. 2014). The studies additionally revealed that culture supernatants of both *Bifidobacterium infantis* and *Lactobacillus acidophilus* protected against IL-1 beta-induced disruption of barrier function (Guo et al. 2017). Yu et al. studied the therapeutic effect of *Lactobacillus fructosus* C2 in attenuating the enterotoxigenic *E. coli* K88 or *S. typhimurium* SL1344-induced changes to mucosal barrier integrity (Yu et al. 2015). Their findings indicated that *L. fructosus* C2 reduced the paracellular permeability of fluorescein isothiocyanate–dextran and the expression of pro-inflammatory cytokine IL-8, in addition to p-ERK and p-JNK. Overall, probiotics have shown promising results in maintaining epithelial integrity in normal and in disease conditions.

5.3.4 Role of Immunomodulation in Diarrheal Diseases and Beneficial Effects of Probiotics

A common cause of diarrhea is gut mucosal inflammation resulting from the activation of either innate or adaptive immune responses. Triggering of the innate immune system largely leads to acute inflammation characterized by neutrophilic infiltration and vasodilation. This initial response is mediated by receptor families present on both immune and epithelial cells. One of the key members are toll-like receptors (TLRs) found on resident sentinel cells (i.e., macrophages and IECs). TLRs are typically activated by pathogen-associated molecular patterns, including lipopolysaccharide (LPS) commonly found in Gram-negative bacterial membranes (Gomez-Llorente et al. 2010). The ensuing signal cascade is mediated by MyD88 and involves activation of the NF-κB and MAPK signaling pathways. In some cases, probiotics have been shown to play an important role in enhancing the innate immune response to pathogens. In healthy mice, *L. casei* CRL 431 was found to activate TLR4 and may be used as a surveillance mechanism against pathogenic bacteria (Castillo et al. 2011). Activation of TLR4 leads to the induction of pro-inflammatory mediators, an increase in TLR2 expression, and a reduction in its own expression, which leads to the recruitment of inflammatory cells. This same mechanism has been shown to inhibit *Salmonella* infection in vivo (Castillo et al. 2011). In contrast, some probiotics have been found to be beneficial in modulating the immune response and attenuating inflammation. Indeed, *L. casei* was found to attenuate the pro-inflammatory signal cascade by antagonism of NF-κB (Liu et al. 2012). In another study, *L. reuteri* strains DSM 17938 and ATCC PTA 4659 therapeutically prevented NEC in rats via downregulation of interleukin-6 (IL-6), tumor necrosis factor alpha (TNFα), and TLR4. Results also revealed a significantly increased survival rate and reduced severity of NEC in parallel with inhibition of the NF-κB pathway (Liu et al. 2012). These outcomes suggest that probiotics might play a key role in the regulation of the innate immune system and hence can function by mitigating the effects of diarrheal pathogens.

Regulation of the adaptive immune response is essential in the maintenance of a disease-free state. Adaptive immune responses utilize antigen-presenting cells such as macrophages and dendritic cells to activate cytotoxic T-cells. Cells of this system, specifically T-lymphocytes, are found to be both aberrantly and chronically activated in the pathogenesis of autoimmune conditions such as IBD. Probiotics contribute to immune regulation by enhancing the maturation and chemotaxis of CD4 (+) FoxP3 (+) Treg cells, which are responsible for attenuating T-lymphocyte-mediated immune responses (Kwon et al. 2010). Consequently, probiotics are important for intestinal homeostasis and developing tolerance toward the native microbiome. Atarashi et al. demonstrated that 17 strains of *Clostridia* found in the native human intestinal microbiome maintain a transforming growth factor beta-rich environment to help the expansion and differentiation of Treg cells (Atarashi et al. 2013). Furthermore, in a dinitrobenzene sulfonic acid mouse model of colitis, *L. casei* was found to enhance chemotaxis of Treg cells to the intestinal intima, causing inhibition of T-cell-mediated inflammation and cell death (Atarashi et al. 2013).

Another well-studied mechanism of attenuating the adaptive immune response is the modification of cytokine profiles to inhibit inflammation and lymphocyte maturation. *L. plantarum* may promote this process via the production of immunomodulatory peptides, such as enterotoxin *E. coli* heat-stable enterotoxin (Chen et al. 2010a). Al-Hassi et al. demonstrated that this peptide produced an anti-inflammatory cytokine profile in dendritic cells in conjunction with downregulation inflammatory mediators in both UC patients and healthy controls (Al-Hassi et al. 2014). Furthermore, strains of both *Lactobacilli* and *Bifidobacteria* were shown in separate studies of murine colitis models to re-establish a normal physiological ratio of IL-10 to IL-12, a comparison of anti-inflammatory with pro-inflammatory cytokines, respectively. In addition, Guzy et al. found *E. coli* strain Nissle 1917 to inhibit the secretion of CXCL8, a pro-inflammatory chemokine, and to induce apoptosis in $\gamma\delta$ T cells via a Fas/Fas ligand pathway (Guzy et al. 2008). In summary, the effects of probiotics extend to the regulation of both the innate and the adaptive immune system for the modulation of inflammatory diarrhea.

5.3.5 Alterations in Gut Microbiome Composition in Response to Probiotics Use

The human gut microbiome is quite stable over time, but there is variability at the extremes of age and among different individuals. Diet and other environmental factors can also have a profound impact on the composition of the microbiome (Shreiner et al. 2015). Furthermore, significant variation has been found in the composition of the intestinal microbiota between healthy and diseased individuals (Scott et al. 2015). As such, the gut microbiome has been shown to play a crucial role in the pathogenesis of diarrheal diseases, including certain forms of IBD (Halfvarson et al. 2017). For example, in comparison with UC, CD has been shown to be associated with greater dysbiosis and less microbial diversity (Pascal et al. 2017). Probiotics have displayed promise in the treatment of these diseases and one aspect of their efficacy is tied to their potential to influence the resident microbiota of the gastrointestinal system (Scott et al. 2015). To better elucidate the effects of probiotics treatment on the composition of the native gut microbiome, a study by Toscana et al. evaluated the ability of *L. rhamnosus* HN001 and *B. longum* BB536 to colonize the intestinal environment of healthy subjects (20 healthy Italian volunteers) and modify the composition of the endogenous microbiome.

Administration of probiotics *B. longum* and *L. rhamnosus* resulted in an increase in the abundance of both bacteria in the pre- and post-prandial groups. At the level of the phyla, a significant reduction was discovered in Firmicutes and Proteobacteria (Toscano et al. 2017). In a clinical study of celiac disease, probiotics were also shown to be effective in restoring the physiological gut microbiome. Quagliariello et al. found supplementation of *Bifidobacterium breve* strains B632 and BR03 essential in reducing the abnormal Firmicutes/Bacteroidetes ratios found in the gut of 40 patients with celiac disease. As such, a clinically significant modality of probiotics is their capability to therapeutically modify the composition of the gut microbiome (Quagliariello et al. 2016).

Utilizing in-vivo mouse models, it has been shown that probiotics can restore gut microbiome altered by lifestyle and diet. Park et al. supplemented *Lactobacillus curvatus* HY7601 and *L. plantarum* KY1032 in diet-induced obese mice. Significant alterations were found in the gut microbiota after probiotic supplementation. However, the two most abundant phyla, Firmicutes and Bacteroidetes, did not show considerable changes. Conversely, there was a significant reduction in key species, including Verrucomicrobia and Proteobacteria, indicating that probiotics can also alter the composition of the endogenous gut microbiome of mice (Park et al. 2013). In another study, Zubiria et al. administered probiotic species *L. kefiri* in a mouse model with a fructose-rich diet. Results revealed that *L. kefiri* supplementation exerts compositional changes in gut microbiota, particularly in Bacteroidetes and Firmicutes profiles, while simultaneously attenuating weight gain and epididymal adipose tissue expansion.

These probiotic-induced alterations of the gut microbiome correlated innately with their therapeutic capacity to inhibit pathogenic bacteria. Indeed, research by Mañes-Lázaro et al. revealed that administration of *L. johnsonii* F19785 to chickens significantly altered the gut microbiome of the chickens and inhibited the capacity of pathogenic bacteria *Campylobacter jejuni* to colonize the gut (Manes-Lazaro et al. 2017). Although the current research is promising, detailed studies have been limited in exploring the therapeutic effects of probiotics-mediated gut microbiome alterations and their significance in treating diarrheal diseases. Further mechanistic and clinical studies should provide evidence for probiotics as a therapeutic modality for a wide range of gut microbiome-associated diseases, including diarrhea.

Treg response is directly related to the maintenance of intestinal homeostasis and the development of tolerance toward the resident microbiota (Toscano et al. 2017). Probiotics inducing the Treg response are of paramount importance, notably within the framework of IBD and other inflammatory diseases.

5.3.6 Probiotic Modulation of Vectorial Ion Transport

Although considered to be multifactorial, diarrhea commonly results from decreased intestinal absorption and/or increased secretion of fluid and electrolytes. Therefore, a detailed understanding of the mechanisms of dysregulated ion transport in inflammatory or infectious diarrhea and the role of probiotics in counteracting these effects are likely to define novel therapeutic targets for intervention. Indeed, several reports in recent years demonstrated pro-absorptive effects of probiotics to enhance intestinal absorption of ions and solutes by directly modulating the expression/activity of the transporters involved. In addition, there are reports to show the anti-secretory effects of probiotics, which should also explain the antidiarrheal effects of certain probiotics.

5.3.6.1 Effects on NaCl Absorption

The major route of NaCl absorption in the mammalian intestine is via a coupled operation of Na^+/H^+ exchanger 3 (NHE3, SLC9A3) and the Cl^-/HCO_3^- exchanger down-regulated in adenoma (DRA, SLC26A3). NHE2 and SLC26A6 (PAT-1) expressed on the apical surface also mediate Na^+/H^+ and Cl^-/HCO_3^- exchange respectively. The significance of NHE3 in electrolyte and fluid homeostasis is demonstrated by reduced Na^+ absorption in NHE3 knockout mice and its inhibition, leading to diarrhea (Schultheis et al. 1998; Priyamvada et al. 2015). Similarly, mutations in the *DRA* gene have been shown to cause congenital Cl^- diarrhea, a rare disorder exhibiting metabolic alkalosis, impaired Cl^-/HCO_3^- exchange, and high fecal Cl^- concentration (Schweinfest et al. 2006). Further, $DRA^{-/-}$ mice exhibited a substantial diarrheal phenotype with serum electrolyte imbalances.

As in congenital chloride diarrhea patients, increasing evidence indicates that disturbances in NHE3/DRA function or expression play a major role in the pathophysiology of diarrheal diseases (Schweinfest et al. 2006). For example, previous reports have shown that short-term infection of IECs with enteropathogenic *E. coli* (EPEC), a food-borne pathogen, significantly inhibited Na^+/H^+ and Cl^-/HCO_3^- exchange activities, defining the pathophysiology of EPEC-induced early diarrhea (Gill et al. 2007). Therefore, agents that stimulate luminal NaCl absorption and/or counteract inhibition of NaCl absorption in inflammation/infection could be promising antidiarrheals to be further explored for their therapeutic potential. In this regard, recent studies have demonstrated probiotics-mediated stimulation of NaCl absorption via their direct effects on the expression/activity of NHE3 and DRA. These studies have shown that *L. acidophilus* (LA) or its culture supernatant (LA-CS) stimulates Cl^-/HCO_3^- exchange activity via its short-term effects to increase apical membrane DRA levels involving a PI-3 kinase-dependent mechanism (Borthakur et al. 2008).

Along the same lines, LA showed the long-term effects of stimulating DRA function by increasing DRA expression via transcriptional mechanisms and counteracted the downregulation of DRA expression and function in in-vitro and in-vivo models of inflammation, or in response to the infection of FVB/N mice by *Citrobacter rodentium* (Kumar et al. 2016; Raheja et al. 2010). Additionally, LA/LA-CS stimulated NHE3 expression/function in vitro and in vivo and counteracted *C. rodentium* infection-induced inhibition of NHE3 expression (Kumar et al. 2016; Singh et al. 2012). Furthermore, LA alleviated *C. rodentium*-induced downregulation of carbonic anhydrase I and IV, which are needed to maintain the pool of bicarbonate required for DRA function (Cl^-/HCO_3^- exchange activity).

Studies from the same group further showed that the bacteria-free culture supernatant of *Bifidobacterium* species (*B. breve*, *B. infantis*, *B. bifidum*), important commensals in the healthy human colon, enhanced DRA expression in Caco-2 cells via ERK1/2 MAPK pathway-dependent transactivation of DRA promoter, whereas administration of the live bacteria increased expression of DRA mRNA and protein levels in mouse colon (Kumar et al. 2012).

These studies significantly enhanced our mechanistic insights into the antidiarrheal potential of *Lactobacilli* and *Bifidobacteria*. A key outcome of these studies showing that soluble factor(s) present in the culture supernatant *of L. acidophilus and Bifidobacterium species* mimic the stimulatory effects of live bacteria on NaCl absorption highlights the critical importance of probiotics-derived molecules in developing novel therapeutic strategies for treating diarrhea. A model of the proposed mechanisms of the effects of probiotics on NaCl absorption is depicted in Fig. 5.1. A brief outline of mechanisms of action of probiotics in diarrheal disorders are summarized in Tables 5.2 and 5.3.

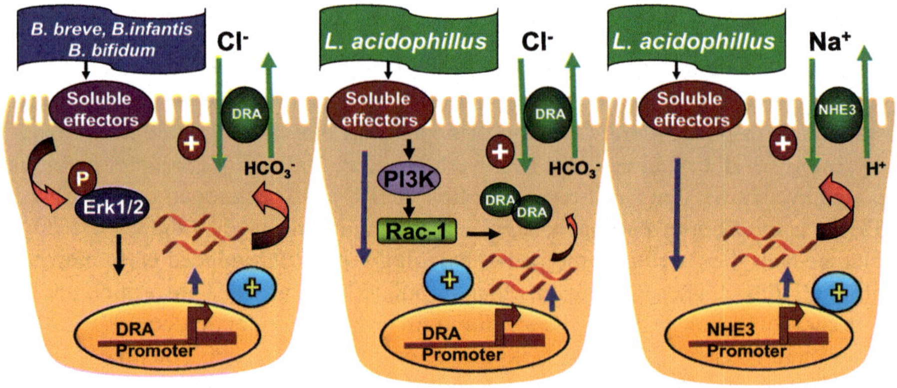

Fig. 5.1 Proposed mechanisms of modulation of NaCl absorption by probiotics. As shown in the schematic diagram, probiotics mediate upregulation of the expression and/or activity of down-regulated in adenoma (DRA) and Na^+/H^+ exchanger 3 (NHE3). (*Left to right*) *1*. The bacteria-free culture supernatant (soluble factors) of *Bifidobacteria* species (*B. breve*, *B. infantis*, *B. bifidum*) transactivated the DRA promoter in Caco-2 cells via the ERK1/2 MAPK pathway, which led to an increase in DRA mRNA and protein levels and ultimately an increase in function. This correlated with an increase in DRA mRNA and protein expression in mouse colon by the administration of the live *Bifidobacteria*. *2*. *L. acidophilus* (LA) or its culture supernatant (LA-CS) stimulated Cl^-/HCO_3^- exchange activity via its short-term effects to increase apical membrane DRA levels involving a PI-3 kinase and RAC-1-dependent mechanism. Similarly, LA also showed long-term effects on DRA by increasing its function and expression via transcriptional mechanisms. *3*. LA/LA-CS also stimulated NHE3 promoter, which led to an increase in NHE3 mRNA and protein expression and function in intestinal epithelial cells

Table 5.2 Genomic Mechanisms

Genus	Species	Protein product	Functional implications	Citation
Bifidobacterium	*Bifidum*	DRA ⇑	Enhanced electroneutral NaCl absorption	Kumar et al. (2012)
	Breve	DRA ⇑	Enhanced electroneutral NaCl absorption	Kumar et al. (2012)
	Infantis	DRA ⇑	Enhanced electroneutral NaCl absorption	Kumar et al. (2012)
Escherichia coli	Nissle	TcpC ⇑	Maintenance of gut epithelial barrier function	Alvarez et al. (2016)
Lactobacillus	*Acidophilus*	DRA ⇑	Enhanced electroneutral NaCl absorption	Raheja et al. (2010)
	Acidophilus	NHE3 ⇑	Enhanced electroneutral NaCl absorption	Singh et al. (2012)
	Gasseri	APF1 ⇑	Pathogen exclusion	Nishiyama et al.
	Johnsonii	SGLT1 ⇑	Enhanced Na+/glucose co-transport	Rooj et al. (2010)
	Plantarum	PepT1 ⇑	Enhanced oligopeptide transport	Chen et al. (2010a)
	Plantarum	STp ⇑	Immunomodulation	Chen et al. (2010a)
	Rhamnosus	SERT ⇑	Enhanced serotonin uptake	Wang et al. (2015)
Saccharomyces	*Boulardii*	SGLT1 ⇑	Enhanced Na+/glucose co-transport	Buts et al. (1999)

DRA Down-Regulated in Adenoma, *TcpC* Toll/IL-1 Receptor-Containing (TIR-containing) Protein C, *NHE3* Na^+/H^+ Exchanger 3, *APF* Aggregation-Promoting Factor, *SGLT1* Sodium-dependent Glucose Cotransporter 1, *PepT1* Peptide Transporter 1, *STp E. coli* heat-stable enterotoxin, *SERT* Serotonin Transporter

5.3.6.2 Effects of Probiotics on CFTR and NKCCL1

Activation of the cystic fibrosis transmembrane conductance regulator (CFTR) chloride channel is another pathway that drives fluid secretion into the intestinal lumen. The CFTR channel is activated in cholera and in diarrhea caused by other bacterial enterotoxins. In cystic fibrosis patients, altered microbial diversity was also observed, with an increase in Firmicutes and a reduction in Bacteroidetes compared with controls, suggesting that gut microbiota might play an important role in cystic fibrosis and its management (Burke et al. 2017) as well.

Another study by Heuvelin et al. analyzed the potency of the probiotic strain *Bifidobacterium breve* C50 (*Bb* C50) and *Lactobacillus rhamnosus 10893* to modulate epithelial Cl^- secretion (Heuvelin et al. 2010). Carbachol- or forskolin (FSK)-induced Cl^- secretion was inhibited in a dose-dependent manner by *B. breve* C50 associated with decreased PKC activity. In addition, FSK-induced Cl^- secretion was also inhibited by *Lactobacillus rhamnosus 10893* (Heuvelin et al. 2010).

Table 5.3 Nongenomic Mechanisms

Genus	Species	Effect	Functional implications	Citation
Bacillus	*Cereus* var. *toyoi*	Counteract CFTR activation	Attenuated Cl^- secretion	Ma et al. (2017)
Bifidobacterium	*Brevis*	Counteract CFTR activation	Attenuated Cl^- secretion	Heuvelin et al. (2010)
	CA1	Bacteriocin ⇑	Antimicrobial factor	Lievin et al. (2000)
	Infantis	Tryptophan ⇑	Enhanced serotonin production	Desbonnet et al. (2008)
	F9	Bacteriocin ⇑	Antimicrobial factor	Lievin et al. (2000)
	Thermophilus	Under investigation	Pathogen exclusion	Gagnon et al. (2016)
	Thetaiotaomicron	Inhibit IFN	Immunomodulation	Resta-Lenert and Barrett (2006)
Escherichia coli	Nissle	Induce T-cell apoptosis	Immunomodulation	Guzy et al. (2008)
Enterococcus	*Faecium*	Counteract CFTR activation	Attenuated Cl^- secretion	Klingspor et al. (2013)
Lactobacilli	*Acidophilus*	Apical membrane DRA ⇑	Enhanced electroneutral NaCl absorption	Borthakur et al. (2008)
	Acidophilus	Apical membrane MCT1⇑	Enhanced short-chain fatty acid absorption	Kumar et al. (2015)
	Acidophilus	Counteract CFTR activation	Attenuated Cl^- secretion	Resta-Lenert and Barrett (2002)
	Acidophilus	MCT1 function ⇑	Enhanced short-chain fatty acid absorption	Kumar et al. (2015)
	Acidophilus	NHE3 function ⇑	Enhanced electroneutral NaCl absorption	Singh et al. (2012)
	Acidophilus	Surface layer protein	Pathogen exclusion	Prado Acosta et al. (2016)
	Amylovorus	SGLT1 function ⇑	Enhanced Na+/glucose co-transport	Rooj et al. (2010)
	Brevis	Surface layer protein	Pathogen exclusion	Prado Acosta et al. (2016)
	Casei	NF-κB antagonism	Immunomodulation	Liu et al. (2012)
	Casei	TLR4 activation	Enhanced Immunosurveillance	Tien et al. (2006)
	Casei	Treg chemotaxis ⇑	Immunomodulation	Atarashi et al. (2013)
	Casei	Under investigation	Pathogen exclusion	Ingrassia et al. (2005)
	Curvatus	DN317 bacteriocin ⇑	Antimicrobial factor	Zommiti et al. (2016)
	Fructosus	IL-8, p-ERK, and p-JNK ⇓	Immunomodulation	Yu et al. (2015)
	Fructosus	Paracellular permeability ⇓	Maintenance of gut epithelial barrier function	Yu et al. (2015)
	Gallinarum	SGLT1 function ⇑	Enhanced Na+/glucose co-transport	Rooj et al. (2010)
	Gasseri	Aggregation promoting factors	Pathogen exclusion	Nishiyama et al. (2015)
	Gasseri	SGLT1 function ⇑	Enhanced Na+/glucose co-transport	Rooj et al. (2010)

	Helveticus	Surface layer protein	Pathogen exclusion	Prado Acosta et al. (2016)
	Johnsonii	SGLT1 function ⇑	Enhanced Na+/glucose co-transport	Rooj et al. (2010)
	Kefiri	Surface layer protein	Pathogen exclusion	Prado Acosta et al. (2016)
	Plantarum	SMCT1 function ⇑	Enhanced short chain fatty acid absorption	Borthakur et al. (2010)
	Plantarum	TNF-alpha ⇓	Immunomodulation	Borthakur et al. (2010)
	Reuteri	IL-6, TNF-alpha, and TLR4 ⇓	Immunomodulation	Liu et al. (2012)
	Rhamnosus	Counteract CFTR activation	Attenuated Cl^- secretion	Heuvelin et al. (2010)
	Casei	Lactic acid production	Antimicrobial factor	Castillo et al. (2011)
	Rhamnosus	Mucus binding pili	Pathogen exclusion	Tytgat et al. (2016)
	Rhamnosus	p40 and p75 secretion	Maintenance of gut epithelial barrier function	Seth et al. (2008)
	Salivarius	Counteract CFTR activation	Attenuated Cl^- secretion	Lomasney et al. (2014)
Streptococcus	*Thermophilus*	Counteract CFTR activation	Attenuated Cl^- secretion	Resta-Lenert and Barrett (2002)
Mixed	*Clostridia*	TGF-beta ⇑	Immunomodulation	Atarashi et al. (2013)
	LA/LP/LS/BI	TNF-alpha ⇓	Immunomodulation	Resta-Lenert and Barrett (2003)
	LA/ *Thermophilus*	Inhibit IFN	Immunomodulation	Resta-Lenert and Barrett (2002)
	VSL #3	IFN-γ ⇓	Immunomodulation	Krishnan et al. (2016)

⇑ indicates an increase and ⇓ indicates a decrease

LA L. acidophilus, *LP L. plantarum*, *LS L. salivarius*, *BI B. lactis*

CFTR Cystic Fibrosis transmembrane conductance Regulator, *IFN* Interferon, *MCT1* Monocarboxylate Transporter 1, *TLR4* toll-like receptor 4, *SMCT1* Sodium-Coupled MCT1, *TGF* Transforming Growth Factor, *TNF* Tumor Necrosis Factor

Similarly, another study demonstrated that the probiotic strains *Enterococcus faecium* NCIMB 10415 and *Bacillus cereus var. toyoi* increased prostaglandin E2-induced short circuit current (Isc) (Klingspor et al. 2013; Ma et al. 2017). Moreover, a recent study showed that *L. salivarius* UCC118 selectively inhibited neurally evoked ion secretion in mice colonic tissues (Lomasney et al. 2014). Furthermore, *Streptococcus thermophilus* (ST) and *Lactobacillus acidophilus* (LA), or the commensal, *Bacteroides thetaiotaomicron* (BT) were able to block the inhibitory effects of IFN on agonist-stimulated chloride secretion, trans-epithelial resistance (TER), and epithelial permeability (Resta-Lenert and Barrett 2006). ST/LA pretreatment resulted in the reversal of the inhibitory effects of IFN on CFTR and the Na-K-Cl co-transporter NKCC1 (Resta-Lenert and Barrett 2006).

Resta-Lenert et al. also demonstrated that probiotics protect against entero-invasive *E. coli* (EIEC) and *S. dublin* (SD) infection in T-84 and HT29/cl.19A cells. Wild-type ST/LA, LA, or ST were able to reverse the effects of EIEC and SD on chloride secretion in IECs. In addition, they showed significantly altered CFTR and NKCC1 function and cellular distribution after treatment with probiotics following EIEC and SD infection (Resta-Lenert and Barrett 2002). These findings also supported the beneficial role of probiotics in decreasing chloride secretion and diarrhea by regulating kinase signaling and levels of ion transport proteins CFTR and NKCC1 in IECs.

5.3.6.3 Effects of Probiotics on Short Chain Fatty Acid Absorption

Short chain fatty acids (SCFAs) are produced in the colonic lumen by bacterial fermentation of dietary fiber or undigested carbohydrates. Butyrate, a key SCFA, serves as the primary fuel for colonocytes, ameliorates mucosal inflammation, maintains epithelial integrity, and exerts antidiarrheal effects by stimulating NaCl absorption. For these intracellular effects, however, optimal absorption of butyrate by the colonic epithelial cells is important (Thibault et al. 2007, 2010). Monocarboxylate transporter 1 (MCT1) plays an important role in the absorption of colonic SCFA. Indeed, previous studies have shown extensive downregulation of MCT1 in intestinal inflammation, leading to decreased availability of butyrate for oxidation in the colonic epithelial cells (Thibault et al. 2007). Therefore, agents that upregulate butyrate absorption are likely to have therapeutic value in counteracting intestinal inflammation and associated diarrhea.

Earlier reports have shown decreased butyrate uptake in Caco-2 cells in response to EPEC infection that was attributed to an EPEC-induced decrease in MCT1 protein level at the apical cell surface (Borthakur et al. 2011). Recent studies from the same laboratory showed that *L. acidophilus* and its CS not only increased MCT1-mediated butyrate uptake, but also alleviated EPEC inhibition of butyrate uptake by blocking EPEC-induced internalization of apical cell surface MCT1 (Kumar et al. 2015). Additionally, *L. plantarum*, another species of *Lactobacilli* that showed no effects on MCT1 activity, upregulated sodium-coupled MCT1 (SMCT1, another SCFA transporter) in rat IEC-6 cells (Borthakur et al. 2010). Also, *L. plantarum* counteracted cytokine-induced inhibition of SMCT1 expression and Na^+-dependent butyrate uptake in IEC-6 cells.

5.3.6.4 Effects of Probiotics on Other Solute Transporters

Characterization of probiotic efficacy in the treatment of diarrheal diseases is not complete without mentioning several areas of emerging study. Recent investigations have found specific probiotic species to alter the expression or function of oligopeptide, serotonin, and glucose transporters in different models of diarrheal diseases. These novel effects correlated with amelioration of the diarrheal phenotype.

Lactobacillus plantarum has recently been shown to enhance the function of the peptide transporter PepT1 in a murine colitis model (Chen et al. 2010b). Utilizing interleukin-10 knockout mice to model spontaneous colitis, Chen et al. observed a marked decrease in PepT1 function in comparison to wild-type mice (Chen et al. 2010a). *L. plantarum* treatment was found to not only mitigate this effect in IL-$10^{-/-}$ mice, but also to exhibit increased transporter function. Immunofluorescence and western blot analysis showed unaltered expression and localization of PepT1 in both WT and IL-$10^{-/-}$ mice. Further investigation revealed that diminished PKC activity in IL-$10^{-/-}$ mice compared with WT led to enhanced PepT1 function.

Lactobacillus rhamnosus GG has been implicated in the treatment of diarrhea associated with IBS (Aragon et al. 2010). IBS is hypothesized to have a psychogenic etiology where brain–gut interactions are fundamentally altered with minimal inflammation. Probiotics have recently emerged as a clinically effective treatment for this condition. To better ascertain the relationship between commensal gut bacteria and IBS, *Bifidobacterium infantis* was observed to significantly upregulate plasma tryptophan, a precursor to the neurotransmitter serotonin (Desbonnet et al. 2008). Furthermore, Wang et al. observed *L. rhamnosus* GG, a probiotic found in commercial formulations, to upregulate serotonin transporter mRNA and protein in HT-29, Caco-2, and mouse IECs (Wang et al. 2015). These novel results suggested that probiotic-mediated modulation of neurotransmitter transport might be a possible mechanism for attenuating IBS-related diarrhea.

Several species of probiotics have also been implicated in the upregulation of glucose transport across the intestinal brush border membranes. In addition, this may be achieved by enhancing function or expression of sodium-dependent glucose co-transporter (SGLT-1) or glucose transporter 2 (Roder et al. 2014). Similar to oral rehydration therapy, probiotics-mediated enhancement of these transporters could counteract the electrolyte loss common in diarrhea of all causes. Early studies showed *Saccharomyces boulardii* to upregulate SGLT-1 function in both pig and rat models (Buts et al. 1999). A recent study by Rooj et al. demonstrated supernatant from cultures of probiotic species *L. gasseri*, *L. amylovorus*, *L. gallinarum*, and *L. johnsonii* to enhance glucose transport by a nongenomic mechanism (Rooj et al. 2010). Putative explanations implicate metabolites produced by these bacterial species to underlie this effect. Mechanisms for the functional enhancement of these nutrient transporters are not well understood and require extensive studies.

5.4 Conclusion

As described above, clinical data indicate that probiotics play a beneficial role in several diarrheal disorders. However, inconclusive evidence for several major etiologies of diarrhea, including CD, traveler's diarrhea, and NEC, highlights the need for more extensive investigations. Furthermore, a key limitation in the current collection of meta-analyses and systematic reviews is the heterogeneity of the study design. Variation in the selection of probiotics formulations, dosage, duration of therapy, and subject selection underlie the difficulties in evaluating the efficacy of probiotics in clinical settings. Also, clinical investigations in the past have overwhelmingly utilized probiotic formulations containing multiple probiotic species or strains. Results of such trials do not pinpoint the effects of individual probiotic species and are difficult to translate to in-vitro or in-vivo experimental models. Therefore, there is a strong need in probiotic clinical trials to obtain high-quality data demonstrating the efficacy of specific individual strains of bacteria.

It should be noted that when comparing current models of treating diarrheal diseases, one avenue, fecal microbiota transplant (FMT), is currently under intense investigation, specifically for *C. difficile*-associated diarrhea. However, The FMT procedure is likely to introduce risks of transmitting various pathogens and metabolic disorders. In addition, FMT has been found to marginally worsen symptoms of IBD, whereas traditional dietary supplementation of probiotics has been shown to be safe for such conditions. Another weakness of FMT is the very concept of the procedure, which some patients find to be rather unpleasant. In such conditions, probiotics may be the preferred treatment for diarrheal pathological conditions.

In agreement with many clinical studies, several studies supported by the in-vitro and in-vivo animal models on vectorial ion and nutrient transporters show promising results in diarrheal diseases. These studies are of critical importance in increasing our understanding of the molecular basis of the beneficial effects of probiotics and probiotics-derived effector molecules in diarrheal diseases. Additionally, the efficacy of probiotics or their secreted soluble factors to counteract inflammation, maintain gut epithelial integrity, mediate pathogen exclusion, and secrete antimicrobial peptides highlights the novel therapeutic potential of different probiotic species in diarrheal diseases.

With increasing evidence indicating the therapeutic role of probiotics in diarrheal diseases, future in-depth investigations are needed to identify bacterial species that are most effective among current probiotic regimens, in addition to optimizing dosages to achieve superior therapeutic effects. Furthermore, in-vitro and in-vivo studies have indicated the importance of probiotics-secreted soluble factors such as bacteriocins, S-layer proteins, and other similar unidentified compounds of low molecular weight. The potential of isolating these soluble factors for further study in in-vitro, in-vivo, and clinical settings presents a presently untapped reservoir of therapeutics for diarrheal diseases. Indeed, such investigations would be beneficial for developing therapeutic interventions for those who have poor gut barrier integrity or who are severely immunocompromised and may not tolerate live probiotic bacterial administration.

In summary, our mechanistic understanding of probiotic effects in the human gut has developed significantly since the turn of the century. Further laboratory and clinical studies may help to build on these advances by leveraging strains of probiotics bacteria and/or their products as novel therapeutics for diarrheal diseases.

Acknowledgments Supported by Department of Veterans Affairs Merit Award # BX002011 (P.K. Dudeja), Merit Award # BX000152 (W.A. Alrefai), VA Senior Research Career Scientist Award (P.K. Dudeja), Research Career Scientist Award (W.A. Alrefai), and the NIDDK grants DK54016, DK81858, and DK92441 (P.K. Dudeja) and DK 109709 (W.A. Alrefai).

References

Alakomi HL, Skytta E, Saarela M, Mattila-Sandholm T, Latva-Kala K, Helander IM (2000) Lactic acid permeabilizes gram-negative bacteria by disrupting the outer membrane. Appl Environ Microbiol 66(5):2001–2005

Al-Hassi HO, Mann ER, Sanchez B, English NR, Peake ST, Landy J et al (2014) Altered human gut dendritic cell properties in ulcerative colitis are reversed by Lactobacillus plantarum extracellular encrypted peptide STp. Mol Nutr Food Res 58(5):1132–1143

Alvarez CS, Badia J, Bosch M, Gimenez R, Baldoma L (2016) Outer membrane vesicles and soluble factors released by probiotic Escherichia coli Nissle 1917 and commensal ECOR63 enhance barrier function by regulating expression of tight junction proteins in intestinal epithelial cells. Front Microbiol 7:1981

Aragon G, Graham DB, Borum M, Doman DB (2010) Probiotic therapy for irritable bowel syndrome. Gastroenterol Hepatol (NY) 6(1):39–44

Atarashi K, Tanoue T, Oshima K, Suda W, Nagano Y, Nishikawa H et al (2013) Treg induction by a rationally selected mixture of Clostridia strains from the human microbiota. Nature 500(7461):232–236

Borthakur A, Gill RK, Tyagi S, Koutsouris A, Alrefai WA, Hecht GA et al (2008) The probiotic Lactobacillus acidophilus stimulates chloride/hydroxyl exchange activity in human intestinal epithelial cells. J Nutr 138(7):1355–1359

Borthakur A, Anbazhagan AN, Kumar A, Raheja G, Singh V, Ramaswamy K et al (2010) The probiotic Lactobacillus plantarum counteracts TNF-{alpha}-induced downregulation of SMCT1 expression and function. Am J Physiol Gastrointest Liver Physiol 299(4):G928–G934

Borthakur A, Kumar A, Dudeja PK (2011) Mechanisms underlying Lactobacillus acidophilus-mediated modulation of intestinal butyrate absorption. Abstract presented at Summer Research Conference at Arizona

Burke DG, Fouhy F, Harrison MJ, Rea MC, Cotter PD, O'Sullivan O et al (2017) The altered gut microbiota in adults with cystic fibrosis. BMC Microbiol 17(1):58

Buts JP, De Keyser N, Marandi S, Hermans D, Sokal EM, Chae YH et al (1999) Saccharomyces boulardii upgrades cellular adaptation after proximal enterectomy in rats. Gut 45(1):89–96

Castillo NA, Perdigon G, de Moreno de Leblanc A (2011) Oral administration of a probiotic Lactobacillus modulates cytokine production and TLR expression improving the immune response against Salmonella enterica serovar Typhimurium infection in mice. BMC Microbiol 11:177

Chen HQ, Yang J, Zhang M, Zhou YK, Shen TY, Chu ZX et al (2010a) Lactobacillus plantarum ameliorates colonic epithelial barrier dysfunction by modulating the apical junctional complex and PepT1 in IL-10 knockout mice. Am J Physiol Gastrointest Liver Physiol 299(6): G1287–G1297

Chen HQ, Shen TY, Zhou YK, Zhang M, Chu ZX, Hang XM et al (2010b) Lactobacillus plantarum consumption increases PepT1-mediated amino acid absorption by enhancing protein kinase C activity in spontaneously colitic mice. J Nutr 140(12):2201–2206

Collins JW, Keeney KM, Crepin VF, Rathinam VA, Fitzgerald KA, Finlay BB et al (2014) Citrobacter rodentium: infection, inflammation and the microbiota. Nat Rev Microbiol 12(9):612–623

Conte MP, Longhi C, Marazzato M, Conte AL, Aleandri M, Lepanto MS et al (2014) Adherent-invasive Escherichia coli (AIEC) in pediatric Crohn's disease patients: phenotypic and genetic pathogenic features. BMC Res Notes 7:748

Culligan EP, Hill C, Sleator RD (2009) Probiotics and gastrointestinal disease: successes, problems and future prospects. Gut Pathog 1(1):19

Desbonnet L, Garrett L, Clarke G, Bienenstock J, Dinan TG (2008) The probiotic Bifidobacteria infantis: an assessment of potential antidepressant properties in the rat. J Psychiatr Res 43(2):164–174

Gagnon M, Vimont A, Darveau A, Fliss I, Jean J (2016) Study of the ability of bifidobacteria of human origin to prevent and treat rotavirus infection using colonic cell and mouse models. PLoS One 11(10):e0164512

Gallo A, Passaro G, Gasbarrini A, Landolfi R, Montalto M (2016) Modulation of microbiota as treatment for intestinal inflammatory disorders: an uptodate. World J Gastroenterol 22(32):7186–7202

Gibson PR, Barrett JS (2010) The concept of small intestinal bacterial overgrowth in relation to functional gastrointestinal disorders. Nutrition 26(11):1038–1043

Gill RK, Borthakur A, Hodges K, Turner JR, Clayburgh DR, Saksena S et al (2007) Mechanism underlying inhibition of intestinal apical Cl/OH exchange following infection with enteropathogenic E. coli. J Clin Invest 117(2):428–437

Gomez-Llorente C, Munoz S, Gil A (2010) Role of Toll-like receptors in the development of immunotolerance mediated by probiotics. Proc Nutr Soc 69(3):381–389

Guo S, Gillingham T, Guo Y, Meng D, Zhu W, Walker WA et al (2017) Secretions of Bifidobacterium infantis and Lactobacillus acidophilus protect intestinal epithelial barrier function. J Pediatr Gastroenterol Nutr 64(3):404–412

Guzy C, Paclik D, Schirbel A, Sonnenborn U, Wiedenmann B, Sturm A (2008) The probiotic Escherichia coli strain Nissle 1917 induces gammadelta T cell apoptosis via caspase- and FasL-dependent pathways. Int Immunol 20(7):829–840

Halfvarson J, Brislawn CJ, Lamendella R, Vazquez-Baeza Y, Walters WA, Bramer LM et al (2017) Dynamics of the human gut microbiome in inflammatory bowel disease. Nat Microbiol 2:17004

Hassan M, Kjos M, Nes I, Diep D, Lotfipour F (2012) Natural antimicrobial peptides from bacteria: characteristics and potential applications to fight against antibiotic resistance. J Appl Microbiol 113(4):723–736

Hecht AL, Casterline BW, Earley ZM, Goo YA, Goodlett DR, Wardenburg JB (2016) Strain competition restricts colonization of an enteric pathogen and prevents colitis. EMBO Rep 17((9):e201642282

Helander IM, Mattila-Sandholm T (2000) Permeability barrier of the gram-negative bacterial outer membrane with special reference to nisin. Int J Food Microbiol 60(2–3):153–161

Heuvelin E, Lebreton C, Bichara M, Cerf-Bensussan N, Heyman M (2010) A Bifidobacterium probiotic strain and its soluble factors alleviate chloride secretion by human intestinal epithelial cells. J Nutr 140(1):7–11

Hirano J, Yoshida T, Sugiyama T, Koide N, Mori I, Yokochi T (2003) The effect of Lactobacillus rhamnosus on enterohemorrhagic Escherichia coli infection of human intestinal cells in vitro. Microbiol Immunol 47(6):405–409

Ingrassia I, Leplingard A, Darfeuille-Michaud A (2005) Lactobacillus casei DN-114 001 inhibits the ability of adherent-invasive Escherichia coli isolated from Crohn's disease patients to adhere to and to invade intestinal epithelial cells. Appl Environ Microbiol 71(6):2880–2887

Johnson-Henry KC, Nadjafi M, Avitzur Y, Mitchell DJ, Ngan BY, Galindo-Mata E et al (2005) Amelioration of the effects of Citrobacter rodentium infection in mice by pretreatment with probiotics. J Infect Dis 191(12):2106–2117

Klingspor S, Martens H, Caushi D, Twardziok S, Aschenbach JR, Lodemann U (2013) Characterization of the effects of Enterococcus faecium on intestinal epithelial transport properties in piglets. J Anim Sci 91(4):1707–1718

Kostic AD, Xavier RJ, Gevers D (2014) The microbiome in inflammatory bowel disease: current status and the future ahead. Gastroenterology 146(6):1489–1499

Krishna Rao R, Samak G (2013) Protection and restitution of gut barrier by probiotics: nutritional and clinical implications. Curr Nutr Food Sci 9(2):99–107

Krishnan M, Penrose HM, Shah NN, Marchelletta RR, McCole DF (2016) VSL#3 probiotic stimulates T-cell protein tyrosine phosphatase-mediated recovery of IFN-gamma-induced intestinal epithelial barrier defects. Inflamm Bowel Dis 22(12):2811–2823

Kumar A, Hecht C, Priyamvada S, Anbazhagan AN, Alakkam A, Borthakur A et al (2012) Probiotic Bifidobacterium species stimulate human SLC26A3 gene function and expression in in vitro and in vivo models. Gastroenterology (Abstract) 142(5):S-542

Kumar A, Alrefai WA, Borthakur A, Dudeja PK (2015) Lactobacillus acidophilus counteracts enteropathogenic E. coli-induced inhibition of butyrate uptake in intestinal epithelial cells. Am J Physiol Gastrointest Liver Physiol 309(7):G602–G607

Kumar A, Anbazhagan AN, Coffing H, Chatterjee I, Priyamvada S, Gujral T et al (2016) Lactobacillus acidophilus counteracts inhibition of NHE3 and DRA expression and alleviates diarrheal phenotype in mice infected with Citrobacter rodentium. Am J Physiol Gastrointest Liver Physiol 311(5):G817–GG26

Kwon HK, Lee CG, So JS, Chae CS, Hwang JS, Sahoo A et al (2010) Generation of regulatory dendritic cells and CD4+Foxp3+ T cells by probiotics administration suppresses immune disorders. Proc Natl Acad Sci USA 107(5):2159–2164

Landy J, Ronde E, English N, Clark SK, Hart AL, Knight SC et al (2016) Tight junctions in inflammatory bowel diseases and inflammatory bowel disease associated colorectal cancer. World J Gastroenterol 22(11):3117–3126

Lievin V, Peiffer I, Hudault S, Rochat F, Brassart D, Neeser J et al (2000) Bifidobacterium strains from resident infant human gastrointestinal microflora exert antimicrobial activity. Gut 47(5):646–652

Liu Y, Fatheree NY, Mangalat N, Rhoads JM (2012) Lactobacillus reuteri strains reduce incidence and severity of experimental necrotizing enterocolitis via modulation of TLR4 and NF-kappaB signaling in the intestine. Am J Physiol Gastrointest Liver Physiol 302(6):G608–G617

Lomasney KW, Cryan JF, Hyland NP (2014) Converging effects of a Bifidobacterium and Lactobacillus probiotic strain on mouse intestinal physiology. Am J Physiol Gastrointest Liver Physiol 307(2):G241–G247

Ma X, Wang X, Kang N, Chen T, Ji H, Lv L et al (2017) The effect of Tong-Xie-Yao-Fang on intestinal mucosal mast cells in postinfectious irritable bowel syndrome rats. Evid Based Complement Alternat Med 2017:9086034

Manes-Lazaro R, Van Diemen PM, Pin C, Mayer MJ, Stevens MP, Narbad A (2017) Administration of Lactobacillus johnsonii FI9785 to chickens affects colonisation by Campylobacter jejuni and the intestinal microbiota. Br Poult Sci 58:1–9

Mardini HE, Grigorian AY (2014) Probiotic mix VSL#3 is effective adjunctive therapy for mild to moderately active ulcerative colitis: a meta-analysis. Inflamm Bowel Dis 20(9):1562–1567

Martinez FA, Balciunas EM, Converti A, Cotter PD, de Souza Oliveira RP (2013) Bacteriocin production by Bifidobacterium spp. A review. Biotechnol Adv 31(4):482–488

Miele E, Pascarella F, Giannetti E, Quaglietta L, Baldassano RN, Staiano A (2009) Effect of a probiotic preparation (VSL#3) on induction and maintenance of remission in children with ulcerative colitis. Am J Gastroenterol 104(2):437–443

Nébot-Vivinus M, Harkat C, Bzioueche H, Cartier C, Plichon-Dainese R, Moussa L et al (2014) Multispecies probiotic protects gut barrier function in experimental models. World J Gastroenterol 20(22):6832

Nishiyama K, Nakazato A, Ueno S, Seto Y, Kakuda T, Takai S et al (2015) Cell surface-associated aggregation-promoting factor from Lactobacillus gasseri SBT2055 facilitates host colonization and competitive exclusion of Campylobacter jejuni. Mol Microbiol 98(4):712–726

Park DY, Ahn YT, Park SH, Huh CS, Yoo SR, Yu R et al (2013) Supplementation of Lactobacillus curvatus HY7601 and Lactobacillus plantarum KY1032 in diet-induced obese mice is associated with gut microbial changes and reduction in obesity. PLoS One 8(3):e59470

Pascal V, Pozuelo M, Borruel N, Casellas F, Campos D, Santiago A et al (2017) A microbial signature for Crohn's disease. Gut 66(5):813–822

Prado Acosta M, Ruzal SM, Cordo SM (2016) S-layer proteins from Lactobacillus sp. inhibit bacterial infection by blockage of DC-SIGN cell receptor. Int J Biol Macromol 92:998–1005

Priyamvada S, Gomes R, Gill RK, Saksena S, Alrefai WA, Dudeja PK (2015) Mechanisms underlying dysregulation of electrolyte absorption in inflammatory bowel disease-associated diarrhea. Inflamm Bowel Dis 21(12):2926–2935

Quagliariello A, Aloisio I, Bozzi Cionci N, Luiselli D, D'Auria G, Martinez-Priego L et al (2016) Effect of Bifidobacterium breve on the intestinal microbiota of coeliac children on a gluten free diet: a pilot study. Nutrients 8(10):660

Raheja G, Singh V, Ma K, Boumendjel R, Borthakur A, Gill RK et al (2010) Lactobacillus acidophilus stimulates the expression of SLC26A3 via a transcriptional mechanism. Am J Physiol Gastrointest Liver Physiol 298(3):G395–G401

Resta-Lenert S, Barrett KE (2002) Enteroinvasive bacteria alter barrier and transport properties of human intestinal epithelium: role of iNOS and COX-2. Gastroenterology 122(4):1070–1087

Resta-Lenert S, Barrett KE (2003) Live probiotics protect intestinal epithelial cells from the effects of infection with enteroinvasive Escherichia coli (EIEC). Gut 52(7):988–997

Resta-Lenert S, Barrett KE (2006) Probiotics and commensals reverse TNF-alpha- and IFN-gamma-induced dysfunction in human intestinal epithelial cells. Gastroenterology 130(3):731–746

Ritchie ML, Romanuk TN (2012) A meta-analysis of probiotic efficacy for gastrointestinal diseases. PLoS One 7(4):e34938

Roder PV, Geillinger KE, Zietek TS, Thorens B, Koepsell H, Daniel H (2014) The role of SGLT1 and GLUT2 in intestinal glucose transport and sensing. PLoS One 9(2):e89977

Rooj AK, Kimura Y, Buddington RK (2010) Metabolites produced by probiotic Lactobacilli rapidly increase glucose uptake by Caco-2 cells. BMC Microbiol 10:16

Russell JB, Diez-Gonzalez F (1997) The effects of fermentation acids on bacterial growth. Adv Microb Physiol 39:205–234

Salari P, Nikfar S, Abdollahi M (2012) A meta-analysis and systematic review on the effect of probiotics in acute diarrhea. Inflamm Allergy Drug Targets 11(1):3–14

Sang LX, Chang B, Zhang WL, Wu XM, Li XH, Jiang M (2010) Remission induction and maintenance effect of probiotics on ulcerative colitis: a meta-analysis. World J Gastroenterol 16(15):1908–1915

Schmitz H, Fromm M, Bentzel CJ, Scholz P, Detjen K, Mankertz J et al (1999) Tumor necrosis factor-alpha (TNFalpha) regulates the epithelial barrier in the human intestinal cell line HT-29/B6. J cell Sci 112(Pt 1):137–146

Schroeder B, Duncker S, Barth S, Bauerfeind R, Gruber AD, Deppenmeier S et al (2006) Preventive effects of the probiotic Escherichia coli strain Nissle 1917 on acute secretory diarrhea in a pig model of intestinal infection. Dig Dis Sci 51(4):724–731

Schultheis PJ, Clarke LL, Meneton P, Miller ML, Soleimani M, Gawenis LR et al (1998) Renal and intestinal absorptive defects in mice lacking the NHE3 Na+/H+ exchanger. Nat Genet 19(3):282–285

Schweinfest CW, Spyropoulos DD, Henderson KW, Kim JH, Chapman JM, Barone S et al (2006) slc26a3 (dra)-deficient mice display chloride-losing diarrhea, enhanced colonic proliferation, and distinct up-regulation of ion transporters in the colon. J Biol Chem 281(49):37962–37971

Scott KP, Antoine JM, Midtvedt T, van Hemert S (2015) Manipulating the gut microbiota to maintain health and treat disease. Microb Ecol Health Dis 26:25877

Seth A, Yan F, Polk DB, Rao RK (2008) Probiotics ameliorate the hydrogen peroxide-induced epithelial barrier disruption by a PKC- and MAP kinase-dependent mechanism. Am J Physiol Gastrointest Liver Physiol 294(4):G1060–G1069

Shen J, Zuo Z-X, Mao A-P (2014) Effect of probiotics on inducing remission and maintaining therapy in ulcerative colitis, Crohn's disease, and pouchitis: meta-analysis of randomized controlled trials. Inflamm Bowel Dis 20(1):21–35

Shreiner AB, Kao JY, Young VB (2015) The gut microbiome in health and in disease. Curr Opin Gastroenterol 31(1):69–75

Singh V, Raheja G, Borthakur A, Kumar A, Gill RK, Alakkam A et al (2012) Lactobacillus acidophilus upregulates intestinal NHE3 expression and function. Am J Physiol Gastrointest Liver Physiol 303(12):G1393–G1401

Thibault R, De Coppet P, Daly K, Bourreille A, Cuff M, Bonnet C et al (2007) Down-regulation of the monocarboxylate transporter 1 is involved in butyrate deficiency during intestinal inflammation. Gastroenterology 133(6):1916–1927

Thibault R, Blachier F, Darcy-Vrillon B, de Coppet P, Bourreille A, Segain JP (2010) Butyrate utilization by the colonic mucosa in inflammatory bowel diseases: a transport deficiency. Inflamm Bowel Dis 16(4):684–695

Tien MT, Girardin SE, Regnault B, Le Bourhis L, Dillies MA, Coppee JY et al (2006) Anti-inflammatory effect of Lactobacillus casei on Shigella-infected human intestinal epithelial cells. J Immunol 176(2):1228–1237

Toscano M, De Grandi R, Stronati L, De Vecchi E, Drago L (2017) Effect of Lactobacillus rhamnosus HN001 and Bifidobacterium longum BB536 on the healthy gut microbiota composition at phyla and species level: a preliminary study. World J Gastroenterol 23(15):2696–2704

Tytgat HL, Douillard FP, Reunanen J, Rasinkangas P, Hendrickx AP, Laine PK et al (2016) Lactobacillus rhamnosus GG Outcompetes Enterococcus faecium via Mucus-Binding Pili: evidence for a novel and heterospecific probiotic mechanism. Appl Environ Microbiol 82(19):5756–5762

Wang YM, Ge XZ, Wang WQ, Wang T, Cao HL, Wang BL et al (2015) Lactobacillus rhamnosus GG supernatant upregulates serotonin transporter expression in intestinal epithelial cells and mice intestinal tissues. Neurogastroenterol Motil 27(9):1239–1248

Yu Q, Yuan L, Deng J, Yang Q (2015) Lactobacillus protects the integrity of intestinal epithelial barrier damaged by pathogenic bacteria. Front Cell Infect Microbiol 5:26

Zanfardino A, Criscuolo G, Di Luccia B, Pizzo E, Ciavatta ML, Notomista E et al (2017) Identification of a new small bioactive peptide from Lactobacillus gasseri supernatant. Benefic Microbes 8(1):133–141

Zommiti M, Almohammed H, Ferchichi M (2016) Purification and characterization of a novel anti-campylobacter bacteriocin produced by Lactobacillus curvatus DN317. Probiotics Antimicrob Proteins 8(4):191–201

Chapter 6
The Influence of Microbiota on Gastrointestinal Motility

Shreya Raja, Vivek Batra, and Shanthi Srinivasan

Abstract In recent years there has been a significant advancement in the role of gut microbiota in regulating gastrointestinal motility. The bidirectional cross talk between the host and gut microbiota has been implicated in the regulation of both physiological and pathophysiological conditions. Intestinal dysbiosis or alteration in the composition of intestinal microbiota can result in impaired host intestinal permeability, immune response, and metabolism, leading to a proinflammatory state. In this review, we focus on the role of the gut microbiome in regulating gastrointestinal motility and shaping the enteric nervous system. We highlight the mechanisms of microbial metabolites in regulating intestinal motility. Several host factors such as diet and genetic predisposition can influence the gut microbial diversity and ultimately contribute to dysbiosis. Intestinal dysbiosis can contribute to the pathophysiology of disorders such as irritable bowel syndrome and chronic intestinal pseudo-obstruction. Manipulation of the gut microbiome is a promising therapeutic target for the treatment of motility disorders. Modification of gut microbiota through diet, antibiotics, probiotics, prebiotics, and fecal microbiota transplantation are all promising strategies for the treatment of gastrointestinal motility disorders that are currently under investigation.

Keywords Motility disorders • Transplantation • Fecal microbiota • Enteric neurons • Irritable bowel syndrome • Amyotrophic lateral sclerosis

---

S. Raja • S. Srinivasan (✉)
Division of Digestive Diseases, Emory University School of Medicine, Atlanta, GA, USA

Atlanta VAMC, Decatur, GA, USA
e-mail: ssrini2@emory.edu

V. Batra
Department of Medical Oncology, Sidney Kimmel Cancer Center, Thomas Jefferson University, Philadelphia, PA, USA

J. Sun, P.K. Dudeja (eds.), *Mechanisms Underlying Host-Microbiome Interactions in Pathophysiology of Human Diseases*, Physiology in Health and Disease,
https://doi.org/10.1007/978-1-4939-7534-1_6

6.1 Introduction

There is evolving evidence that interplay exists between the gut microbiota and the enteric nervous system (ENS), resulting in altered gastrointestinal motility. The conventional wisdom was that motor patterns influence the size, location, and diversity of the microbiota (Quigley 2011). Recent literature and studies have challenged this view, and several studies have demonstrated that the gut microbiota may exert significant control over the enteral nervous system of the host. The bidirectional cross talk between the gut microbiota and the host has been established; yet, the exact dynamics of the relationship remain an active field of investigation.

Several gastrointestinal disorders have been linked to alterations in gut microbiota, including irritable bowel syndrome (IBS), acute diarrhea, inflammatory bowel disease (IBD), and small intestinal bacterial overgrowth syndrome (SIBO) (Aziz et al. 2013). Alterations in gastrointestinal flora with the use of antibiotics can lead to *Clostridium difficile*-infected diarrhea (Kelly and LaMont 2008). Alterations in intestinal bacteria, and the subsequent production of carcinogens may contribute to the pathogenesis of colon cancer (Rowland 2009).

In a newborn, the gastrointestinal tract is sterile at birth. It is colonized within the first few days to weeks after birth by the microbiota. This microbiota is essential for survival as it confers a symbiotic and commensal advantage to the host. Antibiotics use in the early stages of life can predispose to gastrointestinal disorders later in life. The human microbiome contains nearly 100 trillion cells, about tenfold the number of human cells, and encodes about 150 times more genes than human cells (Ley et al. 2006). Metagenomic sequencing of fecal samples of 124 Europeans found that 99% of the genes in the human intestinal microbiome are bacterial. There are 1000–1150 prevalent bacterial species and each individual has at least 160 such species, which are largely shared (Qin et al. 2010). Most of the bacteria found in the human GI tract belong to the Bacteroidetes, Firmicutes, Actinobacteria, and Proteobacteria phyla (Gill et al. 2006). The gut microbiota performs three broad functions, which are well described and include metabolic, immunological, and trophic functions. The metabolic function of bacteria is to extract calories from complex oligosaccharides and promote the absorptive capacity of the intestinal epithelium (Sekirov et al. 2010). The concentration of bacteria varies with age and location in the gastrointestinal tract. The distal colon has the maximum concentration of bacteria, approaching almost 10^{11} bacteria per gram. The colon is populated mainly by anaerobic bacteria owing to the low oxygen concentration in the colon, including *Bacteroides*, *Bifidobacterium*, *Clostridium*, *Lactobacillus*, and *Porphyromonas*. The stomach and proximal intestine have relatively small numbers of bacteria because of the acidic environment. Gram-negative and anaerobic bacteria populate the terminal ileum (Mackie et al. 1999). In this chapter, we first describe the influence of microbiota in shaping the ENS. Next, we discuss how microbial metabolites can regulate intestinal motility. Host factors can affect gut microbial diversity, and we demonstrate how dysbiosis can lead to motility disorders such as IBS and chronic intestinal pseudo-obstruction (CIPO).

6.2 Microbiota Effects on the Enteric Nervous System

6.2.1 Early Development of the ENS

The effects of the microbiota on the development of the ENS are best illustrated in studies performed by colonizing germ-free animals with microbiota. Structural and functional differences have been found in the jejunal and ileal myenteric plexus of postnatal germ-free mice compared with specific pathogen-free mice. Postnatal germ-free mice also have decreased nerve density, decreased neurons per ganglion, and an increased ratio of nitrergic neurons compared with specific pathogen-free mice. In addition, decreased jejunal and ileal contractility has been observed in postnatal germ-free mice compared with specific pathogen-free mice (Collins et al. 2014). A recent animal experiment performed by McVey Neufeld et al. (2013) showed that the microbiome is essential for normal gut intrinsic afferent neuron excitability in the mouse. The investigators examined the electrophysiological properties of myenteric plexus neurons in germ-free mice, specific pathogen-free mice, and germ-free mice conventionalized with intestinal bacteria. Significant findings from this study included decreased neuronal excitability in the myenteric afterhyperpolarization neurons and prolonged post-action potential slow afterhyperpolarization in germ-free mice compared with the other two groups of mice. Another animal study examined small bowel myoelectrical activity in germ-free rats colonized with specific bacterial strains. Interestingly, colonization with anaerobic strains including *Clostridium tabificum* alone, and *Lactobacillus acidophilus* and *Bifidobacterium bifidum* in combination stimulated small intestinal transit by stimulating phase III of the migrating motor complex (MMC) sooner, whereas aerobic strains such as *Micrococcus luteus* and *Escherichia coli* suppressed or had no significant effect on the initiation of phase III of the MMC respectively (Husebye et al. 2001). The mechanism by which the microbiota modulates the MMC is unclear; however, the results of this study suggest that changes that arise in the local environment as a result of anaerobic metabolism may influence small intestinal motility. These landmark animal experiments provide evidence that the microbiota are necessary for the normal development and physiological function of the ENS.

6.2.2 Neurohormones

Serotonin (5-HT) functions as both a neurotransmitter and a local hormone, which is present in the gastrointestinal tract and in the central nervous system. It plays diverse roles in digestion, including initiation of intestinal secretion and peristalsis. Serotonin activates both intrinsic excitatory and inhibitory enteric motor neurons. It can stimulate cholinergic neurons to release acetylcholine, which results in smooth muscle contraction, or it can stimulate inhibitory nitrergic neurons to release nitric oxide, which results in smooth muscle relaxation (Sikander et al. 2009). Several small studies suggest that gut microbiota might modulate the biosynthesis and release of 5-HT in the gastrointestinal tract. The gut microbiota produces short chain fatty acids (SCFAs) via fermentation of carbohydrates, and SCFAs modulate the release of 5-HT from enterochromaffin cells in vivo (Fukumoto et al. 2003). The presence of the cholera toxin induces release of 5-HT in the rat jejunum (Bearcroft et al. 1996). In-vitro studies suggest that infection with enteropathogenic *Escherichia coli* might decrease the activity of the serotonin transporter in intestinal epithelial cells, resulting in decreased concentrations and uptake of 5-HT (Esmaili et al. 2009).

6.2.3 Microbial Metabolism

Short chain fatty acids such as butyrate, propionate, and acetate are produced as by-products of enteric bacterial fermentation of resistant starches. The proportions of each SCFA produced vary based on the presence of a particular microbiota and dietary fiber intake. Gram-positive anaerobic bacteria from the Firmicutes phylum, such as *Faecalibacterium prausnitzii* and *Eubacterium rectale/Roseburia* spp., are significant producers of SCFAs. The most widely studied SCFA is butyrate, which is essential for enterocyte function. Butyrate provides a source of energy for enterocytes, mediates intestinal epithelial cell turnover, modulates production of inflammatory cytokines, and alters gastrointestinal motility (Canani et al. 2011). The motor effects of SCFAs differ based on chain length. In-vitro and in-vivo rat studies have shown that butyrate increases the proportion of ascending excitatory cholinergic neurons and increases colonic contractile activity; however, these findings were not demonstrated for propionate or acetate (Soret et al. 2010). In a study performed on guinea pigs, butyrate increased the frequency of full-length propagations in the proximal colon and increased the velocity of propagation in the distal colon. In contrast, propionate blocked full and short propagations and had a biphasic effect on nonpropagating contractions, and acetate decreased short and total propagations (Hurst et al. 2014). The presence of SCFAs may also increase the proportion of ascending excitatory cholinergic neurons and colonic contractile activity (Soret et al. 2010). The mechanisms by which SCFAs affect gastrointestinal motility are an active area of study.

Low-grade inflammation has been observed in patients with IBS. Immunohistochemical studies of intestinal biopsy specimens from IBS patients have shown increased lymphocyte and mast cell infiltration in addition to myenteric neurodegeneration in patients with IBS compared with controls (Chadwick et al. 2002; Tornblom et al. 2002; Barbara et al. 2004). Increased cytokines such as IL-1β,

IL-6, IL-8, LPS-induced IL-6, and TNF-α have also been observed in IBS patients (Liebregts et al. 2007). The etiology of the inflammation is thought to be multifactorial, and disruption of the intestinal epithelial barrier resulting in increased permeability has been proposed as a possible mechanism. Lipopolysaccharide (LPS), a component of cell walls of Gram-negative bacteria, does not permeate the intestinal epithelial barrier under normal conditions. However, LPS crosses the epithelial barrier in settings of increased permeability, such as inflammation and infection, and activates toll-like receptors (TLRs) found on macrophages and dendritic cells. Activation of TLRs serves as protection against further injury. The mechanism by which specific TLRs distinguish between virulent and commensal bacteria is under study (Rakoff-Nahoum et al. 2004).

The host innate immune system is also able to detect the presence of microbial metabolites through pattern recognition receptors and immune-sensing complexes known as inflammasomes. Specifically, the NLRP-6 inflammasome plays a role in colonic homeostasis and can be influenced by microbiota-associated metabolites, including taurine, histamine, and spermine in a mouse model. These metabolites alter the secretion of IL-8 from colonic epithelial cells, which subsequently affects the downstream production of antimicrobial peptides. Disruption of the NLRP-6/IL-8/anti-microbial peptide axis results in increased susceptibility to inflammation in the host (Levy et al. 2015, 2017).

Bile acids serve many functions in the gastrointestinal tract. They are best known for their role in facilitating the absorption of lipids and fat-soluble vitamins; however, they also have antimicrobial properties, increase mucosal permeability, enhance intestinal secretion of water and electrolytes, and participate in signaling pathways in the ENS that modulate motility and sensation. The gut microbiota metabolically alters bile acids through the process of deconjugation. Bile acid malabsorption results in diarrhea, and is commonly seen in patients with ileal resection, Crohn's disease, and diarrhea-predominant IBS. Several causative mechanisms for bile acid malabsorption have been suggested, including a deficiency of fibroblast growth factor 19 and genetic alterations of receptors involved in bile acid synthesis and transport. Established modalities for treating bile acid diarrhea include bile acid sequestrants such as cholestyramine, colestipol, and colesevelam. Bile acid receptor agonists that bind Farnesoid X receptor, such as obeticholic acid, are under development (Camilleri 2015).

Gut bacteria ferment carbohydrates and produce various gaseous by-products, such as carbon dioxide, hydrogen, methane, and hydrogen sulfide. The microbes generate energy via this process. Interestingly, there is an association between excessive gas production and clinical disorders. Excess hydrogen production has been implicated in SIBO and IBS. SIBO is defined by the presence of excessive bacteria in the small intestine and can be diagnosed using a glucose or lactulose hydrogen breath test or by culturing small intestinal aspirate. During a lactulose hydrogen breath test, lactulose is ingested, which is subsequently digested by the gut bacteria. Excess production of hydrogen is observed as a result of poor transit and the resulting overgrowth of bacteria in the gut (Pimentel et al. 2013).

Methane has been linked to constipation, constipation-predominant IBS, and obesity. Methane gas is a common by-product of fermentation produced by intestinal bacteria and results in slowed small intestinal transit. In animal models, infusion of methane has resulted in 59% slowing of small intestinal transit (Pimentel et al. 2006). Hydrogen sulfide is a toxic, pungent gas that is produced by both microbes and mammalian tissues, and studies have shown that it may play a role in gastrointestinal motility by indirectly mediating secretion, nociception, and smooth muscle relaxation. It has been linked to ulcerative colitis, although the mechanism of action remains unclear (Medani et al. 2011).

6.2.4 Host Factors

The effect of diet and genetics on shaping the gut microbiota has been demonstrated by a comparative study in children from Europe and rural Africa. This study compared the fecal microbiota of European children with the that of African children from a village in Burkina Faso. The African diet was high-fiber, low-fat, low-protein, and vegetarian, whereas the European diet consisted of mostly animal protein, sugars, starch, and low-fiber foods. Analysis of microbiota from the African children showed increased predominance of Bacteroidetes and decreased concentrations of Firmicutes. Specifically, Bacteroidetes such as *Prevotella*- and *Xylanibacter*-containing bacterial genes for cellulose and xylan digestion were found in the microbiota of the African children, but not in European children. In contrast, European children were found to have higher concentrations of Firmicutes and Enterobacteriaceae than African children. The concentration of fecal SCFAs also differed between the two groups, with the African cohort having higher concentrations of SCFAs noted in their stool samples. The findings of this study suggest that exposure to diets high in polysaccharides may influence the evolution of the gut microbiota that can best utilize this diet to produce energy (De Filippo et al. 2010).

Obesity has significantly increased in the last few decades, especially in the USA. The gut microbiota has been shown to be different in obese individuals; a decreased concentration of Bacteroidetes has been noted. The pathophysiology is poorly understood, but it is hypothesized that changes in gut flora might lead to altered proinflammatory molecules and changes in host–gene expression, which affect the gut epithelial and endocrine function and have an impact on insulin resistance and adiposity (Ley 2010).

6.3 Microbiota in Neurological Diseases

The bidirectional interaction between the nervous system and gastrointestinal tract is further illustrated by the role of the gut microbiota in the pathogenesis of neurological conditions, particularly Parkinson's disease (PD) and amyotrophic lateral sclerosis (ALS). Altered microbial composition has also been observed in multiple sclerosis, autism spectrum disorder, and Alzheimer's disease (Scheperjans et al. 2015; Zhang et al. 2017; Chen et al. 2016; Luna et al. 2017; Shen et al. 2017).

6.3.1 Parkinson's Disease

Parkinson's disease is a neurodegenerative condition resulting from the accumulation of a specific amyloid protein, α-synuclein (ASO), within neurons. Motor deficits seen clinically in PD arise from the deposition of ASO in the dopaminergic neurons of the substantia nigra pars compacta in the midbrain. Gastrointestinal symptoms such as constipation may precede neurological symptoms by many years and arise from ASO deposition in enteric neurons.

Intestinal inflammation is present in patients with PD. Increased serum markers of endotoxins such as LPS binding protein, increased intestinal permeability, and increased mucosal staining of Gram-negative bacteria, ASO, and oxidative stress markers such as nitrotyrosine have been observed in a small human study (Forsyth et al. 2011). Differences in fecal microbiota composition have been identified in PD patients compared with healthy controls. A relative increase in Enterobacteriaceae and a decrease in Prevotellaceae have been observed in PD (Scheperjans et al. 2015; Unger et al. 2016). Decreased concentrations of stool SCFAs have been observed in PD patients compared with healthy controls, and may contribute to the development of gastrointestinal dysmotility (Unger et al. 2016).

Gene–environment interactions are thought to play a role in the pathogenesis of PD. An elegant experiment performed by Sampson et al. (2016) demonstrated that germ-free transgenic mice overexpressing ASO have reduced motor deficits, reduced microglial activation, and fewer ASO inclusions compared with ASO mice colonized with complex microbiota. These findings suggest that the presence of a microbiota might be required for the development of PD. Further experiments by this group have shown that treating germ-free ASO mice with SFCAs induces motor deficits, suggesting a possible mechanism by which microbiota modulate microglia. Colonization of germ-free ASO mice with microbiota obtained from patients with PD also induces clinical symptoms when compared with colonization with healthy donors. Therefore, it is plausible that the microbiota might influence the development of PD in genetically predisposed hosts.

6.3.2 Amyotrophic Lateral Sclerosis

Dysbiosis has been implicated in the development of ALS. ALS is a rapidly progressive, fatal neuromuscular condition affecting motor neurons associated with mutations in the Cu/Zn superoxide dismutase gene (SOD1). The G93A transgenic mouse model, which contains the human SOD1 gene, has been developed to study ALS.

Analysis of the fecal microbiota composition of G93A mice has shown decreased concentrations of butyrate-producing bacteria such as *Butyrivibrio fibrisolvens*, *Escherichia coli*, and *Fermicus* compared with wild-type mice. A loss of integrity of tight junctions and increased permeability of the intestinal epithelium has been observed in G93A mice, suggesting that a disruption of intestinal homeostasis might be involved in the pathogenesis of ALS (Wu et al. 2015). A recent landmark study from Zhang et al. demonstrated that oral butyrate treatment delayed progression of ALS symptoms, and significantly prolonged life span in the G93A mouse. Interestingly, treatment with butyrate also increased the proportion of butyrate-producing bacteria. Improvements in the structural integrity of the murine gut with restoration of tight junction proteins, decreased intestinal permeability, and decreased aggregation of SOD1 mutant proteins were also observed following butyrate treatment. Manipulation of the microbiome and its metabolites may emerge as a new therapeutic target for ALS (Zhang et al. 2017).

6.4 Dysbiosis in Motility Disorders

6.4.1 Small Intestinal Bacterial Overgrowth

Small intestinal bacterial overgrowth is defined by the presence of greater than 10^5 colony-forming units per ml of bacteria in jejunal aspirate. Clinical manifestations of SIBO include bloating, distention, diarrhea, abdominal pain, and weight loss. Host mechanisms to prevent SIBO include secretion of gastric acid, bile acid and pancreatic fluids, normal gastrointestinal motility, a competent ileocecal valve, and IgA production on the mucosal surface of the intestines (Miazga et al. 2015). The acidity of the stomach prevents bacterial growth in the small intestine. Reduced gastric acid production may occur as a result of increased aging or from exposure to *Helicobacter pylori* and may lead to SIBO (Dukowicz et al. 2007). The use of proton pump inhibitors has been considered a risk factor for SIBO; however, previous studies have yielded inconsistent results. A large study from the Mayo Clinic including 1,191 patients showed that the results of glucose hydrogen breath testing did not differ between PPI users and non-users. This study also showed that risk factors for a positive breath test included older age and the presence of diarrhea (Ratuapli et al. 2012).

Dysmotility likely plays a role in the pathogenesis of bacterial overgrowth. A small study by Vantrappen et al. (1977) compared the motor complexes seen on manometry tracings from normal subjects and those with SIBO, as defined by a positive bile acid breath test and found that the patients with SIBO had absent or disordered motor complexes. The function of the intestinal motor complexes is thought to be clearance of secretions, desquamated cells and nutrients. Impairment of this mechanism creates an environment that promotes bacterial overgrowth. Another study performed by Stotzer et al. (1996) also showed differences in antroduodenojejunal pressure tracings in healthy subjects compared with patients with SIBO, confirmed by duodenal aspirate and glucose hydrogen breath testing. The most significant finding in this study was the loss of phase III activity in the antrum and small intestine of SIBO patients compared with healthy subjects. Physiologically, phase III of the motor complex functions to clear stomach and intestinal contents during the fasting state. A recent Norwegian study examined intestinal motor patterns in patients with late radiation enteropathy by analyzing small intestinal manometry tracings, gastric pH, and bacterial counts. Interestingly, abnormalities in the MMC were found to be predictors of increased gram-negative bacilli concentrations in patients with late radiation enteropathy (Husebye et al. 1995).

Delayed small intestinal transit time has been demonstrated in patients with SIBO. A recent retrospective study examined 72 patients who underwent wireless motility capsule testing and lactulose hydrogen breath testing. Subjects with positive lactulose hydrogen breath tests were found to have longer small bowel and whole gut transit times than those with normal lactulose hydrogen breath tests. Based on these results, the authors concluded that delayed small intestinal transit contributes to the pathogenesis of SIBO (Roland et al. 2015).

6.4.2 Chronic Intestinal Pseudo-Obstruction

Chronic intestinal pseudo-obstruction clinically presents with the symptoms of obstruction in the absence of a mechanical blockage. CIPO is challenging to identify, and unfortunately many patients may experience recurrent episodes before diagnosis. In CIPO, the intestinal smooth muscle does not effectively contract and propagate luminal contents (De Giorgio et al. 2011).

The pathogenesis of CIPO is not very well understood. Inflammatory and neurodegenerative processes can damage the myenteric ganglia and the interstitial cells of Cajal. Viruses that have been implicated in CIPO include Herpesviridae and John Cunningham virus. Secondary causes of CIPO include neurological disorders such as PD, collagen vascular disease, and endocrine diseases. Chronic alcohol abuse has also been linked to CIPO. Ongoing studies are being performed to identify genetic causes of CIPO. Derangements in gut microbiota occur as a result of altered intestinal motility, and this population frequently presents with symptoms of bacterial overgrowth (Gabbard and Lacy 2013).

6.4.3 Irritable Bowel Syndrome

Dysbiosis has been observed in patients with functional bowel disorders. Functional bowel disorders include IBS, functional diarrhea, chronic idiopathic constipation, and functional bloating. IBS is the most common functional bowel disorder with an estimated 10–20% prevalence in the Western world (Saito et al. 2002). Diagnosis is symptom-based as described by the Rome Criteria. The pathogenesis of functional bowel disorders is multifactorial, with both biological and psychosocial circumstances influencing the course of the disease. Physiological studies suggest that the fecal microbiota might alter the brain–gut axis in functional bowel disorders, which results in aberrant sensorimotor and enteroendocrine function, loss of integrity of the intestinal epithelial barrier, and increased intestinal inflammation (Ringel 2017).

The composition of the fecal microbiota in patients with IBS differs from healthy controls and may also vary by IBS subtype. In a study conducted Rajilic-Stojanovic et al. (2011), phylogenetic microarray analysis of fecal samples show that Bacteroidetes are decreased and Firmicutes are increased in patients with IBS compared with healthy controls. Interestingly, *Faecalibacterium* species were the only Firmicutes found in lower concentrations in IBS patients compared with healthy controls; *Faecalibacterium prausnitzii* is known to have anti-inflammatory properties based on previous studies performed in animal models of colitis. Pathogenic species in the Firmicutes phylum including *Streptococcus* species were found in higher concentrations in IBS patients.

It has been argued that the development of post-infectious IBS (PI-IBS) provides compelling evidence that the microbiota is involved in the pathogenesis of IBS. PI-IBS is a condition presenting with classic symptoms of IBS following acute gastrointestinal infection. The odds ratios of developing PI-IBS after acute gastroenteritis are six- to sevenfold based on previous studies (Thabane et al. 2007; Halvorson et al. 2006). Risk factors for PI-IBS include younger age, female gender, co-existing psychiatric comorbidities such as anxiety and depression, and prolonged infection and fever. Jalanka-Tuovinen et al. (2014) performed phylogenetic microarray analysis of fecal samples from patients with IBS-D, PI-IBS, and healthy subjects and found that the composition of the microbiota of patients with IBS-D and PI-IBS was similar, and that both differed from healthy controls. However, it should be noted that this study performed by Jalanka-Tuovinen et al. (2014) found increased Bacteroidetes and decreased Firmicutes in the IBS groups, which conflicts with the study discussed earlier by Rajilic-Stojanovic et al. (2011). The lack of consistency between studies suggests that although dysbiosis is observed in IBS, the specific microbiota involved in the pathogenesis is still unclear.

6.5 Conclusion

Manipulation of the gut microbiome has emerged as a therapeutic target for the treatment of motility disorders. Modification of the gut microbiota through diet, antibiotics, probiotics, prebiotics, and fecal microbiota transplantation (FMT) is currently under investigation. Recently, the efficacy of the low fermentable oligo-, di-, monosaccharides, and polyols (FODMAP) diet has been demonstrated in the treatment of IBS. FODMAPs are osmotically active, poorly absorbed carbohydrates that are highly susceptible to bacterial fermentation. A recent meta-analysis shows that IBS symptom severity is reduced on a low-FODMAP diet (Marsh et al. 2016). Although the low-FODMAP diet appears promising, its long-term safety has not been assessed.

Studies examining the use of prebiotics and probiotics are currently underway. Prebiotics are foods ingested by the host that cannot be digested in the small intestine and are subsequently fermented in the colon. Prebiotics selectively promote the growth and proliferation of species of commensal bacteria such as *Lactobacillus* and *Bifidobacterium* with known health benefits. Studies addressing the efficacy of prebiotics are limited, but there are data to suggest that ingestion of inulin-type fructans might be beneficial (Quigley and Quera 2006). In contrast to prebiotics, probiotics are live microorganisms ingested by the host. There are many varieties of probiotics available commercially, and clinical efficacy appears to be strain-specific (Whelan 2011). A systematic review of 16 randomized controlled trials demonstrates improvement in IBS symptoms after treatment with *Bifidobacterium infantis* 35624 (Brenner et al. 2009).

Non-absorbable antibiotics such as rifaximin are the mainstay of treatment for SIBO, and have been recently approved for the treatment of diarrhea-predominant IBS (Pimentel et al. 2011). Early antibiotics exposure, specifically to macrolides and tetracyclines, has been linked to the subsequent development of functional bowel disorders, and treatment with antibiotics such as clindamycin and cephalosporins has been connected to the development of *C. difficile* infection (Villarreal et al. 2012). These seemingly conflicting findings of antibiotics being both causative and therapeutic suggests that the microbiome might be selectively targeted by antibiotics that may be beneficial or harmful to commensal bacteria.

Fecal microbiota transplantation is a process in which stool from healthy donors is transplanted to patients with the hope of curing an underlying gastrointestinal disorder. Restoration of a healthy gut microbiome to a diseased host underlines the importance of the microbiota in modulating normal gastrointestinal function. The efficacy of FMT has been established for recurrent *C. difficile* infection, and is currently FDA-approved for this indication (Rossen et al. 2015). Ongoing studies of FMT in the treatment of IBD and motility disorders are limited but promising. A preliminary Chinese study examining 9 patients with CIPO demonstrated an improvement in symptoms of bloating and abdominal pain in addition to the ability to tolerate enteral feeds following FMT via a nasojejunal tube (Gu et al. 2017). A randomized controlled trial examining the efficacy of FMT compared with laxative therapy in patients with slow-transit constipation showed 30% improvement in patients with FMT, although the rate of treatment-related adverse events was higher in the FMT group (Tian et al. 2017). A recently published open label study on FMT in 10 Japanese patients with IBS (8 with IBS-D, 1 with IBS-C, and 1 with IBS-M) demonstrated clinical improvement in symptoms in 6 patients at 4 weeks after FMT. Analysis of stool microbiota composition showed increased diversity following FMT. Higher concentrations of *Bifidobacterium* species were noted in donor stools from the patients who responded to FMT (Mizuno et al. 2017). Although these early studies are encouraging, FMT for the treatment of motility disorders remains experimental.

References

Aziz Q, Dore J, Emmanuel A, Guarner F, Quigley EM (2013) Gut microbiota and gastrointestinal health: current concepts and future directions. Neurogastroenterol Motil 25(1):4–15

Barbara G, Stanghellini V, De Giorgio R, Cremon C, Cottrell GS, Santini D et al (2004) Activated mast cells in proximity to colonic nerves correlate with abdominal pain in irritable bowel syndrome. Gastroenterology 126(3):693–702

Bearcroft CP, Perrett D, Farthing MJ (1996) 5-hydroxytryptamine release into human jejunum by cholera toxin. Gut 39(4):528–531

Brenner DM, Moeller MJ, Chey WD, Schoenfeld PS (2009) The utility of probiotics in the treatment of irritable bowel syndrome: a systematic review. Am J Gastroenterol 104 (4):1033–1049

Camilleri M (2015) Bile acid diarrhea: prevalence, pathogenesis, and therapy. Gut Liver 9 (3):332–339

Canani RB, Costanzo MD, Leone L, Pedata M, Meli R, Calignano A (2011) Potential beneficial effects of butyrate in intestinal and extraintestinal diseases. World J Gastroenterol 17 (12):1519–1528

Chadwick VS, Chen W, Shu D, Paulus B, Bethwaite P, Tie A et al (2002) Activation of the mucosal immune system in irritable bowel syndrome. Gastroenterology 122(7):1778–1783

Chen J, Chia N, Kalari KR, Yao JZ, Novotna M, Soldan MM et al (2016) Multiple sclerosis patients have a distinct gut microbiota compared to healthy controls. Sci Rep 6:28484

Collins J, Borojevic R, Verdu EF, Huizinga JD, Ratcliffe EM (2014) Intestinal microbiota influence the early postnatal development of the enteric nervous system. Neurogastroenterol Motil 26(1):98–107

De Filippo C, Cavalieri D, Di Paola M, Ramazzotti M, Poullet JB, Massart S et al (2010) Impact of diet in shaping gut microbiota revealed by a comparative study in children from Europe and rural Africa. Proc Natl Acad Sci USA 107(33):14691–14696

De Giorgio R, Cogliandro RF, Barbara G, Corinaldesi R, Stanghellini V (2011) Chronic intestinal pseudo-obstruction: clinical features, diagnosis, and therapy. Gastroenterol Clin North Am 40 (4):787–807

Dukowicz AC, Lacy BE, Levine GM (2007) Small intestinal bacterial overgrowth: a comprehensive review. Gastroenterol Hepatol (NY) 3(2):112–122

Esmaili A, Nazir SF, Borthakur A, Yu D, Turner JR, Saksena S et al (2009) Enteropathogenic Escherichia coli infection inhibits intestinal serotonin transporter function and expression. Gastroenterology 137(6):2074–2083

Forsyth CB, Shannon KM, Kordower JH, Voigt RM, Shaikh M, Jaglin JA et al (2011) Increased intestinal permeability correlates with sigmoid mucosa alpha-synuclein staining and endotoxin exposure markers in early Parkinson's disease. PLoS One 6(12):e28032

Fukumoto S, Tatewaki M, Yamada T, Fujimiya M, Mantyh C, Voss M et al (2003) Short-chain fatty acids stimulate colonic transit via intraluminal 5-HT release in rats. Am J Physiol Regul Integr Comp Physiol 284(5):R1269–R1276

Gabbard SL, Lacy BE (2013) Chronic intestinal pseudo-obstruction. Nutr Clin Pract 28 (3):307–316

Gill SR, Pop M, Deboy RT, Eckburg PB, Turnbaugh PJ, Samuel BS et al (2006) Metagenomic analysis of the human distal gut microbiome. Science 312(5778):1355–1359

Gu L, Ding C, Tian H, Yang B, Zhang X, Hua Y et al (2017) Serial frozen fecal microbiota transplantation in the treatment of chronic intestinal pseudo-obstruction: a preliminary study. J Neurogastroenterol Motil 23(2):289–297

Halvorson HA, Schlett CD, Riddle MS (2006) Postinfectious irritable bowel syndrome—a meta-analysis. Am J Gastroenterol 101(8):1894–1899

Hurst NR, Kendig DM, Murthy KS, Grider JR (2014) The short chain fatty acids, butyrate and propionate, have differential effects on the motility of the guinea pig colon. Neurogastroenterol Motil 26(11):1586–1596

Husebye E, Skar V, Hoverstad T, Iversen T, Melby K (1995) Abnormal intestinal motor patterns explain enteric colonization with gram-negative bacilli in late radiation enteropathy. Gastroenterology 109(4):1078–1089

Husebye E, Hellstrom PM, Sundler F, Chen J, Midtvedt T (2001) Influence of microbial species on small intestinal myoelectric activity and transit in germ-free rats. Am J Physiol Gastrointest Liver Physiol 280(3):G368–G380

Jalanka-Tuovinen J, Salojarvi J, Salonen A, Immonen O, Garsed K, Kelly FM et al (2014) Faecal microbiota composition and host-microbe cross-talk following gastroenteritis and in postinfectious irritable bowel syndrome. Gut 63(11):1737–1745

Kelly CP, LaMont JT (2008) Clostridium difficile—more difficult than ever. N Engl J Med 359 (18):1932–1940

Levy M, Thaiss CA, Zeevi D, Dohnalova L, Zilberman-Schapira G, Mahdi JA et al (2015) Microbiota-modulated metabolites shape the intestinal microenvironment by regulating NLRP6 inflammasome signaling. Cell 163(6):1428–1443

Levy M, Blacher E, Elinav E (2017) Microbiome, metabolites and host immunity. Curr Opin Microbiol 35:8–15

Ley RE (2010) Obesity and the human microbiome. Curr Opin Gastroenterol 26(1):5–11

Ley RE, Peterson DA, Gordon JI (2006) Ecological and evolutionary forces shaping microbial diversity in the human intestine. Cell 124(4):837–848

Liebregts T, Adam B, Bredack C, Roth A, Heinzel S, Lester S et al (2007) Immune activation in patients with irritable bowel syndrome. Gastroenterology 132(3):913–920

Luna RA, Oezguen N, Balderas M, Venkatachalam A, Runge JK, Versalovic J et al (2017) Distinct microbiome-neuroimmune signatures correlate with functional abdominal pain in children with autism spectrum disorder. Cell Mol Gastroenterol Hepatol 3(2):218–230

Mackie RI, Sghir A, Gaskins HR (1999) Developmental microbial ecology of the neonatal gastrointestinal tract. Am J Clin Nutr 69(5):1035S–1045S

Marsh A, Eslick EM, Eslick GD (2016) Does a diet low in FODMAPs reduce symptoms associated with functional gastrointestinal disorders? A comprehensive systematic review and meta-analysis. Eur J Nutr 55(3):897–906

McVey Neufeld KA, Mao YK, Bienenstock J, Foster JA, Kunze WA (2013) The microbiome is essential for normal gut intrinsic primary afferent neuron excitability in the mouse. Neurogastroenterol Motil 25(2):183–e88

Medani M, Collins D, Docherty NG, Baird AW, O'Connell PR, Winter DC (2011) Emerging role of hydrogen sulfide in colonic physiology and pathophysiology. Inflamm Bowel Dis 17 (7):1620–1625

Miazga A, Osinski M, Cichy W, Zaba R (2015) Current views on the etiopathogenesis, clinical manifestation, diagnostics, treatment and correlation with other nosological entities of SIBO. Adv Med Sci 60(1):118–124

Mizuno S, Masaoka T, Naganuma M, Kishimoto T, Kitazawa M, Kurokawa S et al (2017) Bifidobacterium-rich fecal donor may be a positive predictor for successful fecal microbiota transplantation in patients with irritable bowel syndrome. Digestion 96(1):29–38

Pimentel M, Lin HC, Enayati P, van den Burg B, Lee HR, Chen JH et al (2006) Methane, a gas produced by enteric bacteria, slows intestinal transit and augments small intestinal contractile activity. Am J Physiol Gastrointest Liver Physiol 290(6):G1089–G1095

Pimentel M, Lembo A, Chey WD, Zakko S, Ringel Y, Yu J et al (2011) Rifaximin therapy for patients with irritable bowel syndrome without constipation. N Engl J Med 364(1):22–32

Pimentel M, Mathur R, Chang C (2013) Gas and the microbiome. Curr Gastroenterol Rep 15 (12):356

Qin J, Li R, Raes J, Arumugam M, Burgdorf KS, Manichanh C et al (2010) A human gut microbial gene catalogue established by metagenomic sequencing. Nature 464(7285):59–65

Quigley EM (2011) Microflora modulation of motility. J Neurogastroenterol Motil 17(2):140–147

Quigley EM, Quera R (2006) Small intestinal bacterial overgrowth: roles of antibiotics, prebiotics, and probiotics. Gastroenterology 130(2 Suppl 1):S78–S90

Rajilic-Stojanovic M, Biagi E, Heilig HG, Kajander K, Kekkonen RA, Tims S et al (2011) Global and deep molecular analysis of microbiota signatures in fecal samples from patients with irritable bowel syndrome. Gastroenterology 141(5):1792–1801

Rakoff-Nahoum S, Paglino J, Eslami-Varzaneh F, Edberg S, Medzhitov R (2004) Recognition of commensal microflora by toll-like receptors is required for intestinal homeostasis. Cell 118 (2):229–241

Ratuapli SK, Ellington TG, O'Neill MT, Umar SB, Harris LA, Foxx-Orenstein AE et al (2012) Proton pump inhibitor therapy use does not predispose to small intestinal bacterial overgrowth. Am J Gastroenterol 107(5):730–735

Ringel Y (2017) The gut microbiome in irritable bowel syndrome and other functional bowel disorders. Gastroenterol Clin North Am 46(1):91–101

Roland BC, Ciarleglio MM, Clarke JO, Semler JR, Tomakin E, Mullin GE et al (2015) Small intestinal transit time is delayed in small intestinal bacterial overgrowth. J Clin Gastroenterol 49(7):571–576

Rossen NG, MacDonald JK, de Vries EM, D'Haens GR, de Vos WM, Zoetendal EG et al (2015) Fecal microbiota transplantation as novel therapy in gastroenterology: a systematic review. World J Gastroenterol 21(17):5359–5371

Rowland IR (2009) The role of the gastrointestinal microbiota in colorectal cancer. Curr Pharm Des 15(13):1524–1527

Saito YA, Schoenfeld P, Locke GR 3rd (2002) The epidemiology of irritable bowel syndrome in North America: a systematic review. Am J Gastroenterol 97(8):1910–1915

Sampson TR, Debelius JW, Thron T, Janssen S, Shastri GG, Ilhan ZE et al (2016) Gut microbiota regulate motor deficits and neuroinflammation in a model of Parkinson's disease. Cell 167 (6):1469–1480.e12

Scheperjans F, Aho V, Pereira PA, Koskinen K, Paulin L, Pekkonen E et al (2015) Gut microbiota are related to Parkinson's disease and clinical phenotype. Mov Disord 30(3):350–358

Sekirov I, Russell SL, Antunes LC, Finlay BB (2010) Gut microbiota in health and disease. Physiol Rev 90(3):859–904

Shen L, Liu L, Ji HF (2017) Alzheimer's disease histological and behavioral manifestations in transgenic mice correlate with specific gut microbiome state. J Alzheimers Dis 56(1):385–390

Sikander A, Rana SV, Prasad KK (2009) Role of serotonin in gastrointestinal motility and irritable bowel syndrome. Clin Chim Acta 403(1–2):47–55

Soret R, Chevalier J, De Coppet P, Poupeau G, Derkinderen P, Segain JP et al (2010) Short-chain fatty acids regulate the enteric neurons and control gastrointestinal motility in rats. Gastroenterology 138(5):1772–1782

Stotzer PO, Bjornsson ES, Abrahamsson H (1996) Interdigestive and postprandial motility in small-intestinal bacterial overgrowth. Scand J Gastroenterol 31(9):875–880

Thabane M, Kottachchi DT, Marshall JK (2007) Systematic review and meta-analysis: the incidence and prognosis of post-infectious irritable bowel syndrome. Aliment Pharmacol Ther 26(4):535–544

Tian H, Ge X, Nie Y, Yang L, Ding C, McFarland LV et al (2017) Fecal microbiota transplantation in patients with slow-transit constipation: a randomized, clinical trial. PLoS One 12(2): e0171308

Tornblom H, Lindberg G, Nyberg B, Veress B (2002) Full-thickness biopsy of the jejunum reveals inflammation and enteric neuropathy in irritable bowel syndrome. Gastroenterology 123 (6):1972–1979

Unger MM, Spiegel J, Dillmann KU, Grundmann D, Philippeit H, Burmann J et al (2016) Short chain fatty acids and gut microbiota differ between patients with Parkinson's disease and age-matched controls. Parkinsonism Relat Disord 32:66–72

Vantrappen G, Janssens J, Hellemans J, Ghoos Y (1977) The interdigestive motor complex of normal subjects and patients with bacterial overgrowth of the small intestine. J Clin Invest 59 (6):1158–1166

Villarreal AA, Aberger FJ, Benrud R, Gundrum JD (2012) Use of broad-spectrum antibiotics and the development of irritable bowel syndrome. WMJ 111(1):17–20

Whelan K (2011) Probiotics and prebiotics in the management of irritable bowel syndrome: a review of recent clinical trials and systematic reviews. Curr Opin Clin Nutr Metab Care 14 (6):581–587

Wu S, Yi J, Zhang YG, Zhou J, Sun J (2015) Leaky intestine and impaired microbiome in an amyotrophic lateral sclerosis mouse model. Physiol Rep 3(4) pii: e12356

Zhang YG, Wu S, Yi J, Xia Y, Jin D, Zhou J et al (2017) Target intestinal microbiota to alleviate disease progression in amyotrophic lateral sclerosis. Clin Ther 39(2):322–336

Chapter 7
Altered Microbiota and Their Metabolism in Host Metabolic Diseases

Beng San Yeoh and Matam Vijay-Kumar

Abstract The mammalian intestine harbors trillions of bacteria collectively known as the gut microbiota. This menagerie of gut microbes performs diverse metabolic roles, many of which are prerequisites to maintaining their symbiotic relationship with the host. Recent years have seen a surge in studies underscoring the profound consequences of microbiota dysregulation and dysbiosis in promoting metabolic disorders. This chapter examines several key concepts and potential mechanisms that accentuate the link between gut microbiome and metabolic diseases. Accumulated data from a variety of animal and human studies indicate that a dysbiotic microbiota can play a key role in the instigation of metabolic diseases via the following potential mechanisms: increasing calorie extraction; producing obesogenic metabolites; causing metabolic endotoxemia-induced, low-grade, chronic inflammation; and reprogramming the host inflammatory/metabolic responses to favor the development of metabolic syndrome.

7.1 Gut Microbiota: A Neglected Organ Within an Organ

The consortium of microbes inhabiting our body, including the skin and mucosal surfaces, is collectively known as the "microbiota." Out of the 52 bacterial phyla known to exist on Earth (Rappe and Giovannoni 2003), only five to seven phyla—predominantly Firmicutes, Bacteroidetes, Proteobacteria, and Actinobacteria—are capable of colonizing the human gut. The most prevalent bacterial phyla in the gut are the Bacteroidetes and Firmicutes (composed of mostly obligate anaerobes that are difficult to culture), whereas Proteobacteria constitutes only a minor community, despite the latter's predominance in most terrestrial environments (Costello et al. 2009; Human Microbiome Project 2012). These microbes colonize the gut

B.S. Yeoh • M. Vijay-Kumar (✉)
Department of Nutritional Sciences, The Pennsylvania State University, University Park, PA, USA
e-mail: mvk13@psu.edu

J. Sun, P.K. Dudeja (eds.), *Mechanisms Underlying Host-Microbiome Interactions in Pathophysiology of Human Diseases*, Physiology in Health and Disease,
https://doi.org/10.1007/978-1-4939-7534-1_7

rapidly after birth and accrue to tens of trillions, a sum that was previously thought to vastly outnumber the host cells by at least tenfold (Hooper et al. 2002). After accounting for the host hematopoietic cells, recent estimates now estimate the ratio between bacteria (~3.8×10^{13}) and host cells (~3.0×10^{13}) to be much closer to 1:1 (Sender et al. 2016a, b; Bianconi et al. 2013). Yet, compared with the human genome, the collective genome of the gut microbiota is enumerated to be 150 times larger, which has led it to be considered our "second genome" (Grice and Segre 2012). The discovery that the sequences in bacterial 16S rRNA genes can be exploited to discern microbial phylogeny (Zuckerkandl and Pauling 1965; Woese 1987; Woese and Fox 1977) has resulted in the development of next-generation sequencing technologies capable of cataloging bacteria taxonomically from phyla to species. Aided by the more sophisticated metagenomics sequencing techniques (von Mering et al. 2007; Riesenfeld et al. 2004), studies to date have characterized more than 1000 bacterial species from the human gut microbiota (Qin et al. 2010).

Despite the heterogeneity in the gut microbiota composition among individuals, a collage of dissimilar species converges to form a "core" microbiome, which performs metabolic functions that are essential to the host (Turnbaugh et al. 2009; Qin et al. 2010; Human Microbiome Project 2012). Such interplay between the host and its microbiota goes beyond simple symbiosis, with terms such as "holobiont," "composite organism," and "superorganism" now used to describe their metabolic inter-dependence. The events preceding the development of a functional "core" microbiota are not well-elucidated; the current paradigm asserts that the neonatal gut is first colonized by pioneer species, particularly *Escherichia coli* from the phylum Proteobacteria, shortly after birth (Koren et al. 2012; Palmer et al. 2007). Consumption of oxygen by these initial settlers would establish an anaerobic niche, thus heralding the subsequent colonization by strict anaerobes from the phyla Firmicutes and Bacteroidetes (Secher et al. 2016). Contrary to long-standing dogma, human milk is not sterile, but is a rich source of diverse bacteria (predominantly *Staphylococcus* spp., *Streptococcus* spp. and *Pseudomonas* spp.) for infants. Based on an estimated median bacterial load of 10^6 cells/ml in healthy women, it is projected that exclusively breast-fed infants consume about 7–8 billion bacteria per day (Boix-Amoros et al. 2016). The composition of the neonatal microbiota is relatively plastic and amenable to changes during early development, but gradually stabilizes and is thought to maintain many of its characteristics throughout the life of the host. Even so, the composition of the adult microbiota can be rapidly altered in response to changes in dietary habits and aging (David et al. 2014; Faith et al. 2011; Wu et al. 2011; Claesson et al. 2012; Yatsunenko et al. 2012).

Substantial alteration of the microbial ecology, however, could result in the loss of functional diversity within the core microbiota and give rise to so-called "dysbiosis." Microbiota dysbiosis denotes any considerable imbalance in microbial ecology that leads to a negative host response, including geographical dislocation (e.g., growth of colonic bacteria in the small intestine) and over-representation or depletion of a particular species, family or phyla; although not all altered microbiota can cause a phenotypic effect. For instance, antibiotics use during the first 6 months of life in human infants can perturb the maturation of the gut microbiota and has been associated with the risk of developing childhood obesity (Trasande et al. 2013). Early-life exposure to antibiotics in mice has also been shown to alter the host–microbe metabolic programming into favoring weight gain and adiposity later in life (Cox et al. 2014; Cho et al. 2012). These provocative findings echo the agricultural use of antibiotics to promote growth in livestock (Gaskins et al. 2002), which further underpins the profound effects of gut dysbiosis on host metabolism.

7.2 Gut Dysbiosis: Tipping the Balance Toward Metabolic Syndrome

Metabolic syndrome denotes abnormalities in parameters such as hypertension, hyperlipidemia, and dysglycemia, which lead to the development of type II diabetes and cardiovascular disease (CVD). The epidemic of metabolic syndrome and obesity is growing in many developed and developing countries. In the USA, the prevalence of obesity in adults has increased more than 75% since 1980, with more than half of the US population classified as overweight (Flegal et al. 2010). It is estimated that 86% of the US population will be either obese or overweight by 2030 (Wang et al. 2008). The obesity crisis has been largely ascribed to overnutrition and a sedentary lifestyle; it was not until the recent two decades that studies began to recognize the gut microbiota as a new player in the development of metabolic diseases. Compared with the stability of the genome, the microbiome is vulnerable and can be altered much more rapidly and substantially (Yatsunenko et al. 2012) owing to the influence of modern dietary habits, food processing, hygienic practices, and antibiotics use. Population-wide alterations in microbiota composition, with the slow disappearance of *H. pylori* being one of many examples of this phenomenon, may represent a plausible factor in the epidemic increase in metabolic and other microbiota-associated diseases (Cover and Blaser 2009).

The pioneering studies by Gordon and colleagues have unraveled significant differences in the microbiota composition between lean and obese humans, with the latter displaying increases in Firmicutes and a concomitant decrease in Bacteroidetes (Ley et al. 2006). Feeding mice with an obesogenic high-fat, low-fiber diet recapitulates a similar increase in the Firmicutes/Bacteroidetes ratio, a phenomenon now described as a shift in the "obese-type" microbiota (Turnbaugh et al. 2006). The disparity between the presumed "obese" and "lean" microbiota is further accentuated by a distinctive reduction in bacterial richness and diversity (Le Chatelier et al. 2013) and an expansion of Proteobacteria (Shin et al. 2015) in the "obese" microbiota. One of the key pieces of evidence supporting the existence of such an obesogenic microbiota came from a seminal study comparing the gut microbiota from twins discordant for obesity (Ridaura et al. 2013).

The transfer of fecal microbiota from obese twins, but not their lean counterparts, into germ-free (GF) mice was sufficient to transmit the obesity phenotype, notably the increase in adiposity (Ridaura et al. 2013). However, the increase in adiposity in mice harboring the obese twins' microbiota was prevented upon co-housing with mice that had the microbiota from the lean twins; this observation correlated strongly with the transfer of several species of Bacteroidetes from the "lean" to the "obese" group of mice (Ridaura et al. 2013).

Recent studies have begun to elucidate the extent to which gut dysbiosis can be promoted by various societal changes, especially those pertaining to intake of processed foods. Pivotal work by Gewirtz and colleagues demonstrates that feeding dietary emulsifiers, namely carboxymethylcellulose and polysorbate-80, to mice substantially alters their gut microbiota and reduces its distance from the epithelia by more than twofold (Chassaing et al. 2015a). Emulsifiers are detergent-like chemicals widely thought to be generally safe as artificial food preservatives; yet, their unexpected property in promoting the breakdown of the mucus layer and provoking microbiota encroachment and dysbiosis implies that such chemicals might not be entirely harmless. The emulsifiers-fed mice developed a metabolic syndrome that is microbiota-dependent and transmissible to non-emulsifier-fed mice via microbiota transplant (Chassaing et al. 2015a). Another food additive, saccharin (an artificial sweetener), has likewise been shown to promote gut dysbiosis and microbiota-dependent glucose intolerance in both humans and mice (Suez et al. 2014). Although the dysbiosis-inducing effects of saccharin are not well-understood, Suez et al. documented that the obesity in saccharin-fed mice is associated with an increase in gut bacterial glycan-degrading pathways (Suez et al. 2014), which may enhance the production of short-chain fatty acids (SCFAs) and provide the mice with extra calories (Turnbaugh et al. 2006).

Gut dysbiosis may be perceived to be pathological; however, this is not a foregone conclusion. A study demonstrated that mice deficient in dual-specificity phosphatase 6 (DUSP6) harbored a distinct set of gut microbiota; yet, such altered microbiota turned out to be beneficial in conferring protection to the host against high-fat diet (HFD)-induced obesity (Ruan et al. 2016). The anti-obesogenic microbiota from DUSP6-deficient (*Dusp6*KO) mice is transmissible via fecal transplants and mediates its effects by significantly increasing the energy expenditure in the recipient GF wild-type (WT) mice. In another instance, a study documents that a "dysbiotic" microbiota, despite its elevated Firmicutes and low diversity, may have beneficial effects against metabolic diseases (Nicolas et al. 2017). Instead of using GF mice as recipients, Nicolas et al. opted to use conventional mice and demonstrated that these mice developed resistance to diet-induced obesity upon acquiring the "dysbiotic" microbiota from obese mice. Such counter-intuitive findings underscore that gut dysbiosis is not necessarily pathological, but may be contextually dependent on the beneficial/adverse metabolic adaptation of the dysbiotic microbiota and the host.

Studies described herein are by no means an exhaustive example of studies implicating the role of gut microbiota in metabolic disorders. Nonetheless, these studies and many others in the field provide key insights into the potential mechanisms, which could explain, in part or wholly, the way in which gut microbes can modulate the host's metabolic well-being. Potential mechanisms by which the microbiota influences host metabolism can be broadly categorized as either: the consequences of microbial metabolism (e.g., increased calorie extraction from ingested foods, production of obesogenic metabolites) or host–bacteria interaction mediated via activation of pathogen recognition receptors (PRR; e.g., TLR2, TLR4, TLR5) by microbial ligands resulting in altered inflammatory and metabolic responses (e.g., low-grade inflammation, metabolic endotoxemia, microbial encroachment; Fig. 7.1). This chapter covers several key concepts that are prevalent in the field, and provides examples of the extent to which specific bacteria and/or their metabolites have been shown to affect host metabolism.

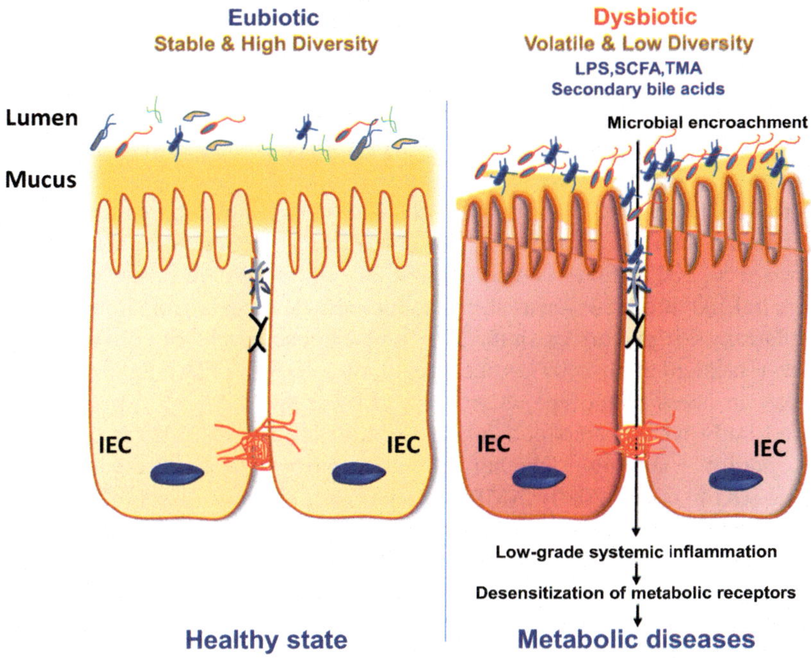

Fig. 7.1 Potential mechanisms by which gut microbial dysbiosis can result in metabolic diseases. A healthy gut is presumed to be capable of restricting the localization of gut microbiota to the luminal side of the intestine. However, substantial alteration in the gut ecology may lead to gut microbiotal dysbiosis, which is characterized by reduced species richness and diversity, and increased microbial encroachment on the mucin layer. Increased translocation of bacteria and exposure to microbial product [e.g., lipopolysaccharide (LPS), short-chain fatty acids (SCFAs), trimethylamine (TMA), secondary bile acids] could promote low-grade chronic inflammation and/or reprogram the host metabolic state into favoring the development of metabolic diseases

7.3 Microbiota-Mediated Alterations of the Energy Harvest, Expenditure, and Storage in the Host

The exposure to subclinical or low doses of antibiotics, as discussed earlier, could result in dysbiosis, which in turn drives the development of metabolic disorders. Yet, mice raised under GF conditions are seemingly resistant to HFD-induced obesity (Backhed et al. 2007). These paradigms are not contradictory, but in fact highlight the notion that complete and partial depletion of the gut microbes differentially program the host metabolism. Backhed et al. postulated that the resistance to diet-induced obesity observed in GF mice could be attributable to the constitutive activation of adenosine monophosphate-activated protein kinase (AMPK) in the liver and muscles (Backhed et al. 2007). As the rheostat for sensing energy status, the elevated AMPK activates ATP-generating catabolic pathways, promotes fatty acid oxidation in mitochondria, and reduces hepatic glycogen synthase activity (Hardie et al. 1998). The introduction of gut microbiota into GF mice abrogated their anti-obesity phenotype, and this is correlated with the decrease in AMPK activity in the liver (Backhed et al. 2007). Moreover, GF mice colonized with "obese" microbiota displayed increases in total body fat and calorie intake, and a greater capacity for energy harvest (Turnbaugh et al. 2006).

Microbiota transplantation into GF mice was reported to suppress their production of fasting-induced adipocyte factor (FIAF; alias angiopoietin-like protein 4) (Backhed et al. 2007). FIAF is a potent inhibitor of lipoprotein lipase, whereby the decrease in FIAF production was shown to increase de novo hepatic lipogenesis and accumulation of triglycerides in the liver and adipose tissues of conventionalized GF mice (Backhed et al. 2004). Accordingly, mice lacking FIAF have been shown to display an obese phenotype, characterized by reduced energy expenditure, and increased body weight and food intake (Kim et al. 2010). FIAF also modulates AMPK activity, although whether FIAF increases (Backhed et al. 2007) or decreases (Kim et al. 2010) AMPK activity remains a subject of contention. These studies nevertheless implicated the active role of gut microbiota in modulating host energy extraction, and glucose and lipid metabolism. However, the microbiota-associated factor(s) and the pathway(s) that influence FIAF and AMPK still need to be further investigated.

Recent evidence suggests that gut microbiota could also modulate host energy metabolism by impeding the emergence of beige adipocytes. Unlike the fat-storing white adipocytes, the beige and brown adipocytes contain more mitochondria and uncoupling proteins to dissociate oxidative phosphorylation and burn fat and glucose to generate heat (Cannon and Nedergaard 2004). When compared to conventional mice, both GF mice and antibiotics-treated mice have substantially more beige adipose tissue, which is correlated with improved glucose tolerance, insulin sensitivity, and resistance to diet-induced obesity (Suarez-Zamorano et al. 2015). This phenomenon is reversible, as re-colonizing the GF mice with gut microbiota induces the whitening of beige adipose tissue, abrogates the anti-obesity phenotype of GF mice, and reverts their metabolic phenotype (Suarez-Zamorano et al. 2015). The potential mechanisms by which the absence of gut microbiota alters the white/brown adipocyte ratio are likely to involve interactions between the host and the metabolites derived from gut bacteria. Such host-bacteria interaction may be reciprocal, as one study suggests that the induction of beige/brown adipose tissue activity due to cold exposure can also shape the gut microbiota that, in turn, helps to promote adaptive thermogenesis (Worthmann et al. 2017). Taken together, these studies collectively lend support to the prevailing hypothesis that gut microbiota are modulators of their host energy harvest, storage, and expenditure.

7.4 Microbiota Generates Metabolites Associated with Metabolic Diseases

A great deal of researches over the past decades has shed light on the various contributions of the gut microbiota to host nutrition. For instance, the gut microbiota constantly provides many important nutrients such as thiamine, riboflavin, biotin, vitamin K, vitamin B6 and vitamin B12 to their host (Said 2011). Dubbed as the "virtual endocrine organ", the gut microbiota is also a major source of hormones and secondary metabolites, including serotonin, tryptophan, indoles, dopamine, norepinephrine, histamine, gamma-aminobutyric acid and extracellular ATP (Lyte 2013; Wall et al. 2014; Madsbad 2014; Samuel et al. 2008; Iwase et al. 2010). The fermentation of dietary fibers by gut microbes generates large quantities of SCFAs (Cummings et al. 1987) that can provide an additional source of calories to the host. SCFAs could also affect the host's ability to store energy as fat and to respond to energy intake by promoting the release of gut hormones, such as peptide YY (PYY) (Samuel et al. 2008; Cani and Delzenne 2009) and the incretin glucagon-like peptide (GLP)-1 (Cani et al. 2007b, 2009; Barrera et al. 2011). The bidirectional communication and inter-relationship between host and microbes has been referred to as the gut–brain axis (Field et al. 2010).

In silico techniques have identified a total of 3449 distinct metabolic reactions occurring in the gut, of which 1267 are unique to the gut microbiota, 940 are unique to the host, and 1142 are present in both (Sridharan et al. 2014). Metabolomics assessment revealed that 77 out of the 179 metabolites detected in luminal contents were enriched in mice with an intact microbiota, when compared to GF mice (Matsumoto et al. 2012). Gut bacteria-derived metabolites can traverse into systemic circulation and are estimated to contribute ~10% of the total pool of metabolites present in mammalian blood (Wikoff et al. 2009). Therefore, it is not surprising that microbiota substantially contribute to host metabolism and thus any significant microbial alterations or dysbiosis would result in metabolic diseases. Such outcomes have been associated with the capacity of the microbiota to facilitate the biotransformation of environmental chemicals into a number of obesogenic and diabetogenic compounds (Snedeker and Hay 2012), several of which will be discussed below.

7.4.1 Metabolic Functions of SCFAs

The major SCFAs derived from microbial fermentation of dietary fibers are acetate, propionate and butyrate (Topping and Clifton 2001; Wong et al. 2006). The Bacteroidetes mainly produces acetate and propionate, whereas the Firmicutes produces butyrate as the primary metabolic end product (den Besten et al. 2013b; Macfarlane and Macfarlane 2003); although, the acetate-producing and butyrate-producing bacteria are also capable of cross-feeding one another (Wrzosek et al. 2013). In a prior study, SCFAs were shown to protect against diet-induced obesity; concomitant dietary supplementation of propionate and butyrate to mice completely blocked HFD-induced weight gain, while acetate moderately suppressed the excess weight gain (Lin et al. 2012). In accord with this notion, a study from our group also found that the lack of fermentable fiber (and thus SCFAs) in Western diets may, in part, underlie their obesogenic and inflammatory effects (Chassaing et al. 2015b). On a similar note, we observed the inclusion of dietary soluble fiber substantially protected mice against HFD-induced metabolic syndrome and obesity (Chassaing et al. 2015b).

The beneficial effects of SCFAs are, in part, facilitated by the SCFA receptors G-protein-coupled receptor (GPR) 41 and GPR43 (alias free fatty acid receptor [FFAR] 3 and FFAR2 respectively), which are expressed on enteroendocrine cells in the gut (Nohr et al. 2013). Intriguingly, genetic deficiency of GPR41 promotes obesity in mice and correlates with a reduction in energy expenditure (Bellahcene et al. 2013). GPR43-deficient mice likewise displayed an obese phenotype, whereas GPR43-overexpressing mice remained lean even when fed an HFD (Kimura et al. 2013). The interaction between SCFAs and their receptors is thought to exert an anorexic effect in mice by inducing the release of appetite-regulating gut hormones, namely GLP-1, PYY, and amylin (Samuel et al. 2008; Cani and Delzenne 2009; Cani et al. 2007b, 2009; Barrera et al. 2011). Other potential mechanisms have also been discussed, including the ability of SCFAs to activate peroxisome proliferator-activated receptor gamma and induce the switch from lipogenesis to fat oxidation (den Besten et al. 2015). The "anti-obesity" properties of propionate and butyrate have also been associated with their inhibition of lipolysis and insulin-stimulated de novo lipogenesis in rats (Sekirov et al. 2010).

Despite their acclaimed anti-obesity properties, SCFAs have also been shown to contribute to the development of metabolic disorders, akin to a double-edged sword. It should be noted that SCFAs themselves can serve as substrates for host energy metabolism (den Besten et al. 2013b). In humans, the SCFAs provide approximately 10% of the daily calorie requirements, considering that the average daily diet in Western societies yields approximately 300–600 mmol SCFAs/day (Bergman 1990; Fernandes et al. 2014; Royall et al. 1990). Most SCFAs are metabolized by intestinal colonocytes (Zambell et al. 2003), although a fraction of SCFAs can be transported into portal vein circulation and then to the liver (den Besten et al. 2013b; Bloemen et al. 2009). By using stable isotopes of SCFAs, den Besten et al. (2013a) demonstrated that colonic SCFAs can indeed reach the liver and participate in hepatic de novo lipogenesis. Their study further suggests that acetate and butyrate are lipogenic, whereas propionate is gluconeogenic. This notion is consistent with another study, which suggests that rectal-infused acetate and propionate might be metabolized into serum triglycerides and cholesterol in humans (Wolever et al. 1989).

In our study with *Tlr5*KO mice, which display microbiota-dependent development of metabolic syndrome (Vijay-Kumar et al. 2010), we observed that these mice exhibited decreased cecal oligosaccharides and elevated SCFAs compared with their WT littermates (Singh et al. 2015). Such observations were not seen in antibiotics-treated or GF *Tlr5*KO mice, suggesting that the increased SCFAs in the conventional *Tlr5*KO mice might be derived from the gut microbiota. Intriguingly, administration of SCFAs to *Tlr5*KO mice exacerbated their metabolic syndrome, i.e., increased body weight and fat pad mass, dysglycemia, insulin resistance, and elevation of serum total cholesterol and triglycerides. By feeding ^{13}C-labeled acetate, we demonstrated that gut-derived SCFAs were able to reach the liver and become incorporated into the hepatic and plasma palmitate (C16:0) and triglycerides to a greater extent in *Tlr5*KO mice than in WT mice. We hypothesized that the utilization of SCFAs might be likely to be dependent on the de novo lipogenesis machinery, such as the hepatic acetyl CoA carboxylase (ACC) and stearoyl-CoA desaturase-1 (SCD1), which are elevated in *Tlr5*KO mice. In accordance with this notion, liver-specific deletion of SCD1 was sufficient to prevent most of the indices of metabolic syndrome. This study (Singh et al. 2015) lends support to the hypothesis that increased calorie extraction by gut microbes, via provision of SCFAs, could be one of the mechanisms by which the gut microbiota can promote metabolic syndrome in susceptible hosts.

7.4.1.1 Acetate

Acetate produced in the gut can be subjected to the following metabolic fates:

(1) Locally utilized as a substrate for de novo lipogenesis by colonocytes (Zambell et al. 2003),
(2) Converted into butyrate by gut bacteria (den Besten et al. 2013a),
(3) Oxidized via the tricarboxylic acid (TCA) cycle in the liver, or
(4) Used as a substrate for synthesis of cholesterol, ketone bodies, and long-chain fatty acids (den Besten et al. 2013a). Under starving conditions, the liver can also metabolize fatty acid stores to produce ketone bodies and acetate. Regardless of whether it is derived from the gut or endogenously produced in the liver, acetate can be delivered to extrahepatic tissue via systemic circulation, where it can be utilized in various metabolic processes, including the TCA cycle, fatty acid, and cholesterol biosynthesis (Ballard 1972; Skutches et al. 1979; Yamashita et al. 2007; Shoaie et al. 2013). Apart from its use in lipogenic processes, acetate also performs various functions via its interaction with GPR41 and GPR43. Activation of GPR41 and GPR43 by acetate, for instance,

could result in increased local formation of pancreatic islets, but this compensatory expansion can lead to beta cell dysfunction in the obese and type 2 diabetic mice (Tang et al. 2015).

In one recent study on rats, Shulman and colleagues uncover a novel neurological link between acetate and increases in appetite via the parasympathetic nervous system (Perry et al. 2016). The rats fed an HFD displayed a substantial increase in whole-body turnover of acetate, and had higher concentrations of plasma and fecal acetate, which correlated with higher glucose-stimulated insulin secretion (GSIS; a variant of the glucose tolerance test). The infusion of acetate into rats fed the control diet increased their GSIS to levels comparable with rats fed an HFD. However, such acetate-induced GSIS can be prevented by either surgically severing the vagus nerves or by administering atropine, an agent that inhibits the parasympathetic nervous system. Chronic infusion of acetate in rats over 10 days sustained their increase in acetate turnover and GSIS, culminating into a positive feedback loop promoting the secretion of insulin and ghrelin (alias the "hunger" hormone) that resulted in insulin resistance and hyperphagia respectively. These findings provide strong mechanistic evidence for linking gut microbiota and the onset of obesity, along the gut–brain axis.

7.4.1.2 Propionate

Propionate is another major microbial fermentation metabolite in the human gut with putative health effects (Hosseini et al. 2011). Unlike acetate, propionate is thought to lower lipogenesis and serum cholesterol in both hepatic and nonhepatic tissues in humans (Hosseini et al. 2011), and also reduce the fasting blood glucose and hepatic cholesterol in obese rats (Berggren et al. 1996; Boillot et al. 1995). Studies performed by using radio-labeled propionate demonstrated that propionate is the preferred substrate for hepatic gluconeogenesis (den Besten et al. 2013a). Propionate is first converted into propionyl-CoA by propionate-CoA ligase, and subsequently propionyl-CoA is converted to succinyl-CoA via propionyl-CoA carboxylase, methylmalonyl-CoA epimerase, and methylmalonyl-CoA mutase. Ultimately, succinyl-CoA enters the TCA cycle and is converted to oxaloacetate, the precursor of gluconeogenesis (Bloemen et al. 2010; den Besten et al. 2013a). However, limited data are available to demonstrate the extent to which propionate is partaking in gluconeogenesis and energy metabolism in humans. Concentrations of propionate in portal blood and hepatic venous blood suggest that around 30% of propionate is taken up by the liver (Cummings et al. 1987). In another study, it is estimated that humans use 50% of the propionate as a substrate for hepatic gluconeogenesis (den Besten et al. 2013a).

7.4.1.3 Butyrate

Similar to acetate, butyrate is a major substrate for de novo lipogenesis in colonocytes (Zambell et al. 2003). Butyrate in the gut, at least in part, comes from the inter-conversion between acetate and butyrate mediated by microbial butyryl-CoA:acetate-CoA transferase (den Besten et al. 2013a). Most of the gut-derived butyrate is rapidly metabolized by the colonocytes into the TCA cycle and de novo lipogenesis, in which butyrate is estimated to provide more than 70% energy source (Donohoe et al. 2011). Because of its role as a potent inhibitor of histone deacetylase, butyrate exerts epigenetic effects that can be detrimental to certain cell types, such as the intestinal stem cells, whose proliferation could be inhibited upon exposure (Kaiko et al. 2016). Hence, the rapid utilization of butyrate in the gut can be appropriately regarded as an adaptive metabolic response to protect the proliferative cells sequestered in colonic crypts. The butyrate that reaches the liver, via the hepatic portal vein, is mostly oxidized by the hepatocytes (Bloemen et al. 2010; den Besten et al. 2013a).

7.4.2 Trimethylamine N-oxide

Choline is an essential nutrient that is present in most dietary sources as phosphatidylcholine (e.g., lecithin). Dietary choline is generally metabolized in the liver and incorporated into various biological processes, including the synthesis of lipoproteins, the neurotransmitter acetylcholine, and the phospholipids on cell membranes (Vance 2008). The gut microbiota can convert dietary choline into trimethylamine (TMA; a noxious metabolite known for its strong ammonia-like or "fishy" odor) via the enzymatic activity of microbial TMA lyases (Craciun and Balskus 2012). Microbial-derived TMA is absorbed and delivered to the liver, where TMA can be rapidly detoxified by hepatic flavin monooxygenase 3 (FMO3) to generate trimethylamine-N-oxide (TMAO) for excretion (Bennett et al. 2013). Excessive conversion of dietary choline into TMA by the gut microbiota, however, may reduce the bioavailability of choline to the host, which could impair very low-density lipoprotein secretion, increase accumulation of triglycerides in the liver, and promote hepatic steatosis (Dumas et al. 2006).

In a series of studies, Hazen and colleagues elegantly demonstrated the profound, yet detrimental, effects of TMAO in promoting CVD. Wang et al. reported that dietary supplementation of TMAO or its precursors (choline, betaine, L-carnitine) resulted in the elevation of macrophage receptors CD36 and SR-A1 in the atherosclerosis-prone apolipoprotein E-deficient mice (Wang et al. 2012; Koeth et al. 2013). These receptors potently suppress the reverse cholesterol transport in macrophages, increase cholesterol accumulation, and promote the formation of foam cells, which is one of the earliest cellular hallmarks of atherogenesis (Bremer 1983; Wang et al. 2012). Metabolism of dietary L-carnitine from red meat into

TMAO was likewise suggested to accelerate atherosclerosis in both mice and humans (Koeth et al. 2013). The increase in plasma TMAO and foam cell formation in mice can be prevented by either ablating the gut microbiota (Wang et al. 2011) or inhibiting bacterial TMA lyases (Wang et al. 2015), thus affirming the link between microbial choline metabolism and risk for CVD. It was suggested that high plasma TMAO levels might be associated with an increased risk for major adverse CVD events (Tang et al. 2013, 2014), including enhancing platelet hyperactivity and thrombosis (Zhu et al. 2017).

Despite this, the extent to which choline-rich foods, such as eggs (Miller et al. 2014) and red meat (Koeth et al. 2013), could contribute to TMAO levels and CVD risks remains highly contentious (Meyer et al. 2016). Such controversy is further complicated by the notion that seafoods, which serve as a rich and direct source of TMAO, have not in any way to date been associated with CVD (Landfald et al. 2017). As an alternative explanation to the TMAO-CVD dispute, it was suggested that the hepatic enzyme FMO3 might play a prominent role in promoting CVD in a fashion that is partially independent of TMAO. Warrier et al. found that FMO3 suppresses nonbiliary macrophage reverse cholesterol transport, but promotes increased levels of biliary cholesterol and its absorption from the intestines (Warrier et al. 2015). The knockdown of FMO3 in mice was reported to stimulate macrophage reverse cholesterol transport and improve hepatic cholesterol balance. However, this also resulted in promoting endoplasmic reticulum stress and inflammation, which were postulated to be due to the dampening of liver X receptor activation. Intriguingly, the administration of broad-spectrum antibiotics, but not supplementation of TMAO, was able to correct the imbalance in the cholesterol metabolism of FMO3-deficient mice. These findings suggest that, although the observed metabolic phenotype is microbiota-dependent, the underlying mechanisms may involve microbial metabolites other than TMAO. Warrier et al. asserted that microbial-derived TMA, whose levels accumulated because of the loss of FMO3, may be a likely candidate that could be further studied for its role in promoting FMO3/TMAO-associated CVD risks.

More recently, the gut microbiota-driven TMA/FMO3/TMAO pathway has been linked to obesity and energy metabolism. Schugar et al. reported a positive correlation between high plasma TMAO levels with increased body weight, fat mass, and adiposity in mice fed a high-fat and high-sucrose diet (Schugar et al. 2017). In agreement with animal studies, humans with type 2 diabetes mellitus also displayed significantly higher levels of plasma TMAO than healthy controls (Tang et al. 2017). Intriguingly, the genetic loss of FMO3 substantially protected mice from diet-induced body and adipose weight gain. Further analysis revealed that the absence of FMO3 activity stimulates the energy-storing white adipose tissue to undergo "beiging" and transition into energy-burning beige/brown adipose tissue. Supplementation of TMAO to mice lacking FMO3 reversed the expression of some of the genes involved in the beiging of adipose tissue, although it was not sufficient to abrogate their resistance to diet-induced obesity. These observations are nonetheless consistent with those of another study that reported that depletion of the gut microbiota promotes the beiging of adipose tissue and reduces obesity in mice

(Suarez-Zamorano et al. 2015). Taken together, these key findings provide yet another mechanistic insight into the extent to which gut microbiota can influence obesity by exerting metabolic reprogramming on the adipose tissue.

7.4.3 Microbiota-Derived Secondary Bile Acids

Bile acids are synthesized from cholesterol in the liver, stored in the gallbladder, and released into the small intestine via bile secretion. In addition to their role in facilitating intestinal absorption of dietary lipids and fat-soluble vitamins, bile acids also serve as potent signaling molecules upon re-absorption via the enterohepatic circulation. The primary bile acids synthesized in humans are cholate and chenodeoxycholate (or β-muricholate in mice), and are generally conjugated with glycine or taurine. A fraction of the primary bile acids released into the gut can be deconjugated by bacterial bile acid hydrolases and further metabolized into secondary bile acids (deoxycholate and lithocholate) (Ridlon et al. 2006; Jones et al. 2008). These microbial-derived bile acids have increased hydrophobicity, which enhances the intestinal absorption of dietary lipids and fat-soluble vitamins (Ridlon et al. 2006). Both primary and secondary bile acids can serve as ligands for farnesoid X receptor (FXR; a nuclear receptor) (Makishima et al. 1999; Parks et al. 1999; Wang et al. 1999) and G protein-coupled receptor TGR5 (a cell-surface receptor) (Maruyama et al. 2002; Kawamata et al. 2003). Interestingly, deoxycholate and lithocholate were identified as the more potent agonists for TGR5 compared with the primary bile acids.

The outcome of FXR activation by bile acids is organ-specific: liver FXR increases hepatic conjugation and excretion of bile acids (Pircher et al. 2003; Moschetta et al. 2004), whereas intestinal FXR promotes expression of bile acids binding protein (Grober et al. 1999) and transporters (Landrier et al. 2006; Lee et al. 2006). Activation of intestinal FXR also results in the production of fibroblast growth factor 19 (FGF19 in humans; Fgf15 in mice) as a negative feedback mechanism to inhibit bile acid synthesis in the liver (Holt et al. 2003; Inagaki et al. 2005). The loss of FXR in mice increases hepatic and circulating levels of cholesterol and triglycerides, and reduces the bile acid pools and their excretion (Sinal et al. 2000). The other bile acid receptor, TGR5, plays an equally important role in host metabolism; its activation induces energy expenditure in brown adipose tissue (Watanabe et al. 2006) and GLP-1 production in enteroendocrine cells (Thomas et al. 2009). Accordingly, studies with mice have shown that TGR5 activation helps to confer resistance to obesity (Watanabe et al. 2006; Thomas et al. 2009), whereas deficiency in TGR5 augments the susceptibility to diet-induced obesity (Maruyama et al. 2006).

Tauro-β-muricholic acid (TβMCA) is one such primary bile acid that has been shown in mice to be negatively-associated with obesity because of its feature as an antagonist for intestinal FXR. TβMCA competes with taurocholate to bind with

FXR in the distal gut (Sayin et al. 2013). Yet, unlike taurocholate, TβMCA does not activate FXR (Sayin et al. 2013). The disruption of FXR signaling due to excess TβMCA was shown to be beneficial in mitigating obesity in mice fed an HFD (Li et al. 2013). Genetic loss of FXR in mice also conferred a pronounced resistance against diet-induced obesity (Li et al. 2013). These findings lend support to the emerging paradigm that modulation of the primary/secondary bile acid pool may be one of the myriad mechanisms by which microbiota can influence hepatic cholesterogenesis (via FXR) and energy metabolism (via TGR5), thus possibly contributing to the promotion/mitigation of obesity.

On the other hand, the secondary bile acid deoxycholate was reported to promote the development of obesity-associated hepatocellular carcinoma (HCC) in mice (Yoshimoto et al. 2013). The study employed a chemically induced HCC model, whereby neonatal mice were treated once with the hepatocarcinogen 7,12-dimethylbenz(a)anthracene and then placed on an HFD to potentiate HCC. At the age of 30 weeks, these mice displayed a substantial increase in serum deoxycholate (Yoshimoto et al. 2013), which can be explained by the elevated abundance of *Clostridium* cluster XI populations, which are capable of performing 7-α-dehydroxylation on primary bile acid to generate deoxycholate in the gut (Ridlon et al. 2006). Remarkably, the levels of deoxycholate correlated positively with the severity of HCC, which can be modulated via inhibiting bacterial 7α-dehydroxylation activity or dietary supplementation of deoxycholate (Yoshimoto et al. 2013). Their follow-up study revealed that deoxycholate mediates its pro-tumorigenic effects in synergy with the pro-inflammatory effects of lipoteichoic acid (a cell wall component of Gram-positive bacteria, a ligand for TLR2) to induce cellular senescence and TLR2-dependent inflammation, thus promoting the progression of an obesity-associated HCC.

It is not clear whether the capacity of bile acids to modulate obesity and hepatic inflammation are mutually exclusive, although one recent study was able to recapitulate both phenotypes in mice fed on dietary guar gum (Janssen et al. 2017). Mice fed an HFD supplemented with guar gum (a soluble fiber) displayed an elevated total bile acid pool in the plasma, with substantial increases in TβMCA, taurocholate, and deoxycholate compared with mice fed the control HFD (Janssen et al. 2017). The changes in the bile acid profile correlated well with an increased resistance to diet-induced obesity (i.e., improved body weight, adiposity, glucose tolerance, and reduced hepatic steatosis); yet, at the same time, these mice exhibited enhanced hepatic inflammation and fibrosis (Janssen et al. 2017). Their observation that dietary supplementation with taurocholate could induce hepatic fibrosis in conventional chow-fed mice, but not in microbiota-ablated mice, further implicates the adverse role played by microbiota-driven bile acid dysmetabolism in the liver (Janssen et al. 2017). Although the underlying molecular mechanisms require further study, these findings support the emerging paradigm that microbiota-mediated alterations in the host bile acid pool could partly explain the extent to which gut microbiota can influence hepatic cholesterogenesis, contribute to obesity, and possibly promote the development of pathological liver conditions.

7.5 Loss of Gut Homeostasis Results in Low-Grade Inflammation and Metabolic Dysregulation

The enormous microbial biomass in the gut is enriched with diverse pathogen-associated molecular patterns (microbial ligands recognizable by host PRR) that are likely to elicit adverse inflammatory responses if not regulated appropriately. Accordingly, the host has evolved to compartmentalize the gut bacteria as a strategy to promote tolerance and avert excessive PRR activation in the gut. For instance, the single layer of epithelial lining in the mammalian intestine plays the crucial role of separating the host from the gut microbiota, while allowing selective uptake of nutrients and electrolytes. PRRs, such as the toll-like receptors, are expressed mostly at the basolateral side of the epithelia (Abreu 2010; Gewirtz et al. 2001) to prevent the PRR from being indiscriminately activated by bacteria residing on the apical side of the epithelia. The frontier demarcating the epithelia and the fecal stream is additionally fortified with mucins (most notably MUC2, and to a lesser extent MUC5, MUC6, MUC7, and MUC19) secreted by the goblet cells (McGuckin et al. 2011). These gel-forming mucins form two distinct mucus layers, composed of a sterile thin inner layer and a partially colonized outer layer, whose thickness increases in line with bacterial load in the intestine (Johansson et al. 2008; Atuma et al. 2001). Furthermore, the secretion of trefoil factor 3 and resistin-like molecule β (Relmβ) by goblet cells modulates epithelial restitution (Mashimo et al. 1996) and confers protection against parasitic nematodes that feed on the epithelia (Artis et al. 2004; Herbert et al. 2009), respectively.

The production of antimicrobial factors, primarily by the Paneth cells in the small intestine, further deters gut bacteria from breaching the inner mucus layer. The host antimicrobial arsenal in the gut includes defensins, cathelicidin, lysozymes, and phospholipase A2, which disrupts bacterial cell wall components (Mukherjee et al. 2008). Antimicrobial proteins such as lipocalin 2 and calprotectin can withhold key nutrients (particularly iron, calcium and zinc) from bacteria (Hood and Skaar 2012). Additional antimicrobial peptides in the gut include the ribonuclease Ang4 (which hydrolyzes bacterial RNA) and RegIIIγ (a C-type lectin that limits association between Gram-positive bacteria and the mucosal surface) (Hooper et al. 2003; Vaishnava et al. 2011). Bacterial adhesion to the epithelia is also substantially prevented by the non-specific immunoglobulin A (IgA) present in the enteric mucus, which promotes the clumping of bacteria (Mantis et al. 2011).

The notion that gut microbial encroachment across the enteric mucus is contributing to obesity has gained much traction in recent studies. The thinning of the mucus layer increases the proximity between the bacteria and the host epithelia, resulting in the translocation of bacteria and their products, which, in turn, promotes low-grade inflammation and metabolic syndrome in mice (Chassaing et al. 2017b). To further explore whether microbiota encroachment is also a feature of metabolic syndrome in humans, Chassaing et al. analyzed the colonic biopsies from obese individuals and found that their microbiota was indeed located much closer to the epithelia than those observed in biopsies from healthy individuals (Chassaing et al.

2017a). The microbiota encroachment in humans was found to be strongly correlated with insulin resistance-associated dysglycemia, but not dyslipidemia (Chassaing et al. 2017a). Other similar themes, such as "small intestinal bacterial overgrowth" (SIBO) and "leaky guts" (impaired gut permeability), have also been explored as potential mechanisms by which gut microbiota could have an impact on host physiology (Wigg et al. 2001). These mechanisms are not likely to be mutually exclusive and may occur in tandem to permit translocation of intestinal bacteria and metabolites into the hepatic portal vein.

7.5.1 Metabolic Endotoxemia: A Mechanistic Link Between Dysregulated Inflammatory Responses and Microbiota-Induced Obesity

Among the gut metabolites associated with metabolic disorders, the bacterial lipopolysaccharide (LPS; a component of the outer membrane of Gram-negative bacteria; a TLR4 ligand) is perhaps the most immunologically potent in inducing chronic inflammation and liver injury. LPS is routinely translocated from the gut into the liver, where it is detoxified by hepatic lipase (Shao et al. 2007) or alkaline phosphatase (Koyama et al. 2002). In pathological conditions, the liver may fail to mediate efficient clearance of LPS, which then gives rise to low-grade inflammation that precedes the steatosis in non-alcoholic fatty liver disease. Mechanistically, this endotoxin binds to LPS-binding protein, MD2, and CD14, forming a complex that activates the TLR4 expressed on Kupffer cells in the liver. Activated TLR4 then upregulates the expression of pro-inflammatory cytokines, such as TNF-α (a well-established proinflammatory cytokine), which induces insulin resistance and inhibits lipoprotein lipase. In addition, the activation of LPS-TLR4 signaling on hepatic stellate cells promotes TGF-β-mediated fibrogenesis, which leads to liver fibrosis and then cirrhosis (Seki et al. 2007). The translocation of sufficient LPS into the systemic circulation to promote metabolic dysregulation has been defined as "metabolic endotoxemia."

By showing that an HFD can induce metabolic endotoxemia, Cani et al. provided one of the most compelling pieces of evidence in support of the causal role of the gut microbiota in promoting metabolic diseases. Their seminal study demonstrated that high-fat feeding increases intestinal Gram-negative bacterial load, thus increasing the amount of luminal LPS available to leak into the systemic circulation (Cani et al. 2007a). The high-fat feeding also resulted in classical symptoms of metabolic inflammation, including elevated macrophage infiltration into adipose tissue, body weight gain, and diabetes. Four weeks of continuous subcutaneous infusion of LPS (from *E. coli* 055:B5) recapitulated the hallmarks of metabolic syndrome (increased fasting glucose, insulin resistance, obesity, and steatosis) observed in HFD-fed mice (Cani et al. 2007a). The obesity-promoting effects of metabolic endotoxemia are dependent on the capacity of the host to respond to LPS, as CD14-deficient mice challenged with HFD are seemingly resistant to obesity and metabolic syndrome. In addition, mice deficient in TLR4 do not exhibit insulin resistance and are protected against microbiota-dependent inflammation and metabolic disorders (Csak et al. 2011; Poggi et al. 2007; Tsukumo et al. 2007; Jia et al. 2014; Kim et al. 2012), thus affirming the involvement of LPS and TLR4 signaling in promoting metabolic diseases. Despite this, it should be noted that not all bacterial LPS are immunogenic; for instance, the LPS derived from *Bacteroides dorei* (belonging to the phylum Bacteroidetes) lacks any detectable endotoxin activity and exhibits a structure that is distinct from LPS from *E. coli* (Vatanen et al. 2016).

The subsequent study by Cani et al. demonstrates that metabolic endotoxemia is microbiota-dependent and is associated with HFD-induced intestinal permeability (Cani et al. 2008). HFD feeding for 4 weeks significantly reduced the *Lactobacillus* and *Bacteroides* population in the gut, while concomitantly increasing the intestinal load of Gram-negative bacteria in WT mice. This change in microbiota composition in HFD-fed mice is accompanied by increased cecal and plasma LPS and enhanced intestinal permeability, presumably because of the decreased expression of the tight junction protein. The administration of antibiotics, however, significantly reduced cecal and plasma LPS levels, mitigated the expression of inflammatory markers, and improved metabolic syndrome in WT mice fed an HFD. A similar outcome was observed when antibiotics were administered to the hyperphagic *ob*/*ob* mice. Yet, microbiota ablation improved the insulin sensitivity in *ob*/*ob* mice without altering their body weight. The processes that facilitate the translocation of LPS are not well-understood, although two potential mechanisms have been proposed: luminal LPS absorption via lipid-rich chylomicrons (Vreugdenhil et al. 2003; Ghoshal et al. 2009) and/or leakage of LPS through defects in the tight junctions of the epithelial monolayer (Cani et al. 2008). These mechanisms are likely not mutually exclusive, but may function in parallel to facilitate metabolic endotoxemia in susceptible hosts.

7.5.2 Microbiota-Mediated Metabolic Disorders Due to Innate Immune Deficiency

Studies from our laboratory have shown a paradigmatically similar association of metabolic syndrome with microbiota-mediated low-grade inflammation in mice lacking toll-like receptor 5 (TLR5; an innate immune receptor that senses the bacterial flagellin) (Vijay-Kumar et al. 2010). The absence of TLR5 in mice induces substantial alteration in their gut microbiota, resulting in increased bacterial load and overabundance of Proteobacteria compared with their WT littermates (Carvalho et al. 2012). A fraction of *Tlr5*KO mice were prone to developing spontaneous gut inflammation (Vijay-Kumar et al. 2007); yet, those that did not display signs of colitis eventually developed pronounced hallmarks of metabolic syndrome (Vijay-Kumar et al. 2010). We have demonstrated that noncolitic *Tlr5*KO mice gained 15–20% more body weight and 2.5-fold larger abdominal fat pads than their WT littermates by 20 weeks of age. This increase in fat mass correlated with substantial increases in serum triglycerides, total cholesterol, and blood pressure. Furthermore, *Tlr5*KO mice exhibited hyperglycemia, loss of glycemic control, hyperinsulinemia, and insulin resistance. In accordance with their hyperinsulinemia, *Tlr5*KO mice exhibited an increase in the size and number of functional pancreatic islets. The loss of glycemic control in these mice is likely driven by insulin resistance, which is compensated for by the increase in insulin production; these conditions resemble the metabolic syndrome typically seen in humans. *Tlr5*KO mice exhibited hyperphagia and restricting their food consumption to that of WT littermates prevented most aspects of their metabolic syndrome except insulin resistance. This suggests that the insulin resistance in *Tlr5*KO mice might not be a consequence of increased food consumption or adiposity, but could be driven by inflammation instead. Feeding an HFD to *Tlr5*KO mice further exacerbated their metabolic syndrome, induced insulitis (inflammatory infiltrate in pancreatic islets), and promoted hepatic steatosis.

Studies that sought to determine the etiology of metabolic syndrome in *Tlr5*KO mice reported that the disorder is predominantly driven by the acquired gut dysbiosis, but not because of the deficiency of TLR5 per se (Zhang et al. 2016; Ubeda et al. 2012). This notion is substantiated by the absence of metabolic syndrome in *Tlr5*KO mice maintained under GF conditions (Singh et al. 2015) or in other facilities (Letran et al. 2011). We observed that microbiota ablation via administration of broad-spectrum antibiotics to *Tlr5*KO mice was indeed effective in correcting their metabolic syndrome phenotype. The transplantation of cecal microbiota from *Tlr5*-KO mice into GF WT mice was sufficient to recapitulate most aspects of metabolic syndrome in the recipient mice. However, the genetic deletion of either TLR4 or TLR2 failed to mitigate the features of metabolic syndrome in *Tlr5*kO mice, thus suggesting that the microbiota might mediate their obesogenic effects independently of TLR5, TLR4, and TLR2 (Vijay-Kumar et al. 2010).

A study from Caricilli et al. demonstrated that *Tlr2*KO mice also developed a phenotype (Caricilli et al. 2011) that is reminiscent of the metabolic syndrome associated with *Tlr5*KO mice. Specifically, their study demonstrated that the loss of TLR2 in mice resulted in a threefold increase in Firmicutes and a slight increase in Bacteroidetes compared with controls (i.e., an increase in the Firmicutes/Bacteroidetes ratio). The changes in their gut microbiota were associated with metabolic endotoxemia, low-grade inflammation, insulin resistance, glucose intolerance, and obesity. The transfer of the gut microbiota from *Tlr2*KO mice to GF WT mice conferred many features of metabolic syndrome to the recipients, but this was reversible by treating the recipient mice with antibiotics. Similar microbiota-dependent obesity, metabolic dysfunction, and gut dysbiosis were also observed in mice lacking the innate immune protein Nod2 (Rodriguez-Nunez et al. 2017). Despite this, it should be noted that immune dysregulation does not necessarily lead to adverse microbial alterations that potentiate diseases; instead, down-regulation of some components of the host immunity may be counter-intuitively beneficial. MyD88 is the universal adapter protein downstream of TLRs (TLR1-9; except TLR3); therefore, its absence may serve to downregulate TLR pathways. The loss of MyD88 completely abrogates TLR5 and TLR2 signaling; yet, MyD88-deficient mice did not develop metabolic syndrome, as was observed in *Tlr5*KO and *Tlr2*KO mice (Vijay-Kumar et al. 2010; Caricilli et al. 2011; Everard et al. 2014). Collectively, these studies emphasize that an altered immune response to the gut microbiota could have a substantial impact on obesity, diabetes, and inflammation.

7.5.3 Loss of Inflammasome Signaling Contributes to Metabolic Diseases

Inflammasomes, another component of innate immunity, are also implicated in the development of metabolic syndrome that is dependent on the gut microbiota. Inflammasomes are multi-protein complexes and include Nod-like receptors (e.g., NLRP1, NLRP3, and NLRC4), composed of proteins with leucine-rich repeats and nucleotide-binding domains that sense microbial ligands (including LPS) and damage-associated molecular patterns. Upon activation, inflammasomes induce caspase 1-mediated cleavage of inactive proIL-1β and proIL-18 to their active forms. Increased activation in the hepatic NLRP3 inflammasome has been associated with liver fibrosis and injury (Watanabe et al. 2009; Imaeda et al. 2009). Conversely, the genetic ablation of NLRP3 improves hepatic insulin signaling and protects against obesity (Vandanmagsar et al. 2011).

Yet, in another study, Henao-Mejia et al. demonstrated that mice deficient in either NLRP3 or NLRP6 developed alterations in their gut microbiota that were associated with the severity of diet-induced liver disease and metabolic syndrome (Henao-Mejia et al. 2012). The microbiota dysbiosis in NLRP3- and NLRP6-deficient mice resulted in an increased translocation of bacterial metabolites into the hepatic portal vein, enhanced TNF-α expression, and exacerbated steatosis (Henao-Mejia et al. 2012). Signaling through TLR4 and TLR9 was necessary, as genetic ablation of either TLR prevented severe manifestations of diet-induced metabolic diseases. Consistent with the phenotype observed in mice lacking TLR5 and TLR2 (Vijay-Kumar et al. 2010; Caricilli et al. 2011), the metabolic syndrome in NLRP3- and NLRP6-deficient mice is microbiota-dependent and can be ameliorated by depleting the gut microbiota (Henao-Mejia et al. 2012).

7.5.4 Microbiota-Mediated Metabolic Diseases in the Absence of Adaptive Immunity

A comprehensive understanding of the microbiota's influence over the host metabolism is further complicated by the participation of the adaptive immunity, which cooperates with innate immunity to maintain gut homeostasis (Slack et al. 2009). One such adaptive immune factor, IgA (the most abundant antibody in the gut), is mostly directed against the microbiota and therefore plays a major role in shaping gut homeostasis. The study by Shulzhenko et al. demonstrated that IgA deficiency indirectly impairs metabolic functions of the intestinal epithelium (Shulzhenko et al. 2011). Specifically, they demonstrated that mice deficient in B cells displayed significantly lowered expression of lipid metabolizing genes with a concomitant upregulation of inflammation-related genes, compared with WT mice. Although no change in food intake was observed, B cell-deficient mice exhibited reduced fat and cholesterol absorption that were reflected in decreased perigonadal fat pad mass. The transfer of microbiota from the B cell-deficient mice into the GF WT mice was sufficient to confer the phenotype from the former to the latter.

The intestinal epithelial cells in B cell-deficient mice upregulated their interferon-inducible genes to compensate for IgA deficiency, while simultaneously repressing GATA4-related genes. This resulted in a significant decrease in cholesterol and fat absorption and reduced perigonadal fat in B cell-deficient mice, but had no effect on food intake (Shulzhenko et al. 2011). Rederiving B cell-deficient mice in GF conditions completely abrogated the differences in their gene expression in comparison with WT mice. Colonizing GF B cell-deficient mice with an intact microbiota recapitulated gene expression to levels seen in conventional B cell-deficient mice, irrespective of the source of the colonizing microbiota. Further, stimulating the gut epithelia in vitro with heat-killed *E. coli* or LPS suppresses the expression of metabolic-related genes, but induces expression of immune-related genes (Shulzhenko et al. 2011). It was postulated that, in the absence of luminal IgA, the intestinal epithelium starts responding directly to the microbiota through the action of this metabolic-to-inflammatory transcriptional "switch," resulting in a reduction in body fat. This study uncovers the longstanding enigmatic association among low weight gain, muscle wasting, and lipid malabsorption in individuals with immunodeficiency syndromes, and demonstrates that the microbiota is capable of promoting either an increase or decrease in adiposity, depending on specific conditions.

The interactions between PRR and their microbial ligands have also been implicated in the pathogenesis of type I diabetes, despite it being an autoimmune rather than an inflammatory disease. A study in rats revealed that suppressing the load of PRR ligands in the gut via antibiotics could reduce insulitis and protect the β-cells from autoimmune-mediated destruction (Brugman et al. 2006). However, the lack of PRR signaling may also promote type I diabetes. A study by Chevonsky and colleagues observed that non-obese diabetic (NOD) mice deficient in MyD88 are protected against the development of type I diabetes, whereas GF MyD88KO/NOD mice develop robust diabetes (Wen et al. 2008). Colonization of GF MyD88KO/NOD mice with altered Schaedler bacteria significantly reduced the incidence of diabetes, whereas administration of broad-spectrum antibiotics increased diabetes incidence in these mice. These results suggest that some degree of innate immune activation might be necessary to develop regulatory T cells capable of preventing autoimmunity.

7.6 Manipulating the Gut Microbiota to Treat Metabolic Diseases: A New Clinical Frontier

Established treatments for metabolic diseases have recently been found to involve the gut microbiota. For example, the therapeutic effects of the antidiabetic drug metformin were shown to be, in part, mediated by its ability to influence the gut microbiota and restore butyrate production (Forslund et al. 2015). Similarly, the metabolic benefits of bariatric surgery are also thought to be mediated, in part, by the microbiota (Tremaroli et al. 2015). These findings strongly suggest that interventions that could alter the gut microbiota might be a promising strategy in correcting metabolic diseases. Accordingly, many researchers have begun to explore other means, such as the use of fecal microbiota transplant, probiotics, and prebiotics, in addition to directly targeting certain bacteria and their metabolites as potential therapeutic strategies to reshape the gut microbiota without harming the gut ecosystem.

7.6.1 Fecal Microbiota Transplantation

The concept that an individual's microbiota can be replaced with another via consumption of fecal material was conceived many centuries ago (de Groot et al. 2017). It was not until quite recently that modern medicine unraveled the efficacy of fecal microbiota transplantation (FMT) in treating *Clostridium difficile* infection (CDI) (van Nood et al. 2013). FMT treatment substantially altered the patient's microbiota composition (Li et al. 2016), restored normal bile acids composition (Weingarden et al. 2014), and re-established the microbiota-mediated colonization resistance against *C. difficile*. Based on its successful application in treating CDI, many studies now seek to establish FMT as a viable treatment option for microbiota-driven metabolic diseases, CVDs, inflammatory bowel diseases, and autoimmune diseases (Gupta et al. 2016). Transplanting fecal microbiota from lean donors to obese recipients improved insulin resistance in patients with metabolic syndrome (Vrieze et al. 2012). However, the use of FMT is currently limited because of its notable risks, including the possible transfer of endotoxins or infectious agents (Schwartz et al. 2013) or the appearance of new gastrointestinal complications (De Leon et al. 2013). Instead of instilling the microbiota from anonymous donors, a rational alternative would be to transplant only a defined group of bacteria (e.g., altered Schaedler flora) (Shen et al. 2015).

7.6.2 Probiotics and Prebiotics

Oral delivery of viable strains of probiotic bacteria capable of integrating into the existing gut microbiota is a promising approach to treating metabolic disorders. The administration of probiotic *Lactobacillus* strains, for instance, has been shown to decrease fat mass, alleviate insulin resistance and reduce the risk of type II diabetes in humans (Andreasen et al. 2010; Kadooka et al. 2010). The concomitant administration of probiotic formula VSL#3, which contains a mixture of *Lactobacillus* and *Bifidobacterium* strains, and the prebiotic oligofructose, has also been demonstrated to reduce liver injury and improve liver functions (Ma et al. 2013). The conception, feasibility, and potential mechanisms underlying the efficacy of probiotics and prebiotics have been elegantly reviewed (Delzenne et al. 2011; Verspreet et al. 2016).

7.6.3 Suppressing the Proteobacteria Bloom

Members of the Proteobacteria phylum tend to be present in low numbers in a healthy gut, but often bloom during various pathological conditions, including inflammatory bowel diseases and metabolic disorders (Shin et al. 2015). One clinical strain of Proteobacteria, the *Enterobacter cloacae* B29 isolated from a morbidly obese individual, was found to be particularly obesogenic; its mono-colonization into GF mice was sufficient to potentiate obesity and insulin resistance (Fei and Zhao 2013). Similar to the LPS from *E. coli* (Cani et al. 2007a), the LPS purified from the B29 strain also displayed a strong endotoxin activity capable of potentiating low-grade inflammation in their host (Fei and Zhao 2013). Given the pathological roles associated with *E. coli* (and other members of Proteobacteria), it would be interesting to explore whether gut dysbiosis and metabolic syndrome could be mitigated by specifically depleting *E. coli* via intervention with phage therapy (Sarker et al. 2016), antimicrobial fecal miRNA (Liu et al. 2016), or using FimH inhibitor (Spaulding et al. 2017). The administration of intestinal alkaline phosphatase (IAP), which neutralizes LPS by de-phosphorylating its lipid A moiety (Economopoulos et al. 2016; Fawley and Gourlay 2016), may also be considered a potential strategy to mitigate the metabolic endotoxemia and inflammation associated with Proteobacteria.

7.6.4 Exploiting the Anti-obesogenic Akkermansia muciniphila

Akkermansia muciniphila is a Gram-negative, mucin-degrading bacterium that was previously thought to be a strict anaerobe (Derrien et al. 2004), but was recently shown to be an aerotolerant anaerobe (Ouwerkerk et al. 2016; Reunanen et al. 2015). *A. muciniphila* thrives in the outer mucus layer, where it depends on mucin as its energy source (Derrien et al. 2004). Its prominent feature as anti-obesogenic bacteria was documented in seminal studies that demonstrated that oral gavage of the bacteria protected mice against diet-induced obesity (Everard et al. 2013) and type 2 diabetes (Shin et al. 2014). The mechanisms underlying its anti-obesogenic property are thought to involve its capacity to counteract metabolic endotoxemia (Everard et al. 2013) and modulate the host immune responses and gut barrier function (Ottman et al. 2017). A recent study by Cani and colleagues found that many of the health-promoting effects of *A. muciniphila* can be recapitulated by orally administering either the pasteurized bacterium or its purified membrane protein Amuc_1100 (thus obviating the use of live bacterium) (Plovier et al. 2017), which further highlights their tremendous potential as novel therapeutics for treating obesity.

7.7 Conclusion and Future Directions

Over the last two decades, our understanding of the gut microbiota and its association with various metabolic disorders, namely obesity, diabetes, atherosclerosis, cardiovascular and hepatic diseases, has increased exponentially. Such knowledge was made possible because of the advent of high-throughput multi-"omics" technologies (e.g., metagenomics, metatranscriptomics, metaproteomics, metabolomics) that enable in-depth analyses into the extent to which gut bacteria communicate not only with the host, but also with other bacteria in the gut. Yet, there are still many unanswered questions, for instance:

(1) What defines an ideal/normal/healthy microbiota? Does it vary from person to person?
(2) What is the relative contribution of host genetics (immunodeficiencies) in addition to dietary habits?
(3) What kind of dysbiosis is fixable?
(4) What if dysbiosis is an adaptation to a person's lifestyle or to a pre-existing disorder?

It has been hypothesized that it might not be the composition per se, but rather microbial genetics/metabolism in the gut that predominantly influence host metabolism. Identifying the microbial metabolic pathways in obese and non-obese individuals may help in developing promising therapeutic bacterial formulations. Harnessing the power of a non-obese microbiota to affect weight loss and/or prevent weight gain, however, is both a potential therapeutic approach and a major challenge. Given the tremendous inter-individual heterogeneity and variability in the composition of the gut microbiota, it is thus unlikely that any bacteria consortia developed to treat metabolic disease would be universally applicable. Although an optimal "one-size-fits-all" bacterial consortia might not exist, the approaches that optimize personalized nutrition and medicine (Zeevi et al. 2015) by accounting for individual microbiota disposition may be more effective in treating the various microbiota-associated metabolic disorders.

References

Abreu MT (2010) Toll-like receptor signalling in the intestinal epithelium: how bacterial recognition shapes intestinal function. Nat Rev Immunol 10(2):131–144. https://doi.org/10.1038/nri2707

Andreasen AS, Larsen N, Pedersen-Skovsgaard T, Berg RM, Moller K, Svendsen KD, Jakobsen M, Pedersen BK (2010) Effects of Lactobacillus acidophilus NCFM on insulin sensitivity and the systemic inflammatory response in human subjects. Br J Nutr 104 (12):1831–1838. https://doi.org/10.1017/S0007114510002874

Artis D, Wang ML, Keilbaugh SA, He W, Brenes M, Swain GP, Knight PA, Donaldson DD, Lazar MA, Miller HR, Schad GA, Scott P, Wu GD (2004) RELMbeta/FIZZ2 is a goblet cell-specific immune-effector molecule in the gastrointestinal tract. Proc Natl Acad Sci USA 101 (37):13596–13600. https://doi.org/10.1073/pnas.0404034101

Atuma C, Strugala V, Allen A, Holm L (2001) The adherent gastrointestinal mucus gel layer: thickness and physical state in vivo. Am J Physiol Gastrointest Liver Physiol 280(5):G922–G929

Backhed F, Ding H, Wang T, Hooper LV, Koh GY, Nagy A, Semenkovich CF, Gordon JI (2004) The gut microbiota as an environmental factor that regulates fat storage. Proc Natl Acad Sci 101(44):15718–15723. https://doi.org/10.1073/pnas.0407076101

Backhed F, Manchester JK, Semenkovich CF, Gordon JI (2007) Mechanisms underlying the resistance to diet-induced obesity in germ-free mice. Proc Natl Acad Sci USA 104(3):979–984

Ballard FJ (1972) Supply and utilization of acetate in mammals. Am J Clin Nutr 25(8):773–779

Barrera JG, Sandoval DA, D'Alessio DA, Seeley RJ (2011) GLP-1 and energy balance: an integrated model of short-term and long-term control. Nat Rev Endocrinol 7(9):507–516. https://doi.org/10.1038/nrendo.2011.77

Bellahcene M, O'Dowd JF, Wargent ET, Zaibi MS, Hislop DC, Ngala RA, Smith DM, Cawthorne MA, Stocker CJ, Arch JR (2013) Male mice that lack the G-protein-coupled receptor GPR41 have low energy expenditure and increased body fat content. Br J Nutr 109(10):1755–1764

Bennett BJ, de Aguiar Vallim TQ, Wang Z, Shih DM, Meng Y, Gregory J, Allayee H, Lee R, Graham M, Crooke R, Edwards PA, Hazen SL, Lusis AJ (2013) Trimethylamine-N-oxide, a metabolite associated with atherosclerosis, exhibits complex genetic and dietary regulation. Cell Metab 17(1):49–60. https://doi.org/10.1016/j.cmet.2012.12.011

Berggren AM, Nyman EM, Lundquist I, Bjorck IM (1996) Influence of orally and rectally administered propionate on cholesterol and glucose metabolism in obese rats. Br J Nutr 76 (2):287–294

Bergman EN (1990) Energy contributions of volatile fatty acids from the gastrointestinal tract in various species. Physiol Rev 70(2):567–590

Bianconi E, Piovesan A, Facchin F, Beraudi A, Casadei R, Frabetti F, Vitale L, Pelleri MC, Tassani S, Piva F, Perez-Amodio S, Strippoli P, Canaider S (2013) An estimation of the number of cells in the human body. Ann Hum Biol 40(6):463–471. https://doi.org/10.3109/03014460.2013.807878

Bloemen JG, Venema K, van de Poll MC, Olde Damink SW, Buurman WA, Dejong CH (2009) Short chain fatty acids exchange across the gut and liver in humans measured at surgery. Clin Nutr 28(6):657–661

Bloemen JG, Olde Damink SW, Venema K, Buurman WA, Jalan R, Dejong CH (2010) Short chain fatty acids exchange: is the cirrhotic, dysfunctional liver still able to clear them? Clin Nutr 29(3):365–369. https://doi.org/10.1016/j.clnu.2009.10.002

Boillot J, Alamowitch C, Berger AM, Luo J, Bruzzo F, Bornet FR, Slama G (1995) Effects of dietary propionate on hepatic glucose production, whole-body glucose utilization, carbohydrate and lipid metabolism in normal rats. Br J Nutr 73(2):241–251

Boix-Amoros A, Collado MC, Mira A (2016) Relationship between milk microbiota, bacterial load, macronutrients, and human cells during lactation. Front Microbiol 7:492. https://doi.org/10.3389/fmicb.2016.00492

Bremer J (1983) Carnitine—metabolism and functions. Physiol Rev 63(4):1420–1480

Brugman S, Klatter FA, Visser JT, Wildeboer-Veloo AC, Harmsen HJ, Rozing J, Bos NA (2006) Antibiotic treatment partially protects against type 1 diabetes in the bio-breeding diabetes-prone rat. Is the gut flora involved in the development of type 1 diabetes? Diabetologia 49 (9):2105–2108. https://doi.org/10.1007/s00125-006-0334-0

Cani PD, Delzenne NM (2009) The role of the gut microbiota in energy metabolism and metabolic disease. Curr Pharm Des 15(13):1546–1558

Cani PD, Amar J, Iglesias MA, Poggi M, Knauf C, Bastelica D, Neyrinck AM, Fava F, Tuohy KM, Chabo C, Waget A, Delmee E, Cousin B, Sulpice T, Chamontin B, Ferrieres J, Tanti JF, Gibson GR, Casteilla L, Delzenne NM, Alessi MC, Burcelin R (2007a) Metabolic endotoxemia initiates obesity and insulin resistance. Diabetes 56(7):1761–1772. https://doi.org/10.2337/db06-1491

Cani PD, Hoste S, Guiot Y, Delzenne NM (2007b) Dietary non-digestible carbohydrates promote L-cell differentiation in the proximal colon of rats. Br J Nutr 98(1):32–37. https://doi.org/10.1017/S0007114507691648

Cani PD, Bibiloni R, Knauf C, Waget A, Neyrinck AM, Delzenne NM, Burcelin R (2008) Changes in gut microbiota control metabolic endotoxemia-induced inflammation in high-fat diet-induced obesity and diabetes in mice. Diabetes 57(6):1470–1481. https://doi.org/10.2337/db07-1403

Cani PD, Lecourt E, Dewulf EM, Sohet FM, Pachikian BD, Naslain D, De Backer F, Neyrinck AM, Delzenne NM (2009) Gut microbiota fermentation of prebiotics increases satietogenic and incretin gut peptide production with consequences for appetite sensation and glucose response after a meal. Am J Clin Nutr 90(5):1236–1243. https://doi.org/10.3945/ajcn.2009.28095

Cannon B, Nedergaard J (2004) Brown adipose tissue: function and physiological significance. Physiol Rev 84(1):277–359. https://doi.org/10.1152/physrev.00015.2003

Caricilli AM, Picardi PK, de Abreu LL, Ueno M, Prada PO, Ropelle ER, Hirabara SM, Castoldi A, Vieira P, Camara NO, Curi R, Carvalheira JB, Saad MJ (2011) Gut microbiota is a key modulator of insulin resistance in TLR 2 knockout mice. PLoS Biol 9(12):e1001212. https://doi.org/10.1371/journal.pbio.1001212

Carvalho FA, Koren O, Goodrich JK, Johansson ME, Nalbantoglu I, Aitken JD, Su Y, Chassaing B, Walters WA, Gonzalez A, Clemente JC, Cullender TC, Barnich N, Darfeuille-Michaud A, Vijay-Kumar M, Knight R, Ley RE, Gewirtz AT (2012) Transient inability to manage proteobacteria promotes chronic gut inflammation in TLR5-deficient mice. Cell Host Microbe 12(2):139–152. https://doi.org/10.1016/j.chom.2012.07.004

Chassaing B, Koren O, Goodrich JK, Poole AC, Srinivasan S, Ley RE, Gewirtz AT (2015a) Dietary emulsifiers impact the mouse gut microbiota promoting colitis and metabolic syndrome. Nature 519(7541):92–96. https://doi.org/10.1038/nature14232

Chassaing B, Miles-Brown J, Pellizzon M, Ulman E, Ricci M, Zhang L, Patterson AD, Vijay-Kumar M, Gewirtz AT (2015b) Lack of soluble fiber drives diet-induced adiposity in mice. Am J Physiol Gastrointest Liver Physiol 309(7):G528–G541. https://doi.org/10.1152/ajpgi.00172.2015

Chassaing B, Raja SM, Lewis JD, Srinivasan S, Gewirtz AT (2017a) Colonic microbiota encroachment correlates with dysglycemia in humans. Cell Mol Gastroenterol Hepatol 4 (2):205–221. https://doi.org/10.1016/j.jcmgh.2017.04.001

Chassaing B, Van de Wiele T, De Bodt J, Marzorati M, Gewirtz AT (2017b) Dietary emulsifiers directly alter human microbiota composition and gene expression ex vivo potentiating intestinal inflammation. Gut Microbes. https://doi.org/10.1136/gutjnl-2016-313099

Cho I, Yamanishi S, Cox L, Methe BA, Zavadil J, Li K, Gao Z, Mahana D, Raju K, Teitler I, Li H, Alekseyenko AV, Blaser MJ (2012) Antibiotics in early life alter the murine colonic microbiome and adiposity. Nature 488(7413):621–626. https://doi.org/10.1038/nature11400

Claesson MJ, Jeffery IB, Conde S, Power SE, O'Connor EM, Cusack S, Harris HM, Coakley M, Lakshminarayanan B, O'Sullivan O, Fitzgerald GF, Deane J, O'Connor M, Harnedy N, O'Connor K, O'Mahony D, van Sinderen D, Wallace M, Brennan L, Stanton C, Marchesi JR, Fitzgerald AP, Shanahan F, Hill C, Ross RP, O'Toole PW (2012) Gut microbiota composition correlates with diet and health in the elderly. Nature 488(7410):178–184. https://doi.org/10.1038/nature11319

Costello EK, Lauber CL, Hamady M, Fierer N, Gordon JI, Knight R (2009) Bacterial community variation in human body habitats across space and time. Science 326(5960):1694–1697. https://doi.org/10.1126/science.1177486

Cover TL, Blaser MJ (2009) Helicobacter pylori in health and disease. Gastroenterology 136 (6):1863–1873. https://doi.org/10.1053/j.gastro.2009.01.073

Cox LM, Yamanishi S, Sohn J, Alekseyenko AV, Leung JM, Cho I, Kim SG, Li H, Gao Z, Mahana D, Zarate Rodriguez JG, Rogers AB, Robine N, Loke P, Blaser MJ (2014) Altering the intestinal microbiota during a critical developmental window has lasting metabolic consequences. Cell 158(4):705–721. https://doi.org/10.1016/j.cell.2014.05.052

Craciun S, Balskus EP (2012) Microbial conversion of choline to trimethylamine requires a glycyl radical enzyme. Proc Natl Acad Sci USA 109(52):21307–21312. https://doi.org/10.1073/pnas.1215689109

Csak T, Velayudham A, Hritz I, Petrasek J, Levin I, Lippai D, Catalano D, Mandrekar P, Dolganiuc A, Kurt-Jones E, Szabo G (2011) Deficiency in myeloid differentiation factor-2 and toll-like receptor 4 expression attenuates nonalcoholic steatohepatitis and fibrosis in mice. Am J Physiol Gastrointest Liver Physiol 300(3):G433–G441

Cummings JH, Pomare EW, Branch WJ, Naylor CP, Macfarlane GT (1987) Short chain fatty acids in human large intestine, portal, hepatic and venous blood. Gut 28(10):1221–1227

David LA, Maurice CF, Carmody RN, Gootenberg DB, Button JE, Wolfe BE, Ling AV, Devlin AS, Varma Y, Fischbach MA, Biddinger SB, Dutton RJ, Turnbaugh PJ (2014) Diet rapidly and reproducibly alters the human gut microbiome. Nature 505(7484):559–563. https://doi.org/10.1038/nature12820

De Groot PF, Frissen MN, de Clercq NC, Nieuwdorp M (2017) Fecal microbiota transplantation in metabolic syndrome: history, present and future. Gut Microbes 8(3):253–267. https://doi.org/10.1080/19490976.2017.1293224

De Leon LM, Watson JB, Kelly CR (2013) Transient flare of ulcerative colitis after fecal microbiota transplantation for recurrent Clostridium difficile infection. Clin Gastroenterol Hepatol Off Clin Prac J Am Gastroenterol Assoc 11(8):1036–1038. https://doi.org/10.1016/j.cgh.2013.04.045

Delzenne NM, Neyrinck AM, Backhed F, Cani PD (2011) Targeting gut microbiota in obesity: effects of prebiotics and probiotics. Nat Rev Endocrinol 7(11):639–646. https://doi.org/10.1038/nrendo.2011.126

Den Besten G, Lange K, Havinga R, van Dijk TH, Gerding A, van Eunen K, Muller M, Groen AK, Hooiveld GJ, Bakker BM, Reijngoud DJ (2013a) Gut-derived short-chain fatty acids are vividly assimilated into host carbohydrates and lipids. Am J Physiol Gastrointest Liver Physiol. https://doi.org/10.1152/ajpgi.00265.2013

Den Besten G, van Eunen K, Groen AK, Venema K, Reijngoud DJ, Bakker BM (2013b) The role of short-chain fatty acids in the interplay between diet, gut microbiota, and host energy metabolism. J Lipid Res 54(9):2325–2340. https://doi.org/10.1194/jlr.R036012

Den Besten G, Bleeker A, Gerding A, van Eunen K, Havinga R, van Dijk TH, Oosterveer MH, Jonker JW, Groen AK, Reijngoud DJ, Bakker BM (2015) Short-chain fatty acids protect against high-fat diet-induced obesity via a PPARgamma-dependent switch from lipogenesis to fat oxidation. Diabetes 64(7):2398–2408

Derrien M, Vaughan EE, Plugge CM, de Vos WM (2004) Akkermansia muciniphila gen. nov., sp. nov., a human intestinal mucin-degrading bacterium. Int J Syst Evol Microbiol 54 (Pt 5):1469–1476. https://doi.org/10.1099/ijs.0.02873-0

Donohoe DR, Garge N, Zhang X, Sun W, O'Connell TM, Bunger MK, Bultman SJ (2011) The microbiome and butyrate regulate energy metabolism and autophagy in the mammalian colon. Cell Metab 13(5):517–526. https://doi.org/10.1016/j.cmet.2011.02.018

Dumas ME, Barton RH, Toye A, Cloarec O, Blancher C, Rothwell A, Fearnside J, Tatoud R, Blanc V, Lindon JC, Mitchell SC, Holmes E, McCarthy MI, Scott J, Gauguier D, Nicholson JK (2006) Metabolic profiling reveals a contribution of gut microbiota to fatty liver phenotype in insulin-resistant mice. Proc Natl Acad Sci USA 103(33):12511–12516

Economopoulos KP, Ward NL, Phillips CD, Teshager A, Patel P, Mohamed MM, Hakimian S, Cox SB, Ahmed R, Moaven O, Kaliannan K, Alam SN, Haller JF, Goldstein AM, Bhan AK, Malo MS, Hodin RA (2016) Prevention of antibiotic-associated metabolic syndrome in mice by intestinal alkaline phosphatase. Diabetes Obes Metab 18(5):519–527. https://doi.org/10.1111/dom.12645

Everard A, Belzer C, Geurts L, Ouwerkerk JP, Druart C, Bindels LB, Guiot Y, Derrien M, Muccioli GG, Delzenne NM, de Vos WM, Cani PD (2013) Cross-talk between Akkermansia muciniphila and intestinal epithelium controls diet-induced obesity. Proc Natl Acad Sci USA 110(22):9066–9071. https://doi.org/10.1073/pnas.1219451110

Everard A, Geurts L, Caesar R, Van Hul M, Matamoros S, Duparc T, Denis RG, Cochez P, Pierard F, Castel J, Bindels LB, Plovier H, Robine S, Muccioli GG, Renauld JC, Dumoutier L, Delzenne NM, Luquet S, Backhed F, Cani PD (2014) Intestinal epithelial MyD88 is a sensor switching host metabolism towards obesity according to nutritional status. Nat Commun 5:5648

Faith JJ, McNulty NP, Rey FE, Gordon JI (2011) Predicting a human gut microbiota's response to diet in gnotobiotic mice. Science 333(6038):101–104. https://doi.org/10.1126/science.1206025

Fawley J, Gourlay DM (2016) Intestinal alkaline phosphatase: a summary of its role in clinical disease. J Surg Res 202(1):225–234. https://doi.org/10.1016/j.jss.2015.12.008

Fei N, Zhao L (2013) An opportunistic pathogen isolated from the gut of an obese human causes obesity in germfree mice. ISME J 7(4):880–884. https://doi.org/10.1038/ismej.2012.153

Fernandes J, Su W, Rahat-Rozenbloom S, Wolever TM, Comelli EM (2014) Adiposity, gut microbiota and faecal short chain fatty acids are linked in adult humans. Nutr Diabetes 4:e121

Field BC, Chaudhri OB, Bloom SR (2010) Bowels control brain: gut hormones and obesity. Nat Rev Endocrinol 6(8):444–453. https://doi.org/10.1038/nrendo.2010.93

Flegal KM, Carroll MD, Ogden CL, Curtin LR (2010) Prevalence and trends in obesity among US adults, 1999–2008. JAMA 303(3):235–241. https://doi.org/10.1001/jama.2009.2014

Forslund K, Hildebrand F, Nielsen T, Falony G, Le Chatelier E, Sunagawa S, Prifti E, Vieira-Silva S, Gudmundsdottir V, Pedersen HK, Arumugam M, Kristiansen K, Voigt AY, Vestergaard H, Hercog R, Costea PI, Kultima JR, Li J, Jorgensen T, Levenez F, Dore J, Meta HIT, Nielsen HB, Brunak S, Raes J, Hansen T, Wang J, Ehrlich SD, Bork P, Pedersen O (2015) Disentangling type 2 diabetes and metformin treatment signatures in the human gut microbiota. Nature 528(7581):262–266. https://doi.org/10.1038/nature15766

Gaskins HR, Collier CT, Anderson DB (2002) Antibiotics as growth promotants: mode of action. Anim Biotechnol 13(1):29–42. https://doi.org/10.1081/ABIO-120005768

Gewirtz AT, Navas TA, Lyons S, Godowski PJ, Madara JL (2001) Cutting edge: bacterial flagellin activates basolaterally expressed TLR5 to induce epithelial proinflammatory gene expression. J Immunol 167(4):1882–1885

Ghoshal S, Witta J, Zhong J, de Villiers W, Eckhardt E (2009) Chylomicrons promote intestinal absorption of lipopolysaccharides. J Lipid Res 50(1):90–97. https://doi.org/10.1194/jlr.M800156-JLR200

Grice EA, Segre JA (2012) The human microbiome: our second genome. Annu Rev Genomics Hum Genet 13:151–170. https://doi.org/10.1146/annurev-genom-090711-163814

Grober J, Zaghini I, Fujii H, Jones SA, Kliewer SA, Willson TM, Ono T, Besnard P (1999) Identification of a bile acid-responsive element in the human ileal bile acid-binding protein gene. Involvement of the farnesoid X receptor/9-cis-retinoic acid receptor heterodimer. J Biol Chem 274(42):29749–29754

Gupta S, Allen-Vercoe E, Petrof EO (2016) Fecal microbiota transplantation: in perspective. Ther Adv Gastroenterol 9(2):229–239. https://doi.org/10.1177/1756283X15607414

Hardie DG, Carling D, Carlson M (1998) The AMP-activated/SNF1 protein kinase subfamily: metabolic sensors of the eukaryotic cell? Annu Rev Biochem 67:821–855. https://doi.org/10.1146/annurev.biochem.67.1.821

Henao-Mejia J, Elinav E, Jin C, Hao L, Mehal WZ, Strowig T, Thaiss CA, Kau AL, Eisenbarth SC, Jurczak MJ, Camporez JP, Shulman GI, Gordon JI, Hoffman HM, Flavell RA (2012) Inflammasome-mediated dysbiosis regulates progression of NAFLD and obesity. Nature 482 (7384):179–185. https://doi.org/10.1038/nature10809

Herbert DR, Yang JQ, Hogan SP, Groschwitz K, Khodoun M, Munitz A, Orekov T, Perkins C, Wang Q, Brombacher F, Urban JF Jr, Rothenberg ME, Finkelman FD (2009) Intestinal epithelial cell secretion of RELM-beta protects against gastrointestinal worm infection. J Exp Med 206(13):2947–2957. https://doi.org/10.1084/jem.20091268

Holt JA, Luo G, Billin AN, Bisi J, McNeill YY, Kozarsky KF, Donahee M, Wang DY, Mansfield TA, Kliewer SA, Goodwin B, Jones SA (2003) Definition of a novel growth factor-dependent signal cascade for the suppression of bile acid biosynthesis. Genes Dev 17(13):1581–1591. https://doi.org/10.1101/gad.1083503

Hood MI, Skaar EP (2012) Nutritional immunity: transition metals at the pathogen-host interface. Nat Rev Microbiol 10(8):525–537. https://doi.org/10.1038/nrmicro2836

Hooper LV, Midtvedt T, Gordon JI (2002) How host-microbial interactions shape the nutrient environment of the mammalian intestine. Annu Rev Nutr 22:283–307. https://doi.org/10.1146/annurev.nutr.22.011602.092259

Hooper LV, Stappenbeck TS, Hong CV, Gordon JI (2003) Angiogenins: a new class of microbicidal proteins involved in innate immunity. Nat Immunol 4(3):269–273. https://doi.org/10.1038/ni888

Hosseini E, Grootaert C, Verstraete W, Van de Wiele T (2011) Propionate as a health-promoting microbial metabolite in the human gut. Nutr Rev 69(5):245–258

Human Microbiome Project C (2012) Structure, function and diversity of the healthy human microbiome. Nature 486(7402):207–214. https://doi.org/10.1038/nature11234

Imaeda AB, Watanabe A, Sohail MA, Mahmood S, Mohamadnejad M, Sutterwala FS, Flavell RA, Mehal WZ (2009) Acetaminophen-induced hepatotoxicity in mice is dependent on Tlr9 and the Nalp3 inflammasome. J Clin Invest 119(2):305–314

Inagaki T, Choi M, Moschetta A, Peng L, Cummins CL, McDonald JG, Luo G, Jones SA, Goodwin B, Richardson JA, Gerard RD, Repa JJ, Mangelsdorf DJ, Kliewer SA (2005) Fibroblast growth factor 15 functions as an enterohepatic signal to regulate bile acid homeostasis. Cell Metab 2(4):217–225. https://doi.org/10.1016/j.cmet.2005.09.001

Iwase T, Shinji H, Tajima A, Sato F, Tamura T, Iwamoto T, Yoneda M, Mizunoe Y (2010) Isolation and identification of ATP-secreting bacteria from mice and humans. J Clin Microbiol 48(5):1949–1951. https://doi.org/10.1128/JCM.01941-09

Janssen AWF, Houben T, Katiraei S, Dijk W, Boutens L, van der Bolt N, Wang Z, Brown JM, Hazen SL, Mandard S, Shiri-Sverdlov R, Kuipers F, Willems van Dijk K, Vervoort J,

Stienstra R, Hooiveld G, Kersten S (2017) Modulation of the gut microbiota impacts nonalcoholic fatty liver disease: a potential role for bile acids. J Lipid Res 58(7):1399–1416. https://doi.org/10.1194/jlr.M075713

Jia L, Vianna CR, Fukuda M, Berglund ED, Liu C, Tao C, Sun K, Liu T, Harper MJ, Lee CE, Lee S, Scherer PE, Elmquist JK (2014) Hepatocyte Toll-like receptor 4 regulates obesity-induced inflammation and insulin resistance. Nat Commun 5:3878. https://doi.org/10.1038/ncomms4878

Johansson ME, Phillipson M, Petersson J, Velcich A, Holm L, Hansson GC (2008) The inner of the two Muc2 mucin-dependent mucus layers in colon is devoid of bacteria. Proc Natl Acad Sci USA 105(39):15064–15069. https://doi.org/10.1073/pnas.0803124105

Jones BV, Begley M, Hill C, Gahan CG, Marchesi JR (2008) Functional and comparative metagenomic analysis of bile salt hydrolase activity in the human gut microbiome. Proc Natl Acad Sci USA 105(36):13580–13585. https://doi.org/10.1073/pnas.0804437105

Kadooka Y, Sato M, Imaizumi K, Ogawa A, Ikuyama K, Akai Y, Okano M, Kagoshima M, Tsuchida T (2010) Regulation of abdominal adiposity by probiotics (Lactobacillus gasseri SBT2055) in adults with obese tendencies in a randomized controlled trial. Eur J Clin Nutr 64 (6):636–643. https://doi.org/10.1038/ejcn.2010.19

Kaiko GE, Ryu SH, Koues OI, Collins PL, Solnica-Krezel L, Pearce EJ, Pearce EL, Oltz EM, Stappenbeck TS (2016) The colonic crypt protects stem cells from microbiota-derived metabolites. Cell 165(7):1708–1720. https://doi.org/10.1016/j.cell.2016.05.018

Kawamata Y, Fujii R, Hosoya M, Harada M, Yoshida H, Miwa M, Fukusumi S, Habata Y, Itoh T, Shintani Y, Hinuma S, Fujisawa Y, Fujino M (2003) A G protein-coupled receptor responsive to bile acids. J Biol Chem 278(11):9435–9440. https://doi.org/10.1074/jbc.M209706200

Kim HK, Youn BS, Shin MS, Namkoong C, Park KH, Baik JH, Kim JB, Park JY, Lee KU, Kim YB, Kim MS (2010) Hypothalamic Angptl4/Fiaf is a novel regulator of food intake and body weight. Diabetes 59(11):2772–2780. https://doi.org/10.2337/db10-0145

Kim KA, Gu W, Lee IA, Joh EH, Kim DH (2012) High fat diet-induced gut microbiota exacerbates inflammation and obesity in mice via the TLR4 signaling pathway. PloS One 7(10): e47713. https://doi.org/10.1371/journal.pone.0047713

Kimura I, Ozawa K, Inoue D, Imamura T, Kimura K, Maeda T, Terasawa K, Kashihara D, Hirano K, Tani T, Takahashi T, Miyauchi S, Shioi G, Inoue H, Tsujimoto G (2013) The gut microbiota suppresses insulin-mediated fat accumulation via the short-chain fatty acid receptor GPR43. Nat Commun 4:1829. https://doi.org/10.1038/ncomms2852

Koeth RA, Wang Z, Levison BS, Buffa JA, Org E, Sheehy BT, Britt EB, Fu X, Wu Y, Li L, Smith JD, DiDonato JA, Chen J, Li H, Wu GD, Lewis JD, Warrier M, Brown JM, Krauss RM, Tang WH, Bushman FD, Lusis AJ, Hazen SL (2013) Intestinal microbiota metabolism of L-carnitine, a nutrient in red meat, promotes atherosclerosis. Nat Med 19(5):576–585. https://doi.org/10.1038/nm.3145

Koren O, Goodrich JK, Cullender TC, Spor A, Laitinen K, Backhed HK, Gonzalez A, Werner JJ, Angenent LT, Knight R, Backhed F, Isolauri E, Salminen S, Ley RE (2012) Host remodeling of the gut microbiome and metabolic changes during pregnancy. Cell 150(3):470–480. https://doi.org/10.1016/j.cell.2012.07.008

Koyama I, Matsunaga T, Harada T, Hokari S, Komoda T (2002) Alkaline phosphatases reduce toxicity of lipopolysaccharides in vivo and in vitro through dephosphorylation. Clin Biochem 35(6):455–461

Landfald B, Valeur J, Berstad A, Raa J (2017) Microbial trimethylamine-N-oxide as a disease marker: something fishy? Microb Ecol Health Dis 28(1):1327309. https://doi.org/10.1080/16512235.2017.1327309

Landrier JF, Eloranta JJ, Vavricka SR, Kullak-Ublick GA (2006) The nuclear receptor for bile acids, FXR, transactivates human organic solute transporter-alpha and -beta genes. Am J Physiol Gastrointest Liver Physiol 290(3):G476–G485. https://doi.org/10.1152/ajpgi.00430.2005

Le Chatelier E, Nielsen T, Qin J, Prifti E, Hildebrand F, Falony G, Almeida M, Arumugam M, Batto JM, Kennedy S, Leonard P, Li J, Burgdorf K, Grarup N, Jorgensen T, Brandslund I, Nielsen HB, Juncker AS, Bertalan M, Levenez F, Pons N, Rasmussen S, Sunagawa S, Tap J, Tims S, Zoetendal EG, Brunak S, Clement K, Dore J, Kleerebezem M, Kristiansen K, Renault P, Sicheritz-Ponten T, de Vos WM, Zucker JD, Raes J, Hansen T, Meta HIT, Bork P, Wang J, Ehrlich SD, Pedersen O (2013) Richness of human gut microbiome correlates with metabolic markers. Nature 500(7464):541–546. https://doi.org/10.1038/nature12506

Lee H, Zhang Y, Lee FY, Nelson SF, Gonzalez FJ, Edwards PA (2006) FXR regulates organic solute transporters alpha and beta in the adrenal gland, kidney, and intestine. J Lipid Res 47 (1):201–214. https://doi.org/10.1194/jlr.M500417-JLR200

Letran SE, Lee SJ, Atif SM, Flores-Langarica A, Uematsu S, Akira S, Cunningham AF, McSorley SJ (2011) TLR5-deficient mice lack basal inflammatory and metabolic defects but exhibit impaired CD4 T cell responses to a flagellated pathogen. J Immunol 186(9):5406–5412. https://doi.org/10.4049/jimmunol.1003576

Ley RE, Turnbaugh PJ, Klein S, Gordon JI (2006) Microbial ecology: human gut microbes associated with obesity. Nature 444(7122):1022–1023. https://doi.org/10.1038/4441022a

Li F, Jiang C, Krausz KW, Li Y, Albert I, Hao H, Fabre KM, Mitchell JB, Patterson AD, Gonzalez FJ (2013) Microbiome remodelling leads to inhibition of intestinal farnesoid X receptor signalling and decreased obesity. Nat Commun 4:2384. https://doi.org/10.1038/ncomms3384

Li SS, Zhu A, Benes V, Costea PI, Hercog R, Hildebrand F, Huerta-Cepas J, Nieuwdorp M, Salojarvi J, Voigt AY, Zeller G, Sunagawa S, de Vos WM, Bork P (2016) Durable coexistence of donor and recipient strains after fecal microbiota transplantation. Science 352 (6285):586–589. https://doi.org/10.1126/science.aad8852

Lin HV, Frassetto A, Kowalik EJ Jr, Nawrocki AR, Lu MM, Kosinski JR, Hubert JA, Szeto D, Yao X, Forrest G, Marsh DJ (2012) Butyrate and propionate protect against diet-induced obesity and regulate gut hormones via free fatty acid receptor 3-independent mechanisms. PloS One 7(4):e35240

Liu S, da Cunha AP, Rezende RM, Cialic R, Wei Z, Bry L, Comstock LE, Gandhi R, Weiner HL (2016) The host shapes the gut microbiota via fecal microRNA. Cell Host Microbe 19 (1):32–43. https://doi.org/10.1016/j.chom.2015.12.005

Lyte M (2013) Microbial endocrinology in the microbiome-gut-brain axis: how bacterial production and utilization of neurochemicals influence behavior. PLoS Pathog 9(11):e1003726

Ma YY, Li L, Yu CH, Shen Z, Chen LH, Li YM (2013) Effects of probiotics on nonalcoholic fatty liver disease: a meta-analysis. World J Gastroenterol 19(40):6911–6918

Macfarlane S, Macfarlane GT (2003) Regulation of short-chain fatty acid production. Proc Nutr Soc 62(1):67–72

Madsbad S (2014) The role of glucagon-like peptide-1 impairment in obesity and potential therapeutic implications. Diabetes Obes Metab 16(1):9–21

Makishima M, Okamoto AY, Repa JJ, Tu H, Learned RM, Luk A, Hull MV, Lustig KD, Mangelsdorf DJ, Shan B (1999) Identification of a nuclear receptor for bile acids. Science 284(5418):1362–1365

Mantis NJ, Rol N, Corthesy B (2011) Secretory IgA's complex roles in immunity and mucosal homeostasis in the gut. Mucosal Immunol 4(6):603–611. https://doi.org/10.1038/mi.2011.41

Maruyama T, Miyamoto Y, Nakamura T, Tamai Y, Okada H, Sugiyama E, Nakamura T, Itadani H, Tanaka K (2002) Identification of membrane-type receptor for bile acids (M-BAR). Biochem Biophys Res Commun 298(5):714–719

Maruyama T, Tanaka K, Suzuki J, Miyoshi H, Harada N, Nakamura T, Miyamoto Y, Kanatani A, Tamai Y (2006) Targeted disruption of G protein-coupled bile acid receptor 1 (Gpbar1/M-Bar) in mice. J Endocrinol 191(1):197–205. https://doi.org/10.1677/joe.1.06546

Mashimo H, Wu DC, Podolsky DK, Fishman MC (1996) Impaired defense of intestinal mucosa in mice lacking intestinal trefoil factor. Science 274(5285):262–265

Matsumoto M, Kibe R, Ooga T, Aiba Y, Kurihara S, Sawaki E, Koga Y, Benno Y (2012) Impact of intestinal microbiota on intestinal luminal metabolome. Sci Rep 2:233. https://doi.org/10.1038/srep00233

McGuckin MA, Linden SK, Sutton P, Florin TH (2011) Mucin dynamics and enteric pathogens. Nat Rev Microbiol 9(4):265–278. https://doi.org/10.1038/nrmicro2538

Meyer KA, Benton TZ, Bennett BJ, Jacobs DR Jr, Lloyd-Jones DM, Gross MD, Carr JJ, Gordon-Larsen P, Zeisel SH (2016) Microbiota-dependent metabolite trimethylamine N-oxide and coronary artery calcium in the coronary artery risk development in young adults study (CARDIA). J Am Heart Assoc 5(10). https://doi.org/10.1161/JAHA.116.003970

Miller CA, Corbin KD, da Costa KA, Zhang S, Zhao X, Galanko JA, Blevins T, Bennett BJ, O'Connor A, Zeisel SH (2014) Effect of egg ingestion on trimethylamine-N-oxide production in humans: a randomized, controlled, dose-response study. Am J Clin Nutr 100(3):778–786. https://doi.org/10.3945/ajcn.114.087692

Moschetta A, Bookout AL, Mangelsdorf DJ (2004) Prevention of cholesterol gallstone disease by FXR agonists in a mouse model. Nat Med 10(12):1352–1358. https://doi.org/10.1038/nm1138

Mukherjee S, Vaishnava S, Hooper LV (2008) Multi-layered regulation of intestinal antimicrobial defense. Cell Mol Life Sci 65(19):3019–3027. https://doi.org/10.1007/s00018-008-8182-3

Nicolas S, Blasco-Baque V, Fournel A, Gilleron J, Klopp P, Waget A, Ceppo F, Marlin A, Padmanabhan R, Iacovoni JS, Terce F, Cani PD, Tanti JF, Burcelin R, Knauf C, Cormont M, Serino M (2017) Transfer of dysbiotic gut microbiota has beneficial effects on host liver metabolism. Mol Syst Biol 13(3):921. 10.15252/msb.20167356

Nohr MK, Pedersen MH, Gille A, Egerod KL, Engelstoft MS, Husted AS, Sichlau RM, Grunddal KV, Poulsen SS, Han S, Jones RM, Offermanns S, Schwartz TW (2013) GPR41/FFAR3 and GPR43/FFAR2 as cosensors for short-chain fatty acids in enteroendocrine cells vs FFAR3 in enteric neurons and FFAR2 in enteric leukocytes. Endocrinology 154(10):3552–3564. https://doi.org/10.1210/en.2013-1142

Ottman N, Reunanen J, Meijerink M, Pietila TE, Kainulainen V, Klievink J, Huuskonen L, Aalvink S, Skurnik M, Boeren S, Satokari R, Mercenier A, Palva A, Smidt H, de Vos WM, Belzer C (2017) Pili-like proteins of Akkermansia muciniphila modulate host immune responses and gut barrier function. PloS One 12(3):e0173004. https://doi.org/10.1371/journal.pone.0173004

Ouwerkerk JP, van der Ark KC, Davids M, Claassens NJ, Robert Finestra T, de Vos WM, Belzer C (2016) Adaptation of Akkermansia muciniphila to the oxic-anoxic interface of the mucus layer. Appl Environ Microbiol. https://doi.org/10.1128/AEM.01641-16

Palmer C, Bik EM, DiGiulio DB, Relman DA, Brown PO (2007) Development of the human infant intestinal microbiota. PLoS Biol 5(7):e177. https://doi.org/10.1371/journal.pbio.0050177

Parks DJ, Blanchard SG, Bledsoe RK, Chandra G, Consler TG, Kliewer SA, Stimmel JB, Willson TM, Zavacki AM, Moore DD, Lehmann JM (1999) Bile acids: natural ligands for an orphan nuclear receptor. Science 284(5418):1365–1368

Perry RJ, Peng L, Barry NA, Cline GW, Zhang D, Cardone RL, Petersen KF, Kibbey RG, Goodman AL, Shulman GI (2016) Acetate mediates a microbiome-brain-beta-cell axis to promote metabolic syndrome. Nature 534(7606):213–217. https://doi.org/10.1038/nature18309

Pircher PC, Kitto JL, Petrowski ML, Tangirala RK, Bischoff ED, Schulman IG, Westin SK (2003) Farnesoid X receptor regulates bile acid-amino acid conjugation. J Biol Chem 278(30):27703–27711. https://doi.org/10.1074/jbc.M302128200

Plovier H, Everard A, Druart C, Depommier C, Van Hul M, Geurts L, Chilloux J, Ottman N, Duparc T, Lichtenstein L, Myridakis A, Delzenne NM, Klievink J, Bhattacharjee A, van der Ark KC, Aalvink S, Martinez LO, Dumas ME, Maiter D, Loumaye A, Hermans MP, Thissen JP, Belzer C, de Vos WM, Cani PD (2017) A purified membrane protein from Akkermansia muciniphila or the pasteurized bacterium improves metabolism in obese and diabetic mice. Nat Med 23(1):107–113. https://doi.org/10.1038/nm.4236

Poggi M, Bastelica D, Gual P, Iglesias MA, Gremeaux T, Knauf C, Peiretti F, Verdier M, Juhan-Vague I, Tanti JF, Burcelin R, Alessi MC (2007) C3H/HeJ mice carrying a toll-like receptor 4 mutation are protected against the development of insulin resistance in white adipose tissue in response to a high-fat diet. Diabetologia 50(6):1267–1276. https://doi.org/10.1007/s00125-007-0654-8

Qin J, Li R, Raes J, Arumugam M, Burgdorf KS, Manichanh C, Nielsen T, Pons N, Levenez F, Yamada T, Mende DR, Li J, Xu J, Li S, Li D, Cao J, Wang B, Liang H, Zheng H, Xie Y, Tap J, Lepage P, Bertalan M, Batto JM, Hansen T, Le Paslier D, Linneberg A, Nielsen HB, Pelletier E, Renault P, Sicheritz-Ponten T, Turner K, Zhu H, Yu C, Li S, Jian M, Zhou Y, Li Y, Zhang X, Li S, Qin N, Yang H, Wang J, Brunak S, Dore J, Guarner F, Kristiansen K, Pedersen O, Parkhill J, Weissenbach J, Meta HITC, Bork P, Ehrlich SD, Wang J (2010) A human gut microbial gene catalogue established by metagenomic sequencing. Nature 464 (7285):59–65. https://doi.org/10.1038/nature08821

Rappe MS, Giovannoni SJ (2003) The uncultured microbial majority. Annu Rev Microbiol 57:369–394. https://doi.org/10.1146/annurev.micro.57.030502.090759

Reunanen J, Kainulainen V, Huuskonen L, Ottman N, Belzer C, Huhtinen H, de Vos WM, Satokari R (2015) Akkermansia muciniphila adheres to enterocytes and strengthens the integrity of the epithelial cell layer. Appl Environ Microbiol 81(11):3655–3662. https://doi.org/10.1128/AEM.04050-14

Ridaura VK, Faith JJ, Rey FE, Cheng J, Duncan AE, Kau AL, Griffin NW, Lombard V, Henrissat B, Bain JR, Muehlbauer MJ, Ilkayeva O, Semenkovich CF, Funai K, Hayashi DK, Lyle BJ, Martini MC, Ursell LK, Clemente JC, Van Treuren W, Walters WA, Knight R, Newgard CB, Heath AC, Gordon JI (2013) Gut microbiota from twins discordant for obesity modulate metabolism in mice. Science 341(6150):1241214. https://doi.org/10.1126/science.1241214

Ridlon JM, Kang DJ, Hylemon PB (2006) Bile salt biotransformations by human intestinal bacteria. J Lipid Res 47(2):241–259. https://doi.org/10.1194/jlr.R500013-JLR200

Riesenfeld CS, Schloss PD, Handelsman J (2004) Metagenomics: genomic analysis of microbial communities. Annu Rev Genet 38:525–552. https://doi.org/10.1146/annurev.genet.38.072902.091216

Rodriguez-Nunez I, Caluag T, Kirby K, Rudick CN, Dziarski R, Gupta D (2017) Nod2 and Nod2-regulated microbiota protect BALB/c mice from diet-induced obesity and metabolic dysfunction. Sci Rep 7(1):548. https://doi.org/10.1038/s41598-017-00484-2

Royall D, Wolever TM, Jeejeebhoy KN (1990) Clinical significance of colonic fermentation. Am J Gastroenterol 85(10):1307–1312

Ruan JW, Statt S, Huang CT, Tsai YT, Kuo CC, Chan HL, Liao YC, Tan TH, Kao CY (2016) Dual-specificity phosphatase 6 deficiency regulates gut microbiome and transcriptome response against diet-induced obesity in mice. Nat Microbiol 2:16220. https://doi.org/10.1038/nmicrobiol.2016.220

Said HM (2011) Intestinal absorption of water-soluble vitamins in health and disease. Biochem J 437(3):357–372

Samuel BS, Shaito A, Motoike T, Rey FE, Backhed F, Manchester JK, Hammer RE, Williams SC, Crowley J, Yanagisawa M, Gordon JI (2008) Effects of the gut microbiota on host adiposity are modulated by the short-chain fatty-acid binding G protein-coupled receptor, Gpr41. Proc Natl Acad Sci USA 105(43):16767–16772. https://doi.org/10.1073/pnas.0808567105

Sarker SA, Sultana S, Reuteler G, Moine D, Descombes P, Charton F, Bourdin G, McCallin S, Ngom-Bru C, Neville T, Akter M, Huq S, Qadri F, Talukdar K, Kassam M, Delley M, Loiseau C, Deng Y, El Aidy S, Berger B, Brussow H (2016) Oral phage therapy of acute bacterial diarrhea with two coliphage preparations: a randomized trial in children from Bangladesh. EBioMedicine 4:124–137. https://doi.org/10.1016/j.ebiom.2015.12.023

Sayin SI, Wahlstrom A, Felin J, Jantti S, Marschall HU, Bamberg K, Angelin B, Hyotylainen T, Oresic M, Backhed F (2013) Gut microbiota regulates bile acid metabolism by reducing the levels of tauro-beta-muricholic acid, a naturally occurring FXR antagonist. Cell Metab 17 (2):225–235. https://doi.org/10.1016/j.cmet.2013.01.003

Schugar RC, Shih DM, Warrier M, Helsley RN, Burrows A, Ferguson D, Brown AL, Gromovsky AD, Heine M, Chatterjee A, Li L, Li XS, Wang Z, Willard B, Meng Y, Kim H, Che N, Pan C, Lee RG, Crooke RM, Graham MJ, Morton RE, Langefeld CD, Das SK, Rudel LL, Zein N, McCullough AJ, Dasarathy S, Tang WHW, Erokwu BO, Flask CA, Laakso M, Civelek M, Naga Prasad SV, Heeren J, Lusis AJ, Hazen SL, Brown JM (2017) The TMAO-producing enzyme flavin-containing monooxygenase 3 regulates obesity and the beiging of white adipose tissue. Cell Rep 19(12):2451–2461. https://doi.org/10.1016/j.celrep.2017.05.077

Schwartz M, Gluck M, Koon S (2013) Norovirus gastroenteritis after fecal microbiota transplantation for treatment of Clostridium difficile infection despite asymptomatic donors and lack of sick contacts. Am J Gastroenterol 108(8):1367. https://doi.org/10.1038/ajg.2013.164

Secher T, Brehin C, Oswald E (2016) Early settlers: which *E. coli* strains do you not want at birth? Am J Physiol Gastrointest Liver Physiol 311(1):G123–G129. https://doi.org/10.1152/ajpgi.00091.2016

Seki E, De Minicis S, Osterreicher CH, Kluwe J, Osawa Y, Brenner DA, Schwabe RF (2007) TLR4 enhances TGF-beta signaling and hepatic fibrosis. Nat Med 13(11):1324–1332

Sekirov I, Russell SL, Antunes LC, Finlay BB (2010) Gut microbiota in health and disease. Physiol Rev 90(3):859–904

Sender R, Fuchs S, Milo R (2016a) Are we really vastly outnumbered? Revisiting the ratio of bacterial to host cells in humans. Cell 164(3):337–340. https://doi.org/10.1016/j.cell.2016.01.013

Sender R, Fuchs S, Milo R (2016b) Revised estimates for the number of human and bacteria cells in the body. PLoS Biol 14(8):e1002533. https://doi.org/10.1371/journal.pbio.1002533

Shao B, Lu M, Katz SC, Varley AW, Hardwick J, Rogers TE, Ojogun N, Rockey DC, Dematteo RP, Munford RS (2007) A host lipase detoxifies bacterial lipopolysaccharides in the liver and spleen. J Biol Chem 282(18):13726–13735

Shen TC, Albenberg L, Bittinger K, Chehoud C, Chen YY, Judge CA, Chau L, Ni J, Sheng M, Lin A, Wilkins BJ, Buza EL, Lewis JD, Daikhin Y, Nissim I, Yudkoff M, Bushman FD, Wu GD (2015) Engineering the gut microbiota to treat hyperammonemia. J Clin Invest 125 (7):2841–2850. https://doi.org/10.1172/JCI79214

Shin NR, Lee JC, Lee HY, Kim MS, Whon TW, Lee MS, Bae JW (2014) An increase in the Akkermansia spp. population induced by metformin treatment improves glucose homeostasis in diet-induced obese mice. Gut 63(5):727–735. https://doi.org/10.1136/gutjnl-2012-303839

Shin NR, Whon TW, Bae JW (2015) Proteobacteria: microbial signature of dysbiosis in gut microbiota. Trends Biotechnol 33(9):496–503. https://doi.org/10.1016/j.tibtech.2015.06.011

Shoaie S, Karlsson F, Mardinoglu A, Nookaew I, Bordel S, Nielsen J (2013) Understanding the interactions between bacteria in the human gut through metabolic modeling. Sci Rep 3:2532

Shulzhenko N, Morgun A, Hsiao W, Battle M, Yao M, Gavrilova O, Orandle M, Mayer L, Macpherson AJ, McCoy KD, Fraser-Liggett C, Matzinger P (2011) Crosstalk between B lymphocytes, microbiota and the intestinal epithelium governs immunity versus metabolism in the gut. Nat Med 17(12):1585–1593. https://doi.org/10.1038/nm.2505

Sinal CJ, Tohkin M, Miyata M, Ward JM, Lambert G, Gonzalez FJ (2000) Targeted disruption of the nuclear receptor FXR/BAR impairs bile acid and lipid homeostasis. Cell 102(6):731–744

Singh V, Chassaing B, Zhang L, San Yeoh B, Xiao X, Kumar M, Baker MT, Cai J, Walker R, Borkowski K, Harvatine KJ, Singh N, Shearer GC, Ntambi JM, Joe B, Patterson AD, Gewirtz AT, Vijay-Kumar M (2015) Microbiota-dependent hepatic lipogenesis mediated by Stearoyl CoA Desaturase 1 (SCD1) promotes metabolic syndrome in TLR5-deficient mice. Cell Metab 22(6):983–996. https://doi.org/10.1016/j.cmet.2015.09.028

Skutches CL, Holroyde CP, Myers RN, Paul P, Reichard GA (1979) Plasma acetate turnover and oxidation. J Clin Invest 64(3):708–713

Slack E, Hapfelmeier S, Stecher B, Velykoredko Y, Stoel M, Lawson MA, Geuking MB, Beutler B, Tedder TF, Hardt WD, Bercik P, Verdu EF, McCoy KD, Macpherson AJ (2009)

Innate and adaptive immunity cooperate flexibly to maintain host-microbiota mutualism. Science 325(5940):617–620. https://doi.org/10.1126/science.1172747

Snedeker SM, Hay AG (2012) Do interactions between gut ecology and environmental chemicals contribute to obesity and diabetes? Environ Health Perspect 120(3):332–339. https://doi.org/10.1289/ehp.1104204

Spaulding CN, Klein RD, Ruer S, Kau AL, Schreiber HL, Cusumano ZT, Dodson KW, Pinkner JS, Fremont DH, Janetka JW, Remaut H, Gordon JI, Hultgren SJ (2017) Selective depletion of uropathogenic *E. coli* from the gut by a FimH antagonist. Nature 546(7659):528–532. https://doi.org/10.1038/nature22972

Sridharan GV, Choi K, Klemashevich C, Wu C, Prabakaran D, Pan LB, Steinmeyer S, Mueller C, Yousofshahi M, Alaniz RC, Lee K, Jayaraman A (2014) Prediction and quantification of bioactive microbiota metabolites in the mouse gut. Nat Commun 5:5492. https://doi.org/10.1038/ncomms6492

Suarez-Zamorano N, Fabbiano S, Chevalier C, Stojanovic O, Colin DJ, Stevanovic A, Veyrat-Durebex C, Tarallo V, Rigo D, Germain S, Ilievska M, Montet X, Seimbille Y, Hapfelmeier S, Trajkovski M (2015) Microbiota depletion promotes browning of white adipose tissue and reduces obesity. Nat Med 21(12):1497–1501. https://doi.org/10.1038/nm.3994

Suez J, Korem T, Zeevi D, Zilberman-Schapira G, Thaiss CA, Maza O, Israeli D, Zmora N, Gilad S, Weinberger A, Kuperman Y, Harmelin A, Kolodkin-Gal I, Shapiro H, Halpern Z, Segal E, Elinav E (2014) Artificial sweeteners induce glucose intolerance by altering the gut microbiota. Nature 514(7521):181–186. https://doi.org/10.1038/nature13793

Tang WH, Wang Z, Levison BS, Koeth RA, Britt EB, Fu X, Wu Y, Hazen SL (2013) Intestinal microbial metabolism of phosphatidylcholine and cardiovascular risk. N Engl J Med 368(17):1575–1584. https://doi.org/10.1056/NEJMoa1109400

Tang WH, Wang Z, Fan Y, Levison B, Hazen JE, Donahue LM, Wu Y, Hazen SL (2014) Prognostic value of elevated levels of intestinal microbe-generated metabolite trimethylamine-N-oxide in patients with heart failure: refining the gut hypothesis. J Am Coll Cardiol 64(18):1908–1914. https://doi.org/10.1016/j.jacc.2014.02.617

Tang C, Ahmed K, Gille A, Lu S, Grone HJ, Tunaru S, Offermanns S (2015) Loss of FFA2 and FFA3 increases insulin secretion and improves glucose tolerance in type 2 diabetes. Nat Med 21(2):173–177

Tang WH, Wang Z, Li XS, Fan Y, Li DS, Wu Y, Hazen SL (2017) Increased trimethylamine N-oxide portends high mortality risk independent of glycemic control in patients with type 2 diabetes mellitus. Clin Chem 63(1):297–306. https://doi.org/10.1373/clinchem.2016.263640

Thomas C, Gioiello A, Noriega L, Strehle A, Oury J, Rizzo G, Macchiarulo A, Yamamoto H, Mataki C, Pruzanski M, Pellicciari R, Auwerx J, Schoonjans K (2009) TGR5-mediated bile acid sensing controls glucose homeostasis. Cell Metab 10(3):167–177. https://doi.org/10.1016/j.cmet.2009.08.001

Topping DL, Clifton PM (2001) Short-chain fatty acids and human colonic function: roles of resistant starch and nonstarch polysaccharides. Physiol Rev 81(3):1031–1064

Trasande L, Blustein J, Liu M, Corwin E, Cox LM, Blaser MJ (2013) Infant antibiotic exposures and early-life body mass. Int J Obes 37(1):16–23. https://doi.org/10.1038/ijo.2012.132

Tremaroli V, Karlsson F, Werling M, Stahlman M, Kovatcheva-Datchary P, Olbers T, Fandriks L, le Roux CW, Nielsen J, Backhed F (2015) Roux-en-Y gastric bypass and vertical banded gastroplasty induce long-term changes on the human gut microbiome contributing to fat mass regulation. Cell Metab 22(2):228–238. https://doi.org/10.1016/j.cmet.2015.07.009

Tsukumo DM, Carvalho-Filho MA, Carvalheira JB, Prada PO, Hirabara SM, Schenka AA, Araujo EP, Vassallo J, Curi R, Velloso LA, Saad MJ (2007) Loss-of-function mutation in Toll-like receptor 4 prevents diet-induced obesity and insulin resistance. Diabetes 56(8):1986–1998. https://doi.org/10.2337/db06-1595

Turnbaugh PJ, Ley RE, Mahowald MA, Magrini V, Mardis ER, Gordon JI (2006) An obesity-associated gut microbiome with increased capacity for energy harvest. Nature 444(7122):1027–1031. https://doi.org/10.1038/nature05414

Turnbaugh PJ, Hamady M, Yatsunenko T, Cantarel BL, Duncan A, Ley RE, Sogin ML, Jones WJ, Roe BA, Affourtit JP, Egholm M, Henrissat B, Heath AC, Knight R, Gordon JI (2009) A core gut microbiome in obese and lean twins. Nature 457(7228):480–484. https://doi.org/10.1038/nature07540

Ubeda C, Lipuma L, Gobourne A, Viale A, Leiner I, Equinda M, Khanin R, Pamer EG (2012) Familial transmission rather than defective innate immunity shapes the distinct intestinal microbiota of TLR-deficient mice. J Exp Med 209(8):1445–1456. https://doi.org/10.1084/jem.20120504

Vaishnava S, Yamamoto M, Severson KM, Ruhn KA, Yu X, Koren O, Ley R, Wakeland EK, Hooper LV (2011) The antibacterial lectin RegIIIgamma promotes the spatial segregation of microbiota and host in the intestine. Science 334(6053):255–258. https://doi.org/10.1126/science.1209791

Van Nood E, Vrieze A, Nieuwdorp M, Fuentes S, Zoetendal EG, de Vos WM, Visser CE, Kuijper EJ, Bartelsman JF, Tijssen JG, Speelman P, Dijkgraaf MG, Keller JJ (2013) Duodenal infusion of donor feces for recurrent Clostridium difficile. N Engl J Med 368(5):407–415. https://doi.org/10.1056/NEJMoa1205037

Vance DE (2008) Role of phosphatidylcholine biosynthesis in the regulation of lipoprotein homeostasis. Curr Opin Lipidol 19(3):229–234

Vandanmagsar B, Youm YH, Ravussin A, Galgani JE, Stadler K, Mynatt RL, Ravussin E, Stephens JM, Dixit VD (2011) The NLRP3 inflammasome instigates obesity-induced inflammation and insulin resistance. Nat Med 17(2):179–188

Vatanen T, Kostic AD, d'Hennezel E, Siljander H, Franzosa EA, Yassour M, Kolde R, Vlamakis H, Arthur TD, Hamalainen AM, Peet A, Tillmann V, Uibo R, Mokurov S, Dorshakova N, Ilonen J, Virtanen SM, Szabo SJ, Porter JA, Lahdesmaki H, Huttenhower C, Gevers D, Cullen TW, Knip M, Group DS, Xavier RJ (2016) Variation in microbiome LPS immunogenicity contributes to autoimmunity in humans. Cell 165(4):842–853. https://doi.org/10.1016/j.cell.2016.04.007

Verspreet J, Damen B, Broekaert WF, Verbeke K, Delcour JA, Courtin CM (2016) A critical look at prebiotics within the dietary fiber concept. Annu Rev Food Sci Technol 7:167–190. https://doi.org/10.1146/annurev-food-081315-032749

Vijay-Kumar M, Sanders CJ, Taylor RT, Kumar A, Aitken JD, Sitaraman SV, Neish AS, Uematsu S, Akira S, Williams IR, Gewirtz AT (2007) Deletion of TLR5 results in spontaneous colitis in mice. J Clin Invest 117(12):3909–3921. https://doi.org/10.1172/JCI33084

Vijay-Kumar M, Aitken JD, Carvalho FA, Cullender TC, Mwangi S, Srinivasan S, Sitaraman SV, Knight R, Ley RE, Gewirtz AT (2010) Metabolic syndrome and altered gut microbiota in mice lacking Toll-like receptor 5. Science 328(5975):228–231. https://doi.org/10.1126/science.1179721

Von Mering C, Hugenholtz P, Raes J, Tringe SG, Doerks T, Jensen LJ, Ward N, Bork P (2007) Quantitative phylogenetic assessment of microbial communities in diverse environments. Science 315(5815):1126–1130. https://doi.org/10.1126/science.1133420

Vreugdenhil AC, Rousseau CH, Hartung T, Greve JW, van't Veer C, Buurman WA (2003) Lipopolysaccharide (LPS)-binding protein mediates LPS detoxification by chylomicrons. J Immunol 170(3):1399–1405

Vrieze A, Van Nood E, Holleman F, Salojarvi J, Kootte RS, Bartelsman JF, Dallinga-Thie GM, Ackermans MT, Serlie MJ, Oozeer R, Derrien M, Druesne A, Van Hylckama Vlieg JE, Bloks VW, Groen AK, Heilig HG, Zoetendal EG, Stroes ES, de Vos WM, Hoekstra JB, Nieuwdorp M (2012) Transfer of intestinal microbiota from lean donors increases insulin sensitivity in individuals with metabolic syndrome. Gastroenterology 143(4):913–916.e917. https://doi.org/10.1053/j.gastro.2012.06.031

Wall R, Cryan JF, Ross RP, Fitzgerald GF, Dinan TG, Stanton C (2014) Bacterial neuroactive compounds produced by psychobiotics. Adv Exp Med Biol 817:221–239

Wang H, Chen J, Hollister K, Sowers LC, Forman BM (1999) Endogenous bile acids are ligands for the nuclear receptor FXR/BAR. Molecular Cell 3(5):543–553

Wang Y, Beydoun MA, Liang L, Caballero B, Kumanyika SK (2008) Will all Americans become overweight or obese? Estimating the progression and cost of the US obesity epidemic. Obesity (Silver Spring) 16(10):2323–2330. https://doi.org/10.1038/oby.2008.351

Wang Z, Klipfell E, Bennett BJ, Koeth R, Levison BS, Dugar B, Feldstein AE, Britt EB, Fu X, Chung YM, Wu Y, Schauer P, Smith JD, Allayee H, Tang WH, DiDonato JA, Lusis AJ, Hazen SL (2011) Gut flora metabolism of phosphatidylcholine promotes cardiovascular disease. Nature 472(7341):57–63. https://doi.org/10.1038/nature09922

Wang D, Xia M, Yan X, Li D, Wang L, Xu Y, Jin T, Ling W (2012) Gut microbiota metabolism of anthocyanin promotes reverse cholesterol transport in mice via repressing miRNA-10b. Circ Res 111(8):967–981. https://doi.org/10.1161/CIRCRESAHA.112.266502

Wang Z, Roberts AB, Buffa JA, Levison BS, Zhu W, Org E, Gu X, Huang Y, Zamanian-Daryoush M, Culley MK, DiDonato AJ, Fu X, Hazen JE, Krajcik D, DiDonato JA, Lusis AJ, Hazen SL (2015) Non-lethal inhibition of gut microbial trimethylamine production for the treatment of atherosclerosis. Cell 163(7):1585–1595. https://doi.org/10.1016/j.cell.2015.11.055

Warrier M, Shih DM, Burrows AC, Ferguson D, Gromovsky AD, Brown AL, Marshall S, McDaniel A, Schugar RC, Wang Z, Sacks J, Rong X, Vallim TA, Chou J, Ivanova PT, Myers DS, Brown HA, Lee RG, Crooke RM, Graham MJ, Liu X, Parini P, Tontonoz P, Lusis AJ, Hazen SL, Temel RE, Brown JM (2015) The TMAO-generating enzyme flavin monooxygenase 3 is a central regulator of cholesterol balance. Cell Rep. https://doi.org/10.1016/j.celrep.2014.12.036

Watanabe M, Houten SM, Mataki C, Christoffolete MA, Kim BW, Sato H, Messaddeq N, Harney JW, Ezaki O, Kodama T, Schoonjans K, Bianco AC, Auwerx J (2006) Bile acids induce energy expenditure by promoting intracellular thyroid hormone activation. Nature 439(7075):484–489. https://doi.org/10.1038/nature04330

Watanabe A, Sohail MA, Gomes DA, Hashmi A, Nagata J, Sutterwala FS, Mahmood S, Jhandier MN, Shi Y, Flavell RA, Mehal WZ (2009) Inflammasome-mediated regulation of hepatic stellate cells. Am J Physiol Gastrointest Liver Physiol 296(6):G1248–G1257

Weingarden AR, Chen C, Bobr A, Yao D, Lu Y, Nelson VM, Sadowsky MJ, Khoruts A (2014) Microbiota transplantation restores normal fecal bile acid composition in recurrent Clostridium difficile infection. Am J Physiol Gastrointest Liver Physiol 306(4):G310–G319

Wen L, Ley RE, Volchkov PY, Stranges PB, Avanesyan L, Stonebraker AC, Hu C, Wong FS, Szot GL, Bluestone JA, Gordon JI, Chervonsky AV (2008) Innate immunity and intestinal microbiota in the development of Type 1 diabetes. Nature 455(7216):1109–1113. https://doi.org/10.1038/nature07336

Wigg AJ, Roberts-Thomson IC, Dymock RB, McCarthy PJ, Grose RH, Cummins AG (2001) The role of small intestinal bacterial overgrowth, intestinal permeability, endotoxaemia, and tumour necrosis factor alpha in the pathogenesis of non-alcoholic steatohepatitis. Gut 48(2):206–211

Wikoff WR, Anfora AT, Liu J, Schultz PG, Lesley SA, Peters EC, Siuzdak G (2009) Metabolomics analysis reveals large effects of gut microflora on mammalian blood metabolites. Proc Natl Acad Sci USA 106(10):3698–3703. https://doi.org/10.1073/pnas.0812874106

Woese CR (1987) Bacterial evolution. Microbiol Rev 51(2):221–271

Woese CR, Fox GE (1977) Phylogenetic structure of the prokaryotic domain: the primary kingdoms. Proc Natl Acad Sci USA 74(11):5088–5090

Wolever TM, Brighenti F, Royall D, Jenkins AL, Jenkins DJ (1989) Effect of rectal infusion of short chain fatty acids in human subjects. Am J Gastroenterol 84(9):1027–1033

Wong JM, de Souza R, Kendall CW, Emam A, Jenkins DJ (2006) Colonic health: fermentation and short chain fatty acids. J Clin Gastroenterol 40(3):235–243

Worthmann A, John C, Ruhlemann MC, Baguhl M, Heinsen FA, Schaltenberg N, Heine M, Schlein C, Evangelakos I, Mineo C, Fischer M, Dandri M, Kremoser C, Scheja L, Franke A, Shaul PW, Heeren J (2017) Cold-induced conversion of cholesterol to bile acids in mice shapes the gut microbiome and promotes adaptive thermogenesis. Nat Med. https://doi.org/10.1038/nm.4357

Wrzosek L, Miquel S, Noordine ML, Bouet S, Joncquel Chevalier-Curt M, Robert V, Philippe C, Bridonneau C, Cherbuy C, Robbe-Masselot C, Langella P, Thomas M (2013) Bacteroides thetaiotaomicron and Faecalibacterium prausnitzii influence the production of mucus glycans and the development of goblet cells in the colonic epithelium of a gnotobiotic model rodent. BMC Biol 11:61. https://doi.org/10.1186/1741-7007-11-61

Wu GD, Chen J, Hoffmann C, Bittinger K, Chen YY, Keilbaugh SA, Bewtra M, Knights D, Walters WA, Knight R, Sinha R, Gilroy E, Gupta K, Baldassano R, Nessel L, Li H, Bushman FD, Lewis JD (2011) Linking long-term dietary patterns with gut microbial enterotypes. Science 334(6052):105–108. https://doi.org/10.1126/science.1208344

Yamashita H, Fujisawa K, Ito E, Idei S, Kawaguchi N, Kimoto M, Hiemori M, Tsuji H (2007) Improvement of obesity and glucose tolerance by acetate in type 2 diabetic Otsuka Long-Evans Tokushima Fatty (OLETF) rats. Biosci Biotechnol Biochem 71(5):1236–1243

Yatsunenko T, Rey FE, Manary MJ, Trehan I, Dominguez-Bello MG, Contreras M, Magris M, Hidalgo G, Baldassano RN, Anokhin AP, Heath AC, Warner B, Reeder J, Kuczynski J, Caporaso JG, Lozupone CA, Lauber C, Clemente JC, Knights D, Knight R, Gordon JI (2012) Human gut microbiome viewed across age and geography. Nature 486 (7402):222–227. https://doi.org/10.1038/nature11053

Yoshimoto S, Loo TM, Atarashi K, Kanda H, Sato S, Oyadomari S, Iwakura Y, Oshima K, Morita H, Hattori M, Honda K, Ishikawa Y, Hara E, Ohtani N (2013) Obesity-induced gut microbial metabolite promotes liver cancer through senescence secretome. Nature 499 (7456):97–101. https://doi.org/10.1038/nature12347

Zambell KL, Fitch MD, Fleming SE (2003) Acetate and butyrate are the major substrates for de novo lipogenesis in rat colonic epithelial cells. J Nutr 133(11):3509–3515

Zeevi D, Korem T, Zmora N, Israeli D, Rothschild D, Weinberger A, Ben-Yacov O, Lador D, Avnit-Sagi T, Lotan-Pompan M, Suez J, Mahdi JA, Matot E, Malka G, Kosower N, Rein M, Zilberman-Schapira G, Dohnalova L, Pevsner-Fischer M, Bikovsky R, Halpern Z, Elinav E, Segal E (2015) Personalized nutrition by prediction of glycemic responses. Cell 163 (5):1079–1094. https://doi.org/10.1016/j.cell.2015.11.001

Zhang W, Hartmann R, Tun HM, Elson CO, Khafipour E, Garvey WT (2016) Deletion of the toll-like receptor 5 gene per se does not determine the gut microbiome profile that induces metabolic syndrome: environment trumps genotype. PloS One 11(3):e0150943. https://doi.org/10.1371/journal.pone.0150943

Zhu W, Wang Z, Tang WHW, Hazen SL (2017) Gut microbe-generated trimethylamine N-oxide from dietary choline is prothrombotic in subjects. Circulation 135(17):1671–1673. https://doi.org/10.1161/CIRCULATIONAHA.116.025338

Zuckerkandl E, Pauling L (1965) Molecules as documents of evolutionary history. J Theor Biol 8 (2):357–366

Chapter 8
The Influence of the Microbiota on the Etiology of Colorectal Cancer

Melissa C. Kordahi and R. William DePaolo

Abstract The microbiome of the gastrointestinal (GI) tract is estimated to comprise 39 trillion organisms that act in a symbiotic relationship with the surrounding tissue cells to maintain homeostasis. Constituents of the gut microbiota occupy either a planktonic niche within the fecal stream, are adherent to the gut mucosa, or are associated with the mucous layer. Alterations in the gut microbiota at any of these levels, caused by the genetics of an individual or by environmental factors, can disturb this homeostatic relationship and promote disease such as colorectal cancer (CRC). CRC is the third most common form of cancer in both men and women and the second leading cause of cancer-related death in the USA, representing a considerable disease burden. The intimate association between the microbiota and the cells of the colon sets the stage for a number of interactions that may contribute to carcinogenesis. Although only a few specific commensal species may play a direct causal role in CRC, more general shifts in the composition may promote local inflammation through the engagement of innate immune receptors encoded within the colonic tissue. Changes in gene expression within the microbiota may also be important as virulence factors are altered and metabolites are produced that may have detrimental effects on the tissue. In this chapter, we explore the theoretical bodyworks through which certain members of the microbiota are believed to cause CRC, the sensing of microbiota-associated molecular patterns by innate immune receptors known as toll-like-receptors (TLRs) and the various strategies aimed at manipulating the microbiota and targeting the TLRs, in the hope of developing new treatment approaches.

Keywords Colon cancer • Microbes • Microbiota • Innate immunity • Toll-like receptors

---

M.C. Kordahi • R. William DePaolo (✉)
Department of Pathology, University of Washington, Seattle, WA, USA
e-mail: wdepaolo@medicine.washington.edu

J. Sun, P.K. Dudeja (eds.), *Mechanisms Underlying Host-Microbiome Interactions in Pathophysiology of Human Diseases*, Physiology in Health and Disease,
https://doi.org/10.1007/978-1-4939-7534-1_8

8.1 Introduction

Colorectal cancer (CRC) is the third most common form of cancer in both men and women and the second leading cause of cancer-related death in the USA (Bardhan and Liu 2013). CRC can be divided into three subtypes: heritable, sporadic, and inflammation-associated (Valle 2014). Family studies have identified mutations in dominant genes such as the adenomatous polyposis coli gene (APC) and mismatch repair genes in only 10–15% of CRC (Valle 2014). These tumors typically develop at earlier ages than the sporadic and inflammation-associated CRC (Valle 2014). Epidemiological studies have shown that the most CRC cases are sporadic, arising from non-shared environmental factors, and rapidly increase in incidence beyond age 50 (Bardhan and Liu 2013). A third subset of CRC, comprising less than 1% of all CRCs, is associated with chronic inflammatory diseases such as Crohn's disease and ulcerative colitis, the two main forms of inflammatory bowel disease (Bardhan and Liu 2013; Karin 2009).

The intestine is home to a large microbial ecosystem that provides protective, structural, and metabolic functions. Due to the proximity of the microbiota to the intestinal epithelium and underlying immune cells, tightly regulated communication must occur to prevent abnormal tissue responses that could lead to chronic inflammation and malignancy. Coordination of intestinal responses are initiated through the recognition of both microbial-associated and host cellular-associated ligands by innate immune receptors, such as the toll-like receptor (TLR) and NOD-like receptor (NLR) families. Thus, mucosal homeostasis in the healthy intestine depends largely on the interplay between commensal microbiota, host genetics, and the immune status of the mucosal tissue. Disruption or imbalances of these signals can lead to uncontrolled inflammation and changes within the microbiota, which play a significant role in tumorigenesis and tumor progression (Grivennikov et al. 2010).

As the intestinal microbiome most likely plays an etiological role in all three subsets of CRC, its importance in CRC has been most clearly illustrated in models using germ-free (GF) mice, which develop less inflammation and fewer tumors than conventionally housed mice (Zackular et al. 2013). Moreover, studies using GF mice colonized with the microbiota from tumor-bearing mice showed a significant increase in tumorigenesis in the colon compared with GF animals colonized with a healthy gut microbiome, further suggesting that the gut microbiome contributes directly to tumorigenesis (Belkaid and Hand 2014). Alterations in the composition of the microbiota have also been identified in both mouse models of cancer and patients with CRC (Sobhani et al. 2011; Chen et al. 2012; Geng et al. 2013). However, it is important to consider that the initiation and progression of colon cancer is likely not due to one unique bacterial species, as many members of the microbiota have been identified as contributors to colon cancer pathogenesis (Sears and Garrett 2014). In that regard, each bacterial species may contribute to carcinogenesis by a distinct microbial signature that could include the production of metabolites and other by-products, stimulation of innate immunity, changes in location, and/or changes in bacterial gene expression.

8.2 Microbiota Associated with CRC

The number of commensal bacteria in a normal healthy gut is equal to the number of our own eukaryotic cells, but there is an even more astonishing amount of genetic diversity that these bacteria contribute to our physiology. It is estimated that for every one of our genes, there are approximately 145 microbial genes. This roughly equals 3.3 million bacterial genes in the gut to the 23,000 in the human genome (Qin et al. 2010). Therefore, although it is important to consider how commensal composition changes in a diseased state, we must also pay close attention to changes in the gene expression of the microbiota. Changes in microbial gene expression may be influenced by intrinsic factors such as polymorphisms in the host genome or the immune status of the mucosal tissue, and they may be influenced by extrinsic factors such as diet, infection, and exposure to xenobiotics. These factors may induce genetic programs in the commensal microbiota, that modulate virulence protein expression, metabolites, genotoxins, and/or carcinogenic molecules, leading to direct neoplastic effects (Sears and Garrett 2014). They may also intensify neoplasia through the induction of local inflammation (Grivennikov et al. 2010). For example, bacterially produced toxins can lead to DNA damage, inhibition of apoptosis or induction of cellular proliferation. In addition to toxin production, bacteria also produce reactive oxygen species (ROS). ROS can directly promote the initiation and progression of carcinogenesis by causing DNA damage or alter cellular signaling and activation pathways, leading to cell survival and proliferation signals (Reuter et al. 2010).

8.2.1 Microbiota Species Most Commonly Correlated with CRC Pathogenesis

Streptococcus gallolyticus Streptococcus gallolyticus is a low-grade pathogen known for its involvement in bacteremia and endocarditis. It belongs to the Firmicutes family, and is a frequent colonizer of the intestinal tract. Interestingly, 25–80% of patients with *S. gallolyticus* in the bloodstream have concomitant colon adenomas (Reynolds et al. 1983). Further studies have shown that *S. gallolyticus* bacteremia is specifically associated with an aggressive form of polyp, the tubular villous adenoma, necessitating careful clinical screening in a certain subset of patients (Hoen et al. 1994). The etiological role of *S. gallolyticus* in CRC is thought to be mediated through specific virulence mechanisms involving adherence and induction of inflammatory factors.

Streptococcus Gallolyticus possesses a pilus protein, encoded by the *pil1* locus, with a collagen-binding domain allowing it to attach to mucosal surfaces and also translocate into systemic circulation. The *pil1* locus is further induced through metabolic processes or promoted by the dysbiosis of the microbiota. This microorganism has been shown to translocate efficiently through a para-cellular epithelial route and promote vasodilatation and capillary permeability, thereby promoting vascularization of neoplasms. *S. gallolyticus* also induces strong inflammatory signals such as cyclooxygenase-2 (COX2), interferon-γ (IFN-γ), interleukin-1β (IL-1β), which may lead to alterations in apoptosis and proliferation, formation of nitric oxide and free radicals that directly cause DNA damage, or production of angiogenic factors such as interleukin-8 IL-8 (Abdulamir et al. 2011; Boleij et al. 2009).

Enterococcus faecalis Another microorganism belonging to the Firmicutes is *E. faecalis*. *E. faecalis* has been linked to CRC pathogenesis because certain strains have the capacity to produce ROS. The high levels of ROS can damage DNA and create genomic instabilities, two events that can lead to transformation in the colon's epithelium. The involvement of certain *E. faecalis* strains in CRC pathogenesis has further been assessed in studies showing that the ROS produced by the microorganism were involved in distal colitis, DNA damage, and cancer in GF IL10−/− mice. On the other hand, *E. faecalis* strains that did not produce ROS induced inflammation, but not tumorigenesis. In addition to ROS, an *E. faecalis* symbiont found in the oral cavity was capable of inducing mucosal macrophages to produce another chromosomal-breaking factor called 4-hydroxy-2-nonenal, a breakdown product of omega-6 polyunsaturated fatty acids. Despite the abundant experimental literature, human studies linking superoxide-producing *E. faecalis* strains to tumorigenesis are lacking (Wang et al. 2012).

Enterotoxigenic Bacteroides fragilis Enterotoxigenic *Bacteroides fragilis* (ETBF) belongs to the Bacteroidetes phylum, and may be considered oncogenic under certain circumstances, because of its virulence factor *B. fragilis* toxin (BFT), or fragilysin. BFT, like ROS, can induce DNA damage in vivo (Sears and Garrett 2014). Further, BFT has been shown to rapidly alter the structure and function of colonic epithelial cells, including the cleavage of the tumor suppressor protein, E-cadherin. E-cadherin is a transmembrane protein confined to epithelial cells and responsible for maintaining the tight junctions between neighboring cells. The extracellular domain of one E-cadherin molecule interacts with E-cadherin molecules on neighboring cells. A prerequisite for intercellular adhesion is the cytoplasmic linkage of E-cadherin to β-catenin. The association of β-catenin at the cell membrane prevents its nuclear translocation and activation of oncogenic signals. The cleavage of E-cadherin by BFT increases cytosolic levels of β-catenin, allowing it to translocate to the nucleus and increase epithelial cell proliferation and expression of proto-oncogenes such as myelocytomatosis viral oncogene (*MYC*) (Sears and Garrett 2014). In ApcMin/+ mice, BFT has been shown to induce colonic hyperplasia and tumor initiation via induction of signal transducer and activator of transcription 3 (STAT3) and a T_H17 inflammatory response (Wu et al. 2009). ApcMin/+ mice develop tumors in the small bowel with limited formation in the colon. However, colonization with ETBF increases tumorigenesis in the distal colon, but not in the small intestine and histological findings showed that colonic adenomas are detectable much faster in mice colonized with ETBF than in ApcMin/+ mice that were ETBF-free. In humans, one study detected ETBF at a significantly higher frequency in the stools of consecutive cases of CRC compared with concurrent hospital-based, age- and gender-matched patients without CRC (Toprak et al. 2006). However, the development of an IL-17 immune response has been linked to a worse prognosis in human CRC, indicating that long-term ETBF colonization may promote colon carcinogenesis in certain predisposed individuals.

Escherichia coli Unlike ETBF, and *E. faecalis*, whose importance in CRC has been identified using mouse models or in preclinical studies, *E. coli* has been isolated from human CRC patients and the importance of this microorganism in the immunopathophysiology of CRC has been verified experimentally (Shen et al. 2010). *E. coli* is a member of the Enterobacteriaceae family and constitutes less than 1% of a healthy individual's fecal microbiota, when performing 16S rDNA sequence analysis. Despite it being found in relatively low numbers compared with other commensal bacteria, *E. coli* is a very common cause of intestinal disease (Apperloo-Renkema et al. 1990). During inflammation, *E. coli* often becomes a dominant member in the gut microbiota after 16S fecal rDNA sequence analysis, a phenotype particularly associated with clinical irritable bowel disease (IBD) and, in animal models, chronic inflammation (Mukhopadhya et al. 2012). Although the molecular mechanism that *E. coli* uses to expand during tumorigenesis is not known, Enterobacteriaceae and other Proteobacteria have evolved a number of strategies to utilize products or by-products formed during an inflammatory response.

Escherichia coli is particularly interesting because, in addition to the changes observed in its abundance, it seems that it can also alter its gene expression in an inflamed gut. Evidence for changes in gene expression come from the analysis of clinical isolates from patients with a chronic disease, such as IBD and CRC. These studies demonstrate that *E. coli* alters its functional characteristics by inducing a more pathogenic phenotype, including an increase in its adherence and invasive abilities (Darfeuille-Michaud et al. 2004). Analysis of *E. coli* strains isolated from IBD and CRC patients has identified a number of genes that encode factors influencing tumorigenesis. The first is cytolethal distending toxin or CDT-V, which can directly cause DNA damage (Nesić et al. 2004). However, only a small number of *E. coli* strains carry this gene. More recently, a natural peptide–polyketide genotoxin called colibactin was identified in *E. coli* isolated from IBD and CRC patients as well (Prorok-Hamon et al. 2014). This genotoxin is encoded by the 54-kb polyketide synthase (PKS) genotoxicity island. The importance of this genotoxicity island was first demonstrated by Jobin and colleagues when a mutant *E. coli* strain harboring a deletion in PKS was still able to induce inflammation, but had less DNA damage, tumor numbers, and bacterial invasion in mice lacking the gene encoding IL-10 (Arthur et al. 2012). Thus, in the context of IL-10 deficiency, carcinogenesis requires expression of a bacterial genotoxin in addition to the genotype-dependent inflammation. PKS may promote oncogenesis via its ability to directly bind DNA and causes double-stranded breaks (Cuevas-Ramos et al. 2010). *E. coli* has been associated with gender-specific differences in CRC development as well. It was demonstrated that hemolytic type I *E. Coli* is significantly associated with adenoma and

CRC in female patients only, upon the analysis of a large number of clinical *E. coli* isolates. This was linked to the activation of the expression of the tumor suppressor BIM by acting in part on hypoxia-induced α-subunit (Jin et al. 2016).

Fusobacterium nucleatum *Fusobacterium Nucleatum* is a fastidious anaerobe, belonging to the Fusobacteria family, that has also been isolated from patients with CRC and its pro-tumorigenic effects have been verified in experimental models. Despite its presence in the colon, *F. nucleatum* is most abundant in the oral cavity where it is associated with dental plaques and gum disease. Although considered an oral bacterium, *F. nucleatum* has been intimately linked to gut inflammation and carcinogenesis, because of its isolation from IBD and CRC patients (Strauss et al. 2011). Similar to *E. coli*, there seem to be pro-tumorigenic effects due to inflammation caused by the expression of the microorganism's own genes. There is a strong correlation between the abundance of *F. nucleatum* and the magnitude of the inflammatory response, especially in terms of tumor necrotic factor-α (TNF-α) and interleukin-10 (IL-10) expression (McCoy et al. 2013). However, the molecular signals that induce these cytokines are not well understood. In fact, the findings to date demonstrate that the induction of pro-inflammatory cytokine responses are likely TLR- and NLR-independent, yet rely on the sensing of *F. nucleatum* by viral-associated innate receptors such as cytoplasmic retinoic acid-inducible gene I (Lee and Tan 2014).

The genetics of *F. nucleatum* are still widely undefined, as this microorganism is difficult to isolate, culture, and manipulate experimentally and clinically. Studies using a periodontal disease-derived *F. nucleatum* strain suggested that the invasive and carcinogenic properties of *F. nucleatum* might be mediated by the activated complex of the FadA adhesin (FadAc), a well-characterized virulence protein (Rubinstein et al. 2013). In vitro colon carcinoma cell-line studies and in vivo tumor xenograft models revealed that FadAc binds to a select extracellular domain of E-cadherin. This binding triggers invasion of the organism and activation of β-catenin/Wnt signaling with stimulation of cell proliferation or tumor growth. Evaluation of tumor tissue from adenoma and adenocarcinoma patients compared with normal colon tissue from nontumorous individuals revealed elevated gene copy numbers of *fadA*, but more interesting was the fact that the highest *fadA* gene copies were detected in cancer tissues and associated with increases in expression of representative Wnt and nuclear factor kappa B (NF-κB) genes (Rubinstein et al. 2013). In another study, *F. nucleatum* was shown to be associated with the promotion of colonic tumor formation in mice through the identification of a host polysaccharide Gal-GalNAc and fusobacterial lectin (*Fap2*) that explained the abundance of the microorganism in CRC. Indeed, *Fap 2* was shown to mediate *F. nucleatum* binding to Gal-GalNAc overexpressed in CRC and targeting host Gal-GalNAc or *Fap2* may provide a way to reduce the *F. nucleatum* drive of CRC (Abed et al. 2016). Anecdotally, *F. nucleatum* was also shown to be more abundant on colon tumors from Spanish individuals compared with tumors from individuals in the USA or Vietnam. This suggests that *F. nucleatum* colonization might vary regionally, indicating that environmental causes may affect the virulence of this microorganism and its involvement in CRC (Kostic et al. 2012).

These five CRC-associated members of the microbiota, *S. gallolyticus*, *E. faecalis*, ETBF, *E. coli*, and *F. nucleatum* have also been suggested to breach the colonic mucus layer and persistently adhere to the mucosa. This mucosal adherence is likely necessary for their oncogenic potential because it allows a more intimate contact with the epithelium. This intimate association could affect the rate of initiation and progression of CRC by promoting inflammation via the stimulation of innate receptors, it could allow direct targeting of factors that may cause DNA damage, or in the context of an individual's genetic makeup and environment, have an impact on cellular turnover (Soler et al. 1999; Grivennikov et al. 2012). Furthermore, alterations in the composition of the microbiome can favor changes in bacterial virulence and metabolism genes, supporting the hypothesis that specific microbes might act sequentially, and in synergy, with certain microbial communities involved in colon carcinogenesis. The findings concerning these microorganisms demonstrate that there are three major mechanisms through which bacteria may contribute to human CRC pathogenesis. The first is that certain members of the colonic microbial community are capable of triggering signaling pathways classically activated during carcinogenesis, such as Wnt and E-cadherin. The second mechanism is through the induction of oxidative stress pathways leading to DNA damage, and the last mechanism is through the activation of signaling pathways leading to an inflammatory cytokine response and the production of interleukin-17 (IL-17), IFN-γ, and TNF-α (Sears and Garrett 2014). However, simplification of these bacterial processes in CRC does not rule out the possibility that additional signaling pathways may be involved, nor does it diminish the possibility that as yet undefined regulatory elements could be targeted by the microbiota. These microorganisms and their genomic potential to induce CRC are summarized in Fig. 8.1.

8.3 The Influence of Microbiota-Derived Metabolites on CRC

As discussed earlier, the microbiota contributes to the immunopathophysiology of CRC via inflammation and modulation of pathways, leading to carcinogenesis. In addition to these pro-tumorigenic effects of the microbiome, accumulating evidence suggests that there might be an influence of the wider microbial community on CRC through its secreted metabolites. Some of the metabolites secreted from the microbiota exert an important beneficial influence on human health, whereas others have been linked to the pathogenesis of cancer by influencing inflammation, DNA damage, and apoptosis, as summarized in Fig. 8.2.

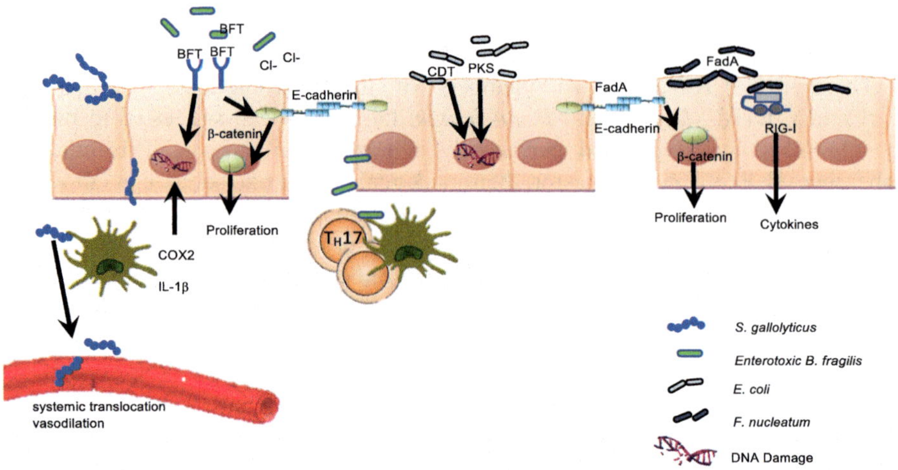

Fig. 8.1 The commensal species most commonly associated with colorectal cancer (CRC) are *Streptococcus gallolyticus* (*blue*) enterotoxigenic *Bacteroides fragilis* (*green*), *Escherichia coli* (*gray*), *Fusobacterium nucleatum* (*purple*) and *Enterococcus faecalis* (not shown). *S. gallolyticus* promotes inflammation via its paracellular translocation and subsequent stimulation of mucosal dendritic cells causing elevated Cox2 and inflammatory cytokines. S. *gallolyticus* also promotes vascularization and vasodilation. Enterotoxigenic *B. fragilis* (ETBF) produces fragilysin or BFT, which can directly cause DNA damage or increase proliferation by freeing intracellular β-catenin from E-cadherin. Chloride ions produced by ETBF and the dissociation of E-cadherin allows bacterial translocation and stimulation of T_H17-mediated immunity. *E. coli* species are predominant during inflammation and in CRC alter gene expression to become more mucosally associated. *E. coli* expresses CDT and PKS, which both have DNA-damaging effects. *F. nucleatum* expresses a virulence factor, FadA, which binds to the extracellular portion of E-cadherin and dissociates β-catenin, allowing activation of proto-oncogenic and proliferative pathways. *F. nucleatum* also contributes to CRC via the induction inflammation after recognition by intracellular retinoic acid-inducible gene I

8.3.1 Metabolites that Contribute to CRC

Nitrogenous Compounds A subset of both Bacteroidetes and some Firmicutes ferment aromatic amino acids from proteins and produce potentially bioactive ammonia and nitrogenous products, particularly *N*-nitroso compounds (NOCs). Both ammonia and NOCs are carcinogenic agents. In the case of NOCs, their carcinogenic nature is due to their ability to alkylate DNA, resulting in genetic mutations. NOCs are positively associated with CRC in Europeans and although some pre-formed NOCs are taken in as part of the diet, they can also be formed via endogenous microbial fermentation that occurs in the colon through the expression of nitro- and nitrate-reductases encoded by Proteobacteria (Loh et al. 2011; Roisin Hughes 2000).

Sulfides Hydrogen sulfide is a major product of the gut and occurs through the reduction of diet-derived sulfate. Sulfides cause both a breakdown in the epithelial barrier and DNA damage via the activation of ROS. An increase in sulfate-reducing

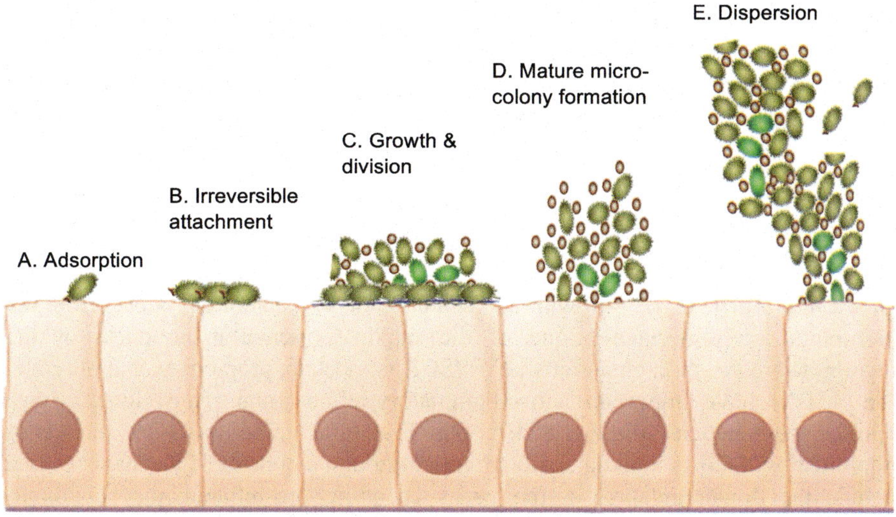

Fig. 8.2 The five stages of bacterial biofilm formation. (**a**) Bacteria reversibly attach to solid support. (**b**) Bacteria become irreversibly attached, and aggregate to form a matrix. (**c**) Maturation phase: cells become layered and the effects of quorum sensing begin. (**d**) Clusters reach maximum thickness. (**e**) Escape of planktonic bacteria from matrix dispersion

bacteria such as *Desulfovibrio* spp. Are not likely driving these pathways in CRC, as they have not been increased in the stool of patients. Therefore, the increase in hydrogen sulfide may be due to changes in bacterial gene expression and activity rather than composition (Reuter et al. 2010).

Bile Acids Gut bacteria are also important contributors to bile acid metabolism and thus may play a role in the biology linking bile acids to colon cancer. For instance, prolonged consumption of red meat and saturated fatty acids increases the risk of CRC. This was observed among descendants of low-risk individuals who moved to developed countries and converted to a western-type diet (Berg et al. 1973). At the gut microbiome level, a high-fat diet may alter the composition of the microbiome by resulting in a bloom of the sulfur-reducing bacterium *Bilophila wadsworthia*, which in turn was shown to exacerbate colitis and inflammation in IL10−/− mice (Devkota et al. 2012) and can also modify the composition of the microbiota owing to their strong antimicrobial activities. Bile acids have been implicated in carcinogenesis and increases in bile acid concentrations have been observed in the stool of patients with CRC (Barrasa et al. 2013). The mechanism by which bile salts contribute to CRC is likely due to the generation of DNA-damaging ROS and reactive nitrogen species, which can lead to increased DNA damage and an increase in the mutation rates (Ajouz et al. 2014).

It is important to note that although certain metabolites such as lithocholic and deoxycholic acid have been shown to be pro-inflammatory and linked to the development of colon cancer, others, such as ursodeoxycholic acid have been shown to have certain health benefits in pre-clinical and clinical studies. *Clostridium*,

Ruminococcus, and *Eubacterium* strains, from the commensal microflora, have the capacity to convert chenodeoxycholic acid to ursodeoxycholic acid. The latter have been shown to be beneficial for CRC prevention in patients with a history of adenomas and IBD in small retrospective studies (Carey and Lindor 2012).

8.3.2 Metabolites that Are Beneficial in CRC

Short Chain Fatty Acids Still with the idea that not all metabolic products from the microbiota are tumorigenic, one of the major fermentation products of the microbiota are the short chain fatty acids (SCFAs) acetate, propionate, and butyrate. The SCFAs have important anti-inflammatory effects and many studies have reported the beneficial role that SCFAs play in colonic homeostasis and models of inflammation. SCFAs modulate inflammation through a number of mechanisms that include the downregulation of the genes encoding pro-inflammatory cytokines through the inhibition of histone deacetylases and the induction of regulatory T cells (Atarashi et al. 2011; Furusawa et al. 2013). Of the three SCFAs, butyrate is mainly produced in the proximal colon and it has been regarded as one of the most important nutrients for colonocytes. Other important anti-tumorigenic effects of butyrate include inhibiting proliferation and selectively inducing apoptosis of CRC cells (Fung et al. 2012). Butyrate also possesses possible anti-carcinogenic effects as it suppresses the expression of COX-2 in cancer cell lines (Zhang et al. 2004). Bacterial pathways of butyrate production have been characterized in *Clostridium acetobutylicum*, and more recently in *Butyrivibrio fibrisolvens*, a butyrate producer that has been found in the gastrointestinal (GI) tract of humans, dogs, and cats, and *Enterococcus durans* (Raz et al. 2007).

8.4 Microbiota, Biofilm Formation, and CRC

The commensal bacteria found within our intestines are not all free-floating entities, but are in a planktonic state. Some of the microbes live in complex communities called biofilms. Biofilms are populations of microbes held to each other, to surfaces or at an interface by microbial-produced polymeric matrices. To form and maintain biofilms, bacteria must induce a different set of genes that aids in increasing the concentration of bacteria, a regulation referred to as quorum-sensing. Bacteria in a biofilm may also differ from their planktonic counterparts in antimicrobial resistance and expression of different virulence genes (Burmølle et al. 2014). The five steps of biofilm formation are illustrated and described in Fig. 8.3.

It is believed that the intestinal microbiota can form biofilms along the mucosal surface in healthy individuals. However, these data come from human sudden death studies and may be an artifact resulting from the embedding time, as healthy subjects lacked these biofilm-like structures (Dejea et al. 2014). Despite their controversial

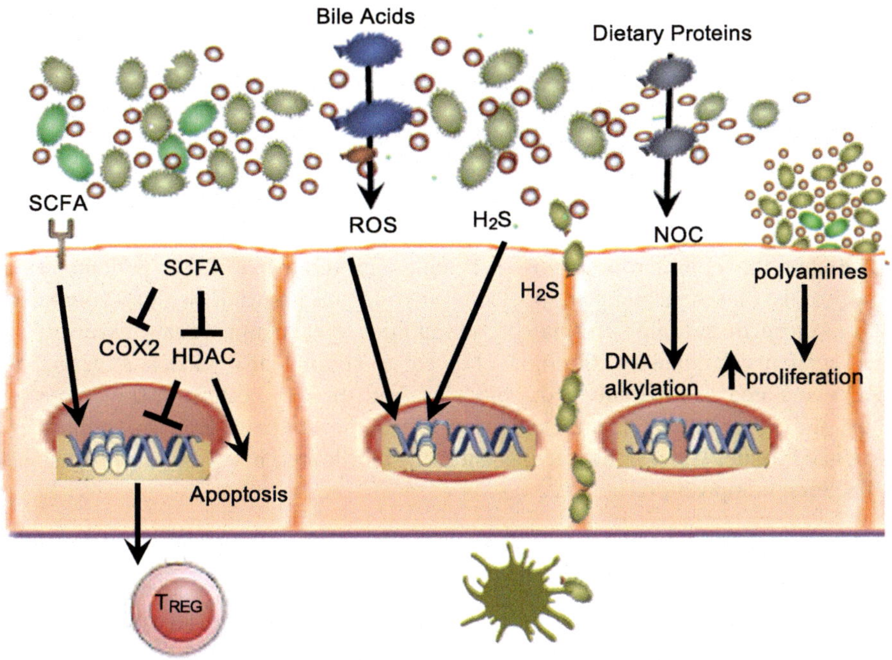

Fig. 8.3 Pro- and anti-tumor effects of bacterial metabolites. Short chain fatty acids (SCFAs) include butyrate, propionate, and acetate. Extracellular or secreted SCFAs can bind to a GP and provides signals to upregulate important genes needed to modulate T regulatory cell (T_{REG}) differentiation. Intracellular SCFAs block COX2. SCFAs also promote T_{REGs} and inhibit apoptosis through modulation of histone deacetylases. Bile acids promote carcinogenesis via the breakdown by certain commensals that generate DNA-damaging ROS. Hydrogen sulfide (H2S) is produced by many bacteria and can damage DNA or cause a breach in the epithelial barrier, allowing translocation of commensals and generating inflammation. Dietary proteins can be metabolized to N-nitroso compounds that alkylate DNA and cause cancer. Biofilms generate polyamines, which can increase cellular proliferation

presence in healthy individuals, biofilms have been associated with nonmalignant pathological conditions such as IBD and with colon-associated malignancy (Dejea et al. 2014; Swidsinski et al. 2005; Johnson et al. 2015). Interestingly, in human CRC, the presence of bacterial biofilms was linked to cancer location with right-sided colon adenomas and cancers possessing biofilms, whereas left-sided cancers were biofilm-negative (Johnson et al. 2015). Siuzdak and colleagues also went on to show a direct correlation between biofilms and the upregulation of a specific polyamine metabolite, N1, N12-diacetylspermine. However, polyamine production was connected to biofilm formation and was not specific to the cancer cells associated with the biofilm, as measurement of up-regulated polyamines was identified in paired normal tissue and in the small number of biofilm-positive left-sided cancers (Johnson et al. 2015). The expression of polyamine is known to enhance eukaryotic cellular proliferation, microbial growth, and cell wall formation (Gerner and Meyskens 2004). Thus, it seems that the presence of biofilms increases polyamines and epithelial proliferation, and creates an environment that more amenable to oncogenic transformation.

With regard to the microbial composition of biofilms, it seems that the organization of the biofilms rather than specific composition may be more important in the pathogenesis of CRC. As Dejea et al. (2014) found that most right-sided tumors are associated with biofilms and that the normal colon tissue from these patients was also biofilm-positive. In contrast, normal tissue from patients with biofilm-negative tumors were always biofilm-negative (Dejea et al. 2014). Furthermore, analysis found that the communities from nonmalignant tissues of CRC patients were significantly closer in composition to those from the tissue of healthy volunteers than to those from tumor-associated communities. Yet, nonmalignant tissues from CRC patients containing biofilms showed a significantly closer structure to CRC-associated tissue than to the biopsies from healthy, cancer-free volunteers. Thus, the observations from examining these tumor-associated and nontumor-associated tissues suggest that the presence of a biofilm might correlate with changes in bacterial composition.

8.5 Microbiota-Dependent Sensing by the Innate Immune System and CRC

The function of the inflammatory response is to protect us against microbes and to help repair and regenerate tissue damage caused by both infectious and non-infectious agents. Thus, it is not surprising that there is an intrinsic link between inflammation and cancer. This is demonstrated by the fact that 15% of worldwide cancers are microbial-associated, and that a number of cancers are associated with chronic inflammatory diseases, such as gastritis, inflammatory bowel disease, and thyroiditis (Selgrad et al. 2008). Our innate immune system imparts our bodies with sensors for tissue damage and infection, with major examples being TLRs and NOD-like receptors. Defective innate immune responses may lead to inadequate pathogen eradication, recurrent tissue injury or failure of anti-inflammatory responses, which can cause chronic inflammation and support tumorigenesis. In CRC, the severity of inflammation strongly correlates with the risk of CRC in patients with IBD and we believe that one of the major factors that drives chronic inflammation in genetically susceptible IBD patients is the commensal microbiota (Rutter et al. 2004; Gupta et al. 2007). Thus, it definitely makes sense that many of the genetic polymorphisms identified in large-scale, genome-wide association studies of IBD patients were genes directly or indirectly involved in microbial sensing (e.g., TLR, NLR), microbial handling (e.g., autophagy) or innate inflammation (cytokines, chemokines).

8.5.1 Toll-like Receptor Biology

Toll-like receptors are evolutionarily conserved, type I transmembrane pattern recognition receptors (PRRs) that sense conserved microbial motifs also called pathogen-associated molecular patterns (Pasare and Medzhitov 2004). Currently, 10 TLRs have been identified (TLR1–TLR10) in humans and 12 in the mouse (TLR1–9, TLR11–13) (Gay and Gangloff 2007). TLR1, -2, -4, -5, and -6 are localized at the cell surface, TLR3, -4, -7, -8, and -9 are present in the intracellular compartment (Akira and Hemmi 2003), and TLR4 can be expressed both intracellularly and extracellularly (Pasare and Medzhitov 2004). TLRs can recognize a variety of ligands such as lipids and lipopeptides (TLR1, -2, -4, -6), bacterial flagellin (TLR5), and fragments of nucleic acids (TLR3, -7, -8, -9). The expression of TLRs is not limited to immune cells as their expression has also been found on non-immune cells. Most of the TLRs are expressed throughout the small and large intestines; however, the localization and function of all the individual TLRs are still unclear.

Activation of TLRs can induce a number of signaling pathways, which results in the up-regulation of genes involved in co-stimulation, inflammation, cellular metabolism, survival, and death. Receptor dimerization of a TLR results in differential recruitment of specific adaptor proteins, including MyD88, MyD88 adaptor-like (Mal, also known as TIR domain-containing adaptor protein), TIR domain-containing adaptor-inducing interferon-β (TRIF), or TRIF-related adaptor molecule, which drive subsequent signaling. This results in the activation of a number of downstream pathways, including NF-κB, mTOR, interferon regulatory factor, PI3K/Akt and mitogen-activated protein kinase (MAPK) pathways that lead to signals affecting inflammation, regeneration, cell survival, and proliferation (Fitzgerald et al. 2001; Kawai and Akira 2011). To expand their PAMP-recognizing capacity, TLRs also have the capacity to form heterodimers and homodimers and bind different accessory proteins, leading to different signal transduction pathways, such as MD-2 and CD14, that form a complex with TLR4 in response to lipopolysaccharide (LPS). Another example is TLR2, which is able to switch its ability to produce pro- and anti-inflammatory responses by dimerization with several co-receptors such as TLR2 itself, TLR1, TLR6, and TLR10. Recent studies suggest that TLR2/TLR6 dimerization might activate a particular signal transduction pathway that induces the transcription of pro-inflammatory molecules, whereas TLR2/TLR1 dimerization promotes the anti-inflammatory pathway that leads to the expression of IL-10 and the trans-differentiation of Th17 and iTreg cells (Melmed et al. 2003).

8.5.2 Toll-like Receptors Associated with CRC

The idea that TLRs might be involved in tumorigenesis came about at the end of the nineteenth century, when William Coley observed that repeated injections of a mixture of bacterial toxins from the Gram-positive bacterium *Streptococcus pneumoniae* and the Gram-negative bacterium *Serratia marcescens* yielded efficient anti-tumor effects, providing proof that microbial products, rather than the infection itself, mediate an anti-tumor effect (Coley 1991). Shear and Turner later discovered that LPS, which is a component of the membrane of Gram-negative bacteria, was the fraction of Coley's toxin that accounted for its anti-tumor effect (Lundin and Checkoway 2009). As LPS stimulates TLR4, these results suggest that the anti-tumor effect of Coley's toxin was a result of TLR activation.

Toll-like receptors play many reported and sometimes conflicting roles in tumorigenesis. The number of potential roles are likely due to the number of different TLRs, the different sets of genetic signatures induced by each TLR, and the differences in expression within the GI tract.

Toll-like receptor signaling may provide activation of antitumor immunity by inducing downstream mediators such as type I interferons. This activation has recently been used to generate TLR agonists as potential candidates for cancer immunotherapy. In contrast, there is much literature on the role of TLRs as tumor promoters through the induction of pathways linked to inflammation, wound healing, tissue regeneration, cell survival, and cell death. Owing to their ability to promote cell survival and proliferation and regulate apoptosis and cell death, the function of a given TLR signal may be highly influenced by many factors such as the tissue microenvironment, the host's genetic signature or the microbiota. TLR sensing of the microbiota is even more complex, as an individual TLR can influence either tumor promotion or antitumor immunity, depending on the context in which it is activated. Regardless of the effect, it is well accepted that TLR recognition of the microbiota plays an influential role in CRC, and this has been demonstrated in a number of studies. For example, the deletion of the TLR adapter molecule Myd88, the absence of the microbiota in GF mice on the APCmin background, or treating mice with broad spectrum antibiotics decreases the incidence and severity of cancer in both sporadic and colitis-associated cancer models (Fitzgerald et al. 2001; Kawai and Akira 2011; Couturier-Maillard et al. 2013; Klimesova et al. 2013).

8.5.3 Specific Contributions of Individual TLRs to CRC

TLR1 Our group has previously shown that genetic deficiency in TLR1 promotes acute enteric infection by *Yersinia enterocolitica*. Examining that model further, we uncovered an altered cellular immune response that promotes the recruitment of neutrophils, which in turn increases metabolism of the respiratory electron acceptor tetrathionate by *Yersinia*. These events drive permanent alterations in anti-commensal immunity, microbiota composition, and chronic inflammation, which persist long after *Yersinia* is cleared (Kamdar et al. 2016). These data demonstrate that acute infection can drive long-term immune and microbiota alterations, leading to chronic inflammatory disease in genetically predisposed individuals and potentially predispose them to cancer. Interestingly, humans express a variant of *TLR1* in which a hydrophobic isoleucine is replaced by a hydrophilic serine at the (I602S) transmembrane domain. When the variant allele is expressed, TLR1 does not localize to the surface membrane, but is rather trapped intracellularly in the cytosol of the cell (Schumann and Tapping 2007). This reduces the TLR1-mediated activation of NF-κB by extracellular ligands. Recently, we have shown that metastatic CRC patients expressing one or both variant alleles of *TLR1* I602S have a better response to treatment in addition to improved progression-free survival, when treated with FOLFIRI (a combination therapy of irinotecan, 5-fluorouracil, and folic acid) and bevacizumab, a humanized monoclonal antibody and angiogenesis inhibitor, suggesting that endogenous TLR2/1 ligands might play a pro-tumorigenic role (Okazaki et al. 2016).

TLR2 The role of TLR2 remains controversial in CRC. In one study, there were no differences in tumor burden between wild-type and TLR2-deficient mice, using a colitis-associated neoplasia model of azoxymethane (AOM) and dextran sulfate sodium (DSS). Increased tumor number and higher IL-6, IL-17A, and phospho-STAT3 levels were identified in TLR2-deficient mice by a different group using a similar AOM-DSS model (Salcedo et al. 2010; Lowe et al. 2010). The different outcomes seen in TLR2 studies may be explained by the fact that TLR2 can form heterodimers with TLR1, TLR6, and, in humans, TLR10. The ability to bind different TLRs expands the number of ligands that TLR2 recognizes, and potentially alters the downstream signaling molecules associated with activation. Moreover, TLR1 and TLR6 allow TLR2 to sense tri-acylated and di-acylated lipoproteins respectively, and, depending upon it binding TLR2, has been shown to induce different MAPK and genetic signatures (Depaolo et al. 2008). Therefore, if the microbiota of an animal colony differed greatly between two groups, with one containing more TLR2/6-activating bacteria and the other more TLR2/1, then even the baseline microenvironment could be different, and knocking out TLR2 may modulate a response in one instance and not in the other.

TLR4 Toll-like receptor 4 can activate both MyD88-dependent and TRIF-dependent signaling pathways, depending upon its cellular expression (Troutman et al. 2012). Using the same AOM/DSS model, TLR4 deletion was shown to strongly reduce inflammation and tumor burden, whereas transgenic mice overexpressing constitutively activated TLR4 in the intestine exhibit a higher sensitivity to colitis-associated neoplasia due to activation of β-catenin signaling pathways (Fukata et al. 2009, 2011; Santaolalla et al. 2013). In contrast, one recent study showed that intestinal overexpression of constitutively activated TLR4 in the APC/Min/+ model reduces tumor load by increasing tumor cell apoptosis (Li et al. 2012, 2014).

An anti-tumor role for TLR4 has also been shown. The release of HMGB1, a TLR4 ligand, by damaged or necrotic tumor cells can trigger TLR4 activation in local immune cells, enhancing antigen presentation and promoting anti-tumor immunity. In humans, a TLR4 loss-of-function allele is associated with less cross-presentation of antigens and results in relapse and an increase in metastasis in patients with breast cancer (Apetoh et al. 2007). Another human variant, TLR4-D299G was recently identified as an aberrant innate immune mediator that may create an auto-inflammatory environment, favoring excessive intestinal epithelial cell (IEC) remodeling and driving tumor progression. This polymorphism seems to compromise the recruitment of the signaling adaptors MyD88 and TRIF, thereby impairing the downstream activation of NF-κB target genes. Instead, activation of STAT3 is likely the principal target of this SNP, thereby promoting malignant tumor progression in human IECs (Cario 2013). Moreover, primary human sporadic adenocarcinomas from patients carrying the TLR4-D299G are more frequently associated with advanced tumor stage.

Lastly, LPS, which can also be derived from Gram-negative commensal bacteria, has been used in phase II clinical trials for the treatment of CRC and was shown to lead to tumor regression when directly injected into adoptively transferred tumors (Otto et al. 1996).

TLR2/6 Lactic acid bacteria are a group of commensal bacterial strains that were shown to have anti-tumor potential in several probiotic studies (see Sect. "Probiotics"). A recent study also found that several strains such as *Lactobacillus plantarum* CCFM634, *L. plantarum* CCFM734, *L. fermentum* CCFM381, *Lactobacillus acidophilus* CCFM137, and *Streptococcus thermophilus* CCFM218 stimulated TLR2/TLR6, providing an insight into lactic acid bacteria-specific host–microbe interactions (Ren et al. 2016).

TLR10 Like TLR1 and TLR6, TLR10 can form heterodimers with TLR2. Although TLR6/2 dimerization allows recognition of di-acylated lipoproteins and lactic acid bacteria, no ligand has yet been found for TLR10 and little is known about TLR10 and CRC.

Toll-like receptor 9 is activated by both bacterial and viral DNA, immunoglobulin–DNA complexes, and synthetic oligodeoxynucleotides (ODNs), which contain unmethylated CpG sequences (Kanzler et al. 2007). Apical expression and activation of TLR9 on epithelial cells by bacterial DNA fragments have been reported to maintain colonic homeostasis. TLR9-induced type 1 interferons have also been shown to mediate the anti-inflammatory effects in experimental colitis (Lee et al. 2006). In CRC, expression of TLR9 has been shown to be higher in adenomatous polyps, but decreased in hyperplastic and villous polyps from patients who developed CRC, suggesting that TLR9 expression might play a protective role against malignant transformation (Rachmilewitz et al. 2002). Recently, much research has focused on antitumor immunity induced by TLR9 antagonists and inhibitory ODNs (inh-ODNs; discussed below).

TLR Adaptor Molecules Similar to TLR2, conflicting results were also found in studies looking at the effect of Myd88 deletion on CRC in the APCmin/+ model. One study demonstrated that APCmin/+Myd88−/− mice develop fewer colonic tumors than APCmin/+ mice, indicating that bacterial signaling contributes to tumorigenesis in the context of APC mutations (Rakoff-Nahoum and Medzhitov 2007). However, a subsequent study found that MyD88-dependent activation of ERK stabilizes β-catenin; in this way, the absence of MyD88 protects against APC-dependent tumors (Lee et al. 2010). Thus, depending on the model used, MyD88-deficiency either protects or increases tumorigenesis.

8.6 Manipulation of the Microbiome as a Treatment for CRC

Specific microbiota species and their products constitute potential targets for modulating colon cancer because of their immunological potential and/or protective effects on colon carcinogenesis. A large body of evidence exists chronicling the influence of certain commensal micro-organisms on both the development of the mucosal immune system and the modulation of innate inflammatory responses to maintain homeostasis. The commensal microbiota has evolved a number of mechanisms that help to modulate the inflammatory response. Indeed, it has been shown to induce anti-inflammatory cytokines such as IL-10 and TGF-β that directly suppress inflammatory signaling, reduce antigen presentation, and can induce regulatory T cells. Other commensal species may directly inhibit colon cancer through the generation of ROS that can, in combination with platinum compounds, greatly enhance the effect of chemotherapies.

In addition to their immune-modulatory effects, commensal microbiota has been shown to reduce cell cycle progression, induce apoptosis of tumor cells, and alter the host's metabolome, by producing specific bacterial enzymes that enhance the production of beneficial or protective metabolites. Moreover, the microbiota and its products may be exploited for diagnosis and detection purposes. For example, comparing the microbial characteristics of patients with CRC with those of IBD patients could begin to dissociate whether the changes observed in the microbiota are inflammation- or cancer-dependent.

8.6.1 *Specific Therapies Targeting the Microbiota to Treat CRC*

Probiotics According to the US Food and Drug Administration (FDA), probiotics are live microorganisms that are intended to have health benefits. Products sold as probiotics include foods, dietary supplements, and products that can be used topically. Although some probiotics have shown promise in research studies, strong scientific evidence to support the specific uses of probiotics for most health conditions is lacking. The FDA has not yet approved any probiotics for preventing or treating any health problems. Some experts have also cautioned that the rapid growth in marketing and use of probiotics may have outpaced scientific research for many of their proposed uses and benefits, as we have yet to define the therapeutic windows for probiotic dosing, the potential side effects, and the accurate stratification of patients most likely to benefit from such therapies.

However, in-vitro and in-vivo studies, in addition to some clinical trials, do suggest that certain microbial species in colon cancer might have a promising beneficial effect. For instance, a particular study showed that the administration of *Lactobacillus rhamnosus* GG (LGG) induced a significant reduction in polyamine biosynthesis in HGC-27 and DLD-1 cancer lines, suggesting that LGG might have the potential to change the host's metabolome and consequently halt the proliferation of tumor cells (Orlando et al. 2009). As mentioned earlier, polyamine is a metabolite that is necessary for cell proliferation. It was also found to be produced by microbial biofilms. Thus, targeting polyamine production and biofilm interactions could be another strategy for treating CRC, for example, by the administration of probiotic strains with adhesive properties, that can prevent the establishment of pathogenic biofilms by competitive exclusion. This ability of probiotics to adhere to GI mucus is of considerable importance in their potential to exert a modulatory effect in situ and the adhesion of probiotic bacteria to epithelial cells has been shown to prevent the establishment of pathogens (Russo et al. 2014). Resta-Lenert and Barret showed that exposure of biofilms to live, but not heat-inactivated, probiotic *S. thermophilus* and *L. acidophilus* strains significantly limited adhesion, invasion, and physiological dysfunction induced by exposure to an entero-invasive strain of *E. coli* (Resta-Lenert and Barrett 2003). A similar effect has been demonstrated for a probiotic strain of *L. plantarum*, which had a protective effect against damage to the integrity of Caco-2 monolayers and the structure and distribution of TJ proteins by enteroinvasive *E. coli* (Qin et al. 2009). Another study showed that Caco-2 cells exposed to *L. plantarum* bacteria significantly induced human beta-defensin 2 mRNA expression and secretion in a dose-dependent manner compared with controls. This was inhibited by anti-TLR2 neutralizing antibodies, suggesting that *L. plantarum* may signal through this microbial PRR and generate an anti-cancer response (Paolillo et al. 2009). In-vivo examples include a study that showed that daily oral administration of microencapsulated *L. acidophilus* in a yogurt formulation to APC+/− mice resulted in significant suppression of colon tumor incidence, tumor multiplicity, and reduced tumor size. Moreover, the treated animals exhibited fewer GI intra-epithelial neoplasia with a lower grade of dysplasia in tumors, that this probiotic might have a potential benefit (Urbanska et al. 2016). In another study, Park et al. fed F344 male rats with *Bacillus polyfermenticus*, showing that these rats displayed significantly lower numbers of aberrant crypt foci than the control group. Supplementation with *B. polyfermenticus* induced less leukocytic DNA damage and plasma lipid peroxidation levels, in addition to a lower plasma total antioxidant potential, suggesting that *B. polyfermenticus* might exert a protective effect on the antioxidant system and the process of colon carcinogenesis (Park et al. 2007).

Last, the potential benefit of certain microorganisms was also tested in some clinical trials involving human subjects. In one study, the administration of *Lactobacillus casei* was evaluated as a method of preventing the occurrence of colorectal tumors. The occurrence of tumors with a grade of moderate atypia or higher was significantly lower in the patient group after 2–4 years of treatment with *L. casei* compared with the control group (Ishikawa et al. 2005). In another 12-week clinical trial completed back in 2007, polypectomized patients were treated with LGG, *Bifidobacterium lactis* Bb12, and oligofructose-enriched inulin. The treatment resulted in significant changes in fecal microbiota of the patients, as the proportion of *Bifidobacterium* and *Lactobacillus* increased and *Clostridium perfringens* decreased. The intervention was also associated with a significant reduction of colorectal epithelial cell proliferation in the patients (Rafter et al. 2007). Another study reported in a prospective trial that in 31 CRC patients, *Lactobacilli johnsonii* (La1), but not *Bifidobacterium longum* (BB536), affects intestinal microbiota by reducing the concentration of pathogens and modulating intestinal dendritic cell (DC) and T-cell specific responses exemplified by a reduction in the release of anti-inflammatory cytokines and a reduction of potentially pathogenic bacteria. Moreover, DCs isolated from patients colonized with La1 had a significantly blunted ability to proliferate, and it is possible that La1 might prevent excessive activation of DCs and the development of Th1-polarized immunity in the intestinal mucosa (Gianotti et al. 2010). Recently, a cohort study with 12 years' follow-up on 45,241 volunteers determined that high yogurt intake was significantly associated with decreased CRC risk, suggesting that the long-term administration of probiotic formulations might reduce the incidence of CRC (Pala et al. 2011). These studies definitely show promising findings, but we have yet to elucidate the mechanisms underlying these potential therapeutic effects. Perhaps unraveling how probiotics may contribute to CRC treatment will uncover new insights into cancer immunotherapy.

Fecal Microbiota Transplantation Another way of manipulating the intestinal microbiota in the hope of enhancing or ameliorating intestinal disease involves fecal microbiota transplantation (FMT) from healthy donors to individuals with specific diseases. Significant clinical effectiveness of FMT has been demonstrated for recurrent *Clostridium difficile* infection and ongoing studies are investigating FMT for other diseases such as IBD (Pamer 2014). Transplantation of autologous microbial populations to recipients likely triggers mucosal immune responses that, depending on the microbiota composition and the recipient's genotype, could range from pro- to anti-inflammatory. However, the impact of FMT on the recipient immune system is complex and unpredictable, and the ongoing discovery of commensal microbes and investigation of their impact on the host will lead to the development of new personalized probiotic agents and microbial consortia that will eventually replace FMT.

TLR Agonists and Antagonists We are beginning to understand the therapeutic potential of individual TLRs in the fight against colon cancer through a number of recent clinical trials and through pre-clinical studies (Hedayat et al. 2012). However, the development of optimal therapeutic strategies for targeting TLR signaling depends on a fuller characterization of the roles of TLR and their downstream adaptors under nondiseased and diseased conditions in the intestine, and through a more thorough understanding of the molecular mechanisms behind their role in the pathophysiology of CRC. TLRs in the gut are constantly exposed to commensal-derived ligands, inducing downstream signaling pathways that contribute to gut mucosal immunity and homeostasis. The pleiotropic nature of TLRs and their ability to activate/modulate many important cellular mechanisms including intestinal permeability, inflammation, cell survival and death, regeneration and repair, autophagy, and tolerance makes them candidates for targeted intervention. Various strategies for modulating TLR signaling exist, including administration of specific commensal species and/or ligands, administration of antibodies to block TLR signaling, and directly targeting the signaling of TLRs on tumor cells to trigger an anti-tumor innate and/or adaptive immune responses.

As mentioned above, TLR9 signaling is not associated with carcinogenesis, but rather plays an important role in generating anti-tumor responses, making it a very promising therapeutic choice. Pre-treatment with the TLR9 ligand CpG DNA was recently shown to ameliorate both chemically induced and spontaneous colitis in mice by inhibiting pro-inflammatory cytokines and chemokines (Rachmilewitz et al. 2002). TLR9 agonists were also shown to suppress tumor growth, and metastasis of CRC cells to the lung and liver, thus increasing overall survival when administered alone or in combination with chemotherapy regimens in CRC xenografts (Fűri et al. 2013; Kim et al. 2012; Zoglmeier et al. 2011). The effects of TLR9 have been observed at the site of drug delivery and at systematic sites, and are dependent upon the presence of MyD88 (Heckelsmiller et al. 2002a, b; Westwood et al. 2009). TLR9 agonists are also being used to boost the anti-tumoral immune response of the host in the context of antigen-loaded DC immunotherapy. The combination of DC and TLR9 agonists have been used in CRC and have demonstrated a stronger immune response, both local and systemic anti-tumoral effects, and the induction of an anti-tumor memory response in the host (Heckelsmiller et al. 2002a).

Toll-like receptor 4 signaling has been shown to perpetuate carcinogenesis through the induction of prostaglandins and through the activation of β-catenin. Thus, patients with CRC may benefit from administration of antagonists that can inhibit TLR4 signaling. A recent study by Wei-Ting Kuo et al. showed that the administration of a synthetic analog of LPS that inhibits TLR4 signaling caused an increase in tumor cell apoptosis and decreased the tumor burden (Kuo et al. 2016). Similar to TLR9, TLR7 agonists seem to promote anti-tumor responses. For example, intravenous administration of the selective TLR7 agonist DSR-29133 led to enhanced anti-tumor effects in a solid tumor model (Dovedi et al. 2016).

Toll-like receptors can also be finely tuned to optimize tumor responsiveness to chemotherapy through their microbial recognition. Experimentally, polyI:C, a TLR3 agonist, was shown to enhance cycloheximide-induced apoptosis of tumor cells, whereas TLR9 agonist administered with chemotherapy regimens increased the overall survival of mice in a CRC xenograft model (Jiang et al. 2008). These studies are opening up a new range of clinical applications for TLR agonists as an adjuvant for chemotherapy (Adams 2009; Bhardwaj et al. 2010). Table 8.1 summarizes information regarding the various human TLRs that have been implicated in colon cancer, their natural and synthetic ligands, their location throughout the large intestine, and their therapeutic potential.

Table 8.1 Toll-like receptor (*TLR*) biology and therapeutic potential in colorectal cancer (*CRC*) (Uematsu and Akira 2008; Yu et al. 2010)

TLR	Ligands	Location in the large intestine	Effects on immunological functions associated with CRC
TLR2/1	Tri-acylated lipopeptides^{m}	Myofibroblasts	TLR1 I602S variant is associated with better response to CRC treatment and progression-free survival
	Pam3CSK4^{s}	Circulating PMNs	
TLR2	PG^{m}, LP^{m}, Zymosan^{m}, GPI^{m}, Tc52^{m}	LP macrophages, myofibroblast, circulating PMNs	TLR2KO had increased tumor number, IL6, IL17A, p-STAT3 in AOM/DSS CRC mouse model
	OspA^{m}, Porin^{m}, LcrV^{m}, HSP60^{h}, HSP70^{h}, HMGB1^{h}, endoplasmin, biglycan^{h}, human cardiac myosin^{h}, hyaluronan^{h}, monosodium urate^{h} crystals^{h}		
TLR3	IMQ^{m}	Epithelial cells	Experimentally, polyI:C, was shown to enhance cycloheximide-induced apoptosis of tumor cells
	Poly(I:C)^{m}	DC, Myofibroblasts	
	Poly(A:U)^{m}		
	mRNA^{h}		
TLR4	LPS^{m}, MPLA^{m}, Mannan^{m}	Endothelial cells	TLR4 signaling enhances carcinogenesis through the induction of STAT3, prostaglandins, and activation of β-catenin
	Glycoinositol phospholipids^{m}	LP macrophages, IEC, DC	Patients with CRC may benefit from administration of antagonists that can inhibit TLR4 signaling
		– Myofibroblasts – Circulating PMNs	Phase II clinical trials using LPS for the treatment of CRC lead to tumor regression when directly injected into adoptively transferred tumors
	Biglycan^{h}, CD138^{h}, β-defensin^{h}, endoplasmin^{h}, fibrinogen^{h}, heparin sulfate^{h}, HMGB1^{h}, HSP22^{h}, HSP60^{h}, HSP70^{h}, HSP72^{h}, hyaluronan^{h}, monosodium urate crystals^{h}, Resistin^{h}, S100 proteins^{h}, surfactant protein A^{h}, tenascin-C^{h}		
TLR5	Flagellin^{m}	LP endothelial cells, LP macrophages, myofibroblasts, circulating PMNs	Lack of MyD88 or TLR5 expression dramatically enhanced tumor growth and inhibited tumor necrosis in mouse xenografts of human colon cancer (Rhee et al. 2008)
TLR2/6	Di-acylated lipopeptides^{m}	Myofibroblasts	No association with CRC
	LTA^{m}	Circulating PMNs	
	GPI anchor^{m}		

(continued)

Table 8.1 (continued)

TLR	Ligands	Location in the large intestine	Effects on immunological functions associated with CRC
TLR7	RSQ 848[s]	Myofibroblasts	IV administration of the selective TLR7 agonist DSR-29133 led to enhanced anti-tumor effects in a solid tumor model
	siRNA[h]	Circulating PMNs	
TLR8	ssRNA40[m]		
	Human cardiac myosin[h]		
TLR9	CpG DNA[m]	Myofibroblasts	Anti-tumor immunity induced by TLR9 antagonists and inhibitory oligodeoxynucleotides
	Hemozoin[m]	Circulating PMNs	TLR9 agonists suppress tumor growth, and metastasis of CRC, increasing overall survival when administered alone or in combination with chemotherapy regimens in CRC xenografts
	HMGB1[h]		DC and TLR9 agonists strengthen local and systemic anti-tumoral effects, and induce anti-tumor memory response
TLR10	None to date		

h human, *m* microbial, *s* synthetic, *PG* peptidoglycans, *LP* lipoproteins, *GPI* glycophosphatidylinositol, *HSP* heat shock protein, *HMGB1* high mobility group box protein 1, *IMQ* imiquimod-R837, *LPS* lipopolysaccharide, *MPLA* monophosphoryl lipid A, *RSQ 848* resiquimod, *CpG* oligodeoxynucleotide, *LP* lamina propria, *PMNs* polymorphonuclear leukocytes

8.7 Conclusion

The gut microbiota possesses a huge genomic and metabolic potential. It can also have an impact on the immune system through microbial PRR receptors that live in close proximity to these commensal microorganisms in the gut. Thus, the gut microbiota constitutes a potential target for modulating colon cancer risk, through diagnostic or treatment strategies, offering a very exciting field of investigation and discoveries for scientists.

References

Abdulamir AS, Hafidh RR, Abu Bakar F (2011) The association of Streptococcus bovis/gallolyticus with colorectal tumors: the nature and the underlying mechanisms of its etiological role. J Exp Clin Cancer Res 30:11

Abed J et al (2016) Fap2 mediates Fusobacterium nucleatum colorectal adenocarcinoma enrichment by binding to tumor-expressed Gal-GalNAc. Cell Host Microbe 20:215–225

Adams S (2009) Toll-like receptor agonists in cancer therapy. Immunotherapy 1:949–964
Ajouz H, Mukherji D, Shamseddine A (2014) Secondary bile acids: an underrecognized cause of colon cancer. World J Surg Oncol 12:164
Akira S, Hemmi H (2003) Recognition of pathogen-associated molecular patterns by TLR family. Immunol Lett 85:85–95
Apetoh L et al (2007) Toll-like receptor 4-dependent contribution of the immune system to anticancer chemotherapy and radiotherapy. Nat Med 13:1050–1059
Apperloo-Renkema HZ, Van der Waaij BD, Van der Waaij D (1990) Determination of colonization resistance of the digestive tract by biotyping of Enterobacteriaceae. Epidemiol Infect 105:355–361
Arthur JC et al (2012) Intestinal inflammation targets cancer-inducing activity of the microbiota. Science 338:120–123
Atarashi K et al (2011) Induction of colonic regulatory T cells by indigenous Clostridium species. Science 331:337–341
Bardhan K, Liu K (2013) Epigenetics and colorectal cancer pathogenesis. Cancers 5:676–713
Barrasa JI, Olmo N, Lizarbe MA, Turnay J (2013) Bile acids in the colon, from healthy to cytotoxic molecules. Toxicol Vitro Int J Publ Assoc BIBRA 27:964–977
Belkaid Y, Hand T (2014) Role of the microbiota in immunity and inflammation. Cell 157:121–141
Berg JW, Howell MA, Silverman SJ (1973) Dietary hypotheses and diet-related research in the etiology of colon cancer. Health Serv Rep 88:915–924
Bhardwaj N, Gnjatic S, Sawhney NB (2010) TLR AGONISTS: are they good adjuvants? Cancer J Sudbury Mass 16:382–391
Boleij A, Schaeps RMJ, Tjalsma H (2009) Association between Streptococcus bovis and colon cancer. J Clin Microbiol 47:516
Burmølle M, Ren D, Bjarnsholt T, Sørensen SJ (2014) Interactions in multispecies biofilms: do they actually matter? Trends Microbiol 22:84–91
Carey EJ, Lindor KD (2012) Chemoprevention of colorectal cancer with ursodeoxycholic acid: cons. Clin Res Hepatol Gastroenterol 36(Suppl 1):S61–S64
Cario E (2013) The human TLR4 variant D299G mediates inflammation-associated cancer progression in the intestinal epithelium. Oncoimmunology 2:e24890
Chen W, Liu F, Ling Z, Tong X, Xiang C (2012) Human intestinal lumen and mucosa-associated microbiota in patients with colorectal cancer. PLoS One 7:e39743
Coley WB (1991) The treatment of malignant tumors by repeated inoculations of erysipelas. With a report of ten original cases. 1893. Clin Orthop 262:3–11
Couturier-Maillard A et al (2013) NOD2-mediated dysbiosis predisposes mice to transmissible colitis and colorectal cancer. J Clin Invest 123:700–711
Cuevas-Ramos G et al (2010) Escherichia coli induces DNA damage in vivo and triggers genomic instability in mammalian cells. Proc Natl Acad Sci USA 107:11537–11542
Darfeuille-Michaud A et al (2004) High prevalence of adherent-invasive Escherichia coli associated with ileal mucosa in Crohn's disease. Gastroenterology 127:412–421
Dejea CM et al (2014) Microbiota organization is a distinct feature of proximal colorectal cancers. Proc Natl Acad Sci USA 111:18321–18326
Depaolo RW et al (2008) Toll-like receptor 6 drives differentiation of tolerogenic dendritic cells and contributes to LcrV-mediated plague pathogenesis. Cell Host Microbe 4:350–361
Devkota S et al (2012) Dietary-fat-induced taurocholic acid promotes pathobiont expansion and colitis in Il10-/- mice. Nature 487:104–108
Dovedi SJ et al (2016) Intravenous administration of the selective toll-like receptor 7 agonist DSR-29133 leads to anti-tumor efficacy in murine solid tumor models which can be potentiated by combination with fractionated radiotherapy. Oncotarget 7:17035–17046
Fitzgerald KA et al (2001) Mal (MyD88-adapter-like) is required for toll-like receptor-4 signal transduction. Nature 413:78–83
Fukata M et al (2009) Innate immune signaling by toll-like receptor-4 (TLR4) shapes the inflammatory microenvironment in colitis-associated tumors. Inflamm Bowel Dis 15:997–1006

Fukata M et al (2011) Constitutive activation of epithelial TLR4 augments inflammatory responses to mucosal injury and drives colitis-associated tumorigenesis. Inflamm Bowel Dis 17:1464–1473
Fung KYC, Cosgrove L, Lockett T, Head R, Topping DL (2012) A review of the potential mechanisms for the lowering of colorectal oncogenesis by butyrate. Br J Nutr 108:820–831
Fűri I et al (2013) Epithelial toll-like receptor 9 signaling in colorectal inflammation and cancer: clinico-pathogenic aspects. World J Gastroenterol 19:4119–4126
Furusawa Y et al (2013) Commensal microbe-derived butyrate induces the differentiation of colonic regulatory T cells. Nature 504:446–450
Gay NJ, Gangloff M (2007) Structure and function of Toll receptors and their ligands. Annu Rev Biochem 76:141–165
Geng J, Fan H, Tang X, Zhai H, Zhang Z (2013) Diversified pattern of the human colorectal cancer microbiome. Gut Pathog 5:2
Gerner EW, Meyskens FL (2004) Polyamines and cancer: old molecules, new understanding. Nat Rev Cancer 4:781–792
Gianotti L et al (2010) A randomized double-blind trial on perioperative administration of probiotics in colorectal cancer patients. World J Gastroenterol 16:167–175
Grivennikov SI, Greten FR, Karin M (2010) Immunity, inflammation, and cancer. Cell 140:883–899
Grivennikov SI et al (2012) Adenoma-linked barrier defects and microbial products drive IL-23/IL-17-mediated tumour growth. Nature 491:254–258
Gupta RB et al (2007) Histologic inflammation is a risk factor for progression to colorectal neoplasia in ulcerative colitis: a cohort study. Gastroenterology 133:1099–1105; quiz 1340–1341
Heckelsmiller K et al (2002a) Combined dendritic cell- and CpG oligonucleotide-based immune therapy cures large murine tumors that resist chemotherapy. Eur J Immunol 32:3235–3245
Heckelsmiller K et al (2002b) Peritumoral CpG DNA elicits a coordinated response of CD8 T cells and innate effectors to cure established tumors in a murine colon carcinoma model. J Immunol 1950 169:3892–3899
Hedayat M, Takeda K, Rezaei N (2012) Prophylactic and therapeutic implications of toll-like receptor ligands. Med Res Rev 32:294–325
Hoen B et al (1994) Tumors of the colon increase the risk of developing Streptococcus bovis endocarditis: case-control study. Clin Infect Dis Off Publ Infect Dis Soc Am 19:361–362
Ishikawa H et al (2005) Randomized trial of dietary fiber and Lactobacillus casei administration for prevention of colorectal tumors. Int J Cancer 116:762–767
Jiang Q, Wei H, Tian Z (2008) Poly I:C enhances cycloheximide-induced apoptosis of tumor cells through TLR3 pathway. BMC Cancer 8:12
Jin Y et al (2016) Hemolytic E. coli promotes colonic tumorigenesis in females. Cancer Res 76:2891–2900
Johnson CH et al (2015) Metabolism links bacterial biofilms and colon carcinogenesis. Cell Metab 21:891–897
Kamdar K et al (2016) Genetic and metabolic signals during acute enteric bacterial infection alter the microbiota and drive progression to chronic inflammatory disease. Cell Host Microbe 19:21–31
Kanzler H, Barrat FJ, Hessel EM, Coffman RL (2007) Therapeutic targeting of innate immunity with toll-like receptor agonists and antagonists. Nat Med 13:552–559
Karin M (2009) NF-κB as a critical link between inflammation and cancer. Cold Spring Harb Perspect Biol 1:a000141
Kawai T, Akira S (2011) Toll-like receptors and their crosstalk with other innate receptors in infection and immunity. Immunity 34:637–650
Kim S et al (2012) Berberine suppresses the TPA-induced MMP-1 and MMP-9 expressions through the inhibition of PKC-α in breast cancer cells. J Surg Res 176:e21–e29

Klimesova K et al (2013) Altered gut microbiota promotes colitis-associated cancer in IL-1 receptor-associated kinase M-deficient mice. Inflamm Bowel Dis 19:1266–1277

Kostic AD et al (2012) Genomic analysis identifies association of Fusobacterium with colorectal carcinoma. Genome Res 22:292–298

Kuo W-T, Lee T-C, Yu LC-H (2016) Eritoran suppresses colon cancer by altering a functional balance in toll-like receptors that bind lipopolysaccharide. Cancer Res 76:4684–4695

Lee P, Tan KS (2014) Fusobacterium nucleatum activates the immune response through retinoic acid-inducible gene I. J Dent Res 93:162–168

Lee J, Rachmilewitz D, Raz E (2006) Homeostatic effects of TLR9 signaling in experimental colitis. Ann NY Acad Sci 1072:351–355

Lee SH et al (2010) ERK activation drives intestinal tumorigenesis in Apc(min/+) mice. Nat Med 16:665–670

Li Y et al (2012) Gut microbiota accelerate tumor growth via c-jun and STAT3 phosphorylation in APCMin/+ mice. Carcinogenesis 33:1231–1238

Li Y et al (2014) Constitutive TLR4 signalling in intestinal epithelium reduces tumor load by increasing apoptosis in APC(Min/+) mice. Oncogene 33:369–377

Loh YH et al (2011) N-Nitroso compounds and cancer incidence: the European prospective investigation into cancer and nutrition (EPIC)-Norfolk study. Am J Clin Nutr 93:1053–1061

Lowe EL et al (2010) Toll-like receptor 2 signaling protects mice from tumor development in a mouse model of colitis-induced cancer. PloS One 5:e13027

Lundin JI, Checkoway H (2009) Endotoxin and cancer. Environ Health Perspect 117:1344–1350

McCoy AN et al (2013) Fusobacterium is associated with colorectal adenomas. PLoS One 8:e53653

Melmed G et al (2003) Human intestinal epithelial cells are broadly unresponsive to toll-like receptor 2-dependent bacterial ligands: implications for host-microbial interactions in the gut. J Immunol 170:1406–1415

Mukhopadhya I, Hansen R, El-Omar EM, Hold GL (2012) IBD—what role do Proteobacteria play? Nat Rev Gastroenterol Hepatol 9:219–230

Nesić D, Hsu Y, Stebbins CE (2004) Assembly and function of a bacterial genotoxin. Nature 429:429–433

Okazaki S et al (2016) Clinical significance of TLR1 I602S polymorphism for patients with metastatic colorectal cancer treated with FOLFIRI plus Bevacizumab. Mol Cancer Ther 15:1740–1745

Orlando A, Messa C, Linsalata M, Cavallini A, Russo F (2009) Effects of Lactobacillus rhamnosus GG on proliferation and polyamine metabolism in HGC-27 human gastric and DLD-1 colonic cancer cell lines. Immunopharmacol Immunotoxicol 31:108–116

Otto F et al (1996) Phase II trial of intravenous endotoxin in patients with colorectal and non-small cell lung cancer. Eur J Cancer 1990 32A:1712–1718

Pala V et al (2011) Yogurt consumption and risk of colorectal cancer in the Italian European prospective investigation into cancer and nutrition cohort. Int J Cancer 129:2712–2719

Pamer EG (2014) Fecal microbiota transplantation: effectiveness, complexities, and lingering concerns. Mucosal Immunol 7:210–214

Paolillo R, Romano Carratelli C, Sorrentino S, Mazzola N, Rizzo A (2009) Immunomodulatory effects of Lactobacillus plantarum on human colon cancer cells. Int Immunopharmacol 9:1265–1271

Park E, Jeon G-I, Park J-S, Paik H-D (2007) A probiotic strain of Bacillus polyfermenticus reduces DMH induced precancerous lesions in F344 male rat. Biol Pharm Bull 30:569–574

Pasare C, Medzhitov R (2004) Toll-like receptors: linking innate and adaptive immunity. Microbes Infect 6:1382–1387

Prorok-Hamon M et al (2014) Colonic mucosa-associated diffusely adherent afaC+ Escherichia coli expressing lpfA and pks are increased in inflammatory bowel disease and colon cancer. Gut 63:761–770

Qin H, Zhang Z, Hang X, Jiang Y (2009) L. plantarum prevents enteroinvasive Escherichia coli-induced tight junction proteins changes in intestinal epithelial cells. BMC Microbiol 9:63

Qin J et al (2010) A human gut microbial gene catalog established by metagenomic sequencing. Nature 464:59–65
Rachmilewitz D et al (2002) Immunostimulatory DNA ameliorates experimental and spontaneous murine colitis. Gastroenterology 122:1428–1441
Rafter J et al (2007) Dietary synbiotics reduce cancer risk factors in polypectomized and colon cancer patients. Am J Clin Nutr 85:488–496
Rakoff-Nahoum S, Medzhitov R (2007) Regulation of spontaneous intestinal tumorigenesis through the adaptor protein MyD88. Science 317:124–127
Raz I, Gollop N, Polak-Charcon S, Schwartz B (2007) Isolation and characterisation of new putative probiotic bacteria from human colonic flora. Br J Nutr 97:725–734
Ren C et al (2016) Identification of TLR2/TLR6 signalling lactic acid bacteria for supporting immune regulation. Sci Rep 6:34561
Resta-Lenert S, Barrett KE (2003) Live probiotics protect intestinal epithelial cells from the effects of infection with enteroinvasive Escherichia coli (EIEC). Gut 52:988–997
Reuter S, Gupta SC, Chaturvedi MM, Aggarwal BB (2010) Oxidative stress, inflammation, and cancer: how are they linked? Free Radic Biol Med 49:1603–1616
Reynolds JG, Silva E, McCormack WM (1983) Association of Streptococcus bovis Bacteremia with bowel disease. J Clin Microbiol 17:696–697
Rhee SH, Im E, Pothoulakis C (2008) Toll-like receptor 5 engagement modulates tumor development and growth in a mouse xenograft model of human colon cancer. Gastroenterology 135:518–528
Roisin Hughes IRR (2000) Metabolic activities of the gut microflora in relation to cancer. Microb Ecol Health Dis 12:179–185
Rubinstein MR et al (2013) Fusobacterium nucleatum promotes colorectal carcinogenesis by modulating E-cadherin/β-catenin signaling via its FadA adhesin. Cell Host Microbe 14:195–206
Russo F, Linsalata M, Orlando A (2014) Probiotics against neoplastic transformation of gastric mucosa: effects on cell proliferation and polyamine metabolism. World J Gastroenterol 20:13258–13272
Rutter MD et al (2004) Cancer surveillance in longstanding ulcerative colitis: endoscopic appearances help predict cancer risk. Gut 53:1813–1816
Salcedo R et al (2010) MyD88-mediated signaling prevents development of adenocarcinomas of the colon: role of interleukin 18. J Exp Med 207:1625–1636
Santaolalla R et al (2013) TLR4 activates the β-catenin pathway to cause intestinal neoplasia. PloS One 8:e63298
Schumann RR, Tapping RI (2007) Genomic variants of TLR1—it takes (TLR-)two to tango. Eur J Immunol 37:2059–2062
Sears CL, Garrett WS (2014) Microbes, microbiota, and colon cancer. Cell Host Microbe 15:317–328
Selgrad M et al (2008) The role of viral and bacterial pathogens in gastrointestinal cancer. J Cell Physiol 216:378–388
Shen XJ et al (2010) Molecular characterization of mucosal adherent bacteria and associations with colorectal adenomas. Gut Microbes 1:138–147
Sobhani I et al (2011) Microbial dysbiosis in colorectal cancer (CRC) patients. PLoS One 6:e16393
Soler AP et al (1999) Increased tight junctional permeability is associated with the development of colon cancer. Carcinogenesis 20:1425–1431
Strauss J et al (2011) Invasive potential of gut mucosa-derived Fusobacterium nucleatum positively correlates with IBD status of the host. Inflamm Bowel Dis 17:1971–1978
Swidsinski A, Weber J, Loening-Baucke V, Hale LP, Lochs H (2005) Spatial organization and composition of the mucosal flora in patients with inflammatory bowel disease. J Clin Microbiol 43:3380–3389
Toprak NU et al (2006) A possible role of Bacteroides fragilis enterotoxin in the aetiology of colorectal cancer. Clin Microbiol Infect 12:782–786

Troutman TD, Bazan JF, Pasare C (2012) Toll-like receptors, signaling adapters and regulation of the pro-inflammatory response by PI3K. Cell Cycle 11:3559–3567
Uematsu S, Akira S (2008) Toll-Like receptors (TLRs) and their ligands. Handb Exp Pharmacol 183:1–20. https://doi.org/10.1007/978-3-540-72167-3_1
Urbanska AM, Bhathena J, Cherif S, Prakash S (2016) Orally delivered microencapsulated probiotic formulation favorably impacts polyp formation in APC (Min/+) model of intestinal carcinogenesis. Artif Cells Nanomed Biotechnol 44:1–11
Valle L (2014) Genetic predisposition to colorectal cancer: where we stand and future perspectives. World J Gastroenterol 20:9828–9849
Wang X et al (2012) 4-hydroxy-2-nonenal mediates genotoxicity and bystander effects caused by Enterococcus faecalis-infected macrophages. Gastroenterology 142:543–551.e7
Westwood JA et al (2009) Toll-like receptor triggering and T-cell costimulation induce potent antitumor immunity in mice. Clin Cancer Res 15:7624–7633
Wu S et al (2009) A human colonic commensal promotes colon tumorigenesis via activation of T helper type 17 T cell responses. Nat Med 15:1016–1022
Yu L, Wang L, Chen S (2010) Endogenous toll-like receptor ligands and their biological significance. J Cell Mol Med 14:2592–2603
Zackular JP et al (2013) The gut microbiome modulates colon tumorigenesis. mBio 4:e00692-613
Zhang Z-H, Ouyang Q, Gan H-T (2004) Targeting cyclooxygenase-2 with sodium butyrate and NSAIDs on colorectal adenoma/carcinoma cells. World J Gastroenterol 10:2954–2957
Zoglmeier C et al (2011) CpG blocks immunosuppression by myeloid-derived suppressor cells in tumor-bearing mice. Clin Cancer Res 17:1765–1775

Chapter 9
Oral Microbiome: Potential Link to Systemic Diseases and Oral Cancer

Adrian A. Vasquez, Jeffrey L. Ram, Milad S. Qazazi, Jun Sun, and Ikuko Kato

Abstract The human oral microbiome comprises more than 2,000 bacterial taxa, including a large number of opportunistic pathogens, and it is considered to be the second most diverse microbial community following stool. However, oral microbiome as a whole in determining human health and diseases has been understudied compared with the gut microbiome. Yet, potential links between oral bacteria and a range of systemic diseases have long been recognized based on their associations with periodontal diseases and surgical dental procedures. These pathological conditions include sepsis/endocarditis, cardiovascular diseases and their established risk factors, chronic kidney disease, Alzheimer's disease, rheumatoid arthritis, and head and neck cancer, and are reviewed here. Although local inflammation and physical interventions from dietary and hygiene habits can facilitate systemic dissemination of oral bacteria, studies have delineated several bacterial virulence factors from oral pathogens that are involved in systemic dissemination, inflammation, immune evasion, and cytotoxicities, and are thus relevant to systemic diseases. Unfortunately, mechanistic information to date has been primarily derived from a few periodontal pathogens. To fully elucidate host–microbial and microbial–microbial interactions pertinent to human health and disease, use of a multi-omics approach, including metagenomics, metabolomics, and transcriptomics, may be required.

A.A. Vasquez • J.L. Ram • M.S. Qazazi
Department of Physiology, Wayne State University School of Medicine, Detroit, MI, USA

J. Sun
Department of Medicine, School of Medicine, University of Illinois at Chicago, Chicago, IL, USA

I. Kato (✉)
Department of Oncology, Wayne State University School of Medicine, Detroit, MI, USA

Department of Pathology, Wayne State University School of Medicine, Detroit, MI, USA
e-mail: katoi@karmanos.org

J. Sun, P.K. Dudeja (eds.), *Mechanisms Underlying Host-Microbiome Interactions in Pathophysiology of Human Diseases*, Physiology in Health and Disease,
https://doi.org/10.1007/978-1-4939-7534-1_9

9.1 Introduction

The link between oral health and systemic disease has been postulated for many centuries. In fact, the focal infection theory resulted in many teeth being extracted without relief for the patient during the nineteenth century (Li et al. 2000). Although treatment methods may have changed, the basic idea of these outdated treatments, that oral health may affect systemic health, is even more strongly supported today. As is reviewed in this chapter, comprehensive studies of the oral microbiome, its metabolic activities, and its potential for systemic pathogenicity have been conducted and the association and mechanisms of interaction of the oral microbiome with the rest of the body are the focus of much current research.

The oral microbiome consists of a wide range of microorganisms, e.g., viruses, fungi, protozoa, archaea, and bacteria, and includes a large number of opportunistic pathogens that are involved in dental, periodontal, respiratory, cardiovascular, and other diseases. In fact, upper respiratory infection (common cold) caused by a range of viruses is the most frequent acute illness diagnosed worldwide (Vos et al. 2015), and most people experience caries and periodontitis caused by oral bacteria in their lifetime (Warinner et al. 2014). Also, Epstein–Barr virus and human papilloma virus have long been recognized to be causal agents for nasopharyngeal and oropharyngeal cancers respectively. This high rate of infections by opportunistic pathogens suggests a state of dysbiosis of the oral microbiome rather than symbiosis, as occurs with other human microbiomes.

Yet, the oral microbiome as a whole in determining human health and diseases has been rather understudied compared with the gut microbiome. Older studies relied on the presence or severity of periodontal diseases and culture from these lesions in assessing the effects of oral bacteria on the risk of various diseases. However, recent progress in genomic technology based on 16S rRNA sequencing has revolutionized this field, providing a culture-independent comprehensive view of the oral bacterial community. Exploiting this rapidly growing information, the chapter primarily focuses on bacteria, with some reservations. The chapter does not address well-established bacterial etiology, i.e., the associations with dental and periodontal diseases, and aspiration pneumonia, and also excludes literature concerning oral pathogens migrating and growing in the lower digestive tract as part of its local flora, because the associations with such bacteria are covered by other chapters. In addition to those chapters, there are excellent reviews on this emerging issue, specifically the association between bacteria originating in the oral cavity (such as Fusobacterium) and colorectal cancer, and their plausible biological pathways (Sun and Kato 2016; Flynn et al. 2016). We further limit the description to the diseases for which evidence suggests involvement of specific bacteria, their toxin/virulence factors, or antibiotic/probiotic treatment, beyond their associations with periodontal diseases or dental caries alone. Here, we summarize current understanding of the potential link between oral bacteria and a range of systemic diseases and oral (head and neck) cancers.

9.2 Oral Microbiome Composition, Diversity, Evolution, Acquisition and Development

The human oral microbiome consists of hundreds of species of bacteria in addition to viruses and fungi. Collectively, this community of organisms contributes to the well-being or dysregulation of the host. Ecologically, the oral microbiome is very diverse and with the advent of DNA sequencing advances, much progress has been made in this area. In fact, the diversity of the oral microbiome is suggested to be among the best known to date (Duran-Pinedo and Frias-Lopez 2015). However, fundamental ecological knowledge about the human oral microbiome is still lacking, and research aimed at answering some of these questions is needed.

The 2,000 or so bacterial taxa that inhabit the oral microbiome give rise to a plethora of diseases (Warinner et al. 2014). Studies of ancient dental plaque revealed a constantly evolving acquisition and turnover of the oral microbiome as human populations expanded, diversified, and fluctuated among hunter–gatherer, agrarian, industrial, and post-industrial (Adler et al. 2013). Their results suggest that recent populations of bacteria in the oral microbiome might be less diverse than ancient populations and that this might result in the disease states seen today.

A recent work summarized the application of ecological principles for studying the oral microbiome (Zaura and Mira 2015). This perspective is important because the oral microbiome can be analyzed as a typical ecological habitat, encompassing both abiotic and biotic factors. Mclean (2014) summarizes present strategies and suggests that future studies might need to focus on pH, an abiotic factor within the ecological perspective.

An insightful review of the acquisition of the oral microbiome analyzed it in the context of ecological succession concepts (Sampaio-Maia and Monteiro-Silva 2014). They indicate that the first colonizers are *Streptococcus* and *Staphylococcus*, and their role in succession can be compared with species that inhabit habitats that are just being made available (Sampaio-Maia and Monteiro-Silva 2014). These species are considered r-selected species that have a high reproductive and growth rate and exploit new territory (Green 1980). Bacteria are further colonized through the process of microbial succession, leading to complex and stable communities (Sampaio-Maia and Monteiro-Silva 2014). The last colonizers include slow-growing bacteria such as Fusobacteria that are very stable and successful once established (Green 1980; Sampaio-Maia and Monteiro-Silva 2014). Other studies previously reported a similar succession of bacteria when oral microbiota of children at various developmental stages were analyzed (Crielaard et al. 2011).

Acquiring the oral microbiome after birth is essential and is considered by some groups as the critical first step to maintaining a normal oral microbiome during adult life (Zaura et al. 2014). If the delivery is by the vagina, newborns show a higher oral microbiome taxonomic diversity comprising *Haemophilus parainfluenzae*, *Streptococcus sanguinis*, *Streptococcus* spp., and *Cardiobacterium hominis* in comparison with the microbiomes of newborns delivered by Cesarean (C)-section (Lif Holgerson et al. 2011). In contrast, oral bacteria in newborns delivered by C-section had higher populations of *Slackia exigua*, *Veillonella* spp., *V. atypica*, and *V. parvula* (Lif Holgerson et al. 2011). As suggested by Holgerson et al. (2013), oral microbiomes also differed in breastfed and formula-fed newborns. The oral microbiome of breastfed newborns included *Lactobacillus* and *Streptococcus*, whereas formula-fed newborns had more anaerobic bacteria and lacked lactobacilli (Holgerson et al. 2013).

With the advent of powerful omics technologies including transcriptomics, proteomics, and metabolomics, we are getting a deeper view of the composition of the oral microbiome. Results from the Human Microbiome Project Consortium gave great insights into the diversity and variation between body habitats and between races, but left many questions unanswered (Human Microbiome Project Consortium 2012). Compared with other body sites such as stool, skin, and vagina the oral microbiome had the highest median alpha diversity operational taxonomic units (OTUs), but one of the lowest beta diversities (Human Microbiome Project Consortium 2012). The NIH Human Microbiome Project states on its online portal that it has been developed to be a "central repository for all Human Microbiome Project data with the aim to characterize microbial communities found at multiple human body sites" (http://hmpdacc.org/). This resource is continually updated and can be consulted to access the most up-to-date literature on the pan microbial communities. The site describes each habitat as having a core resident bacteria population that distinguishes it from other locations. For example, the oral cavity mostly comprises *Streptococcus*, *Haemophilus* (buccal mucosa), *Actinomyces* (supragingival plaque), *Prevotella* (adjacent to the subgingival plaque), and to a lesser degree *Corynebacterium*, *Moraxella*, *Propionibacterium*, and *Veillonella* (Human_Microbiome_Project_Consortium 2012). A study using this public database mined the metagenomics data to construct and portray host genetic variation correlations with oral microbiota (Blekhman et al. 2015). This approach revealed correlations in alpha (within individual) diversity between host variation and body sites such as the oral microbiome (Blekhman et al. 2015). Another interesting result was the alpha diversity correlation between stool and palatine tonsils (Blekhman et al. 2015).

The habitats within the mouth have been described as containing distinct microenvironments that support specific populations of microbes (Warinner et al. 2015). Using fluorescent probes, a recent paper by Mark Welch et al. (2016), demonstrated the spatial distribution of bacteria in the mouth giving insights into the "biogeography" of the mouth in which diverse habitats enable multiple species of bacteria to co-exist. Not only are the bacteria of interest, but also the metabolome that arises, such as the 400,000 protein families from the oral cavity that have not yet been functionally characterized (Human_Microbiome_Project_Consortium 2012).

Studies of the ecology and metabolome of the oral microbiome continue to expand our understanding and appreciation of the integral role that the oral microbiome plays in human health. With continued research on the oral microbiome, our knowledge is likely to expand to the high level that has recently been achieved for the gut microbiome. Advances in preventing and restoring gut microbiome dysfunction have been especially important for human health in recent times, and we expect similar benefits with an increased understanding of the roles of the oral microbiome in human health.

9.3 Associations with Pathological Conditions

9.3.1 Bacteremia and Endocarditis

Bacteria in the mouth can be a source of systemic infection, leading to bacteremia. Bacteria that have mediated such infections include *Streptococcus* spp. and *Lactobacillus* spp., and serious systemic infections are most frequently reported in immunosuppressed or immunocompromised patients, or those with other serious health issues, such as vascular disease or diabetes. In addition, bacteria that are not specifically associated with the oral microbiome, but that may exist there as transient oronasal commensals, such as *S. aureus*, may also become the source of systemic infections. Endocarditis infections are most frequently reported; infections with noncardiac involvement occur as well.

Several studies have observed that the most commonly observed species present in bacteremia following dental procedures are various streptococcal species, including *Streptococcus* spp., *Peptostreptococcus micros* (Bahrani-Mougeot et al. 2008), *S. mitis*, *S. oralis*, and *S. sanguinis* (Forner et al. 2006). *Veillonella dispar* or *V. parvula*, and *Dialister pneumosintes* have also been reported (Bahrani-Mougeot et al. 2008). A brain abscess that occurred following a dental extraction has been attributed to *Arcanobacterium haemolyticum* (Vargas et al. 2006).

Infections of heart valves, a condition known as endocarditis, often require surgery (valve replacement) and are associated with high levels of morbidity and mortality (Pang et al. 2015). While the bacteria that cause endocarditis may sometimes be of cutaneous or unknown origin, about 30% of endocarditis cases appear to have been caused by bacteria of dental origin (Delahaye et al. 2016). Approximately 70% of these putative oral-source infections were caused by viridans streptococci, although 10% were due to a group of species known as HACEK bacteria (*Haemophilus*, *Aggregatibacter*, *Cardiobacterium*, *Eikenella*, *Kingella*), and were usually associated with patients having concurrent dental infections, including periodontal disease and tooth decay (Delahaye et al. 2016). *Streptococcus mutans*, which is most frequently found in dental caries, is the most prevalent bacteria of dental origin found in coronary artery endothelial cells in patients undergoing heart valve surgery (Abranches et al. 2009); molecular identification techniques have confirmed this observation in a variety of cardiac tissues (Oliveira et al. 2015; Nakano et al. 2009).

Although many endocarditis patients have periodontal disease, the decay-causing bacteria *S. mutans* is found more commonly than the periodontal disease pathogen *P. gingivalis* in the cardiac tissue (Oliveira et al. 2015). Various authors have speculated that this might indicate that *S. mutans* may more readily evade immune system responses when translocating from the oral cavity to cardiac tissues. Despite a mortality rate of 5–10% (up to 30% for in-hospital subjects) for patients with endocarditis, this degree of mortality seems more frequently associated with non-oral bacteria (often *S. aureus*) than *Streptococcus* spp. (Pang et al. 2015; Slipczuk et al. 2013). However, although *S. aureus* is often assumed to be of non-oral origin, cases have been reported in which the suspected source is from dental procedures (Kasmi et al. 2014). Nevertheless, despite the seriousness of the condition, endocarditis of oral origin associated with *S. mutans* seems less likely than with *S. aureus* to result in death and also, the global rate of oral *Streptococcus* involvement in infectious endocarditis has decreased in recent decades (Selton-Suty et al. 2001).

9.3.2 Cardiovascular Diseases

Much evidence accumulated over several decades supports an association of cardiovascular diseases (CVDs) with oral bacteria. These studies encompass major cardiovascular events, i.e., ischemic heart disease and stroke, and well-established risk factors for these conditions namely, hypertension, atherosclerosis, and metabolic syndrome consisting of dyslipidemia, abdominal obesity, and high blood sugar. This section summarizes these conditions according to various types of available evidence.

9.3.2.1 The Association with Periodontal Disease and Its Treatments

Several meta-analysis studies examined associations between periodontal disease defined by diverse measures and different types of CVD and its precursors. Nibali et al. summarized 20 studies published through 2011. Using a random effects model, they reported that the presence of metabolic syndrome is associated with the presence of periodontitis in a total of 36,337 subjects (odds ratio [OR] 1.71; 95% confidence interval [CI] 1.42–2.03). When only studies with definite diagnoses were included ($n = 16{,}405$), the magnitude of association increased to an OR of 2.09 (95% CI 1.28–3.44). However, the directionality of this association (i.e., which condition makes the other more susceptible) has been controversial, as most studies have relied on the cross-sectional study design (Watanabe and Cho 2014). In fact, only one group of investigators used a longitudinal study design and demonstrated that the onset of two of more metabolic syndrome components during follow-up is associated with the presence of periodontal pockets at baseline (Morita et al. 2010).

Zeng et al. recently analyzed 15 observational studies concerning periodontal disease and carotid atherosclerosis that were published through February 2015 and involved a total of 17,330 participants (Zeng et al. 2016). They reported a pooled OR of 1.27 (95% CI: 1.14–1.41) with statistically significant heterogeneity. Subgroup analysis adjusted for smoking and diabetes mellitus weakened this association (OR: 1.08; 95% CI: 1.00–1.18). Again, the data from longitudinal studies are limited. A study from Austria failed to show a significant association with the Periodontal Index, but showed significant associations with decayed, missing or filled teeth and with poor oral hygiene (Schillinger et al. 2006).

Blaizot et al. reviewed 29 observational studies published between 1989 and 2007 on various CVD endpoints (including coronary heart disease, stroke, and atherosclerosis), yielding pooled odds ratios of 2.35 (95% CI 1.87–2.96) for case–control and 1.34 (95% CI 1.27–1.42) for cohort studies. Since then a few new studies (both case–control and cohort) have reported stronger associations (OR 4–8) of periodontitis with stroke (Pradeep et al. 2010; Sim et al. 2008; Jimenez et al. 2009). Two of these studies (Sim et al. 2008; Jimenez et al. 2009) suggest that there might be a further increase in risk in younger (<60 or <65 years) populations.

Furthermore, several clinical trials of periodontal treatments have provided additional evidence to reinforce the associations reported in observational studies. In a single arm study, the standard periodontal treatment (scaling and root planing) resulted in an improved blood lipid profile during a 12-month follow-up (Buhlin et al. 2009). Another single arm study using anti-infective periodontal therapy found reduction of carotid intima media thickness after 12 months of the treatment (Piconi et al. 2009). Likewise, a Turkish group reported significant improvement in endothelial function immediately after nonsurgical periodontal therapy (Mercanoglu et al. 2004). As the control group without treatment in this study consisted of individuals without periodontitis, the authors did not completely rule out chronological fluctuations in endothelial function in patients with periodontitis. Finally, a randomized clinical trial comparing the standard versus intensive periodontal therapies revealed that topical antibiotic administration resulted in a significant improvement in the blood lipid profile (D'Aiuto et al. 2006).

9.3.2.2 Serum Antibodies Against Oral Bacteria

Serological assays to detect and quantify antibodies against specific oral pathogens were often used in earlier studies to address the role of specific bacteria. Most investigators used whole bacteria extract as an antigen, applying it to an ELISA, whereas a few other recent studies focused on antibodies against specific bacterial virulence proteins. In these studies, IgG antibody or IgG and IgA antibodies were measured. Most commonly studied are *Porphyromonas gingivalis* and *Aggregatibacter actinomycetemcomitans*. Mustapha et al. performed a meta-analysis including various types of CVD endpoints (Mustapha et al. 2007), reporting a summary odds ratio for coronary heart disease of 1.75 (95% CI 1.32–2.34), combining all bacterial and antibody types together, whereas no increase in risk was observed for stroke. A subsequent cohort study concerning stroke did not yield ORs consistently increased across genders (Pussinen et al. 2007). Conversely, a case–control study measuring antibodies against four periodontal pathogens, including *Tannerella forsythia* and *Treponema denticola*, in addition to the two aforementioned bacteria, confirmed that higher antibody levels in any of these bacteria were associated with an increased risk for myocardial infarction (Lund Håheim et al. 2008). Hyvärinen et al. (2012) reported significantly higher IgA levels against *A. actinomycetemcomitans* in patients with acute coronary syndrome compared with control subjects. The meta-analysis by Mustapha et al.

(2007) also revealed that individuals with elevated. *P. gingivalis* antibodies presented with a more advanced degree of atherosclerosis, measured by carotid intimal media thickness, whereas Beck et al. suggested that an increase in carotid intimal media thickness might have been associated with antibody titers to a wider range of oral bacteria including some commensals such as *S. oralis* (Beck et al. 2005). Jeong et al. specifically investigated auto-immune reactions to a peptide (Pep19) from *P. gingivalis* human heat shock protein 60 homolog, GroEL, and found that patients diagnosed with atherosclerosis had higher antibody titers to this bacterial protein (Jeong et al. 2012). Moreover, two recent cross-sectional studies have described an increased risk for metabolic syndrome being associated with elevated antibody titers to *P. gingivalis* and to *A. actinomycetemcomitans* in Japan (Iwasaki et al. 2016) and in Finland (Hyvärinen et al. 2015) respectively.

9.3.2.3 Bacteria from the Oral Cavity

The presence of antibodies to oral pathogens does not necessarily distinguish past and current active infection. Several recent studies made attempts to detect and quantify oral bacteria that were compared according to clinical phenotypes. Using real-time polymer chain reaction (PCR), Matsushita et al. (2015) reported that a component of metabolic syndrome, abdominal obesity, was associated with the presence of three high-risk periodontal pathogens, *P. gingivalis*, *T. forsythia*, and *T. denticola*, in buccal swab samples. Using pyrosequencing of bacterial 16S rRNA, Koren et al. (2011) discovered associations between dyslipidemia in low-density lipoprotein (LDL) and high-density lipoprotein (HDL) levels, with an abundance of *Fusobacterium* and *Streptococcus* abundance oral swab samples. Desvarieux et al. (2010) applied DNA–DNA hybridization techniques to test 11 bacteria in subgingival plaques and calculated the total bacterial load and the sums of three bacterial groups with different periodontal risk levels. The total bacterial loads were linearly positively associated with both diastolic and systolic blood pressures. The prevalence of hypertension also increased progressively with an increase in the loads from etiological and putative groups of bacteria. Using the same methodology, these authors described significant positive associations of degree of atherosclerosis, measured by carotid intima media thickness, with both cumulative and etiological bacterial loads (Desvarieux et al. 2005). A more recent longitudinal study by the same authors has proved that an increase in etiological bacterial loads over a median 3-year follow-up was associated with progressive thickening of the common carotid artery (Desvarieux et al. 2013). Furthermore, a tenfold increment in salivary *Aggregatibacter actinomycetemcomitans* counts has been noted to increase the risk of acute coronary syndrome 4.31-fold (95CI 1.06–17.5) and that of stable coronary artery disease 7.47-fold (95% CI 1.57–35.5) (Hyvärinen et al. 2012). However, these associations were not replicated when the same group of investigators examined subgingival plaque bacteria (Mäntylä et al. 2013). 16S rRNA pyrosequencing data have provided a different view, pointing to an association of history of myocardial infarction or stroke with a relative abundance of

Anaeroglobus, a new member of the family Veillonellaceae, compared with population controls (Fåk et al. 2015). This study also uncovered 29 additional genera or families differentially distributed between both control groups (population and hospital) combined and patients with a history of myocardial infarction or stroke (Fåk et al. 2015). Most of these 29, however, were not established periodontal pathogens, making the interpretation of these results difficult, until the associations are confirmed by independent studies.

9.3.2.4 Laboratory Studies

Thus far, experimental models to test the effects of periodontal infection on the development of CVDs and their precursors have mostly relied on oral inoculation of *P. gingivalis* alone. However, a study by Blasco-Baque et al. (2012) applied mixed infection of P. gingivalis, Prevotella intermediat, and Fusobacterium nucleatum to C57BL/6 mice fed on a high-fat diet, mimicking the bacteria that occur with human periodontitis. They found that experimental periodontal infection augmented several components of metabolic syndrome, i.e., elevated systolic and diastolic blood pressure and glucose intolerance. A more recent metabolomic study in *P. gingivalis* inoculated C57BL/6 mice not only demonstrated the development of hyperglycemia and hyperinsulinemia, but also revealed changes in metabolomic profiles in the brain, liver and heart, and plasma (Ilievski et al. 2016). Specifically, in the hearts, the upregulation of the glycolytic pathway, indicative of ischemia, was noted, whereas a microbially derived metabolite, 4-ethylphenylsulfate, was systemically elevated. Other investigators used genetically modified mice susceptible to CVD. In *ApoE*-null or deficient mice, oral inoculation of *P. gingivalis* accelerated the growth of atherosclerotic lesions, which was accompanied by increases in several inflammatory markers, such as VCAM, interleukin (IL)-6 and toll-like receptors (TLRs), in the aorta and serum (Lalla et al. 2003; Gibson et al. 2004; Miyamoto et al. 2006; Madan et al. 2007). Furthermore, a study by Gibson et al. (2004) demonstrated that the acceleration of atherosclerosis by experimental *P. gingivalis* infection depended on the presence of fimbriae, as *fimA* mutant strains failed to replicate these phenotypes. According to Turunen et al. (2012), *P. gingivalis* arginine-specific gingipain (Rgp) is responsible for inducing the production of auto-antibodies against oxidized LDL that play a vital role in atherosclerosis. Jeong et al. (2015) further reported that auto-immune reactions to peptides from a *P. gingivalis* human heat shock protein 60 homolog, GroEL, are crucial in the development of arterial plaque and hypercholesteremia in *ApoE*-null mice on a high-fat diet. Importantly, immunization with heat-killed whole bacteria or antibiotic treatment in these mouse models prevents the acceleration of atherosclerosis in addition to inflammatory responses (Gibson et al. 2004; Miyamoto et al. 2006; Madan et al. 2007; Turunen et al. 2012), reinforcing the potential causal association.

9.3.3 Chronic Kidney Disease

Patients with chronic kidney disease (CKD) present decreased function (glomerular filtration rate [GFR] <60 mL/min per 1.73 m^2), which may lead to kidney failure (GFR <15 mL/min per 1.732 m^2) and end-stage renal disease (ESRD). The association between periodontitis and CKD has been documented in both cross-sectional and cohort studies. In community residents in the USA, severe peritonitis is associated with a twofold (OR 2.0; 95% CI 1.23–3.24) increased risk of CKD (Kshirsagar et al. 2005), whereas the presence of periodontitis is associated with an increase in risk of CKD of 1.6-fold (OR 1.62; 95% CI 1.17–2.26) in the National Health and Nutrition Examination Survey (NHANES) III study (Fisher et al. 2011). Furthermore, the latter study clarified that the association with periodontitis was mediated through the development of hypertension and diabetes, which are known risk factors for CKD. A cohort study among native Americans with type 2 diabetes found a five-fold (hazard ratio 1.49; 95% CI 1.4–17.4) increased incidence of ESRD associated with moderate/severe periodontitis compared with no/mild periodontitis (Shultis et al. 2007). Similarly, a longitudinal study in Japan showed that elderly people with a larger inflamed periodontal surface area at the beginning of the study exhibited a significantly greater decline in in kidney function 2 years later (Iwasaki et al. 2012a). Corroboratively, high levels of antibodies against the periodontal pathogens, *P. gingivalis*, *T. denticola*, and *A. actinomycetemcomitans* were associated with CKD, with an odds ratio ranging from 1.6 to 1.8 in USA residents (Kshirsagar et al. 2007), and Japanese elderly people with elevated antibody titers to *P. gingivalis* were 2.6 (95% CI 1.05–3.64) times more likely to present CKD (Iwasaki et al. 2012b). Finally, one small single-arm clinical trial demonstrated that nonsurgical periodontal treatment led to a significant improvement in kidney function, measured by serum cystatin C levels (Graziani et al. 2010). Because there has not been convincing evidence supporting the direct involvement of periodontal pathogens in specific pathological kidney conditions or detection of those pathogens in patients' urine, the observed associations most likely represent the indirect effects of oral pathogens through cardiovascular risk factors as discussed above.

9.3.4 Alzheimer's Disease

Alzheimer's disease (AD) is a type of dementia that occurs commonly in the elderly. This progressive neurodegenerative disease affects regions of the brain associated with memory, speech, and awareness of surroundings (Faizi et al. 2016). Despite numerous studies to elucidate the pathogenesis of this disease, little progress has been made in understanding its causes or effective treatments that can cure the disease (Miklossy 2015). Increases in spending by >$1 trillion in the USA have been predicted by the year 2050 for treating AD in the 13–14 million people who may be suffering from AD by then (Olsen and Singhrao 2015).

The disease starts with short-term memory loss and anxio-depressive symptoms followed by orientation and verbal difficulties (Miklossy 2015). Patients usually survive with the disease for a decade or more, and death could occur from a secondary infection, due to pneumonia or a urinary infection (Miklossy 2015).

Periodontitis is one of the factors that can possibly promote the development of AD, as a sixfold increase in the rate of cognitive decline over a 6-month follow-up period has been reported in AD patients with periodontitis compared with those without (Ide et al. 2016). Periodontitis is a chronic disease, accelerated by the formation of bacteria biofilm that adheres to the surfaces of the tooth to the gingiva (Noble et al. 2014). Periodontitis patients show an elevated pro-inflammatory state, demonstrated by high serum C-reactive protein and pro-inflammatory cytokines (e.g., tumor necrosis factor α [TNFα]) levels, which results in a decrease in the anti-inflammatory capacities (Ide et al. 2016). The increase in serum bacterial lipopolysaccharide (LPS) in patients with periodontitis was detected after chewing and eating (Ide et al. 2016), suggesting bacterial translocation through oral mucosa to the blood stream, and possibly to the brain. Intact bacteria, bacterial virulence factors, and inflammatory mediators can be disseminated to the brain via the blood stream from the oral cavity, which harbors various periodontal pathogens (Olsen and Singhrao 2015). Increases in amyloid beta (Aβ) and tau protein that have been linked to pathological AD conditions may result from the systemic inflammatory products formed by periodontal disease (Watts et al. 2008). The effects of infectious agents and increased inflammatory mediators in the development of AD may be medicated through the APOEε4 genotype. Live bacteria, LPS, and other toxic bacterial products can penetrate the human brain. *APOEε4*, TNFα, and perhaps ephrin type-A receptor 1 are factors that can impair the integrity of the blood–brain barrier (BBB), which allows live bacteria, LPS, and other toxic bacteria to easily penetrate the brain (Olsen and Singhrao 2015). Specifically, *APOEε4* leads to BBB breakdown in brain pericytes by activating the cyclophilin A-matrix metalloproteinase 9 (MMP-9) pathway, which is also involved in periodontal disease. *Chlamydophila pneumoniae* occasionally found in the oral cavity is found in 80–90% of the brain-tissue specimens from late-onset AD patients that showed correlation with *APOEε4* allele expression, as was reported by Balin et al. (cited in Olsen and Singhrao 2015).

A prospective study has demonstrated that when the level of serum IgG against periodontal pathogens (*P. gingivalis*) is increased, the risk of being diagnosed with incident AD increases (Noble et al. 2014). Patients with and without AD can be discriminated based on levels of a mixture of three periodontal bacterial IgG titers (Noble et al. 2014). *Actinomyces naeslundii* are gram-positive rod-shaped bacteria linked to matured dental plaque formation, gingivitis, and dental caries, and it plays an important role in bone loss. A high level of serum *A. naeslundii* IgG was determined to be related to an increased risk of AD (Noble et al. 2014). *Eubacterium nodatum* is also a gram-positive, anaerobic rod-shaped bacterium that is associated with lowering the risk of AD, but its role in oral pathogenesis has not been clearly identified (Noble et al. 2014).

Another follow-up study was carried out to survey patients who were prospectively converted to AD by looking at the serum antibodies to bacteria (Stein et al.

2012). In a comparison with control subjects who did not develop AD or mild cognitive impairment, patients who developed AD displayed a significant increase in the levels of antibodies to *F. nucleatum*, *T. denticola*, and *P. intermedia* (Stein et al. 2012).

Although the connection between peripheral inflammation/infections and brain Aβ, a main pathological feature of AD, has been studied with human periodontal disease, it has still not been determined whether peripheral inflammation or infections facilitates the accumulation of Aβ (Kamer et al. 2015). As discussed above, both clinical evidence and data from animal models clearly indicate that inflammatory components are involved in the pathogenies of AD, but it remains to be elucidated which form of peripheral inflammation or infectious settings plays a significant role in the development of AD (Kamer et al. 2015).

9.3.5 Rheumatoid Arthritis

Rheumatoid arthritis (RA) is an autoimmune chronic inflammatory disease affecting multiple synovial joints. A possible association between RA and periodontitis has been debated for the past 200 years (Rutger Persson 2012). An epidemiological link between these two conditions has been primarily reported in case–control or cross-sectional studies in various populations. A population-based cross-sectional study from the NHANES pointed out that the presence of periodontitis increased the risk of RA by 1.82-fold (95% CI 1.04–3.20) (De Pablo et al. 2008). In a Dutch clinic-based study, the relative risk of RA was 3.6 (95% CI 2.3–5.5) and 3.7 (95% CI 2.4–5.9) for moderate and severe periodontitis respectively (De Smit et al. 2012) and a case–control multicenter study in India reported a 4.28-fold (95% CI 2.35–7.38) increased risk of RA associated with periodontitis (Potikuri et al. 2012). A more recent multicenter study in the USA confirmed that the association was particularly evident for RA positive to antibody against citrullinated protein, which is postulated to be generated by the enzymatic activity of periodontal pathogens (Mikuls et al. 2014).

Porphyromonas gingivalis has been a focus of research interest in this association, because it expresses a unique enzyme named peptidylarginine deiminase (PAD), which catalyzes post-translational modification of arginine to citrulline (Wegner et al. 2010). Citrullinated proteins are known to play a critical role in RA pathogenesis, leading to production of autoantibodies. Several sero-epidemiological studies corroborate this hypothesis, demonstrating antibody titers to this bacterium, but not to other oral pathogens such as *P. intermedia*, *F. nucleatum*, *Eikenella corrodens*, and *A. actinomycetemcomitans*, were elevated in RA patients compared with controls (Okada et al. 2011; Hitchon et al. 2010; Mikuls et al. 2009). This association was also observed even among individuals who were positive to RA-associated autoantibodies, but do not present clinical signs/symptoms of RA, disputing the possibility that periodontal disease is a consequence of RA (Mikuls et al. 2012). Jeong et al. (2012) reported that auto-immune reactions to a peptide (Pep19) from *P. gingivalis*, a human heat shock protein 60 homolog, GroEL, may play an additional role in the development of RA, as serum antibodies against this peptide from *P. gingivalis* in RA patients were elevated compared with those against other nonperiodontal pathogens. There are also supporting data from mouse experimental arthritis models. Pretreatment with oral inoculation of *P. gingivalis* led to the development of severe arthritis (Cantley et al. 2011), whereas immunization with *P. gingivalis* enolase induced autoimmunity to human enolase and clinical arthritis (Kinloch et al. 2011).

More recent studies based on culture-independent, high-throughput sequencing techniques examined the associations of RA with more comprehensive microbial community structure. Scher et al. reported using 16S ribosomal RNA gene pyrosequencing that not *P. gingivalis*, but *Anaeroglobus geminatus* and *Prevotella* and *Leptotrichia* species may play a role in RA pathogenesis (Scher et al. 2012). In a metagenome-wide study, alterations in the dental or saliva microbiome distinguished individuals with RA from healthy controls and correlated with clinical measures of RA. In particular, *Haemophilus* spp. were depleted in individuals with RA and correlated negatively with levels of serum autoantibodies, whereas *Lactobacillus salivarius* was over-represented in individuals with RA and was present in increased amounts in cases of very active RA. This dysbiosis was also partially restored in treated RA patients (Zhang et al. 2015). These studies suggest a possibility that a wider range of oral microbiome or overall community balance might be involved in the pathogenesis of RA.

9.3.6 Oral Cancer

9.3.6.1 Cancer Development

The associations of the presence of periodontal disease and the risk of head and neck cancer have been studied in cross-sectional, case–control, and cohort studies. In the study based on the National Health and Nutrition Examination Survey (NHANES), the severity of periodontal disease was assessed by clinical attachment loss (CAL). More than 15 mm of CAL was strongly associated with the presence of a benign tumor (OR 4.57, 95% CI 2.25–9.30) and precancerous lesions (OR = 1.55, 95% CI 1.06–2.27) (Tezal et al. 2005). The same authors subsequently conducted a case–control study for head and neck cancer (oral, oropharyngeal, and laryngeal cancer) at Rosewell Park Cancer Institute, using alveolar bone loss (ABL) as a measure of periodontitis (Tezal et al. 2009). The authors demonstrated that a 1-mm increase in ABL is associated with a fourfold increase in the risk of head and neck cancer (OR = 4.36, 95% CI 3.16–6.01). The association was strongest for oral, intermediate for oropharyngeal, and weakest for laryngeal cancer. A more recent case–control study conducted in India indicated an approximately twofold increased risk of oral cancer associated with severe gingivitis (OR 2.28, 95% CI 1.18–4.38) and generalized gingival recession (OR 1.74, 95CI 1.15–2.62) (Laprise et al. 2016) with adjustment for a priori selected covariates. Despite these positive findings, a prospective cohort study in US male health professionals failed to confirm the increased risk of oropharyngeal cancer based on self-reported periodontal disease with bone loss (Michaud et al. 2008). Other head and neck cancer sites were not evaluated in this study (Michaud et al. 2008). However, it is important to note that the two well-known periodontal pathogens, *P. gingivalis* and *F. nucleatum*, were recently tested in a new murine chemical carcinogenesis model (Gallimidi et al. 2015). Oral inoculation of these bacteria before and after chemical carcinogen exposure significantly increased tumor size and invasion, through the IL6–STAT3 pathway. Furthermore, co-culture of *P. gingivalis* in immortalized keratinocytes leads to epithelial–mesenchymal transition (Sztukowska et al. 2016) and in oral cancer cells increases cancer stem cell properties and thus the aggressiveness of cancer (Ha et al. 2015).

Porphyromonas gingivalis has been found to colonize on tumor surfaces more densely than on normal mucosa and it can attach and invade gingival epithelial cells through the interaction between fimbriae and host integrin (Yilmaz et al. 2003). It also targets dividing cells in the S phase (Al-Taweel et al. 2016). Potential involvement of *Streptococcus*, which is preferentially found in dental plaques, has also been reported, as it is more frequently detected on oral and pharyngeal cancer sites or in the oral cavity of these patients than in control samples (Sasaki et al. 2005; Tateda et al. 2000). In addition to these studies focusing on specific oral pathogens, others have attempted to characterize the poly-microbial community associated with oral and pharyngeal cancer, employing various techniques, and have yielded highly variable results (Nagy et al. 1998; Mager et al. 2005; Hooper et al. 2006, 2007; Pushalkar et al. 2011, 2012; Schmidt et al. 2014; Sonalika et al. 2012; Hu et al. 2016). However, as summarized in Table 9.1, over-representation of *Streptococcus*, *Fusobacterium*, and *Prevotella* in cancer specimens was noted by multiple studies.

9.3.6.2 Treatment Outcome/Complications

Treatments for head and neck cancer have changed substantially over the past few decades, with increased use of chemoradiation, which has led to improvement in overall survival and local tumor control. Unfortunately, this type of treatment often causes a range of acute and late oral complications (Epstein et al. 2012) and the oral microbiome has been implicated in the development and progression of some of these conditions. Xerostomia are well-known common side effects from radiation therapy for head and neck cancer, which leads to a dramatic shift in the core structure of the oral microbiome (Gao et al. 2015; Hu et al. 2013) and in abundance of pathogenic bacteria and fungi (Almståhl et al. 2015; Tong et al. 2003; Brown et al. 1975). In addition, immunosuppression due to systemic chemotherapy exacerbates these dysbiotic conditions (Panghal et al. 2012).

Treatment-induced oral and oropharyngeal mucositis is an acute debilitating and troublesome condition that profoundly affects patient quality of life. Bacterial

Table 9.1 Summary of poly-microbial studies on oral cancer

References	Microbial methods	Number of cancer cases	Controls	Type of specimens	Increase in tumor	Decrease in tumor	Other findings
Nagy et al. (1998)	Culture	21	Own normal mucosa	Surface biofilm	Total aerobics/anaerobics, *Veillonella*, *Fusobacterium*, *Prevotella*, *Porphyromonas*, *Actinomyces*, *Clostridium*, *Haemophilus*, *Enterobacteriaceae*, *Streptococcus* spp.		
Mager et al. (2005)	DNA–DNA hybridization	45	229 hospital controls	Saliva	*Capnocytophaga gingivalis*, *Prevotella melaninogenica*, *Streptococcus mitis*		
Hooper et al. (2006)	Culture colony 16S rRNA Sanger sequencing	20	Own normal mucosa	Tumor/mucosa tissue	*Micrococcus luteus*		
Hooper et al. (2007)	16S rRNA PCR cloning	10	Own normal mucosa	Tumor/mucosa tissue	*Ralstonia insidiosa*, *Prevotella* sp., *Fusobacterium naviforme*	*Sphingomonas* spp., *Granulicatella adiacens*	
Pushalkar et al. (2011)	16S rRNA PCR cloning, DGGE	10	Own normal mucosa	Tumor/mucosa tissue	*Streptococcus* sp., *Peptostreptococcus stomatis*, *Streptococcus salivarius*, *Streptococcus gordonii*, *Gemella haemolysans*, *Gemella morbillorum*, *Johnsonella ignava*, *Streptococcus parasanguinis*	*Granulicatella adiacens*	

(continued)

Table 9.1 (continued)

References	Microbial methods	Number of cancer cases	Controls	Type of specimens	Increase in tumor	Decrease in tumor	Other findings
Pushalkar et al. (2012)	Pyrosequencing of 16S rRNA amplicon, DGGE	3	Two noncancer controls	Saliva		Richness	Differential DGGE pattern
Sonalika et al. (2012)	Culture	32	72	Saliva	Total anaerobics, Gram-negative anaerobic bacilli, Coliforms		
Schmidt et al. (2014)	Pyrosequencing of 16S rRNA amplicon	25 (16 cancer, 1 in situ, 8 pre-cancer)	Five own normal mucosa and five healthy controls	Tumor/mucosal swab	Fusobacterium	Firmicutes, Actinobacteria	
Hu et al. (2016)	Miseq 16S rRNA sequencing	44 (15 Cancer, 10 leukoplakia)	19 healthy controls	Saliva	Microbial richness/diversity indexes		

DGGT denaturing gradient gel electrophoresis

overgrowth on the affected mucosa and their invasion through ulceration and impaired mucosal barrier also represents a possible source of systemic opportunistic infection or sepsis (Al-Ansari et al. 2015; Meurman et al. 1997). For example, *Staphylococcus aureus* and *S. epidermidis* have been more frequently isolated from both the oral cavity and the blood of patients who received chemotherapy or chemoradiation for oral squamous cell carcinoma than from those who received radiation alone (Panghal et al. 2012). Dysbiosis induced by cytotoxic treatments weakens mucosal protection by resident oral bacteria and leads to enhanced pro-inflammatory reactions to exacerbate and extend mucositis, which initially arises from damaged epithelial cells directly by cytotoxic agents (Donnelly et al. 2003; Stringer and Logan 2015). Using qPCR, Laheij et al. (2012) compared the abundance of selected bacteria in oral rinse samples from cancer patients who developed moderate to severe (ulcerative) mucositis and those who developed no or mild mucositis and found that *P. gingivalis* in particular, in addition to *T. denticola* and *F. nucleatum*, were more abundant and *Parvimonas micra* was less abundant in moderate to severe mucositis patients than in those with no or mild mucositis. More recently, using 16S rRNA amplicon pyrosequencing, Ye et al. (2013) conducted a prospective study among Swedish pediatric cancer patients. They found that, compared with children who did not develop mucositis, children who developed mucositis have higher microbial community diversity (Shannon index) with increased frequencies of *Capnocytophaga*, Peptostreptococcaceae, lactococci, Fusobacteria, and Spirochaetes in their oral mucosa.

Besides these acute toxicities, other important late complications in which the oral microbiome is involved are dental caries and osteoradionecrosis (Buglione et al. 2016). A major microbial shift associated with hyposalivation due to radiation is an increase in cariogenic bacteria, specifically *Streptococcus* and *Lactobacillus* species, which are acid-producing bacteria, as documented by many studies (Almståhl et al. 2015; Tong et al. 2003; Brown et al. 1975). Moreover, hyposalivation leads to loss of buffering capacity, thus further lowering oral pH, and to a decreased supply of enamel remineralization substrates (Deng et al. 2015; Epstein et al. 2012), both of which contribute to the development of dental demineralization and caries. Osteoradionecrosis is a necrotic process resulting from radiation-activated fibrosis and inflammation in irradiated bones (mandible and maxilla). Under this condition, minimal external trauma can cause ulceration, facilitating certain bacterial infection and leading to osteomyelitis (Chrcanovic et al. 2010; Silvestre-Rangil and Silvestre 2011). Several studies have reported the frequent detection of *Actinomyces* in affected bones (Hansen et al. 2006a, b; Støre et al. 2005). Other isolated bacteria are mainly anaerobic, including *P. gingivalis*, *F. nucleatum*, and *Prevotella* (Støre et al. 2005; Aas et al. 2010). Active roles of these bacteria in the pathogenesis of the disease are unclear, but bacterial infection complicates the disease process, impeding successful treatment (Hansen et al. 2006a, b).

Despite these known complications, direct evidence is still lacking as to whether preexisting and treatment-induced oral dysbiosis alters treatment outcome and patient survival in head and neck cancer patients.

9.4 Potential Mechanisms

9.4.1 Experimental Models

9.4.1.1 Animal Models

In addressing mechanistic questions, animal models have distinct advantages over in-vitro studies that take place on plastic surfaces with limited numbers of cell types present. Various animal models for studying the host–bacterial interactions involved in periodontitis have been developed to test bacterial colonization, invasion, induction of host responses, and a repair process (De Molon et al. 2013; Graves et al. 2012). Although large animal models such as nonhuman primates and dogs exist, most studies use rodent (rat and mouse) models. Bacterial inoculation or induction of host responses can be achieved by several modes. One called the "ligature model" physically induces indigenous bacterial accumulation in dental plaques by placing ligatures around teeth. Other models use direct injection of bacterial virulence factors, such as LPS or fimbriae, or bacteria themselves into local tissues, e.g., gingiva and scarp (Graves et al. 2012). All of these are acute models primarily used to study the effects and mechanisms of the development of periodontal disease. The most relevant animal model for studying the chronic effects on the development of systemic diseases is the oral gavage model, where human bacterial cultures are inoculated into an animal oral cavity. This model also allows investigators to examine every step of the model host response (De Molon et al. 2013). A variant of this model is the *A. actinomycetemcomitans* infection model, in which the bacteria are given with the animal feed (Graves et al. 2012). In these models, investigators are able to test mutant bacterial strains that lack specific virulence factors of interest, in addition to human clinical isolates. It should be kept in mind, however, that in both models, the introduction of exogenous bacteria may alter existing native oral flora, unless germ-free animals are used, which may contribute to host responses and pathogenesis. Also, in studies of systemic disease, various genetically modified animal models, such as *ApoE*-deficient mice susceptible to CVD, are often used in conjunction with bacterial exposure (Kebschull et al. 2010), and bacterial inoculation has also been tested to augment the induction of target diseases (e.g., cancer, CVD, arthritis) in established chemical, immunological or dietary pathogenesis models (Gallimidi et al. 2015; Cantley et al. 2011; Kebschull et al. 2010).

9.4.1.2 In-Vitro and Organotypic Models

Although standard monolayer culture of malignant or immortalized oral or gingival cell lines have been used for co-culture, incubation or transfection studies with oral pathogens or their virulence factors, recent advances in 3D culture technology allow investigators to develop new in-vitro models consisting of multiple lineages of organ-specific cells that more closely mimic human organs. To our knowledge, organoids derived from embryonic stem cells or organ-specific adult stem cells, which are standard for other organs, have not yet been available for orodental and pharyngeal structures (Clevers 2016). Instead, several 3D organotypic tissue culture models have been constructed using tissue engineering technology and were originally aimed at clinical application to repair damaged oral tissue (Moharamzadeh et al. 2007, 2012; Kinikoglu et al. 2015). These models have recently have been appreciated for their utility in studying host–bacterial and epithelial–mesenchymal interactions, oral carcinogenesis, and drug efficacy, as a potential alternative to in-vivo testing. Typically, these artificial mucosae consist of three components, epithelial cells, fibroblasts, and natural or synthetic scaffold, but the cells are not from the same origins (not from one stem cell or the same individuals). Earlier models often utilized established cell lines (malignant or immortalized), whereas more recent models have used patients' own keratinocytes and fibroblasts (Mostefaoui et al. 2002; Kinikoglu et al. 2009; Moharamzadeh et al. 2008). Moreover, Peña et al. developed a completely autologous oral mucosa equivalent using a patient's own materials, including fibrin as a scaffold (Pena et al. 2010, 2011). For oral carcinogenesis models, normal epithelial cells are replaced with oral cancer or dysplastic oral keratinocyte cell lines (Colley et al. 2011). More complex models incorporate immunological cells, i.e., monocytes, which may be more suitable for testing host pro- and anti-inflammatory responses (Bao et al. 2015b). In other models, dental tissue is placed on top of the epithelial culture (Gursoy et al. 2012) and bone is attached using a fibrin-based adhesive to engineered mucosa (Almela et al. 2016). These models have been successfully employed to demonstrate variabilities in penetration and intracellular invasion/survival of different strains of periodontal pathogens, such as *F. nucleatum* and *P. gingivalis* (Andrian et al. 2004; Gursoy et al. 2010; Pinnock et al. 2014), internalization of host molecules by *A. actinomycetemcomitans* (Paino et al. 2012), intracellular delivery of antibiotics (Wayakanon et al. 2013), and host immunological responses to multispecies biofilm (Bao et al. 2015a).

9.4.2 *Bacterial Toxin and Virulence Factors*

This section is intended to provide information regarding selected nonspecific bacterial virulence factors that are found ubiquitously in many oral bacteria and species-specific virulence factors that have been well characterized in well-established periodontal pathogens.

9.4.2.1 Non-specific Universal

Lipopolysaccharide

Lipopolysaccharide (LPS) is a major surface component of the bacterial cell wall and is an outer membrane glycolipid that is essential for virtually all Gram-negative bacteria. LPS represents one of the conserved microbial structures that activates the innate immune system (Trent et al. 2006). As a result, LPS is the best studied bacterial universal virulence factor, and in fact one of the commonly used animal models of periodontal disease is based on LPS injection (De Molon et al. 2013). Because of its ubiquitous presence within our environment and in vivo, circulating LPS is not indicative of its origin and is more likely to be contributed by gut bacteria (Moreira et al. 2012). LPS expresses potent bioactivity in extremely small amounts (nanogram to picogram range), causing sepsis (endotoxemia), and modulating a myriad of other host innate inflammatory responses. Specifically, LPS exerts the prototypical stimuli for host activation through myeloid cells (neutrophils, monocytes, macrophages), nonmyeloid cells (fibroblasts, platelets), and other innate host defense mechanisms, such as serum complement, in addition to specific components within the intrinsic coagulation pathway (Dixon and Darveau 2005). LPS is therefore implicated in etiological pathways of wide range of pathological conditions, including inflammatory diseases, cancer, and CVD.

Lipopolysaccharide is composed of three distinct domains—lipid A, core oligosaccharides, and the O-antigen polysaccharide. The lipid A domain, also known as endotoxin, is the bioactive component primarily recognized by host TLR4 (Trent et al. 2006). In most bacteria, lipid A structure is conserved and consists of a mono- or biphosphorylated disaccharide backbone, which is acylated with C12–C14 fatty acids. Investigations of the lipid A structures of various organisms have revealed the presence of considerable diversity, which can be attributed to the action of latent enzymes that modify the canonical lipid A molecule. Variations can range from simple modifications, such as minor alteration in the length of the segment, to dramatic changes in the overall chemical configuration, composition, or attached charge groups, which, in turn, can affect overall structure (Dixon and Darveau 2005).

Potency of LPS components has been traditionally quantified through the use of in-vivo animal models, i.e., the chick embryo lethality test, the rabbit pyrogenicity assay, and the Shwartzman reaction (skin hemorrhagic reaction), whereas in-vitro assays, such as induction or enhancement of IL-1, PGE2, O_2^- production in murine macrophages, and activation of the human complement cascade, have also been introduced. Studies with these models have identified that key determinants of the toxicity of LPS are mediated via the TLR4 pathway and include the number of phosphate groups and the lengths, number, and positions of fatty acids and β, (1–6)-linked D-glucosamine disaccharide backbone in lipid A (Dixon and Darveau 2005; Trent et al. 2006). Interestingly, when lipid As from enteric (e.g., *Escherichia coli*) and oral bacteria (*P. gingivalis*) were compared, investigators found that the potency of oral bacteria was much lower than that of enteric bacteria (Dixon and Darveau

2005; Ogawa et al. 2007). Accumulating evidence indicates, however, that despite the low affinity of *P. gingivalis* LPS to host TLR4, *P. gingivalis* LPS possesses significant biological activity mediated through TLR2 (Ogawa et al. 2007), which has been now acknowledged to be the case for a number of other human pathogens, including *Helicobacter pylori* (Trent et al. 2006). The attenuation in the TLR4 pathway activation may in turn confer advantages for some bacteria in evading host innate immune reactions and thus establishing chronic infections.

GroEL

The bacterial molecule chaperone GroEL belongs to heat shock protein family 60 (HSP60), which represents one of the most conserved proteins in living organisms and is present not only in the cytosol, but also on the cell surface and in the extracellular milieu (Castanié-Cornet et al. 2014). Consequently, GroEL is produced ubiquitously regardless of the Gram positivity of the bacteria. The virulence of this protein among oral microbiome bacteria has been, thus far, primarily studied in Gram-negative bacteria that have been linked to periodontal disease. This molecule exerts virulence through two postulated etiological pathways: direct effects eliciting proinflammatory responses and indirect effects via molecular mimicry. In-vitro studies using various types of human cells, including peripheral blood cells, fibroblasts, and endothelial cells and of bacteria (i.e., *P. gingivalis*, *T. forsythia*, *F. nucleatum* and *A. actinomycetemcomitans*) have shown that co-culture with this protein stimulates secretion of several inflammatory cytokines, mediators, and adhesion molecules, such as IL-6, IL-8, IL-12, MCP1, COX2, ICAM-1 VCAM-1, and E-selectin (Jung et al. 2016; Lee et al. 2012; Nalbant and Saygili 2016; Huang et al. 2016). These effects were mediated by TLR-4 in some bacteria, e.g., *P. gingivalis* (Huang et al. 2016) and *F. nucleatum* (Lee et al. 2012), but not in others, e.g., *T. forsythia* (Jung et al. 2016). A Taiwanese group assessed the direct effects of GroEL on endothelial migratory and angiogenic functions that play important roles in the etiology of both CVD and cancer (Huang et al. 2016; Lin et al. 2015). The results depended on the types of endothelial cells used. Although both capacities were suppressed when coronary artery endothelial cells were exposed to *P. gingivalis* GroEL (Huang et al. 2016), both capacities were increased when endothelial progenitor cells were used (Lin et al. 2015). Both findings were further reinforced by concomitant animal experiments, suggesting their potential involvement in CVD and cancer. Specifically, in the former, *P. gingivalis* GroEL injection accelerated the development of atherogenic lipid profiles (elevated blood LDL and triglyceride) in mice fed a high-cholesterol diet (Huang et al. 2016). In the latter, *P. gingivalis* GroEL led to neovasculogenesis in chicken embryos and accelerated tumor growth in mice (Lin et al. 2015).

Molecular mimicry is caused by the homology between foreign and self-antigens that may elicit cross activation of T and B cells and thereby produce an autoimmune response (Chistiakov et al. 2016; Kebschull et al. 2010). Mammalian HSP60 shares significant homology with bacterial GroEL. Accordingly, anti-GroEL antibodies cross-react with human endothelial HPS60, leading to auto-aggressive destruction through endothelial cytotoxicity (Mayr et al. 1999). HSP-reactive T-cells have indeed been found in the peripheral blood and atherosclerotic plaques of patients with periodontal disease (Yamazaki et al. 2004). Autoantibodies to HPS60 also promoted thrombus formation in a murine chemical thrombosis model (Dieudé et al. 2009). *F. nucleatum* GroEL injection elicited anti-GroEL antibody production in *ApoE* knockout mice, which was accompanied by atherosclerotic plaque formation and increased serum inflammatory markers and LDL (Lee et al. 2012), although direct stimulation rather than autoimmune response may account for some of these results.

9.4.2.2 Bacterial Species-Specific

Porphyromonas Gingivalis

Porphyromonas gingivalis is a nonmotile, Gram-negative obligate anaerobic intracellular bacterium and the best studied oral pathogen, with an array of virulence factors reviewed by others (Zenobia and Hajishengallis 2015; Nakayama 2015). Thus, this section focuses on only well-established virulence factors known to have clinical relevance. Importantly, *P. gingivalis* is considered to be a keystone pathogen, which, even with a very low abundance, can alter commensal microbial composition, leading to a dysbiotic community that further promotes host immune suppression and inflammatory responses (Lamont and Hajishengallis 2015; Hajishengallis et al. 2011). Another distinct feature of *P. gingivalis* virulence is the absence of machineries, e.g., types III and IV, that directly inject effectors into host cells. Rather, this bacterium utilizes a novel type IX secretion system distributed among the Bacteroidetes phylum to secrete virulence proteins into the extracellular milieu (Nakayama 2015).

Porphyromonas gingivalis fimbriae are protein-based filamentous appendages that protrude from the cell surface and are thinner and shorter than a flagellum. Fimbriae facilitate adhesion to host cells, tissues, or to other bacteria, and help mediate bacterial invasion/internalization and thus delivery of other virulence factors (Yoshimura et al. 2009). In addition, *P. gingivalis* fimbriae play a role in the production of outer membrane vesicles that contain other virulence factors (Mantri et al. 2015). *P. gingivalis* expresses two forms of fimbriae, FimA (major/long) and Mfa1 (minor/short), which are genetically distinct, encoded by separate gene clusters, but share similar architecture, comprising five proteins, FimA–E for FimA and Mfa1–5 for Mfa1, designated as one major (FimA/Mfa1), three minor accessory (FimCDE/Mfa345) structural proteins, and one controlling filament length (FimB/Mra2) (Shoji et al. 2004; Yoshimura et al. 2009). Expression and maturation of these proteins also depends on other *P. gingivalis* virulence factors, namely *rgp* and *kgp*, as both *rgp* and *kgp* bind the *fimA* promoter and Rgp cleavages the FimA N-terminal (Xie et al. 2000; Kadowaki et al. 1998). Fimbria proteins themselves exert direct effects on various host cells via adherence to a wide range of molecules (Amano 2010). As a result, both major and minor fimbriae induce expression of proinflammatory cytokines and adhesion molecules on endothelial and hematopoietic cells via TLR2 and TLR4 pathways, while causing immune evasion via CXCR4 and complement receptor 3 pathways (Enersen et al. 2013). The virulence of fimbriae is also determined by amino acid sequence variations in the *fimA* gene, which is grouped into six genotypes, I, Ib, II, III, IV, and V. Genotypes Ib, II, and IV have been designated as high-activity genotypes as these strains lead to elevated inflammatory reactions in vivo and in vitro (Kato et al. 2007; Gao et al. 2012) and are more commonly found in periodontitis patients than in healthy subjects (Nakagawa et al. 2002). Differential distributions in the *fimA* genotypes have also been implicated in CVD risk (Enersen et al. 2013). Finally, several other oral pathogens express fimbriae, but their molecular structures and biological activities have not been well characterized to date (Amano 2010).

Gingipains, a group of cysteine proteinases, represent the most significant virulence factors produced by this bacterium. Because of the asaccharolytic nature of *P. gingivalis*, high proteolytic abilities that degrade proteins from host and other microorganisms are essential for *P. gingivalis* to meet its nutritional requirements (Li and Collyer 2011). Accordingly, this bacterium secretes several proteolytic enzymes besides gingipains, as discussed elsewhere (Sheets et al. 2008). Gingipains (*P. gingivalis* + clostripain) are endopeptidases consisting of arginine-specific (Arg-gingipain [Rgp]) and lysine-specific proteases (Lys-gingipain [Kgp]), which account for 85% of the extracellular proteolytic activity of this bacterium. Rgp and Kgp are encoded by three genes, *rgpA*, *rgpB*, and *kgp* (Potempa et al. 2003). The *rgpA* gene is composed of a pre-pro-fragment, followed by a proteinase domain, and a large hemagglutinin/adhesion (HA) domain at the C-terminus. Although the *kgp* gene product shares a similar size and structure to the *rgpA* gingipain, the rgpB gene is missing the entire section of the HA domain (Potempa et al. 2003; Li and Collyer 2011). Maturation of gingipain proteins requires extensive processing with specific post-translational cleavage and both R and K gingipains are mutually involved in mutual efficient maturation. In addition, a carboxypeptidase (CPG70) processes the C-terminus RgpA and Kgp HA domains, and several other loci (*wpbB*, *porR*, *vimA*, *vimE*, and *vimF*) participate in post-translational carbohydrate modifications of gingipains, resulting in a high heterogeneity of these proteins (Sheets et al. 2008). As far as proteolytic activities are concerned, Rgp exerts indiscriminate powerful proteolytic activities, whereas Kgp exhibits a narrow specificity and weak proteolytic activities. The former is considered to be crucial for this bacterium in nutrient acquisition, but the latter plays more roles in the manipulation of host defense systems (Potempa et al. 2003).

In addition to the role in facilitating the expression of other *P. gingivalis* virulence factors as discussed above for fimbriae, a myriad of virulence activities of gingipains against the host are mediated through the biphasic mechanisms: activation and inactivation of host proteins (Imamura et al. 2003; Guo et al. 2010). Rgps enhanced vascular permeability through prekallikrein activation or direct bradykinin release in combination with Kgp. Rgps also activate protease-activated receptors and induce platelet aggregation, which, together with the coagulation-inducing activity, may explain an emerging link between periodontitis and CVD. Signaling via protease-activated receptors 1 and 2, Raps stimulates oral epithelial cells to produce an array of proinflammatory cytokines, including IL-1alpha, IL-1beta, IL-6, TNFα and IL-8, whereas RgpA and Kgp, but not RgpB, mediate, in a proteolytic activity-independent manner, strong enhancement of the production of proinflammatory cytokines in macrophages (Imamura et al. 2003; Guo et al. 2010). Both gingipains stimulate expression of MMPs and activate secreted latent MMPs that can destroy periodontal tissues and basement membrane, which potentially promotes cancer cell migration and invasion (Whitmore and Lamont 2014). Finally, gingipains degrade various cytokines, components of the complement system and cell surface receptors on immunological cells, epithelial cells and endothelial cells, including C5aR, CD14, CD4, and CD8, thus perturbing the host defense against not only this bacterium, but also other oral pathogens (Imamura et al. 2003; Guo et al. 2010).

Nucleoside diphosphate kinase is an evolutionarily highly conserved enzyme that catalyzes γ-phosphate transfer between nucleoside triphosphates and diphosphates (NTPs/NDPs) (Spooner and Yilmaz 2012). Among a range of oral pathogens, its virulence to human hosts has been exclusively studied in *P. gingivalis* in which *PG1018* encodes the enzyme (Yilmaz et al. 2008). This enzyme functions as an ATPase and thus prevents various ATP-dependent host immune responses mediated through the purinergic receptor P2X7 (Spooner and Yilmaz 2012). This leads to suppression of ATP-dependent host cell apoptosis (Yilmaz et al. 2008) and attenuation of the inflammasome, IL-1β, and HMGB1 extracellular secretion activated by the host eATP-P2X7 signaling pathway (Johnson et al. 2015), which together may contribute to bacterial persistent infection and to tumor proliferation. Interestingly, humans express several isoforms of this enzyme and their overexpression has been reported in surgically resected oral squamous cell carcinoma (Atanasova and Yilmaz 2014). A recent study also indicates that this enzyme inhibits reactive oxygen species (ROS) production in gingival epithelial cells via membrane-bound NADPH–oxidase complex and mitochondrial signaling pathways to counteract host antimicrobial activities, and this mitochondrial exploitation is postulated to affect host energetics (Choi et al. 2013).

The *P. gingivalis* PAD encoded by *ppad* is a putative etiological factor for RA as discussed above. PAD is primarily a cell surface-associated protein transported by T9SS (Nakayama 2015) and it deiminates the guanidino group of carboxyl-terminal arginine residues on a variety of peptides, including the vasoregulatory peptide-hormone bradykinin that controls blood pressure, to yield ammonia and a citrulline residue (Mcgraw et al. 1999). The ability of the enzyme to create altered host epitopes by converting arginine residues is postulated to promote autoimmune reactions by creation of altered host epitopes. Citrullination by prokaryotic PAD (PPAD) is distinct from mammalian PAD which more efficiently citrullinates internal arginine residues, independent of calcium (Mikuls et al. 2012). The dual expression of gingipains and PPAD is deemed to act in concert in RA, the former producing a carboxy-terminal arginine residue that then serves as a target for PPAD (Mikuls et al. 2012). However, a recent transfection experiment using full-length and truncated (secreted) forms of strain W83 PPAD revealed that only the full-length PPAD exhibited the ability to autocitrullinate (Konig et al. 2015), challenging the original hypothesis on RA pathogenesis. As an intracellular pathogen, *P. gingivalis* may still provide host proteins with possibilities to interact with the nonsecreted, full-length PPAD. In addition, PPAD-positive strains have been shown to induce the expression of cyclo-oxygenase 2 (COX-2) and microsomal PGE synthase-1 (mPGES-1), key factors in the prostaglandin synthesis pathway in gingival fibroblasts, augmenting inflammatory reactions (Gawron et al. 2014).

Treponema Denticola

Treponema denticola is a Gram-negative obligate anaerobic bacterium and a key pathogen of periodontitis, belonging to the *Spirochetes* family. Recently, a number of potential virulence factors have been reviewed (Ishihara 2010; Fenno 2012). Among those for which corresponding bacterial genes have been identified are several virulence factors that are specific to this bacterium and possess multiple functions. Major outer sheath protein (MSP) is a 53-kDa protein encoded by *msp* (Fenno et al. 1996), locates within the outer membrane, and demonstrates, instead, that it is predominantly periplasmic, with only limited surface exposure (Caimano et al. 1999). Besides mediating adherence to host cells (Fenno et al. 1996), MSP exerts cytotoxic activity toward human polymorphonuclear leukocytes, suppressing chemotaxis and phagocytosis through disruption of cytoskeleton (Puthengady Thomas et al. 2006), while it activates macrophages through the TLR2-dependent pathway (Nussbaum et al. 2009). More importantly MSP activates host mitogen-activated protein kinase (MAPK) pathways (i.e., ERK1/2), which are key regulators of cell proliferation (Jobin et al. 2007).

Dentilisin, encoded by *prtP*, is an adhesin with pore-forming activity, and an outer membrane-associated chymotrypsin-like protease. Dentilisin has cytotoxic activity, degrading a variety of humoral proteins, including basement membrane components (type IV collagen, laminin, and fibronectin), serum proteins (transferrin, fibrinogen, IgG, IgA, and α1-antitrypsin), and bioactive peptides (cytokines and chemokines) (Ishihara 2010; Fenno 2012). Accordingly, it plays an important role in bacterial invasion through epithelial cells and endothelial cells and also plays a role in immunosuppression. In fact, in-vitro experiments using reconstructed basement membrane (Matrigel) (Grenier et al. 1990) and multi-layer epithelial cell culture (Uitto et al. 1995) have demonstrated that bacterial migration/penetration is potentiated by this enzymatic activity. This bacterial penetration was mediated through the alteration of host epithelial cell tight junctions (Chi et al. 2003). In addition, higher O_2^- production was observed in a dentilisin-positive wild-type strain in comparison with a dentilisin-negative mutant, suggesting another possible mechanism mediating the cytotoxicity of this enzyme (Yamazaki et al. 2006). As a result, in a mouse abscess model, attenuation of lesion formation was observed with the mutant strain, compared with a dentilisin-positive wild-type strain (Ishihara et al. 1998).

Tannerella Forsythia

Tannerella forsythia is an obligate anaerobic Gram-negative member of the Cytophaga-Bacteroides family. Despite the strong association with periodontal disease, this bacterium remains understudied owing to difficult culture conditions, yet a few putative virulence factors have been identified to date (Sharma 2010). The best characterized virulence factor is a surface and secreted protein, BspA (Bacteroides surface protein A), encoded by the *bspA* gene, which belongs to the leucine-rich repeat family (LRR). The presence of LRR domains suggests that the BspA protein might be involved in protein–protein interactions important in mediating host–bacterial or bacterial–bacterial interactions (Sharma 2010). BspA has been shown to trigger the release of pro-inflammatory cytokines from monocytes and chemokines from epithelial cells by activating the TLR2-dependent pathway. TLR1 serves as a coreceptor for TLR2 in this activation, and the BspA LRR domain-1 is involved in activation of TLR2/1 heterodimers (Onishi et al. 2008). In addition, BspA is required for this bacterium to invade epithelial cells (Inagaki et al. 2016), which is mediated through the activation of host PI3K signaling and interactions with host Rac1 GTPase (Mishima and Sharma 2011). In vivo, BspA injection into *ApoE* knockout mice accelerated the development of dyslipidemia and atherosclerosis (Lee et al. 2014). Recently, a gene encoding a novel MMP-like enzyme called karilysin has been sequenced (Karim et al. 2010). This protease has a wide specificity to host proteins and thus is likely to contribute to *T. forsythia* virulence. Karilysin facilitates the shedding of soluble, fully active TNFα from the macrophage surface (Bryzek et al. 2014) and degrades antimicrobial peptide LL-37 (Koziel et al. 2010), a critical component of innate immunity.

Fusobacterium Nucleatum

Fusobacterium nucleatum is a Gram-negative, nonspore-forming, obligate anaerobe of the Fusobacteriaceae family and best appreciated for its role as a component of dental plaque (Han 2015). FadA is a unique adhesin among all adhesins of this bacterium that can bind host cells and is the best-characterized virulence factor, with potential roles in various pathogenic conditions, including CVD and cancer. In fact, FadA is only present in a subset of pathogenic *Fusobacterium* species, i.e., *F. nucleatum* and *F. periodonticum* (Han 2015). FadA exists in two forms, the pre-FadA, consisting of 129 amino-acids and the secreted mature FadA (mFadA), consisting of 111 amino acids (Xu et al. 2007). These two molecules together form an active complex, FadAc, for host–cell binding and invasion in both normal and cancerous cells (Rubinstein et al. 2013; Fardini et al. 2011). FadA binds to various host cell surface proteins, such as cadherins (VE-cadherin on the endothelial cells and E-cadherin on epithelial cells) (Rubinstein et al. 2013; Fardini et al. 2011). FadA binding to VE-cadherin on endothelial cells induces the translocation of VE-cadherin from cell–cell junction to intracellular compartments, increasing the permeability of the endothelium (Fardini et al. 2011). Thus, FadA mediates both direct invasion into the host cells and pericellular invasion via loosened cell–cell junctions, which may facilitate systemic dissemination (Fardini et al. 2011). Furthermore, the increased endothelial permeability allows other bacteria in the vicinity to penetrate through and facilitates both intra- and inter-cellular invasion of other bacterial species (Fardini et al. 2011). Alternatively, FadA binding to E-cadherin on epithelial cells activates β-catenin signaling (Rubinstein et al. 2013), leading to increased transcriptional activity of oncogenes, Wnt, and pro-inflammatory cytokines, in addition to cell proliferation.

FAP2 is another *F. nucleatum* cell surface protein and a component of the type V secretion system (Kaplan et al. 2010), for which a role in virulence has recently been delineated. This protein binds specifically to human immune cells in peripheral blood (Jewett et al. 2000) and in tumor tissue (Gur et al. 2015) on an inhibitory receptor, T cell immunoglobulin and ITIM domain (TIGIT). This binding triggers cell apoptosis, thus leading to immune suppression. Furthermore, this protein has been characterized to be a galactose-sensitive adhesin (Coppenhagen-Glazer et al. 2015) and recognizes a specific host polysaccharide, Gal-GalNAc, which is overexpressed in tumor tissue (Abed et al. 2016). Accordingly, this protein not only enhances the survival of other bacteria, but may also promote tumor progression.

Aggregatibacter actinomycetemcomitans

Aggregatibacter actinomycetemcomitans is a facultative anaerobic, nonmotile, nonsporing, small Gram-negative bacterium that belongs to the class of Gammaproteobacteria and is involved in the pathology of aggressive types of periodontitis (Henderson et al. 2010). This organism expresses two well-established toxins, leukotoxin and cytolethal distending toxin (CDT), which are not unique to this particular organism, because they are also produced by various other bacteria belonging to Proteobacteria (Henderson et al. 2010). Although not all *A. actinomycetemcomitans* strains harbor genetic loci encoding CDT (ranging from 70 to 80%) (Faïs et al. 2016), the expression of leukotoxin is highly dependent on promoter sequences of the corresponding genetic loci (Johansson 2011). Leukotoxin is secreted by a type I secretion system and CDT by a type V secretion system (Zijnge et al. 2012).

The *A. actinomycetemcomitans* leukotoxin operon consists of four genes, *ltxC*, *lltxA*, *ltxB*, and *ltxD* and an upstream promoter. ltxA encodes the structural protein of the toxin of 1,055 amino acids, ltxC for components required for post-translational acylation, and ltxB and D for transport of the toxin to the bacterial outer membrane (Kachlany 2010; Johansson 2011). A 530 bp deletion and an insertion of a DNA-transposable element (IS1301) in this promoter region have been shown to lead to a highly leukotoxic phenotype (Kachlany 2010; Johansson 2011; Henderson et al. 2010). LtxA expressed by *A. actinomycetemcomitans* exhibits a unique specificity against cells of human hematopoietic origin. This restricted cell specificity is mediated through interactions between CD18 molecule of LFA-1 on the host cells and a repeat region of the toxin consisting of 14 tandem repeats of nine amino acids that recognizes and interacts with the host receptor molecule (Johansson 2011; Kachlany 2010). Leukotoxin induces cell death primarily through apoptosis, interfering host mitochondrial respiration pathway, and activating caspase (Henderson et al. 2010). The effects are more potent on neutrophils and monocytes/macrophages than on lymphocytes (Henderson et al. 2010; Johansson 2011). These effects ultimately lead not only to suppression of both host innate and acquired immunity, but also to the release of proteolytic enzymes from neutrophils and of proinflammatory cytokines (e.g., 1L-1β and IL-18) from macrophages (Henderson et al. 2010).

Aggregatibacter actinomycetemcomitans CDT belongs to the AB toxin family and is composed of an active subunit (CdtB) and two binding subunits (CdtA and CdtC) encoded by the *cdtABC* gene cluster (Ahmed et al. 2001). This CDT is distinguished from binary CDT toxin secreted by *Clostridium difficile* in its structure and functions. Importantly, the active CdtB subunit of CDT is functionally and structurally homologous to mammalian DNase I, is translocated into the nucleus, and acts as a genotoxin, leading to DNA damage (Faïs et al. 2016). The CdtB subunit contains putative DNA binding residues (Nesic et al. 2004) and a single nucleotide polymorphism at amino acid 281 (H/R) has been shown to change its biological activity by six log orders (Nishikubo et al. 2006). The high-activity R281 strain was reported to be dominant in clinical isolates (Nishikubo et al. 2006). Through its DNase activity, CDT damages DNA at an extremely low concentration (50 pg/mL), leading to both single-strand and double-strand breaks. These result in cell arrest, followed by DNA repair, cell death or cell senescence (Faïs et al. 2016). Cell fate following CDT exposure depends on cell type, with epithelial and mesenchymal lineages mainly undergoing cell cycle arrest accompanied by cytoplasmic elongation and distension, whereas hematopoietic lineages rapidly move toward apoptosis after a brief cell cycle arrest (Shenker et al. 1999). Loss of hematopoietic lineages may lead to immunosuppression and thus persistent bacterial colonization (Faïs et al. 2016). CDT also inhibits macrophage phagocytosis and is capable of inducing the release of pro-inflammatory cytokines (such as IL-1β, IL-6, and IL-8) from cultured cells (Faïs et al. 2016). Moreover, exposure of gingival tissue or epithelial cells to CDT causes a pronounced increase in the expression and cytosolic distribution of E-cadherin and β-catenin, indicating a potential impact on mucosal barrier function (Damek-Poprawa et al. 2013).

Many properties of this toxin support its involvement in human cancers. CDT-induced DNA damage may be involved in cancer promotion/progression. Chronic intoxication with sublethal doses of CDT can induce DNA damage by increasing the frequency of mutations, resulting in chromosomal instability that may not cause cell death or cell cycle arrest, but that leads to activation of pro-survival signals, enhancing anchorage-independent growth (Guidi et al. 2013). Furthermore, the pro-inflammatory action and immunotoxicity of this toxin may foster tumor survival, proliferation, and progression (via promoting angiogenesis and metastasis) (Faïs et al. 2016). Alternatively, cellular senescence following CDT exposure leads to a senescence-associated secretory phenotype (SASP), which is characterized by secretion of a large number of growth factors and pro-inflammatory cytokines and can promote survival and proliferation of transformed cells (Coppé et al. 2010). Correspondingly, mice infected with another CTD-positive bacterium, *Helicobacter hepaticus*, have been shown to develop hepatic dysplastic lesions (Ge et al. 2007), and the prevalence of CDT-producing *Escherichia coli* was significantly higher in colorectal cancer patients than in diverticulosis patients (Buc et al. 2013). Because *A. actinomycetemcomitans* CDT has been described to be unstable in the absence of sucrose, in contrast from CDT from other bacteria (Yamada et al. 2006), in-vivo carcinogenicity of this bacterial CDT needs to be assessed separately.

9.4.3 Oral–Systemic Bacterial Translocation

Bacteria can enter the circulation to cause infections at remote locations by a number of mechanisms:

(1) Damage to blood vessels during traumatic dental procedures, giving direct exposure of the circulation to bacteria.

(2) Direct entry associated with severe tooth decay.

(3) Ulceration of the epithelium caused by oral infections such as periodontitis.

(4) Uptake of bacteria by phagocytic cells that migrate into the circulation. Some of the most basic procedures of dental hygiene result in bacteria entering the blood stream. For example, although the presence of bacteria of oral origin ordinarily found in circulation is practically nil (tested by culture-based methods, <3%), tooth-brushing resulted in the transient appearance of culturable bacteria of oral origin in 32% of subjects when blood was sampled within 5 min of tooth-brushing (Lockhart et al. 2008). Although brief (back to pre-treatment levels within 20 min), the transient bacteremia associated with tooth-brushing included bacterial species known to be associated with infective endocarditis (23% of subjects) and prosthetic joint infection (12% of subjects) (Mougeot et al. 2015). Probing the periodontal pocket depth in patients with periodontal disease exhibited an increased incidence of post-treatment bacteremia to 20% and 43% in studies by Kinane et al. (2005) and Daly et al. (2001) respectively. Significant increases in transient bacteremia have also been observed in numerous studies on the effects of ultrasonic root scaling (Forner et al. 2006; Heimdahl et al. 1990; Kinane et al. 2005). Given the effects of these relatively non-invasive procedures, it is not surprising that tooth extraction also gives rise to bacteremia, which occurs more frequently and at higher titers if unaccompanied by antibiotic treatment (Barbosa et al. 2015; Heimdahl et al. 1990; Lockhart et al. 2008; Mougeot et al. 2015). A case of post-extraction septicemia caused by *Neisseria meningitidis* has been described as a rare complication of dental extraction (Pedersen et al. 1993). Root canal treatment has also been associated with bacteremia (Heimdahl et al. 1990; Savarrio et al. 2005), being observed more frequently when the instrumentation penetrates through the tooth to several outside the root canal than in treatments in which the root canal reamer remained inside the root canal (Debelian et al. 1995). In many of the cases described above, characterization of the bacteria in blood has been compared with those isolated from the mouth or from the oral area probed or treated and found to be closely matched in type. For example, various species of *Streptococcus* are among the most common oral bacteria found in blood following these dental procedures (Savarrio et al. 2005). Nevertheless, post-dental therapy bacteremia tends

to be of short duration, with bacteria observed within 5 min of therapy but often being cleared from the blood within 30 min, as reported by many of the above references.

Severe tooth infection may also give rise to bacteremia independent of treatment. Although bacterial levels in blood are virtually zero before treatment of simple dental caries, dental infections involving the pulp of the tooth (pulpitis) may be associated with periapical inflammation and systemic bacteremia (Olsen and Van Winkelhoff 2014). Because the infection may often remain local, distant infections of odontogenic origin have occasionally been observed in other organs, such as the liver and brain (Wagner et al. 2006). Also, in rare cases, dental infection has been the initiating cause of necrotizing fasciitis (Bahl et al. 2014).

The epithelium lining the periodontal pockets becomes ulcerated in periodontal disease, providing a potential direct entry pathway for bacteria into the circulation (El Kholy et al. 2015). The dentogingival epithelial surface area, which averages approximately 5 cm^2 in the absence of periodontitis, expands to an average of 20 cm^2 among those individuals with severe enough gum disease to be referred to a periodontist (Hujoel et al. 2001). Numerous studies have shown that periodontal disease increases the bacteremia associated with typical dental therapies such as periodontal probing and root scaling (Olsen 2008). Nevertheless, patients with periodontitis rarely exhibited bacteremia before these treatments (Daly et al. 1997; Reis et al. 2016). Although periodontitis is an underlying condition that may allow bacteria to enter the circulation, bacteremia often occurs only after the gums are manipulated during dental treatment. Prophylactic treatment with antibiotics before such treatments may therefore be helpful in blocking this source of infection.

Although direct entry of bacteria into the circulation may mediate translocation of oral bacteria to sites elsewhere in the body, this pathway immediately exposes the bacteria to various immune defense systems and rapid clearance. However, several instances have been reported in which live bacteria are internalized in cells and thereby carried throughout the body somewhat protected from direct immune attack. For example, dendritic cells (DCs) of the oral mucosa have been reported to carry culturable *P. gingivalis* and *Burkholderia cepacia* and the carriage rate is increased by dental intervention in patients with chronic periodontitis (Carrion et al. 2012). Being carried by DCs enhanced the survival of *P. gingivalis* under aerobic conditions that these anaerobes would encounter in the circulation. Oral streptococci have also been shown to be carried by blood platelets, which the authors have referred to as "inadvertent Trojan horse carriers" of oral bacteria (Deng et al. 2014).

9.4.4 Modulation of Immunological and Inflammatory Parameters

The transition from periodontal health to disease is associated with a dramatic shift from a symbiotic microbial community composed mostly of facultative bacterial genera to a dysbiotic microbial community that is mainly composed of anaerobic genera enriched with virulence factors. First, active bacterial subversion of the host immune response is a critical mechanism that enables the persistence of pathogens and the induction of pathological conditions at systemic sites. Manipulation of the host immune response by a keystone pathogen such as *P. gingivalis* is essential to instigate compositional alterations in the oral microbiota, which can thereby trigger proinflammatory reactions. Although on the one hand, the dysbiotic microbial community needs to evade immune-mediated killing, on the other, it requires sustained inflammation to procure nutrients from host tissue breakdown. Hence, periodontal bacteria manipulate host immune responses through interactions with immunological cells to achieve this goal (Hajishengallis 2015).

A keystone pathogen can subvert neutrophil homeostasis, causing impaired recruitment and chemotaxis, resistance to granule-derived antimicrobial agents and to the oxidative burst, inhibition of phagocytic killing while promoting a nutritionally favorable inflammatory response, and a delay of neutrophil apoptosis (Olsen and Hajishengallis 2016). These responses are mediated through interference with the expression of chemokines and cell adhesion molecules and with cell surface chemotactic peptides, and instigation of C5aR-TLR2 crosstalk and following PI3K activation (Olsen and Hajishengallis 2016). Periodontal pathogens are also capable of blocking complement activation through the degradation of the third complement component (C3) or of key upstream components, such as the mannose-binding lectin, and thus inhibit bacterial opsonization and phagocytosis, recruitment and activation of inflammatory cells, and direct lysis of microorganisms by the C5b–C9 complex (Hajishengallis 2015).

Dysregulated proinflammatory reaction is another important mechanism involved in the pathogenesis of various systemic chronic diseases, such as CVD and cancer. Bacterially induced IL-23 production by innate immune cells, such as DCs and macrophages (Mφ), promotes the survival and expansion of Th17 cells and activates γδ T cells. Acting as a link between innate and adaptive immunity, Th17 cells secrete IL-17, which upregulates expression of granulocyte colony-stimulating factor (G-CSF) and CXC chemokines and the production and release of neutrophils from the bone marrow. Recruited neutrophils, in turn, produce chemokines that selectively recruit more Th17 cells, leading to reinforced feedback for high-level production of IL-17 to maintain the inflammatory state (Hajishengallis 2014).

Another important player in host–bacterial interactions is platelets/thrombocytes, which are now known to play key roles in innate and adaptive immunities and inflammation, in addition to their primary role in blood coagulation (Ferdous and Scott 2015). Bacteria binding to platelets triggers platelet activation. The main platelet receptors that mediate these interactions are glycoprotein (GP)IIb–IIIa, GPIbα, FcγRIIa, complement receptors, and TLRs. This process may involve direct interactions between bacterial proteins and the receptors, or can be mediated by plasma proteins such as fibrinogen, von Willebrand factor, complement, and IgG (Cox et al. 2011). Although promoting blood coagulation shows a clear link between periodontal infection and CVD, platelet activation also induces host immune responses. Platelets have been shown to enhance the phagocytosis of periodontal pathogens by neutrophils. Platelets can also directly phagocytose bacteria, but phagocytosis does not necessarily result in bacterial killing and bacteria can also become trapped in the space between platelets in an aggregate. As a result, phagocytosis of bacteria by platelets can lead to the formation of a pool of viable bacteria, present either intracellularly or within a thrombus, that is protected from the immune system and may serve as a potential reservoir for systemic dissemination (Cox et al. 2011). Human platelet expresses TLR4 to detect TLR4 ligands and induce platelet binding to adherent neutrophils, leading to robust neutrophil activation and formation of neutrophil extracellular traps (NETs), while it directly secretes proinflammatory cytokine and matricidal proteins. Human platelets have also been shown to interact with a triggering receptor expressed on myeloid cells 1 (TREM1) to promote neutrophil-mediated production of ROS and secretion of IL-8 (Ferdous and Scott 2015). Activated platelets express functional CD40 ligand (CD40L/CD154), a molecule of vital importance to the adaptive immune response, and thus are capable of augmenting T-cell responses and promoting B-cell differentiation and immunoglobulin class switching. CD40L expressed on the surface of platelets can also interact with CD40 on endothelial cells to induce adhesion molecules and release of CC-chemokine ligand 2, all of which further promote leukocyte recruitment to inflammatory sites (Ferdous and Scott 2015).

An additional immunological mechanism is molecular mimicry (MM), which was discussed earlier for GroEL. During a viral or bacterial infection, if the organism shares cross-reactive epitopes for B and T cells, immune responses targeting the infectious agent also attack the host, causing autoimmune diseases (Oldstone 2005). Although CVD is not a typical autoimmune disease, molecular mimicry may play an important role in atherogenesis, as structurally homologous human heat shock protein (HSP60) is found in atherosclerotic lesions, whereas autoantibodies against HSP60 may exert endothelial cytotoxicity. Cross-reactive T cells may also lead to the secretion of cytokines, MMPs, and NO by macrophages, thereby fueling inflammatory reactions in the atherosclerotic lesion (Ludewig et al. 2004). On the other hand, MM may act as a protective mechanism for certain neurodegenerative diseases, such as AD. Healthy periodontal bacterium, *S. mutans*, is known to produce amyloid proteins, which in turn elicit the production of autoantibodies against amyloid β (Friedland 2015) that facilitate the clearance of amyloid β (Gu et al. 2014) and thus are inversely associated with temporal lobe atrophy (Kimura et al. 2016).

9.4.5 Microbial Enzymatic Activities and Metabolites

Owing to advancements of sequencing technologies, knowledge of the diversity of the oral microbiome has increased. However, less is understood regarding the oral metabolome, i.e., the metabolic activities of the oral bacterial communities, and how metabolic products of the metabolome may influence systemic health. Metabolites are produced not only by the host, but also by microorganisms, and their actions can be far-reaching (Johnson et al. 2016). Under normal physiological conditions, the effects of metabolites are counteracted by the host to maintain homeostasis (Johnson et al.

2016). Loss of homeostatic control can lead to permanently altered states. Currently, metabolomics emphasizes the need to understand the identities and variation of molecular mechanisms that may one day provide direct targets for treating disease.

In a recent review, Takahashi (2015) identified ethanol and acetaldehydes as among the most prominent known oral bacterial metabolites that may lead to somatic disease and cancer (Takahashi 2015). Ethanol is metabolized by bacteria to acetaldehyde, which may be carcinogenic. The contribution of acetaldehyde generated in the oral bacterial cavity to cancer of the upper digestive tract, which has been postulated for decades, is supported by several recent studies (Salaspuro 2012; Homann et al. 2000; Marttila et al. 2013). Homann et al. reported that oral microbial acetaldehyde production from ethanol plays a major role in alcohol-associated carcinogenesis and also showed that oral acetaldehyde production in smokers was significantly higher than from samples of nonsmokers (Homann et al. 2000). This effect of smoking on acetaldehyde is consistent with the increased presence of squamous cell carcinoma and oral lichenoid disease in smokers (Marttila et al. 2013). Although many species of oral bacteria may be involved, *Neisseria* has been implicated as one microbial genus found in the oral cavity of healthy individuals that can produce acetaldehyde from ethanol and sugar and thereby potentially plays a major role in human carcinogenesis (Moritani et al. 2015; Muto et al. 2000). Overall, Takahashi (2015) emphasized the need for more studies on the metabolome of the oral microbiome.

Acetate, a product of fatty acid breakdown, is also a metabolite that, when altered by the oral microbiome, could lead to diseased states. A recent work by Perry et al. (2016) demonstrated that acetate was sufficient to cause obesity and related sequelae in a gut microbiome study, which was accompanied by an increase in Firmicutes, a decrease in Bacteroidetes, and changes in Proteobacteria abundance (Perry et al. 2016). Whether the oral microbiome may contribute to this was not discussed, but previous work by our laboratory revealed correlations among a diet high in saturated fatty acids (SFAs) and alpha diversity indices in richness, the number of OTUs, and diversity with SFA positively correlated with the relative abundance of Betaproteobacteria (Kato et al. 2016).

The concentration of nitrate, a key player in smooth muscle regulation and cardiovascular health, has also been shown to depend on the commensal microbiota found in the oral cavity (Kapil et al. 2013). A genomic analysis of the nitrate-reducing bacteria populations in the oral cavity revealed several species that may play an important role in maintaining homeostasis (Hyde et al. 2014). Species such as *Neisseria flavescens* and *Haemophilus parainfluenzae* were amongst the highest nitrate-reducers (Hyde et al. 2014). An imbalance in nitrite concentration can be detrimental to human health, especially in smokers, and can lead to cancer (Stepanov et al. 2008).

The overall composition of the metabolome of the human oral microbiome is also affected by external changes to the person. A comparative study of sailors before and after a long sea voyage revealed stark changes in the metabolome and an overall decrease in the diversity of the oral microbiome (Zheng et al. 2015). They report shifts in the metabolome that would result in the sailors being more prone to disease, highlighting the importance of a stable and healthy metabolome for maintaining general health (Zheng et al. 2015). Specifically, microbially derived folate biosynthesis was decreased, along with a concomitant decrease in *Lactobacillus lactis* and a reduction of dihydrofolate reductase and dihydropteroate synthase enzyme activity (Zheng et al. 2015).

9.5 Concluding Remarks

As summarized above and in Fig. 9.1, evidence is growing to support the role of the oral microbiome in etiologies of various systemic conditions beyond dental and periodontal diseases. Although some of the associations have been supported by both observational and experimental studies along with sound biological mechanisms (e.g., CVD, bacteremia), others present rather modest associations with no strong biological mechanisms tested or proposed. It is important to note that an association between a presumed exposure and a disease outcome does not necessarily indicate a causal relationship. Such an association may be caused by close correlation with other known or unknown risk factors of the disease of interest. In the case of oral pathogens, smoking is the greatest concern, as smoking is a known risk factor for almost all the conditions discussed in this chapter, and because

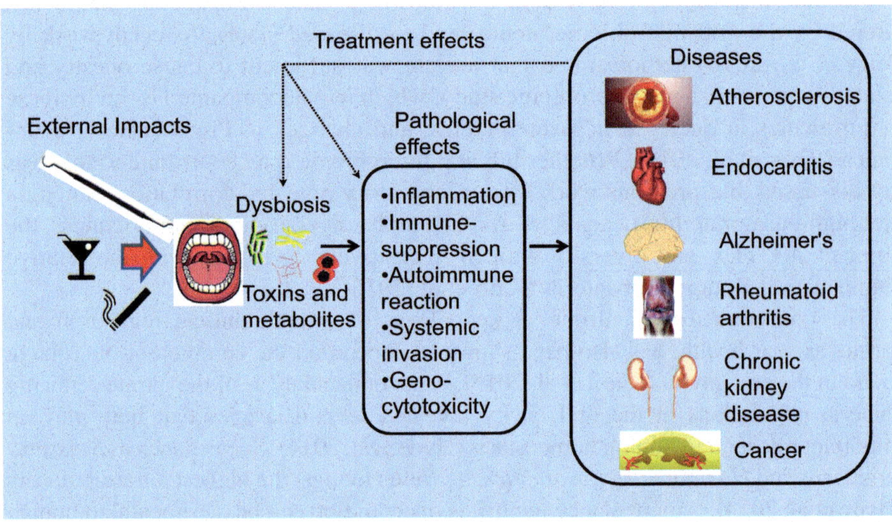

Fig. 9.1 Schematic summary of the mechanistic pathways linking the oral microbiome to various systemic conditions and cancer. Chemical (tobacco, alcohol, etc.) and physical (oral hygiene procedures) may affect oral microbiome composition. Dysbiotic oral microbiome produces more potent toxins (lipopolysaccharide, gingipains, nucleoside diphosphate kinase, cytolethal distending toxin, etc.) and potentially carcinogenic metabolites (acetaldehyde, nitrosamines). These facilitate, together with external stimuli, immune subversion, bacterial invasion, proinflammatory and autoimmune reactions, and cyto-genotoxic damages, eventually resulting in a myriad of disease states listed. Treatments for the resulting diseases such as chemotherapy and radiation may further exacerbate dysbiosis and inflammatory and immune suppressive conditions

smokers have a high prevalence of periodontal disease due to poor oral hygiene behaviors or local/systemic immune suppression (Barbour et al. 1997; Santos et al. 2015). As a result, a shift in the oral microbial structure caused by smoking has been recently acknowledged (Wu et al. 2016; Kato et al. 2017). Therefore, analyses stratified by smoking status or stringent statistical adjustment is necessary to draw valid observations from studies on many conditions still under investigation.

Oral microbiome research to date has provided a wealth of information concerning host–bacterial interaction, using not only epithelial cells, but also immune cells, fibroblasts, and osteoclasts as discussed above. However, a major limitation is that the vast majority of the research was based on *P. gingivalis*, which is a rather uncommon intra-cellular bacterium with very low abundance. Further research is warranted to elucidate virulence mechanisms of other major bacteria in the dysbiotic community members induced by keystone pathogens, e.g., *P. gingivalis*, in addition to protective mechanisms of commensal bacteria. Finally, equally important to host–bacterial interactions are bacterial–bacterial interactions, either in maintaining oral health or in inducing pathological conditions. This is particularly the case for oral bacteria living in a steep oxygen gradient and with a limited nutrient supply that relies on endogenous nutrients from saliva, tissue exudates, crevicular fluids, degenerating host cells or other bacterial metabolites (Wade 2013; Hojo et al. 2009). As a result, metabolic cooperation by consortia of bacteria is necessary for any bacteria to survive, and biofilm development facilitates this cooperation and horizontal gene transfer (Wade 2013; Hojo et al. 2009), which may modulate the expression of bacterial virulence. It is now clear that the behavior of platonic- versus biofilm-associated bacteria is dramatically different (Gabrilska and Rumbaugh 2015). Recent development of in-vitro and in-vivo polymicrobial biofilm models has certainly contributed to knowledge concerning bacterial social behaviors (Gabrilska and Rumbaugh 2015). However, these models do not necessarily replicate the full range of complex natural oral microbiome from obligate aerobics to obligate anaerobics. Perhaps further advances in this area may be facilitated by the introduction of humanized oral microbiomes, as in recent gut microbiome studies.

Acknowledgments This research was supported by a grant from National Institute of Dental and Craniofacial Research (R21 DE023181 to IK). The authors would like to acknowledge the Wikipedia creative commons for several images used in the graphic. We would also like to acknowledge Pixabay as a source of images.

References

Aas JA, Reime L, Pedersen K, Eribe ERK, Abesha-Belay E, Støre G, Olsen I (2010) Osteoradionecrosis contains a wide variety of cultivable and non-cultivable bacteria. J Oral Microbiol 2. https://doi.org/10.3402/jom.v2i0.5072

Abed J, Emgard JE, Zamir G, Faroja M, Almogy G, Grenov A, Sol A et al (2016) Fap2 mediates *Fusobacterium nucleatum* colorectal adenocarcinoma enrichment by binding to tumor-expressed Gal-GalNAc. Cell Host Microbe 20(2):215–225. https://doi.org/10.1016/j.chom.2016.07.006

Abranches J, Zeng L, Belanger M, Rodrigues PH, Simpson-Haidaris PJ, Akin D, Dunn WA, Progulske-Fox A, Burne RA (2009) Invasion of human coronary artery endothelial cells by *Streptococcus mutans* OMZ175. Oral Microbiol Immunol 24(2):141–145. https://doi.org/10.1111/j.1399-302X.2008.00487.x

Adler CJ, Dobney K, Weyrich LS, Kaidonis J, Walker AW, Haak W, Bradshaw CJ et al (2013) Sequencing ancient calcified dental plaque shows changes in oral microbiota with dietary shifts of the neolithic and industrial revolutions. Nat Genet 45(4):450–455. https://doi.org/10.1038/ng.2536

Ahmed HJ, Svensson LA, Cope LD, Latimer JL, Hansen EJ, Ahlman K, Bayat-Turk J, Klamer D, Lagergård T (2001) Prevalence of cdtabc genes encoding cytolethal distending toxin among *Haemophilus ducreyi* and *Actinobacillus actinomycetemcomitans* strains. J Med Microbiol 50(10):860–864. https://doi.org/10.1099/0022-1317-50-10-860

Al-Ansari S, Zecha JAEM, Barasch A, De Lange J, Rozema FR, Raber-Durlacher JE (2015) Oral mucositis induced by anticancer therapies. Curr Oral Health Rep 2(4):202–211. https://doi.org/10.1007/s40496-015-0069-4

Almela T, Brook IM, Moharamzadeh K (2016) Development of three-dimensional tissue engineered bone-oral mucosal composite models. J Mater Sci Mater Med 27(4):1–8. https://doi.org/10.1007/s10856-016-5676-7

Almståhl A, Wikström M, Fagerberg-Mohlin B (2015) Microflora in oral ecosystems and salivary secretion rates—a 3-year follow-up after radiation therapy to the head and neck region. Arch Oral Biol 60(9):1187–1195. https://doi.org/10.1016/j.archoralbio.2015.04.004

Al-Taweel FB, Douglas CWI, Whawell SA (2016) The periodontal pathogen *Porphyromonas gingivalis* preferentially interacts with oral epithelial cells in S phase of the cell cycle. Infect Immun 84(7):1966–1974. https://doi.org/10.1128/iai.00111-16

Amano A (2010) Bacterial adhesins to host components in periodontitis. Periodontol 2000 52(1):12–37. https://doi.org/10.1111/j.1600-0757.2009.00307.x

Andrian E, Grenier D, Rouabhia M (2004) In vitro models of tissue penetration and destruction by *Porphyromonas gingivalis*. Infect Immun 72(8):4689–4698. https://doi.org/10.1128/iai.72.8.4689-4698.2004

Atanasova KR, Yilmaz Ö (2014) Looking in the *Porphyromonas gingivalis*' cabinet of curiosities: the microbium, the host and cancer association. Mol Oral Microbiol 29(2):55–66. https://doi.org/10.1111/omi.12047

Bahl R, Sandhu S, Singh K, Sahai N, Gupta M (2014) Odontogenic infections: microbiology and management. Contemp Clin Dent 5(3):307–311

Bahrani-Mougeot FK, Paster BJ, Coleman S, Ashar J, Barbuto S, Lockhart PB (2008) Diverse and novel oral bacterial species in blood following dental procedures. J Clin Microbiol 46(6):2129–2132. https://doi.org/10.1128/jcm.02004-07

Bao K, Belibasakis GN, Selevsek N, Grossmann J, Bostanci N (2015a) Proteomic profiling of host-biofilm interactions in an oral infection model resembling the periodontal pocket. Sci Rep 5:15999. https://doi.org/10.1038/srep15999

Bao K, Papadimitropoulos A, Akgül B, Belibasakis GN, Bostanci N (2015b) Establishment of an oral infection model resembling the periodontal pocket in a perfusion bioreactor system. Virulence 6(3):265–273. https://doi.org/10.4161/21505594.2014.978721

Barbosa M, Prada-Lopez I, Alvarez M, Amaral B, Casares-De-Cal MD, Tomas I (2015) Post-tooth extraction bacteraemia: a randomized clinical trial on the efficacy of chlorhexidine prophylaxis. PLoS One 10(5):15. https://doi.org/10.1371/journal.pone.0124249

Barbour SE, Nakashima K, Zhang J-B, Tangada S, Hahn C-L, Schenkein HA, Tew JG (1997) Tobacco and smoking: environmental factors that modify the host response (immune system) and have an impact on periodontal health. Crit Rev Oral Biol Med 8(4):437–460. https://doi.org/10.1177/10454411970080040501

Beck JD, Eke P, Lin D, Madianos P, Couper D, Moss K, Elter J, Heiss G, Offenbacher S (2005) Associations between IgG antibody to oral organisms and carotid intima-medial thickness in community-dwelling adults. Atherosclerosis 183(2):342–348. https://doi.org/10.1016/j.atherosclerosis.2005.03.017

Blasco-Baque V, Kémoun P, Loubieres P, Roumieux M, Heymes C, Serino M, Sixou M, Burcelin R (2012) Impact de la maladie parodontale sur la pression artérielle des souris diabétiques. Ann Cardiol Angéiol 61(3):173–177. https://doi.org/10.1016/j.ancard.2012.04.012

Blekhman R, Goodrich JK, Huang K, Sun Q, Bukowski R, Bell JT, Spector TD et al (2015) Host genetic variation impacts microbiome composition across human body sites. Genome Biol 16:191. https://doi.org/10.1186/s13059-015-0759-1

Brown LR, Dreizen S, Handler S, Johnston DA (1975) Effect of radiation-induced xerostomia on human oral microflora. J Dent Res 54(4):740–750. https://doi.org/10.1177/00220345750540040801

Bryzek D, Ksiazek M, Bielecka E, Karim AY, Potempa B, Staniec D, Koziel J, Potempa J (2014) A pathogenic trace of *Tannerella forsythia*—shedding of soluble fully active tumor necrosis factor α from the macrophage surface by karilysin. Mol Oral Microbiol 29(6):294–306. https://doi.org/10.1111/omi.12080

Buc E, Dubois D, Sauvanet P, Raisch J, Delmas J, Darfeuille-Michaud A, Pezet D, Bonnet R (2013) High prevalence of mucosa-associated *E. coli* producing cyclomodulin and genotoxin in colon cancer. PLoS One 8(2):e56964. https://doi.org/10.1371/journal.pone.0056964

Buglione M, Cavagnini R, Di Rosario F, Sottocornola L, Maddalo M, Vassalli L, Grisanti S et al (2016) Oral toxicity management in head and neck cancer patients treated with chemotherapy and radiation: dental pathologies and osteoradionecrosis (part 1) literature review and consensus statement. Crit Rev Oncol Hematol 97:131–142. https://doi.org/10.1016/j.critrevonc.2015.08.010

Buhlin K, Hultin M, Norderyd O, Persson L, Pockley AG, Pussinen PJ, Rabe P, Klinge B, Gustafsson A (2009) Periodontal treatment influences risk markers for atherosclerosis in patients with severe periodontitis. Atherosclerosis 206(2):518–522. https://doi.org/10.1016/j.atherosclerosis.2009.03.035

Caimano MJ, Bourell KW, Bannister TD, Cox DL, Radolf JD (1999) The *Treponema denticola* major sheath protein is predominantly periplasmic and has only limited surface exposure. Infect Immun 67(8):4072–4083

Cantley MD, Haynes DR, Marino V, Bartold PM (2011) Pre-existing periodontitis exacerbates experimental arthritis in a mouse model. J Clin Periodontol 38(6):532–541. https://doi.org/10.1111/j.1600-051X.2011.01714.x

Carrion J, Scisci E, Miles B, Sabino GJ, Zeituni AE, Gu Y, Bear A, Genco CA, Brown DL, Cutler CW (2012) Microbial carriage state of peripheral blood dendritic cells (DCs) in chronic periodontitis influences dc differentiation, atherogenic potential. J Immunol 189(6):3178–3187. https://doi.org/10.4049/jimmunol.1201053

Castanié-Cornet M-P, Bruel N, Genevaux P (2014) Chaperone networking facilitates protein targeting to the bacterial cytoplasmic membrane. Biochim Biophys Acta Mol Cell Res 1843(8):1442–1456. https://doi.org/10.1016/j.bbamcr.2013.11.007

Chi B, Qi M, Kuramitsu HK (2003) Role of dentilisin in *Treponema denticola* epithelial cell layer penetration. Res Microbiol 154(9):637–643. https://doi.org/10.1016/j.resmic.2003.08.001

Chistiakov DA, Orekhov AN, Bobryshev YV (2016) Links between atherosclerotic and periodontal disease. Exp Mol Pathol 100(1):220–235. https://doi.org/10.1016/j.yexmp.2016.01.006

Choi CH, Spooner R, Deguzman J, Koutouzis T, Ojcius DM, Yilmaz Ö (2013) *Porphyromonas gingivalis*-nucleoside-diphosphate-kinase inhibits ATP-induced reactive-oxygen-species via P2X7 receptor/NADPH-oxidase signalling and contributes to persistence. Cell Microbiol 15(6):961–976. https://doi.org/10.1111/cmi.12089

Chrcanovic BR, Reher P, Sousa AA, Harris M (2010) Osteoradionecrosis of the jaws—a current overview—part 1. Oral Maxillofac Surg 14(1):3–16. https://doi.org/10.1007/s10006-009-0198-9

Clevers H (2016) Modeling development and disease with organoids. Cell 165(7):1586–1597. https://doi.org/10.1016/j.cell.2016.05.082

Colley HE, Hearnden V, Jones AV, Weinreb PH, Violette SM, Macneil S, Thornhill MH, Murdoch C (2011) Development of tissue-engineered models of oral dysplasia and early invasive oral squamous cell carcinoma. Br J Cancer 105(10):1582–1592. http://www.nature.com/bjc/journal/v105/n10/suppinfo/bjc2011403s1.html

Coppé J-P, Desprez P-Y, Krtolica A, Campisi J (2010) The senescence-associated secretory phenotype: the dark side of tumor suppression. Ann Rev Pathol 5:99–118. https://doi.org/10.1146/annurev-pathol-121808-102144

Coppenhagen-Glazer S, Sol A, Abed J, Naor R, Zhang X, Han YW, Bachrach G (2015) Fap2 of *Fusobacterium nucleatum* is a galactose-inhibitable adhesin involved in coaggregation, cell adhesion, and preterm birth. Infect Immun 83(3):1104–1113. https://doi.org/10.1128/iai.02838-14

Cox D, Kerrigan SW, Watson SP (2011) Platelets and the innate immune system: mechanisms of bacterial-induced platelet activation. J Thromb Haemost 9(6):1097–1107. https://doi.org/10.1111/j.1538-7836.2011.04264.x

Crielaard W, Zaura E, Schuller AA, Huse SM, Montijn RC, Keijser BJ (2011) Exploring the oral microbiota of children at various developmental stages of their dentition in the relation to their oral health. BMC Med Genomics 4:22. https://doi.org/10.1186/1755-8794-4-22

D'Aiuto F, Parkar M, Nibali L, Suvan J, Lessem J, Tonetti MS (2006) Periodontal infections cause changes in traditional and novel cardiovascular risk factors: results from a randomized controlled clinical trial. Am Heart J 151(5):977–984. https://doi.org/10.1016/j.ahj.2005.06.018

Daly C, Mitchell D, Grossberg D, Highfield J, Stewart D (1997) Bacteraemia caused by periodontal probing. Aust Dent J 42(2):77–80. https://doi.org/10.1111/j.1834-7819.1997.tb00100.x

Daly CG, Mitchell DH, Highfield JE, Grossberg DE, Stewart D (2001) Bacteremia due to periodontal probing: a clinical and microbiological investigation. J Periodontol 72 (2):210–214. https://doi.org/10.1902/jop.2001.72.2.210

Damek-Poprawa M, Korostoff J, Gill R, Dirienzo JM (2013) Cell junction remodeling in gingival tissue exposed to a microbial toxin. J Dent Res 92(6):518–523. https://doi.org/10.1177/0022034513486807

De Molon RS, De Avila ED, Cirelli JA (2013) Host responses induced by different animal models of periodontal disease: a literature review. J Investig Clin Dent 4(4):211–218. https://doi.org/10.1111/jicd.12018

De Pablo P, Dietrich T, McAlindon TE (2008) Association of periodontal disease and tooth loss with rheumatoid arthritis in the US population. J Rheumatol 35(1):70–76

De Smit M, Westra J, Vissink A, Doornbos-Van Der Meer B, Brouwer E, Van Winkelhoff AJ (2012) Periodontitis in established rheumatoid arthritis patients: a cross-sectional clinical, microbiological and serological study. Arthritis Res Ther 14(5):R222. https://doi.org/10.1186/ar4061

Debelian GJ, Olsen I, Tronstad L (1995) Bacteremia in conjunction with endodontic therapy. Endod Dent Traumatol 11(3):142–149

Delahaye F, M'hammedi A, Guerpillon B, De Gevigney G, Boibieux A, Dauwalder O, Bouchiat C, Vandenesch F (2016) Systematic search for present and potential portals of entry for infective endocarditis. J Am Coll Cardiol 67(2):151–158. https://doi.org/10.1016/j.jacc.2015.10.065

Deng LQ, Bensing BA, Thamadilok S, Yu H, Lau K, Chen X, Ruhl S, Sullam PM, Varki A (2014) Oral streptococci utilize a Siglec-like domain of serine-rich repeat adhesins to preferentially target platelet sialoglycans in human blood. PLoS Pathogens 10(12):15. https://doi.org/10.1371/journal.ppat.1004540

Deng J, Jackson L, Epstein JB, Migliorati CA, Murphy BA (2015) Dental demineralization and caries in patients with head and neck cancer. Oral Oncol 51(9):824–831. https://doi.org/10.1016/j.oraloncology.2015.06.009

Desvarieux M, Demmer RT, Rundek T, Boden-Albala B, Jacobs DR, Sacco RL, Papapanou PN (2005) Periodontal microbiota and carotid intima-media thickness: the oral infections and vascular disease epidemiology study (invest). Circulation 111(5):576–582. https://doi.org/10.1161/01.cir.0000154582.37101.15

Desvarieux M, Demmer RT, Jacobs DR, Rundek T, Boden-Albala B, Sacco RL, Papapanou PN (2010) Periodontal bacteria and hypertension: the oral infections and vascular disease epidemiology study (invest). J Hypertens 28(7):1413–1421. https://doi.org/10.1097/HJH.0b013e328338cd36

Desvarieux M, Demmer RT, Jacobs DR, Papapanou PN, Sacco RL, Rundek T (2013) Changes in clinical and microbiological periodontal profiles relate to progression of carotid intima-media thickness: the oral infections and vascular disease epidemiology study. J Am Heart Assoc 2(6). https://doi.org/10.1161/jaha.113.000254

Dieudé M, Gillis MA, Théorêt JF, Thorin E, Lajoie G, Levine JS, Merhi Y, Rauch J (2009) Autoantibodies to heat shock protein 60 promote thrombus formation in a murine model of arterial thrombosis. J Thromb Haemost 7(4):710–719. https://doi.org/10.1111/j.1538-7836.2009.03305.x

Dixon DR, Darveau RP (2005) Lipopolysaccharide heterogeneity: innate host responses to bacterial modification of lipid a structure. J Dent Res 84(7):584–595. https://doi.org/10.1177/154405910508400702

Donnelly JP, Bellm LA, Epstein JB, Sonis ST, Symonds RP (2003) Antimicrobial therapy to prevent or treat oral mucositis. Lancet Infect Dis 3(7):405–412. https://doi.org/10.1016/S1473-3099(03)00668-6

Duran-Pinedo AE, Frias-Lopez J (2015) Beyond microbial community composition: functional activities of the oral microbiome in health and disease. Microbes Infect 17(7):505–516. https://doi.org/10.1016/j.micinf.2015.03.014

El Kholy K, Genco RJ, Van Dyke TE (2015) Oral infections and cardiovascular disease. Trends Endocrinol Metabol 26(6):315–321. https://doi.org/10.1016/j.tem.2015.03.001

Enersen M, Nakano K, Amano A (2013) *Porphyromonas gingivalis* fimbriae. J Oral Microbiol 5. https://doi.org/10.3402/jom.v5i0.20265

Epstein JB, Thariat J, Bensadoun R-J, Barasch A, Murphy BA, Kolnick L, Popplewell L, Maghami E (2012) Oral complications of cancer and cancer therapy. Cancer J Clin 62(6):400–422. https://doi.org/10.3322/caac.21157

Faïs T, Delmas J, Serres A, Bonnet R, Dalmasso G (2016) Impact of CDT toxin on human diseases. Toxins 8(7):220. https://doi.org/10.3390/toxins8070220

Faizi M, Seydi E, Abarghuyi S, Salimi A, Nasoohi S, Pourahmad J (2016) A search for mitochondrial damage in Alzheimer's disease using isolated rat brain mitochondria. Iran J Pharm Res 15:185–195

Fåk F, Tremaroli V, Bergström G, Bäckhed F (2015) Oral microbiota in patients with atherosclerosis. Atherosclerosis 243(2):573–578. https://doi.org/10.1016/j.atherosclerosis.2015.10.097

Fardini Y, Wang X, Témoin S, Nithianantham S, Lee D, Shoham M, Han YW (2011) *Fusobacterium nucleatum* adhesin FadA binds vascular endothelial cadherin and alters endothelial integrity. Mol Microbiol 82(6):1468–1480. https://doi.org/10.1111/j.1365-2958.2011.07905.x

Fenno JC (2012) *Treponema denticola* interactions with host proteins. J Oral Microbiol 4. https://doi.org/10.3402/jom.v4i0.9929

Fenno JC, Müller KH, McBride BC (1996) Sequence analysis, expression, and binding activity of recombinant major outer sheath protein (Msp) of *Treponema denticola*. J Bacteriol 178 (9):2489–2497

Ferdous F, Scott TR (2015) A comparative examination of thrombocyte/platelet immunity. Immunol Lett 163(1):32–39. https://doi.org/10.1016/j.imlet.2014.11.010

Fisher MA, Taylor GW, West BT, McCarthy ET (2011) Bidirectional relationship between chronic kidney and periodontal disease: a study using structural equation modeling. Kidney Int 79(3):347–355. https://doi.org/10.1038/ki.2010.384

Flynn KJ, Baxter NT, Schloss PD (2016) Metabolic and community synergy of oral bacteria in colorectal cancer. mSphere 1(3):e00102–e00116. https://doi.org/10.1128/mSphere.00102-16

Forner L, Larsen T, Kilian M, Holmstrup P (2006) Incidence of bacteremia after chewing, tooth brushing and scaling in individuals with periodontal inflammation. J Clin Periodontol 33 (6):401–407. https://doi.org/10.1111/j.1600-051X.2006.00924.x

Friedland RP (2015) Mechanisms of molecular mimicry involving the microbiota in neurodegeneration. J Alzheimers Dis 45(2):349–362. https://doi.org/10.3233/jad-142841

Gabrilska RA, Rumbaugh KP (2015) Biofilm models of polymicrobial infection. Future Microbiol 10(12):1997–2015. https://doi.org/10.2217/fmb.15.109

Gallimidi AB, Fischman S, Revach B, Bulvik R, Maliutina A, Rubinstein AM, Nussbaum G, Elkin M (2015) Periodontal pathogens *Porphyromonas gingivalis* and *Fusobacterium nucleatum* promote tumor progression in an oral-specific chemical carcinogenesis model. Oncotarget 6 (26):22613–22623

Gao L, Xu Y, Meng S, Wu Y, Huang H, Su R, Zhao L (2012) Identification of the putative specific pathogenic genes of *Porphyromonas gingivalis* with type II fimbriae. DNA and Cell Biol 31 (6):1027–1037. https://doi.org/10.1089/dna.2011.1487

Gao L, Hu Y, Wang Y, Jiang W, He Z, Zhu C, Ma R, Huang Z (2015) Exploring the variation of oral microbiota in supragingival plaque during and after head-and-neck radiotherapy using pyrosequencing. Arch Oral Biol 60(9):1222–1230. https://doi.org/10.1016/j.archoralbio.2015.05.006

Gawron K, Bereta G, Nowakowska Z, Łazarz-Bartyzel K, Łazarz M, Szmigielski B, Mizgalska D et al (2014) Peptidylarginine deiminase from *Porphyromonas gingivalis* contributes to infection of gingival fibroblasts and induction of prostaglandin E2-signaling pathway. Mol Oral Microbiol 29(6):321–332. https://doi.org/10.1111/omi.12081

Ge Z, Rogers AB, Feng Y, Lee A, Xu S, Taylor NS, Fox JG (2007) Bacterial cytolethal distending toxin promotes the development of dysplasia in a model of microbially induced hepatocarcinogenesis. Cell Microbiol 9(8):2070–2080. https://doi.org/10.1111/j.1462-5822.2007.00939.x

Gibson FC, Hong C, Chou H-H, Yumoto H, Chen J, Lien E, Wong J, Attardo Genco C (2004) Innate immune recognition of invasive bacteria accelerates atherosclerosis in apolipoprotein E-deficient mice. Circulation 109(22):2801–2806. https://doi.org/10.1161/01.cir.0000129769.17895.f0

Graves DT, Kang J, Andriankaja O, Wada K, Rossa C (2012) Animal models to study host-bacteria interactions involved in periodontitis. Front Oral Biol 15:117–132. https://doi.org/10.1159/000329675

Graziani F, Cei S, La Ferla F, Vano M, Gabriele M, Tonetti M (2010) Effects of non-surgical periodontal therapy on the glomerular filtration rate of the kidney: an exploratory trial. J Clin Periodontol 37(7):638–643. https://doi.org/10.1111/j.1600-051X.2010.01578.x

Green RF (1980) A note on K-selection. Am Nat 116(2):291–296. https://doi.org/10.1086/283627

Grenier D, Uitto VJ, McBride BC (1990) Cellular location of a *Treponema denticola* chymotrypsinlike protease and importance of the protease in migration through the basement membrane. Infect Immun 58(2):347–351

Gu H, Zhong Z, Jiang W, Du E, Dodel R, Liu J, Farlow MR, Zheng W, Du Y (2014) The role of choroid plexus in IVIG-induced beta-amyloid clearance. Neuroscience 270:168–176. https://doi.org/10.1016/j.neuroscience.2014.04.011

Guidi R, Guerra L, Levi L, Stenerlöw B, Fox JG, Josenhans C, Masucci MG, Frisan T (2013) Chronic exposure to the cytolethal distending toxins of gram-negative bacteria promotes genomic instability and altered DNA damage response. Cell Microbiol 15(1):98–113. https://doi.org/10.1111/cmi.12034

Guo Y, Nguyen KA, Potempa J (2010) Dichotomy of gingipains action as virulence factors: from cleaving substrates with the precision of a surgeon's knife to a meat chopper-like brutal degradation of proteins. Periodontol 2000 54(1):15–44. https://doi.org/10.1111/j.1600-0757.2010.00377.x

Gur C, Ibrahim Y, Isaacson B, Yamin R, Abed J, Gamliel M, Enk J et al (2015) Binding of the Fap2 protein of *Fusobacterium nucleatum* to human inhibitory receptor TIGIT protects tumors from immune cell attack. Immunity 42(2):344–355. https://doi.org/10.1016/j.immuni.2015.01.010

Gursoy UK, Pollanen M, Kononen E, Uitto VJ (2010) Biofilm formation enhances the oxygen tolerance and invasiveness of *Fusobacterium nucleatum* in an oral mucosa culture model. J Periodontol 81(7):1084–1091. https://doi.org/10.1902/jop.2010.090664

Gursoy UK, Pollanen M, Kononen E, Uitto VJ (2012) A novel organotypic dento-epithelial culture model: effect of *Fusobacterium nucleatum* biofilm on B-defensin-2, -3, and LL-37 expression. J Periodontol 83(2):242–247. https://doi.org/10.1902/jop.2011.110177

Ha NH, Woo BH, Kim DJ, Ha ES, Choi JI, Kim SJ, Park BS, Lee JH, Park HR (2015) Prolonged and repetitive exposure to *Porphyromonas gingivalis* increases aggressiveness of oral cancer cells by promoting acquisition of cancer stem cell properties. Tumor Biol 36(12):9947–9960. https://doi.org/10.1007/s13277-015-3764-9

Hajishengallis G (2014) Immunomicrobial pathogenesis of periodontitis: keystones, pathobionts, and host response. Trends Immunol 35(1):3–11. https://doi.org/10.1016/j.it.2013.09.001

Hajishengallis G (2015) Periodontitis: from microbial immune subversion to systemic inflammation. Nat Rev Immunol 15(1):30–44. https://doi.org/10.1038/nri3785

Hajishengallis G, Liang S, Payne MA, Hashim A, Jotwani R, Eskan MA, McIntosh ML et al (2011) A low-abundance biofilm species orchestrates inflammatory periodontal disease through the commensal microbiota and the complement pathway. Cell Host Microbe 10 (5):497–506. https://doi.org/10.1016/j.chom.2011.10.006

Han YW (2015) *Fusobacterium nucleatum*: a commensal-turned pathogen. Curr Opin Microbiol 23:141–147. https://doi.org/10.1016/j.mib.2014.11.013

Hansen T, Kunkel M, Kirkpatrick CJ, Weber A (2006a) Actinomyces in infected osteoradionecrosis—underestimated? Hum Pathol 37(1):61–67. https://doi.org/10.1016/j.humpath.2005.09.018

Hansen T, Wagner W, Kirkpatrick CJ, Kunkel M (2006b) Infected osteoradionecrosis of the mandible: follow-up study suggests deterioration in outcome for patients with actinomyces-positive bone biopsies. Int J Oral Maxillofac Surg 35(11):1001–1004. https://doi.org/10.1016/j.ijom.2006.08.006

Heimdahl A, Hall G, Hedberg M, Sandberg H, Soder PO, Tuner K, Nord CE (1990) Detection and quantitation by lysis-filtration of bacteremia after different oral surgical procedures. J Clin Microbiol 28(10):2205–2209

Henderson B, Ward JM, Ready D (2010) Aggregatibacter (actinobacillus) actinomycetemcomitans: a triple A* periodontopathogen? Periodontol 2000 54(1):78–105. https://doi.org/10.1111/j.1600-0757.2009.00331.x

Hitchon CA, Chandad F, Ferucci ED, Willemze A, Ioan-Facsinay A, Van Der Woude D, Markland J et al (2010) Antibodies to *Porphyromonas gingivalis* are associated with anticitrullinated protein antibodies in patients with rheumatoid arthritis and their relatives. J Rheumatol 37 (6):1105–1112. https://doi.org/10.3899/jrheum.091323

Hojo K, Nagaoka S, Ohshima T, Maeda N (2009) Bacterial interactions in dental biofilm development. J Dent Res 88(11):982–990. https://doi.org/10.1177/0022034509346811

Holgerson PL, Vestman NR, Claesson R, Ohman C, Domellof M, Tanner AC, Hernell O, Johansson I (2013) Oral microbial profile discriminates breast-fed from formula-fed infants. J Pediatr Gastroenterol Nutr 56(2):127–136. https://doi.org/10.1097/MPG.0b013e31826f2bc6

Homann N, Tillonen J, Meurman JH, Rintamaki H, Lindqvist C, Rautio M, Jousimies-Somer H, Salaspuro M (2000) Increased salivary acetaldehyde levels in heavy drinkers and smokers: a microbiological approach to oral cavity cancer. Carcinogenesis 21(4):663–668. https://doi.org/10.1093/carcin/21.4.663

Hooper SJ, Crean SJ, Lewis MAO, Spratt DA, Wade WG, Wilson MJ (2006) Viable bacteria present within oral squamous cell carcinoma tissue. J Clin Microbiol 44(5):1719–1725. https://doi.org/10.1128/jcm.44.5.1719-1725.2006

Hooper SJ, Crean S-J, Fardy MJ, Lewis MAO, Spratt DA, Wade WG, Wilson MJ (2007) A molecular analysis of the bacteria present within oral squamous cell carcinoma. J Med Microbiol 56(12):1651–1659. https://doi.org/10.1099/jmm.0.46918-0

Hu Y-J, Shao Z-Y, Wang Q, Jiang Y-T, Ma R, Tang Z-S, Liu Z, Liang J-P, Huang Z-W (2013) Exploring the dynamic core microbiome of plaque microbiota during head-and-neck radiotherapy using pyrosequencing. PLoS One 8(2):e56343. https://doi.org/10.1371/journal.pone.0056343

Hu X, Zhang Q, Hua H, Chen F (2016) Changes in the salivary microbiota of oral leukoplakia and oral cancer. Oral Oncol 56:e6–e8. https://doi.org/10.1016/j.oraloncology.2016.03.007

Huang CY, Shih CM, Tsao NW, Lin YW, Shih CC, Chiang KH, Shyue SK, Chang YJ, Hsieh CK, Lin FY (2016) The GroEL protein of *Porphyromonas gingivalis* regulates atherogenic phenomena in endothelial cells mediated by upregulating toll-like receptor 4 expression. Am J Transl Res 8(2):384–404

Hujoel PP, White BA, Garcia RI, Listgarten MA (2001) The dentogingival epithelial surface area revisited. J Periodont Res 36(1):48–55. https://doi.org/10.1034/j.1600-0765.2001.00011.x

Human_Microbiome_Project_Consortium (2012) Structure, function and diversity of the healthy human microbiome. Nature 486(7402):207–214. http://www.nature.com/nature/journal/v486/n7402/abs/nature11234.html#supplementary-information

Hyde ER, Andrade F, Vaksman Z, Parthasarathy K, Jiang H, Parthasarathy DK, Torregrossa AC et al (2014) Metagenomic analysis of nitrate-reducing bacteria in the oral cavity: implications for nitric oxide homeostasis. PLoS One 9(3):e88645. https://doi.org/10.1371/journal.pone.0088645

Hyvärinen K, Mäntylä P, Buhlin K, Paju S, Nieminen MS, Sinisalo J, Pussinen PJ (2012) A common periodontal pathogen has an adverse association with both acute and stable coronary artery disease. Atherosclerosis 223(2):478–484. https://doi.org/10.1016/j.atherosclerosis.2012.05.021

Hyvärinen K, Salminen A, Salomaa V, Pussinen PJ (2015) Systemic exposure to a common periodontal pathogen and missing teeth are associated with metabolic syndrome. Acta Diabetol 52(1):179–182. https://doi.org/10.1007/s00592-014-0586-y

Ide M, Harris M, Stevens A, Sussams R, Hopkins V, Culliford D, Fuller J et al (2016) Periodontitis and cognitive decline in Alzheimer's disease. PLoS One 11(3):9. https://doi.org/10.1371/journal.pone.0151081

Ilievski V, Kinchen JM, Prabhu R, Rim F, Leoni L, Unterman TG, Watanabe K (2016) Experimental periodontitis results in prediabetes and metabolic alterations in brain, liver and heart: global untargeted metabolomic analyses. J Oral Biol (Northborough) 3(1). 10.13188/12377-13987X.1000020

Imamura T, Travis J, Potempa J (2003) The biphasic virulence activities of gingipains: activation and inactivation of host proteins. Curr Protein Pept Sci 4(6):443–450

Inagaki S, Kimizuka R, Kokubu E, Saito A, Ishihara K (2016) *Treponema denticola* invasion into human gingival epithelial cells. Microb Pathog 94:104–111. https://doi.org/10.1016/j.micpath.2016.01.010

Ishihara K (2010) Virulence factors of *Treponema denticola*. Periodontol 2000 54(1):117–135. https://doi.org/10.1111/j.1600-0757.2009.00345.x

Ishihara K, Kuramitsu HK, Miura T, Okuda K (1998) Dentilisin activity affects the organization of the outer sheath of *Treponema denticola*. J Bacteriol 180(15):3837–3844

Iwasaki M, Taylor GW, Manz MC, Kaneko N, Imai S, Yoshihara A, Miyazaki H (2012a) Serum antibody to *Porphyromonas gingivalis* in chronic kidney disease. J Dent Res 91(9):828–833. https://doi.org/10.1177/0022034512455063

Iwasaki M, Taylor GW, Nesse W, Vissink A, Yoshihara A, Miyazaki H (2012b) Periodontal disease and decreased kidney function in Japanese elderly. Am J Kidney Dis 59(2):202–209. https://doi.org/10.1053/j.ajkd.2011.08.027

Iwasaki M, Minagawa K, Sato M, Kaneko N, Imai S, Yoshihara A, Miyazaki H (2016) Serum antibody to *Porphyromonas gingivalis* in metabolic syndrome among an older Japanese population. Gerodontology 33(2):193–200. https://doi.org/10.1111/ger.12135

Jeong E, Lee JY, Kim SJ, Choi J (2012) Predominant immunoreactivity of *Porphyromonas gingivalis* heat shock protein in autoimmune diseases. J Periodontal Res 47(6):811–816. https://doi.org/10.1111/j.1600-0765.2012.01501.x

Jeong E, Kim K, Kim JH, Cha GS, Kim S-J, Kang HS, Choi J (2015) *Porphyromonas gingivalis* HSP60 peptides have distinct roles in the development of atherosclerosis. Mol Immunol 63(2):489–496. https://doi.org/10.1016/j.molimm.2014.10.004

Jewett A, Hume WR, Le H, Huynh TN, Han YW, Cheng G, Shi W (2000) Induction of apoptotic cell death in peripheral blood mononuclear and polymorphonuclear cells by an oral bacterium, *Fusobacterium nucleatum*. Infect Immun 68(4):1893–1898. https://doi.org/10.1128/iai.68.4.1893-1898.2000

Jimenez M, Krall EA, Garcia RI, Vokonas PS, Dietrich T (2009) Periodontitis and incidence of cerebrovascular disease in men. Ann Neurol 66(4):505–512. https://doi.org/10.1002/ana.21742

Jobin M-C, Virdee I, McCulloch CA, Ellen RP (2007) Activation of MAPK in fibroblasts by *Treponema denticola* major outer sheath protein. Biochem Biophys Res Commun 356 (1):213–218. https://doi.org/10.1016/j.bbrc.2007.02.111

Johansson A (2011) *Aggregatibacter actinomycetemcomitans* leukotoxin: a powerful tool with capacity to cause imbalance in the host inflammatory response. Toxins 3(3):242–259. https://doi.org/10.3390/toxins3030242

Johnson L, Atanasova KR, Bui PQ, Lee J, Hung S-C, Yilmaz Ö, Ojcius DM (2015) *Porphyromonas gingivalis* attenuates ATP-mediated inflammasome activation and HMGB1 release through expression of a nucleoside-diphosphate kinase. Microbes Infect 17 (5):369–377. https://doi.org/10.1016/j.micinf.2015.03.010

Johnson CH, Ivanisevic J, Siuzdak G (2016) Metabolomics: beyond biomarkers and towards mechanisms. Nat Rev Mol Cell Biol 17(7):451–459. https://doi.org/10.1038/nrm.2016.25

Jung YJ, Choi YJ, An SJ, Lee HR, Jun HK, Choi BK (2016) *Tannerella forsythia* GroEL induces inflammatory bone resorption and synergizes with IL-17. Mol Oral Microbiol. https://doi.org/10.1111/omi.12172

Kachlany SC (2010) *Aggregatibacter actinomycetemcomitans* leukotoxin: from threat to therapy. J Dent Res 89(6):561–570. https://doi.org/10.1177/0022034510363682

Kadowaki T, Nakayama K, Yoshimura F, Okamoto K, Abe N, Yamamoto K (1998) Arg-gingipain acts as a major processing enzyme for various cell surface proteins in *Porphyromonas gingivalis*. J Biol Chem 273(44):29072–29076. https://doi.org/10.1074/jbc.273.44.29072

Kamer AR, Pirraglia E, Tsui W, Rusinek H, Vallabhajosula S, Mosconi L, Yi L et al (2015) Periodontal disease associates with higher brain amyloid load in normal elderly. Neurobiol Aging 36(2):627–633. https://doi.org/10.1016/j.neurobiolaging.2014.10.038

Kapil V, Haydar SM, Pearl V, Lundberg JO, Weitzberg E, Ahluwalia A (2013) Physiological role for nitrate-reducing oral bacteria in blood pressure control. Free Radic Biol Med 55:93–100. https://doi.org/10.1016/j.freeradbiomed.2012.11.013

Kaplan CW, Ma X, Paranjpe A, Jewett A, Lux R, Kinder-Haake S, Shi W (2010) *Fusobacterium nucleatum* outer membrane proteins Fap2 and RadD induce cell death in human lymphocytes. Infect Immun 78(11):4773–4778. https://doi.org/10.1128/iai.00567-10

Karim AY, Kulczycka M, Kantyka T, Dubin G, Jabaiah A, Daugherty PS, Thogersen IB, Enghild JJ, Nguyen KA, Potempa J (2010) A novel matrix metalloprotease-like enzyme (Karilysin) of the periodontal pathogen *Tannerella forsythia* ATCC 43037. Biol Chem 391(1):105–117. https://doi.org/10.1515/bc.2010.009

Kasmi G, Refatllari E, Dumani S, Refatllari A (2014) Early infective endocarditis due to *Staphylococcus aureus* following dental procedures. Clin Lab 60(11):1933–1936. https://doi.org/10.7754/Clin.Lab.2014.140502

Kato T, Kawai S, Nakano K, Inaba H, Kuboniwa M, Nakagawa I, Tsuda K et al (2007) Virulence of *Porphyromonas gingivalis* is altered by substitution of fimbria gene with different genotype. Cell Microbiol 9(3):753–765. https://doi.org/10.1111/j.1462-5822.2006.00825.x

Kato I, Vasquez A, Moyerbrailean G, Land S, Djuric Z, Sun J, Lin H-S, Ram JL (2017) Nutritional correlates of oral microbiome. J Am Coll Nutr 36(2):88–98

Kato I, Vasquez AA, Moyerbrailean G, Land S, Sun J, Lin H-S, Ram JL (2016) Oral microbiome and history of smoking and colorectal cancer. J Epidemiol Res 2(2):92

Kebschull M, Demmer RT, Papapanou PN (2010) "Gum bug, leave my heart alone!"—epidemiologic and mechanistic evidence linking periodontal infections and atherosclerosis. J Dent Res 89(9):879–902. https://doi.org/10.1177/0022034510375281

Kimura A, Takemura M, Saito K, Yoshikura N, Hayashi Y, Inuzuka T (2016) Association between naturally occurring anti-amyloid β autoantibodies and medial temporal lobe atrophy in Alzheimer's disease. J Neurol Neurosurg Psychiatry. https://doi.org/10.1136/jnnp-2016-313476

Kinane DF, Riggio MP, Walker KF, Mackenzie D, Shearer B (2005) Bacteraemia following periodontal procedures. J Clin Periodontol 32(7):708–713. https://doi.org/10.1111/j.1600-051X.2005.00741.x

Kinikoglu B, Auxenfans C, Pierrillas P, Justin V, Breton P, Burillon C, Hasirci V, Damour O (2009) Reconstruction of a full-thickness collagen-based human oral mucosal equivalent. Biomaterials 30(32):6418–6425. https://doi.org/10.1016/j.biomaterials.2009.08.010

Kinikoglu B, Damour O, Hasirci V (2015) Tissue engineering of oral mucosa: a shared concept with skin. J Artif Organs 18(1):8–19. https://doi.org/10.1007/s10047-014-0798-5

Kinloch AJ, Alzabin S, Brintnell W, Wilson E, Barra L, Wegner N, Bell DA, Cairns E, Venables PJ (2011) Immunization with *Porphyromonas gingivalis* enolase induces autoimmunity to mammalian α-enolase and arthritis in DR4-IE-transgenic mice. Arthritis Rheum 63 (12):3818–3823. https://doi.org/10.1002/art.30639

Konig MF, Paracha AS, Moni M, Bingham CO, Andrade F (2015) Defining the role of *Porphyromonas gingivalis* peptidylarginine deiminase (PPAD) in rheumatoid arthritis through the study of PPAD biology. Ann Rheum Dis 74(11):2054–2061. https://doi.org/10.1136/annrheumdis-2014-205385

Koren O, Spor A, Felin J, Fåk F, Stombaugh J, Tremaroli V, Behre CJ et al (2011) Human oral, gut, and plaque microbiota in patients with atherosclerosis. Proc Natl Acad Sci 108(Supplement 1):4592–4598. https://doi.org/10.1073/pnas.1011383107

Koziel J, Karim AY, Przybyszewska K, Ksiazek M, Rapala-Kozik M, Nguyen KA, Potempa J (2010) Proteolytic inactivation of LL-37 by karilysin, a novel virulence mechanism of *Tannerella forsythia*. J Innate Immun 2(3):288–293. https://doi.org/10.1159/000281881

Kshirsagar AV, Moss KL, Elter JR, Beck JD, Offenbacher S, Falk RJ (2005) Periodontal disease is associated with renal insufficiency in the atherosclerosis risk in communities (ARIC) study. Am J Kidney Dis 45(4):650–657. https://doi.org/10.1053/j.ajkd.2004.12.009.

Kshirsagar AV, Offenbacher S, Moss KL, Barros SP, Beck JD (2007) Antibodies to periodontal organisms are associated with decreased kidney function. Blood Purif 25(1):125–132

Laheij AMGA, De Soet JJ, Von Dem Borne PA, Kuijper EJ, Kraneveld EA, Van Loveren C, Raber-Durlacher JE (2012) Oral bacteria and yeasts in relationship to oral ulcerations in hematopoietic stem cell transplant recipients. Support Care Cancer 20(12):3231–3240. https://doi.org/10.1007/s00520-012-1463-2

Lalla E, Lamster IB, Hofmann MA, Bucciarelli L, Jerud AP, Tucker S, Lu Y, Papapanou PN, Schmidt AM (2003) Oral infection with a periodontal pathogen accelerates early atherosclerosis in apolipoprotein e-null mice. Arterioscler Thromb Vasc Biol 23(8):1405–1411. https://doi.org/10.1161/01.atv.0000082462.26258.fe

Lamont RJ, Hajishengallis G (2015) Polymicrobial synergy and dysbiosis in inflammatory disease. Trends Mol Med 21(3):172–183. https://doi.org/10.1016/j.molmed.2014.11.004

Laprise C, Shahul HP, Madathil SA, Thekkepurakkal AS, Castonguay G, Varghese I, Shiraz S et al (2016) Periodontal diseases and risk of oral cancer in southern India: results from the hence life study. Int J Cancer 139(7):1512–1519. https://doi.org/10.1002/ijc.30201

Lee HR, Jun HK, Kim HD, Lee SH, Choi BK (2012) *Fusobacterium nucleatum* GroEL induces risk factors of atherosclerosis in human microvascular endothelial cells and ApoE(-/-) mice. Mol Oral Microbiol 27(2):109–123. https://doi.org/10.1111/j.2041-1014.2011.00636.x

Lee HR, Jun HK, Choi BK (2014) *Tannerella forsythia* BspA increases the risk factors for atherosclerosis in ApoE-/- mice. Oral Dis 20(8):803–808. https://doi.org/10.1111/odi.12214

Li N, Collyer CA (2011) Gingipains from *Porphyromonas gingivalis*—complex domain structures confer diverse functions. Eur J Microbiol Immunol (Bp) 1(1):41–58. https://doi.org/10.1556/EuJMI.1.2011.1.7

Li XJ, Kolltveit KM, Tronstad L, Olsen I (2000) Systemic diseases caused by oral infection. Clin Microbiol Rev 13(4):547. https://doi.org/10.1128/Cmr.13.4.547-558.2000

Lif Holgerson P, Harnevik L, Hernell O, Tanner AC, Johansson I (2011) Mode of birth delivery affects oral microbiota in infants. J Dent Res 90(10):1183–1188. https://doi.org/10.1177/0022034511418973

Lin FY, Huang CY, Lu HY, Shih CM, Tsao NW, Shyue SK, Lin CY et al (2015) The GroEL protein of *Porphyromonas gingivalis* accelerates tumor growth by enhancing endothelial progenitor cell function and neovascularization. Mol Oral Microbiol 30(3):198–216. https://doi.org/10.1111/omi.12083

Lockhart PB, Brennan MT, Sasser HC, Fox PC, Paster BJ, Bahrani-Mougeot FK (2008) Bacteremia associated with toothbrushing and dental extraction. Circulation 117(24):3118–3125. https://doi.org/10.1161/circulationaha.107.758524

Ludewig B, Krebs P, Scandella E (2004) Immunopathogenesis of atherosclerosis. J Leukocyte Biol 76(2):300–306. https://doi.org/10.1189/jlb.1203605

Lund Håheim L, Olsen I, Nafstad P, Schwarze P, Rønningen KS (2008) Antibody levels to single bacteria or in combination evaluated against myocardial infarction. J Clin Periodontol 35 (6):473–478. https://doi.org/10.1111/j.1600-051X.2008.01229.x

Madan M, Bishayi B, Hoge M, Messas E, Amar S (2007) Doxycycline affects diet- and bacteria-associated atherosclerosis in an ApoE heterozygote murine model: cytokine profiling implications. Atherosclerosis 190(1):62–72. https://doi.org/10.1016/j.atherosclerosis.2006.02.026.

Mager D, Haffajee A, Devlin P, Norris C, Posner M, Goodson J (2005) The salivary microbiota as a diagnostic indicator of oral cancer: a descriptive, non-randomized study of cancer-free and oral squamous cell carcinoma subjects. J Transl Med 3(1):27

Mantri CK, Chen C-H, Dong X, Goodwin JS, Pratap S, Paromov V, Xie H (2015) Fimbriae-mediated outer membrane vesicle production and invasion of *Porphyromonas gingivalis*. MicrobiologyOpen 4(1):53–65. https://doi.org/10.1002/mbo3.221

Mäntylä P, Buhlin K, Paju S, Persson GR, Nieminen MS, Sinisalo J, Pussinen PJ (2013) Subgingival *Aggregatibacter actinomycetemcomitans* associates with the risk of coronary artery disease. J Clin Periodontol 40(6):583–590. https://doi.org/10.1111/jcpe.12098

Mark Welch JL, Rossetti BJ, Rieken CW, Dewhirst FE, Borisy GG (2016) Biogeography of a human oral microbiome at the micron scale. Proc Natl Acad Sci USA 113(6):E791–E800. https://doi.org/10.1073/pnas.1522149113

Marttila E, Uittamo J, Rusanen P, Lindqvist C, Salaspuro M, Rautemaa R (2013) Acetaldehyde production and microbial colonization in oral squamous cell carcinoma and oral lichenoid disease. Oral Surg Oral Med Oral Pathol Oral Radiol 116(1):61–68. https://doi.org/10.1016/j.oooo.2013.02.009

Matsushita K, Hamaguchi M, Hashimoto M, Yamazaki M, Yamazaki T, Asai K, Yamori M et al (2015) The novel association between red complex of oral microbe and body mass index in healthy Japanese: a population based cross-sectional study. J Clin Biochem Nutr 57 (2):135–139. https://doi.org/10.3164/jcbn.15-19

Mayr M, Metzler B, Kiechl S, Willeit J, Schett G, Xu Q, Wick G (1999) Endothelial cytotoxicity mediated by serum antibodies to heat shock proteins of *Escherichia coli* and *Chlamydia pneumonia*: immune reactions to heat shock proteins as a possible link between infection and atherosclerosis. Circulation 99(12):1560–1566. https://doi.org/10.1161/01.cir.99.12.1560

McGraw WT, Potempa J, Farley D, Travis J (1999) Purification, characterization, and sequence analysis of a potential virulence factor from *Porphyromonas gingivalis*, peptidylarginine deiminase. Infect Immun 67(7):3248–3256

McLean JS (2014) Advancements toward a systems level understanding of the human oral microbiome. Front Cell Infect Microbiol 4:98. https://doi.org/10.3389/fcimb.2014.00098

Mercanoglu F, Oflaz H, Öz O, Gökbuget AY, Genchellac H, Sezer M, Nişanci Y, Umman S (2004) Endothelial dysfunction in patients with chronic periodontitis and its improvement after initial periodontal therapy. J Periodontol 75(12):1694–1700. https://doi.org/10.1902/jop.2004.75.12.1694

Meurman JH, Pyrhönen S, Teerenhovi L, Lindqvist C (1997) Oral sources of septicaemia in patients with malignancies. Oral Oncol 33(6):389–397. https://doi.org/10.1016/S0964-1955(97)00032-8

Michaud DS, Liu Y, Meyer M, Giovannucci E, Joshipura K (2008) Periodontal disease, tooth loss and cancer risk in a prospective study of male health professionals. Lancet Oncol 9(6):550–558. https://doi.org/10.1016/S1470-2045(08)70106-2

Miklossy J (2015) Historic evidence to support a causal relationship between spirochetal infections and Alzheimer's disease. Front Aging Neurosci 7:12. https://doi.org/10.3389/fnagi.2015.00046

Mikuls TR, Payne JB, Reinhardt RA, Thiele GM, Maziarz E, Cannella AC, Holers VM, Kuhn KA, O'Dell JR (2009) Antibody responses to *Porphyromonas gingivalis* (*P. gingivalis*) in subjects with rheumatoid arthritis and periodontitis. Int Immunopharmacol 9(1):38–42. https://doi.org/10.1016/j.intimp.2008.09.008

Mikuls TR, Thiele GM, Deane KD, Payne JB, O'dell JR, Yu F, Sayles H et al (2012) *Porphyromonas gingivalis* and disease-related autoantibodies in individuals at increased risk of rheumatoid arthritis. Arthritis Rheumatol 64(11):3522–3530. https://doi.org/10.1002/art.34595

Mikuls TR, Payne JB, Yu F, Thiele GM, Reynolds RJ, Cannon GW, Markt J et al (2014) Periodontitis and *Porphyromonas gingivalis* in patients with rheumatoid arthritis. Arthritis Rheumatol 66(5):1090–1100. https://doi.org/10.1002/art.38348

Mishima E, Sharma A (2011) *Tannerella forsythia* invasion in oral epithelial cells requires phosphoinositide 3-kinase activation and clathrin-mediated endocytosis. Microbiology 157(8):2382–2391. https://doi.org/10.1099/mic.0.048975-0

Miyamoto T, Yumoto H, Takahashi Y, Davey M, Gibson FC, Genco CA (2006) Pathogen-accelerated atherosclerosis occurs early after exposure and can be prevented via immunization. Infect Immun 74(2):1376–1380. https://doi.org/10.1128/iai.74.2.1376-1380.2006

Moharamzadeh K, Brook IM, Van Noort R, Scutt AM, Thornhill MH (2007) Tissue-engineered oral mucosa: a review of the scientific literature. J Dent Res 86(2):115–124. https://doi.org/10.1177/154405910708600203

Moharamzadeh K, Brook IM, Van Noort R, Scutt AM, Smith KG, Thornhill MH (2008) Development, optimization and characterization of a full-thickness tissue engineered human oral mucosal model for biological assessment of dental biomaterials. J Mater Sci Mater Med 19(4):1793–1801. https://doi.org/10.1007/s10856-007-3321-1

Moharamzadeh K, Colley H, Murdoch C, Hearnden V, Chai WL, Brook IM, Thornhill MH, Macneil S (2012) Tissue-engineered oral mucosa. J Dent Res 91(7):642–650. https://doi.org/10.1177/0022034511435702

Moreira APB, Texeira TFS, Ferreira AB, Do Carmo Gouveia Peluzio M, De Cássia Gonçalves Alfenas R (2012) Influence of a high-fat diet on gut microbiota, intestinal permeability and metabolic endotoxaemia. Br J Nutr 108(5):801–809. https://doi.org/10.1017/S0007114512001213

Morita T, Yamazaki Y, Mita A, Takada K, Seto M, Nishinoue N, Sasaki Y, Motohashi M, Maeno M (2010) A cohort study on the association between periodontal disease and the development of metabolic syndrome. J Periodontol 81(4):512–519. https://doi.org/10.1902/jop.2010.090594

Moritani K, Takeshita T, Shibata Y, Ninomiya T, Kiyohara Y, Yamashita Y (2015) Acetaldehyde production by major oral microbes. Oral Dis 21(6):748–754. https://doi.org/10.1111/odi.12341

Mostefaoui Y, Claveau I, Ross G, Rouabhia M (2002) Tissue structure, and IL-1β, IL-8, and TNF-α secretions after contact by engineered human oral mucosa with dentifrices. J Clin Periodontol 29(11):1035–1041. https://doi.org/10.1034/j.1600-051X.2002.291109.x

Mougeot FKB, Saunders SE, Brennan MT, Lockhart PB (2015) Associations between bacteremia from oral sources and distant-site infections: tooth brushing versus single tooth extraction. Oral Surg Oral Med Oral Pathol Oral Radiol 119(4):430–435. https://doi.org/10.1016/j.oooo.2015.01.009

Mustapha IZ, Debrey S, Oladubu M, Ugarte R (2007) Markers of systemic bacterial exposure in periodontal disease and cardiovascular disease risk: a systematic review and meta-analysis. J Periodontol 78(12):2289–2302. https://doi.org/10.1902/jop.2007.070140

Muto M, Hitomi Y, Ohtsu A, Shimada H, Kashiwase Y, Sasaki H, Yoshida S, Esumi H (2000) Acetaldehyde production by non-pathogenic Neisseria in human oral microflora: implications for carcinogenesis in upper aerodigestive tract. Int J Cancer 88(3):342–350

Nagy KN, Sonkodi I, Szöke I, Nagy E, Newman HN (1998) The microflora associated with human oral carcinomas. Oral Oncol 34(4):304–308. https://doi.org/10.1016/S1368-8375(98)80012-2

Nakagawa I, Amano A, Ohara-Nemoto Y, Endoh N, Morisaki I, Kimura S, Kawabata S, Hamada S (2002) Identification of a new variant of fimA gene of *Porphyromonas gingivalis* and its distribution in adults and disabled populations with periodontitis. J Periodontal Res 37 (6):425–432. https://doi.org/10.1034/j.1600-0765.2002.01637.x

Nakano K, Nemoto H, Nomura R, Inaba H, Yoshioka H, Taniguchi K, Amano A, Ooshima T (2009) Detection of oral bacteria in cardiovascular specimens. Oral Microbiol Immunol 24 (1):64–68. https://doi.org/10.1111/j.1399-302X.2008.00479.x

Nakayama K (2015) *Porphyromonas gingivalis* and related bacteria: from colonial pigmentation to the type IX secretion system and gliding motility. J Periodontal Res 50(1):1–8. https://doi.org/10.1111/jre.12255

Nalbant A, Saygili T (2016) IL12, IL10, IFNγ and TNFα expression in human primary monocytes stimulated with bacterial heat shock GroEL (HSP64) protein. PLoS One 11(4):e0154085. https://doi.org/10.1371/journal.pone.0154085

Nesic D, Hsu Y, Stebbins CE (2004) Assembly and function of a bacterial genotoxin. Nature 429 (6990):429–433. http://www.nature.com/nature/journal/v429/n6990/suppinfo/nature02532_S1.html

Nishikubo S, Ohara M, Ikura M, Katayanagi K, Fujiwara T, Komatsuzawa H, Kurihara H, Sugai M (2006) Single nucleotide polymorphism in the cytolethal distending toxin B gene confers heterogeneity in the cytotoxicity of actinobacillus actinomycetemcomitans. Infect Immun 74 (12):7014–7020. https://doi.org/10.1128/iai.02063-05

Noble JM, Scarmeas N, Celenti RS, Elkind MSV, Wright CB, Schupf N, Papapanou PN (2014) Serum IgG antibody levels to periodontal microbiota are associated with incident Alzheimer disease. PLoS One 9(12):14. https://doi.org/10.1371/journal.pone.0114959

Nussbaum G, Ben-Adi S, Genzler T, Sela M, Rosen G (2009) Involvement of toll-like receptors 2 and 4 in the innate immune response to *Treponema denticola* and its outer sheath components. Infect Immun 77(9):3939–3947. https://doi.org/10.1128/IAI.00488-09

Ogawa T, Asai Y, Makimura Y, Tamai R (2007) Chemical structure and immunobiological activity of *Porphyromonas gingivalis* lipid A. Front Biosci 12:3795–3812

Okada M, Kobayashi T, Ito S, Yokoyama T, Komatsu Y, Abe A, Murasawa A, Yoshie H (2011) Antibody responses to periodontopathic bacteria in relation to rheumatoid arthritis in Japanese adults. J Periodontol 82(10):1433–1441. https://doi.org/10.1902/jop.2011.110020

Oldstone MB (2005) Molecular mimicry, microbial infection, and autoimmune disease: evolution of the concept. Curr Top Microbiol Immunol 296:1–17

Oliveira FAF, Forte CPF, Silva PGD, Lopes CB, Montenegro RC, Dos Santos A, Sobrinho C, Mota MRL, Sousa FB, Alves A (2015) Molecular analysis of oral bacteria in heart valve of patients with cardiovascular disease by real-time polymerase chain reaction. Medicine 94 (47):5. https://doi.org/10.1097/md.0000000000002067

Olsen I (2008) Update on bacteraemia related to dental procedures. Transfus Apher Sci 39 (2):173–178. https://doi.org/10.1016/j.transci.2008.06.008

Olsen I, Hajishengallis G (2016) Major neutrophil functions subverted by *Porphyromonas gingivalis*. J Oral Microbiol 2016:8. https://doi.org/10.3402/jom.v8.30936

Olsen I, Singhrao SK (2015) Can oral infection be a risk factor for Alzheimer's disease? J Oral Microbiol 7:16. https://doi.org/10.3402/jom.v7.29143

Olsen I, Van Winkelhoff AJ (2014) Acute focal infections of dental origin. Periodontol 2000 65 (1):178–189. https://doi.org/10.1111/prd.12018

Onishi S, Honma K, Liang S, Stathopoulou P, Kinane D, Hajishengallis G, Sharma A (2008) Toll-like receptor 2-mediated interleukin-8 expression in gingival epithelial cells by the *Tannerella forsythia* leucine-rich repeat protein BspA. Infect Immun 76(1):198–205. https://doi.org/10.1128/iai.01139-07

Paino A, Lohermaa E, Sormunen R, Tuominen H, Korhonen J, Pöllänen MT, Ihalin R (2012) Interleukin-1β is internalised by viable *Aggregatibacter actinomycetemcomitans* biofilm and locates to the outer edges of nucleoids. Cytokine 60(2):565–574. https://doi.org/10.1016/j.cyto.2012.07.024

Pang PYK, Sin YK, Lim CH, Tan TE, Lim SL, Chao VTT, Chua YL (2015) Surgical management of infective endocarditis: an analysis of early and late outcomes. Eur J Cardiothorac Surg 47 (5):826–832. https://doi.org/10.1093/ejcts/ezu281

Panghal M, Kaushal V, Kadayan S, Yadav JP (2012) Incidence and risk factors for infection in oral cancer patients undergoing different treatments protocols. BMC Oral Health 12(1):1–12. https://doi.org/10.1186/1472-6831-12-22

Pedersen LM, Madsen OR, Gutschik E (1993) Septicemia caused by an unusual *Neisseria meningitidis* species following dental extraction. Scand J Infect Dis 25(1):137–139

Pena I, Junquera LM, Meana A, Garcia E, Garcia V, De Vicente JC (2010) In vitro engineering of complete autologous oral mucosa equivalents: characterization of a novel scaffold. J Periodontal Res 45(3):375–380. https://doi.org/10.1111/j.1600-0765.2009.01248.x

Pena I, Junquera LM, Meana A, Garcia E, Aguilar C, Fresno MF (2011) In vivo behavior of complete human oral mucosa equivalents: characterization in athymic mice. J Periodontal Res 46(2):214–220. https://doi.org/10.1111/j.1600-0765.2010.01330.x

Perry RJ, Peng L, Barry NA, Cline GW, Zhang D, Cardone RL, Petersen KF, Kibbey RG, Goodman AL, Shulman GI (2016) Acetate mediates a microbiome-brain-beta-cell axis to promote metabolic syndrome. Nature 534(7606):213–217. https://doi.org/10.1038/nature18309

Piconi S, Trabattoni D, Luraghi C, Perilli E, Borelli M, Pacei M, Rizzardini G et al (2009) Treatment of periodontal disease results in improvements in endothelial dysfunction and reduction of the carotid intima-media thickness. FASEB J 23(4):1196–1204. https://doi.org/10.1096/fj.08-119578

Pinnock A, Murdoch C, Moharamzadeh K, Whawell S, Douglas CWI (2014) Characterisation and optimisation of organotypic oral mucosal models to study *Porphyromonas gingivalis* invasion. Microbes Infect 16(4):310–319. https://doi.org/10.1016/j.micinf.2014.01.004

Potempa J, Sroka A, Imamura T, Travis J (2003) Gingipains, the major cysteine proteinases and virulence factors of *Porphyromonas gingivalis*: structure, function and assembly of multidomain protein complexes. Curr Protein Pept Sci 4(6):397–407

Potikuri D, Dannana KC, Kanchinadam S, Agrawal S, Kancharla A, Rajasekhar L, Pothuraju S, Gumdal N (2012) Periodontal disease is significantly higher in non-smoking treatment-naive rheumatoid arthritis patients: results from a case-control study. Ann Rheum Dis 71 (9):1541–1544. https://doi.org/10.1136/annrheumdis-2011-200380

Pradeep AR, Hadge P, Arjun Raju P, Shetty SR, Shareef K, Guruprasad CN (2010) Periodontitis as a risk factor for cerebrovascular accident: a case–control study in the Indian population. J Periodontal Res 45(2):223–228. https://doi.org/10.1111/j.1600-0765.2009.01220.x

Pushalkar S, Mane SP, Ji X, Li Y, Evans C, Crasta OR, Morse D, Meagher R, Singh A, Saxena D (2011) Microbial diversity in saliva of oral squamous cell carcinoma. FEMS Immunol Med Microbiol 61(3):269–277. https://doi.org/10.1111/j.1574-695X.2010.00773.x

Pushalkar S, Ji X, Li Y, Estilo C, Yegnanarayana R, Singh B, Li X, Saxena D (2012) Comparison of oral microbiota in tumor and non-tumor tissues of patients with oral squamous cell carcinoma. BMC Microbiol 12(1):144

Pussinen PJ, Alfthan G, Jousilahti P, Paju S, Tuomilehto J (2007) Systemic exposure to *Porphyromonas gingivalis* predicts incident stroke. Atherosclerosis 193(1):222–228. https://doi.org/10.1016/j.atherosclerosis.2006.06.027

Puthengady Thomas B, Sun CX, Bajenova E, Ellen RP, Glogauer M (2006) Modulation of human neutrophil functions in vitro by *Treponema denticola* major outer sheath protein. Infect Immun 74(3):1954–1957. https://doi.org/10.1128/iai.74.3.1954-1957.2006

Reis LC, Rocas IN, Siqueira JF, De Uzeda M, Lacerda VS, Domingues R, Moraes SR, Saraiva RM (2016) Bacteremia after endodontic procedures in patients with heart disease: culture and molecular analyses. J Endod 42(8):1181–1185. https://doi.org/10.1016/j.joen.2016.05.013

Rubinstein MR, Wang X, Liu W, Hao Y, Cai G, Han YW (2013) *Fusobacterium nucleatum* promotes colorectal carcinogenesis by modulating E-cadherin/β-catenin signaling via its FadA adhesin. Cell Host Microbe 14(2):195–206. https://doi.org/10.1016/j.chom.2013.07.012

Rutger Persson G (2012) Rheumatoid arthritis and periodontitis—inflammatory and infectious connections. Review of the literature. J Oral Microbiol 4. https://doi.org/10.3402/jom.v4i0.11829

Salaspuro M (2012) Interactions of alcohol and tobacco in gastrointestinal cancer. J Gastroenterol Hepatol 27:135–139. https://doi.org/10.1111/j.1440-1746.2012.07017.x

Sampaio-Maia B, Monteiro-Silva F (2014) Acquisition and maturation of oral microbiome throughout childhood: an update. Dent Res J (Isfahan) 11(3):291–301

Santos A, Pascual A, Llopis J, Giner L, Kim DM, Levi P Jr, Ramseierg CA (2015) Self-reported oral hygiene habits in smokers and nonsmokers diagnosed with periodontal disease. Oral Health Prev Dent 13(3):245–251. https://doi.org/10.3290/j.ohpd.a34248

Sasaki M, Yamaura C, Ohara-Nemoto Y, Tajika S, Kodama Y, Ohya T, Harada R, Kimura S (2005) *Streptococcus anginosus* infection in oral cancer and its infection route. Oral Dis 11 (3):151–156. https://doi.org/10.1111/j.1601-0825.2005.01051.x

Savarrio L, Mackenzie D, Riggio M, Saunders WP, Bagg J (2005) Detection of bacteraemias during non-surgical root canal treatment. J Dent 33(4):293–303. https://doi.org/10.1016/j.jdent.2004.09.008

Scher JU, Ubeda C, Equinda M, Khanin R, Buischi Y, Viale A, Lipuma L et al (2012) Periodontal disease and the oral microbiota in new-onset rheumatoid arthritis. Arthritis Rheum 64 (10):3083–3094. https://doi.org/10.1002/art.34539

Schillinger T, Kluger W, Exner M, Mlekusch W, Sabeti S, Amighi J, Wagner O, Minar E, Schillinger M (2006) Dental and periodontal status and risk for progression of carotid atherosclerosis: the inflammation and carotid artery risk for atherosclerosis study dental substudy. Stroke 37(9):2271–2276. https://doi.org/10.1161/01.STR.0000236495.82545.2e

Schmidt BL, Kuczynski J, Bhattacharya A, Huey B, Corby PM, Queiroz ELS, Nightingale K et al (2014) Changes in abundance of oral microbiota associated with oral cancer. PLoS One 9(6): e98741. https://doi.org/10.1371/journal.pone.0098741

Selton-Suty C, Doco-Lecompte T, Juilliere Y, Hoen B (2001) Infectious endocarditis in France today. Arch Des Maladies Du Coeur Des Vaisseaux 94(10):1099–1102

Sharma A (2010) Virulence mechanisms of *Tannerella forsythia*. Periodontol 2000 54 (1):106–116. https://doi.org/10.1111/j.1600-0757.2009.00332.x

Sheets SM, Robles-Price AG, McKenzie RME, Casiano CA, Fletcher HM (2008) Gingipain-dependent interactions with the host are important for survival of *Porphyromonas gingivalis*. Front Biosci 13:3215–3238

Shenker BJ, McKay T, Datar S, Miller M, Chowhan R, Demuth D (1999) Actinobacillus actinomycetemcomitans immunosuppressive protein is a member of the family of cytolethal distending toxins capable of causing a G2 arrest in human T cells. J Immunol 162 (8):4773–4780

Shoji M, Naito M, Yukitake H, Sato K, Sakai E, Ohara N, Nakayama K (2004) The major structural components of two cell surface filaments of *Porphyromonas gingivalis* are matured through lipoprotein precursors. Mol Microbiol 52(5):1513–1525. https://doi.org/10.1111/j.1365-2958.2004.04105.x

Shultis WA, Weil EJ, Looker HC, Curtis JM, Shlossman M, Genco RJ, Knowler WC, Nelson RG (2007) Effect of periodontitis on overt nephropathy and end-stage renal disease in type 2 diabetes. Diabetes Care 30(2):306–311. https://doi.org/10.2337/dc06-1184

Silvestre-Rangil J, Silvestre FJ (2011) Clinico-therapeutic management of osteoradionecrosis: a literature review and update. Med Oral Patol Oral Cir Bucal 16(7):e900–e904

Sim S-J, Kim H-D, Moon J-Y, Zavras AI, Zdanowicz J, Jang S-J, Jin B-H, Bae K-H, Paik D-I, Douglass CW (2008) Periodontitis and the risk for non-fatal stroke in Korean adults. J Periodontol 79(9):1652–1658. https://doi.org/10.1902/jop.2008.080015

Slipczuk L, Codolosa JN, Davila CD, Romero-Corral A, Yun J, Pressman GS, Figueredo VM (2013) Infective endocarditis epidemiology over five decades: a systematic review. PLoS One 8(12):17. https://doi.org/10.1371/journal.pone.0082665

Sonalika WG, Amsavardani Tayaar S, Bhat KG, Patil BR, Muddapur MV (2012) Oral microbial carriage in oral squamous cell carcinoma patients at the time of diagnosis and during radiotherapy–a comparative study. Oral Oncol 48(9):881–886. https://doi.org/10.1016/j.oraloncology.2012.03.018

Spooner R, Yilmaz Ö (2012) Nucleoside-diphosphate-kinase: a pleiotropic effector in microbial colonization under interdisciplinary characterization. Microbes Infect 14(3):228–237. https://doi.org/10.1016/j.micinf.2011.10.002

Stein PS, Steffen MJ, Smith C, Jicha G, Ebersole JL, Abner E, Dawson D (2012) Serum antibodies to periodontal pathogens are a risk factor for Alzheimer's disease. Alzheimers Dement 8 (3):196–203. https://doi.org/10.1016/j.jalz.2011.04.006

Stepanov I, Jensen J, Hatsukami D, Hecht SS (2008) New and traditional smokeless tobacco: comparison of toxicant and carcinogen levels. Nicotine Tob Res 10(12):1773–1782. https://doi.org/10.1080/14622200802443544

Støre G, Eribe ERK, Olsen I (2005) DNA–DNA hybridization demonstrates multiple bacteria in osteoradionecrosis. Int J Oral and Maxillofac Surg 34(2):193–196. https://doi.org/10.1016/j.ijom.2004.06.010

Stringer AM, Logan RM (2015) The role of oral flora in the development of chemotherapy-induced oral mucositis. J Oral Pathol Med 44(2):81–87. https://doi.org/10.1111/jop.12152

Sun J, Kato I (2016) Gut microbiota, inflammation and colorectal cancer. Genes Dis 3(2):130–143. https://doi.org/10.1016/j.gendis.2016.03.004

Sztukowska MN, Ojo A, Ahmed S, Carenbauer AL, Wang Q, Shumway B, Jenkinson HF, Wang H, Darling DS, Lamont RJ (2016) *Porphyromonas gingivalis* initiates a mesenchymal-like transition through ZEB1 in gingival epithelial cells. Cell Microbiol 18 (6):844–858. https://doi.org/10.1111/cmi.12554

Takahashi N (2015) Oral microbiome metabolism: from "who are they?" to "what are they doing?". J Dent Res 94(12):1628–1637. https://doi.org/10.1177/0022034515606045

Tateda M, Shiga K, Saijo S, Sone M, Hori T, Yokoyama J, Matsuura K, Takasaka T, Miyagi T (2000) *Streptococcus anginosus* in head and neck squamous cell carcinoma: implication in carcinogenesis. Int J Mol Med 6(6):699–703

Tezal M, Grossi SG, Genco RJ (2005) Is periodontitis associated with oral neoplasms? J Periodontol 76(3):406–410. https://doi.org/10.1902/jop.2005.76.3.406

Tezal M, Sullivan MA, Hyland A, Marshall JR, Stoler D, Reid ME, Loree TR et al (2009) Chronic periodontitis and the incidence of head and neck squamous cell carcinoma. Cancer Epidemiol Biomarkers Prev 18(9):2406–2412. https://doi.org/10.1158/1055-9965.epi-09-0334

Tong HC, Gao XJ, Dong XZ (2003) Non-mutans streptococci in patients receiving radiotherapy in the head and neck area. Caries Res 37(4):261–266

Trent MS, Stead CM, Tran AX, Hankins JV (2006) Invited review: diversity of endotoxin and its impact on pathogenesis. J Endotoxin Res 12(4):205–223. https://doi.org/10.1177/09680519060120040201

Turunen SP, Kummu O, Harila K, Veneskoski M, Soliymani R, Baumann M, Pussinen PJ, Hörkkö S (2012) Recognition of *Porphyromonas gingivalis* gingipain epitopes by natural IgM binding to malondialdehyde modified low-density lipoprotein. PLoS One 7(4):e34910. https://doi.org/10.1371/journal.pone.0034910

Uitto VJ, Pan YM, Leung WK, Larjava H, Ellen RP, Finlay BB, Mcbride BC (1995) Cytopathic effects of *Treponema denticola* chymotrypsin-like proteinase on migrating and stratified epithelial cells. Infect Immun 63(9):3401–3410

Vargas J, Hernandez M, Silvestri C, Jiménez O, Guevara N, Carballo M, Rojas N et al (2006) Brain abscess due to *Arcanobacterium haemolyticum* after dental extraction. Clin Infect Dis 42 (12):1810–1811. https://doi.org/10.1086/504436

Vos T, Barber RM, Bell B, Bertozzi-Villa A, Biryukov S, Bolliger I, Charlson F et al (2015) Global, regional, and national incidence, prevalence, and years lived with disability for 301 acute and chronic diseases and injuries in 188 countries, 1990–2013: a systematic analysis for the global burden of disease study 2013. Lancet 386(9995):743–800. https://doi.org/10.1016/S0140-6736(15)60692-4

Wade WG (2013) The oral microbiome in health and disease. Pharmacol Res 69(1):137–143. https://doi.org/10.1016/j.phrs.2012.11.006

Wagner KW, Schoen R, Schumacher M, Schmelzeisen R, Schulze D (2006) Case report: brain and liver abscesses caused by oral infection with *Streptococcus intermedius*. Oral Surg Oral Med Oral Pathol Oral Radiol Endodontol 102(4):E21–E23. https://doi.org/10.1016/j.tripleo.2006.02.010

Warinner C, Rodrigues JFM, Vyas R, Trachsel C, Shved N, Grossmann J, Radini A et al (2014) Pathogens and host immunity in the ancient human oral cavity. Nat Genet 46(4):336–344. https://doi.org/10.1038/ng.2906

Warinner C, Speller C, Collins MJ (2015) A new era in palaeomicrobiology: prospects for ancient dental calculus as a long-term record of the human oral microbiome. Philos Trans R Soc Lond B Biol Sci 370(1660):20130376. https://doi.org/10.1098/rstb.2013.0376

Watanabe K, Cho YD (2014) Periodontal disease and metabolic syndrome: a qualitative critical review of their association. Arch Oral Biol 59(8):855–870. https://doi.org/10.1016/j.archoralbio.2014.05.003

Watts A, Crimmins E, Gatz M (2008) Inflammation as a potential mediator for the relationship between oral health and Alzheimer's disease. Gerontologist 48:38–38

Wayakanon K, Thornhill MH, Douglas CWI, Lewis AL, Warren NJ, Pinnock A, Armes SP, Battaglia G, Murdoch C (2013) Polymersome-mediated intracellular delivery of antibiotics to treat *Porphyromonas gingivalis*-infected oral epithelial cells. FASEB J 27(11):4455–4465. https://doi.org/10.1096/fj.12-225219

Wegner N, Wait R, Sroka A, Eick S, Nguyen K-A, Lundberg K, Kinloch A, Culshaw S, Potempa J, Venables PJ (2010) Peptidylarginine deiminase from *Porphyromonas gingivalis* citrullinates human fibrinogen and α-enolase: implications for autoimmunity in rheumatoid arthritis. Arthritis Rheum 62(9):2662–2672. https://doi.org/10.1002/art.27552

Whitmore SE, Lamont RJ (2014) Oral bacteria and cancer. PLoS Pathog 10(3):e1003933. https://doi.org/10.1371/journal.ppat.1003933

Wu J, Peters BA, Dominianni C, Zhang Y, Pei Z, Yang L, Ma Y et al (2016) Cigarette smoking and the oral microbiome in a large study of American adults. ISME J. https://doi.org/10.1038/ismej.2016.37

Xie H, Chung WO, Park Y, Lamont RJ (2000) Regulation of the *Porphyromonas gingivalis* fima (Fimbrillin) gene. Infect Immun 68(12):6574–6579. https://doi.org/10.1128/iai.68.12.6574-6579.2000

Xu M, Yamada M, Li M, Liu H, Chen SG, Han YW (2007) FadA from *Fusobacterium nucleatum* utilizes both secreted and nonsecreted forms for functional oligomerization for attachment and invasion of host cells. J Biol Chem 282(34):25000–25009. https://doi.org/10.1074/jbc.M611567200

Yamada T, Komoto J, Saiki K, Konishi K, Takusagawa F (2006) Variation of loop sequence alters stability of cytolethal distending toxin (CDT): crystal structure of cdt from actinobacillus actinomycetemcomitans. Protein Sci Publ Protein Soc 15(2):362–372. https://doi.org/10.1110/ps.051790506

Yamazaki K, Ohsawa Y, Itoh H, Ueki K, Tabeta K, Oda T, Nakajima T et al (2004) T-cell clonality to *Porphyromonas gingivalis* and human heat shock protein 60s in patients with atherosclerosis and periodontitis. Oral Microbiol Immunol 19(3):160–167. https://doi.org/10.1111/j.0902-0055.2004.00134.x

Yamazaki T, Miyamoto M, Yamada S, Okuda K, Ishihara K (2006) Surface protease of *Treponema denticola* hydrolyzes C3 and influences function of polymorphonuclear leukocytes. Microbes Infect 8(7):1758–1763. https://doi.org/10.1016/j.micinf.2006.02.013

Ye Y, Carlsson G, Barr Agholme M, Wilson JAL, Roos A, Henriques-Normark B, Engstrand L, Modéer T, Pütsep K (2013) Oral bacterial community dynamics in paediatric patients with malignancies in relation to chemotherapy-related oral mucositis: a prospective study. Clin Microbiol Infect 19(12):E559–E567. https://doi.org/10.1111/1469-0691.12287

Yilmaz Ö, Young PA, Lamont RJ, Kenny GE (2003) Gingival epithelial cell signalling and cytoskeletal responses to *Porphyromonas gingivalis* invasion. Microbiology 149 (9):2417–2426. https://doi.org/10.1099/mic.0.26483-0

Yilmaz O, Yao L, Maeda K, Rose TM, Lewis EL, Duman M, Lamont RJ, Ojcius DM (2008) Atp scavenging by the intracellular pathogen *Porphyromonas gingivalis* inhibits p2x7-mediated host-cell apoptosis. Cell Microbiol 10(4):863–875. https://doi.org/10.1111/j.1462-5822.2007.01089.x

Yoshimura F, Murakami Y, Nishikawa K, Hasegawa Y, Kawaminami S (2009) Surface components of *Porphyromonas gingivalis*. J Periodontal Res 44(1):1–12. https://doi.org/10.1111/j.1600-0765.2008.01135.x

Zaura E, Mira A (2015) Editorial: the oral microbiome in an ecological perspective. Front Cell Infect Microbiol 5:39. https://doi.org/10.3389/fcimb.2015.00039

Zaura E, Nicu EA, Krom BP, Keijser BJ (2014) Acquiring and maintaining a normal oral microbiome: current perspective. Front Cell Infect Microbiol 4:85. https://doi.org/10.3389/fcimb.2014.00085

Zeng X-T, Leng W-D, Lam Y-Y, Yan BP, Wei X-M, Weng H, Kwong JSW (2016) Periodontal disease and carotid atherosclerosis: a meta-analysis of 17,330 participants. Int J Cardiol 203:1044–1051. https://doi.org/10.1016/j.ijcard.2015.11.092

Zenobia C, Hajishengallis G (2015) *Porphyromonas gingivalis* virulence factors involved in subversion of leukocytes and microbial dysbiosis. Virulence 6(3):236–243. https://doi.org/10.1080/21505594.2014.999567

Zhang X, Zhang D, Jia H, Feng Q, Wang D, Liang D, Wu X et al (2015) The oral and gut microbiomes are perturbed in rheumatoid arthritis and partly normalized after treatment. Nat Med 21(8):895–905. https://doi.org/10.1038/nm.3914

Zheng W, Zhang Z, Liu C, Qiao Y, Zhou D, Qu J, An H, Xiong M, Zhu Z, Zhao X (2015) Metagenomic sequencing reveals altered metabolic pathways in the oral microbiota of sailors during a long sea voyage. Sci Rep 5:9131. https://doi.org/10.1038/srep09131

Zijnge V, Kieselbach T, Oscarsson J (2012) Proteomics of protein secretion by *Aggregatibacter actinomycetemcomitans*. PLoS One 7(7):e41662. https://doi.org/10.1371/journal.pone.0041662

Chapter 10
Candida albicans Commensalism and Human Diseases

Changbin Chen and Xinhua Huang

Abstract *Candida albicans* is by far the most prevalent commensal yeast and frequently causes opportunistic infections in humans. This polymorphic fungus resides as a lifelong, harmless commensal on mucosal surfaces of the oropharynx, gastrointestinal, and genitourinary tracts in healthy individuals, but causes a frequently fatal disseminated infection responsible for multiple forms of disease in immunocompromised patients, including oral, vaginal, dermal, and disseminated candidiasis. Although *C. albicans* colonization can be detected in humans as early as a few weeks after birth, our understanding of its commensal adaptation has arisen only recently from the most basic level. *C. albicans* commensalism requires a homeostatic interplay among fungus, resident microbiota, and host immunity, and disturbance of the balance can lead to pathogenicity of the yeast. This chapter describes a number of important factors contributing to the maintenance of this interaction and summarizes recent progresses regarding mechanisms underlying the regulation of *C. albicans* colonization in various host niches.

Keywords *Candida albicans* • Commensalism • Candidiasis • Microbiota • Host immunity

10.1 Introduction

The fungus *Candida albicans* was first discovered as the etiological agent of oral thrush in the first half of the nineteenth century by F.T. Berg. Since then, it has been isolated from different parts of the human body and a diverse variety of animals (Odds 1988; Jacobsen et al. 2008). So far, this dimorphic organism has been recognized as the most frequently isolated yeast in humans. One of the most

C. Chen (✉) • X. Huang
Unit of Pathogenic Fungal Infection and Host Immunity, CAS Key Laboratory of Molecular Virology and Immunology, Institut Pasteur of Shanghai, Chinese Academy of Sciences, Shanghai, China
e-mail: cbchen@ips.ac.cn

J. Sun, P.K. Dudeja (eds.), *Mechanisms Underlying Host-Microbiome Interactions in Pathophysiology of Human Diseases*, Physiology in Health and Disease,
https://doi.org/10.1007/978-1-4939-7534-1_10

intriguing features of *C. albicans* is its versatility: this fungus gained the ability to strive in a wide range of different niches following a long coevolution with the host (Calderone 2002). *C. albicans* is a part of the natural microbial microbiota and resides as a commensal of the oral cavity, vagina, and gastrointestinal (GI) tract in healthy humans (Znaidi et al. 2015; Zordan and Cormack 2012). For example, it can be detected in oropharyngeal microbiota in up to 50% of normal individuals and in the vagina in 25 to 30% of asymptomatic and healthy women. Furthermore, *C. albicans* was found to be a predominantly opportunistic fungal pathogen contributing to diverse infectious syndromes, such as oral and oropharyngeal candidiasis (OPC), vulvovaginal candidiasis (VVC), mucocutaneous candidiasis, and disseminated candidiasis. It is able to translocate through the mucosa and gain access to internal organs, causing a systemic disease and organ failure (Mavor et al. 2005). *C. albicans* ranks as the fourth most common infectious agent isolated from blood cultures, responsible for 60% of bloodstream-derived fungal infections, and has a mortality rate of approximately 40% (Kett et al. 2011). Interestingly, although the origin of systemic *C. albicans* infection remains controversial, a burgeoning body of evidence has shown that the GI tract is considered the most frequent place for *C. albicans* colonization in humans and may represent the major source of candidemia, given favorable conditions (Nucci and Anaissie 2001; Odds et al. 2006). In a mammalian host, *C. albicans* transits from being a commensal to a pathogen by adapting itself to diverse physiological niches, which differ with regard to a variety of environmental cues such as pH, temperature, nutrients, and stress challenges. Moreover, the yeast competes with other members of the microbiota and confronts the host immune defenses. Thus, in addition to contributions of virulence factors, pathogenicity of *C. albicans* also relies on a number of fitness attributes (Mayer et al. 2013). To date, numerous studies have been carried out to focus on the pathogenic state of *C. albicans*; our knowledge about its commensalism is comparatively limited, although accumulating evidence has suggested that the most effective strategy to reduce candidemia-associated mortalities might be to prevent infections from occurring. Recent progresses in addressing functional roles of commensalism factors indicate that maintaining *C. albicans* in its commensal form requires a tripartite interaction of fungus, resident microbiota, and host immunity, and involves a subtle balance between first, the host's immune defenses, bacterial microbiota, and other local environmental conditions, and second, the yeast's virulence factors. Disturbance of this balance was found to promote proliferation and colonization of *C. albicans* in different mucosal surfaces before invading the tissues (Odds 1988). A commensal stage of *C. albicans* involves regulation and adaptation to diverse host microenvironments and could be sustained by downregulating the pathogenic potential. Moreover, commensalism may represent a specific stage in the human host evolution that enables it to recognize and deal with a potential fungal invader.

In this chapter, we summarize the latest advances in our understanding of commensalism-related mechanisms, enabling *C. albicans* to colonize different host niches. Key commensalism attributes, including microbiota, mucus layer, antimicrobial peptides (AMPs), and gene regulatory circuits, are discussed and evaluated in terms of their roles in specific host niches including the oral cavity, reproductive tract, and GI tract.

10.2 *C. albicans* Commensalism in the Oral Cavity

Candida albicans is normally a commensal fungus, able to colonize saliva and oral mucosa, where it transits into a pathogen causing OPC or oral thrush (Zhu and Filler 2010; Naglik et al. 2011; Nucci and Anaissie 2001). A typical OPC symptom is displayed as geographically extensive thick white plaques formed on the tongue, buccal mucosa, soft palate, and pharynx. In healthy individuals, *C. albicans* is most commonly isolated from the tongue (the mid-line of the middle and posterior thirds), the cheek, or the palatal mucosa (Arendorf and Walker 1979, 1980; Borromeo et al. 1992). The individual's health is a major predisposing factor for *C. albicans* colonization in the oral cavity, supported by statistics showing that the mean rates of oral *Candida* carriage in healthy or hospitalized individuals are 17.7% and 40.6% respectively (Odds 1988). Normally, oral candidiasis is not life-threatening, but still leads to a significant level of morbidity because of chronic pain or discomfort upon mastication and limited nutrition intake in the elderly or immunodeficient patients. Moreover, OPC is one of the first clinical signs of HIV infection, and commonly detected in neonates, the elderly, patients with xerostomia (dry mouth), and individuals undergoing chemotherapy and radiotherapy for head–neck cancers (Fidel 2011). Severe infection in infants can even lead to malnutrition and a failure to thrive (Fidel 2011; Glocker and Grimbacher 2010). Given increased oral *Candida* infection and carriage titers in OPC patients, it is important to analyze the functional role of some key factors in promoting *C. albicans* colonization in this niche.

10.2.1 Adhesion

Candida albicans normally resides at low levels in the oral microbiota of humans. A successful colonization in the oral cavity requires a stable population of *C. albicans* to be maintained in this niche. The yeast cells have to evolve strategies for adherence to the mucus or epithelial surface to avoid removal by host clearance mechanisms in the oral cavity, a continuous-flow environment. First, *C. albicans* colonization or clearance can be determined from a balance between growth in the oral cavity and removal of cells from the mouth. Yeast cells may be washed out or removed by saliva, swallowing, and oral hygiene. Once the balance is disrupted, *C. albicans* can either be cleared or cause OPC. Second, a number of fungal surface adhesins have been identified by different experimental approaches and reagents, and attested to be crucial for the avid adherence of *C. albicans* to epithelial cells. Some of them, including members of the agglutinin-like sequence (ALS) family,

Hwp1, Eap1, have been well characterized (Zordan and Cormack 2012). Dysfunction of cell-wall regulation undoubtedly has a negative effect on adhesion properties. Moreover, using a *C. albicans* isolate known to be highly adherent to buccal epithelial cells (BECs), Imbert-Bernard et al. (1995) fractionated the yeast cell wall extracts for measuring adhesion activity and identified two proteins, with molecular masses of 38 and 54 kDa respectively, from the protein fraction Fr1, as potential candidates contributing to adherence mechanisms of *C. albicans* to human BECs. *C. albicans CaMNTI* gene encodes a mannosyltransferase involved in O-linked mannosylation and mutant cells lacking *CaMNTI* displayed significantly reduced adherence to BECs (Buurman et al. 1998). The secreted aspartyl proteinases (Saps) also implicated in *C. albicans* adherence to BECs and other substrates (Ghannoum and Abu Elteen 1986; Watts et al. 1998). Third, *C. albicans* adherence also involves plenty of ligands, including epithelial and bacterial cell-surface molecules, extracellular matrix proteins, and dental acrylic. Absorbed saliva molecules, including previously identified basic proline-rich proteins IB-6 (O'Sullivan et al. 1997) and Psi (Babu and Dabbous 1986), are receptors for *C. albicans* adhesion and play important roles in maintaining its commensal behavior. Moreover, the influence of carbohydrates on *C. albicans* adherence to BECs has been assessed through in-vitro analyses. Sugars such as glucose, galactose, sucrose, or mannose, significantly enhance adherence, whereas it is not the case for other carbohydrates such as xylose, ribose, fructose, maltose, lactose, or raffinose (Macura and Tondyra 1989). Finally, a biofilm community is formed in the oral cavity when *C. albicans* cells co-adhere with several species of oral bacteria, including *Streptococcus* spp. and *Actinomyces* spp. Adhesive interactions between yeast and bacteria, such as protein–protein interactions and lectin binding, have been proposed to contribute to *C. albicans* colonization in the oral cavity (Millsap et al. 1998). In-vitro assays have validated that carbohydrates (e.g., rhamnose, glucose, GlcNAc, and galactose) isolated from the *S. gordonii* cell wall act as receptors for *C. albicans* adherence (Holmes et al. 1995). Studies also found that inactivation of *S. gordonii* cell-surface polypeptide-encoding genes, including *CSHA*, *CSHB*, *SSPA*, and/or *SSPB*, results in significant reductions in adherence of *C. albicans* cells (Holmes et al. 1996), indicating co-adherence between the yeast and bacteria is multifactorial. However, in some cases bacteria may interfere with the adherence of *C. albicans* cells. For example, a biofilm of *S. gordonii* was found to reduce the adherence of *C. albicans* to polystyrene (Webb et al. 1995), arguing against a positive role of bacteria in *C. albicans* adherence.

Clearly, adherence of *C. albicans* cells to the oral cavity is controlled by a spectrum of adhesive factors operated through various mechanisms. However, contributions of these factors to colonization still need to be fully deciphered owing to environmental diversity of the oral niche and complex adhesive mechanisms of this yeast.

10.2.2 Acquisition

To maintain a high level of *C. albicans* population in the oral cavity, yeast cells must grow and multiply at a rate at least equal to that of clearance. However, the oral cavity is an environment with extremely limited carbon sources, possibly because of the consumption of a large number of bacteria in saliva (Samaranayake et al. 1986), and this reflects a fact that the growth rate of *C. albicans* in saliva is too low to be measured accurately. To overcome this, *C. albicans* has to manipulate its metabolic activities to facilitate assimilation of carbon and nitrogen from the oral cavity, and these activities have been shown to aid the growth and survival of yeast cells in this niche. Competition with other oral microorganisms for nutrients such as glucose significantly affects *C. albicans* growth rate (Ribeiro et al. 2016). Given that oral bacteria are present at most oral sites at concentrations much higher than *C. albicans*, *Candida* cells must compete with them for adhesion sites and nutrients. Antibiotics treatment, which reduces the number of oral bacteria, turns out to be a predisposing factor for *C. albicans* colonization in the oral cavity (Cannon et al. 1995). Moreover, a recent study suggested that a fungal species, *Pichia*, might exert an antagonistic effect on *C. albicans* colonization and the associated mechanisms include nutrient limitation, and modulation of growth and virulence factors (Mukherjee et al. 2014).

10.2.3 Anti-host Clearance Mechanisms

As stated previously, a successful colonization of *C. albicans* in the oral cavity depends on a balance between the rates of its growth and clearance. A major strategy that yeast cells utilize to influence this balance is to modulate host defenses (Cannon et al. 1995). Accumulating evidence has suggested that immune evasion and immunomodulatory strategies carried out by *C. albicans* might be important not only in defending against immune killing, but also in the maintenance of its commensal state. For example, complement regulator molecules, including fH, FHL-1, and C4bp, can be captured by *C. albicans* and directly bind to the fungal cell surface to block all three complement activating pathways (Zipfel et al. 2007). Lymphocytes appear to be important for modulating *C. albicans* oral colonization, because children with thymic athymic dysplasia and HIV-AIDs patients are more susceptible to oral candidiasis (Cleveland et al. 1968; McCarthy et al. 1991; Scully et al. 1994). Signaling pathways involving IL-17 and IL-17RA are crucial for controlling overgrowth and invasion of *C. albicans* in the oral cavity, supported by a study reporting that mice deficient in different components of the IL-17 pathway (IL-23, IL-17RA and RORγt) develop severe oral candidiasis (Conti et al. 2009). Mitogen-activated protein kinase (MAPK) signaling was also found to be important, as oral epithelial cells are able to discriminate between yeast and hypha via a bi-phasic MAPK response. Compared with the invasive hyphal form, a

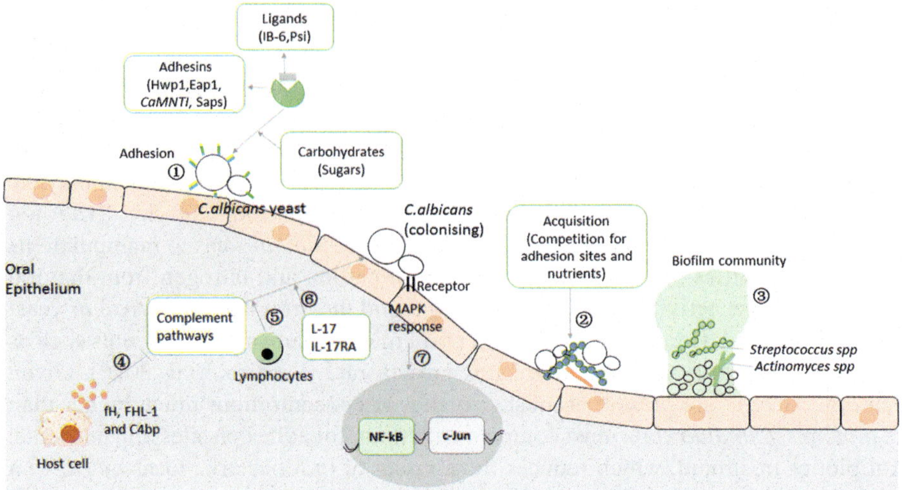

Fig. 10.1 Key factors involved in promoting colonization of *Candida albicans* in the oral cavity. *C. albicans* cells have to evolve strategies facilitating adherence to the mucus or epithelial surface of the oral cavity. The cartoon illustrates some of the important factors contributing to adhesion, acquisition, and anti-host clearance of the yeast, based on a number of publications. For example, factors involved in *1* adhesion (Zordan and Cormack 2012; Buurman et al. 1998; Watts et al. 1998; O'Sullivan et al. 1997); *2* acquisition (Macura and Tondyra 1989; Cannon et al. 1995; Mukherjee et al. 2014; Ribeiro et al. 2016); *3* biofilm (Millsap et al. 1998); *4* host complement (Zipfel et al. 2007); *5* host lymphocytes (Cleveland et al. 1968; McCarthy et al. 1991; Scully et al. 1994); *6* Th17 response (Conti et al. 2009); *7* activation of host MAPK pathways (Moyes et al. 2010; Gow and Hube 2012)

low burden of yeast-form cells favors colonization and can be recognized by an unknown pattern recognition receptor leading to a weak, early, and transient activation of MAPK response followed by induction of NF-κB and c-Jun (Moyes et al. 2010; Gow and Hube 2012). In summary, *C. albicans* has evolved to become a successful commensal on mucosal surfaces of the oral cavity, where its colonization requires actions of different key players of the fungal cell surface, the role of the microbiota, in addition to attributes important to avoiding or overcoming host defense mechanisms (Fig. 10.1).

10.3 *C. albicans* Commensalism in the Reproductive Tract

Fungi are recognized as important components of the vaginal ecosystem in healthy women, although they are vastly outnumbered by their bacterial counterparts in this niche. *C. albicans* colonizes the reproductive tract of 20% of women without causing any overt symptoms, yet it is one of the leading causes of infectious vaginitis. Although vaginal candidiasis is not lethal and rarely has the chance to develop systemic disease, this infection results in significant morbidity, affects a large number of patients, and shows tolerance to conventional therapy. Three quarters of women of reproductive age experience at least one episode of VVC during their lifetime (Sobel 1990; Sobel et al. 1998). *C. albicans* vaginal infection is most frequently detected in young women of childbearing age and 40–50% of women experience a recurrence. Compared with VVC, recurrent VVC (RVVC) is a much more serious clinical condition because of symptoms recurring four or more episodes per year and because of its refractoriness to successful treatment (Cassone 2015). Recent epidemiological studies have found that RVVC prevalence may be higher than previously estimated and that 7–8% of women who experienced the first episode may develop RVVC (Foxman et al. 1998). Clinical signs and symptoms of RVVC include intense pruritus, vaginal discharge, an erythematous vulva, and dyspareunia. Given a poor therapy outcome, recurrent VVC has a severe impact on women and their partners, both physically and psychologically (Mardh et al. 2002; Watson and Calabretto 2007; Aballéa et al. 2013), resulting in considerable suffering and cost, in addition to a markedly negative effect on sexual relations (Foxman et al. 2000).

Although it is far less abundant than bacteria in the vagina, *C. albicans* infection provides a pronounced effect on vaginal health. Despite therapeutic advances, VVC and RVVC remain common problems worldwide, affecting all strata of society. However, understanding of the mechanisms of *C. albicans* commensalism in the vaginal tract has developed slowly.

10.3.1 Estrogen Levels

Predisposition to VVC can be determined by exogenous and endogenous factors. Exogenous factors include the use of antibiotics or oral contraceptives, pregnancy, hormone replacement therapy, and uncontrolled diabetes mellitus (Sobel 1990; Sobel et al. 1998), whereas hormonal levels, such as those of estrogen and progesterone, constitute major endogenous factors (Kalo-Klein and Witkin 1989), and their importance in *C. albicans* vaginal colonization has been addressed. Clinical studies have shown that *C. albicans* is more frequently detected in the vaginal microbiota during pregnancy than at other times (Susic 1988). Women taking oral contraceptive pills are more prone to infection by *C. albicans* during the mid-cycle estrogen surge (Goplerud et al. 1976; Oriel et al. 1972). Besides clinical evidence, the contribution of estrogen to *C. albicans* vaginal infection was further confirmed in animal models, providing an extremely useful tool for identifying factors important in the regulation of susceptibility to *Candida* infection. *C. albicans* is normally absent from the vaginal microbiota in rats or mice and it is impossible to establish *Candida* infection in untreated animals. However, high doses of long-acting estradiol (E_2) promote persistent vaginal colonization of *Candida* in these animals (Cassone and Sobel 2016; Fidel et al. 2000). The reasons for estrogen-promoted *C. albicans* colonization in the vaginal tract have become somewhat clear. *C. albicans* harbors a specific estrogen-binding protein (EBP) and a cytosol receptor or binding system for estrogen has been documented (Skowronski and Feldman 1989). Of significance, vaginal colonization by an isolate of *C. albicans* containing a high-affinity EBP shows an 8.6-fold increase in response to in-vivo treatment with E_2 than with a comparable dose regimen of diethylstilbestrol, an efficacious mammalian estrogen receptor agonist, highlighting that mechanisms for estrogen-sensitive vaginal colonization of *C. albicans* include a functional ligand–EBP interaction within the yeast (Tarry et al. 2005). Moreover, in-vitro studies have shown that estrogen treatment highly induces expression of soluble heat-shock proteins in *C. albicans* and significantly promotes germ tube formation (Burt et al. 2003; White and Larsen 1997). Enhanced *C. albicans* adherence to vaginal epithelial cells was found through estrogen signaling, as a hormone-dependent glycogen production and accumulation from human vaginal epithelial cells significantly contribute to fungal colonization by providing a carbon source for *Candida* growth (Dennerstein and Ellis 2001; McCourtie and Douglas 1981).

10.3.2 Colonization Factors

Similar to its colonization in the oral cavity, *C. albicans* commensalism in the vaginal tract requires yeast adherence to vaginal epithelial cells. So far, only a few factors have been identified and characterized to play roles in regulating vaginal colonization with yeast. A previous study indicates that vaginal *C. albicans* colonization and tissue invasion can be significantly enhanced by germ-tube formation, as a nongerminating mutant fails to induce experimental VVC in vivo (Sobel et al. 1984). Consistent with this result, factors that influence *Candida* germination had a profound effect on symptomatic vaginitis. High-frequency heritable phenotypic switching was also thought be important for spontaneous in-vivo transformation from asymptomatic commensal colonization to symptomatic vaginitis, based on an observation that fresh vaginal *C. albicans* clinical strains isolated during acute vaginitis tend to have high frequency of switching (Soll 1988; Soll et al. 1989).

The observation that production of lactic acid and short-chain fatty acids by vaginal bacteria such as lactobacilli acidifies the vaginal environment suggests that pH regulation might have an impact on *C. albicans* colonization in this niche. Indeed, low pH favors the yeast form, but inhibits an invasive hyphal form of *C. albicans* (Han et al. 2011). Recent in-vitro and in-vivo studies have proposed a model that *C. albicans* co-opts amino acid metabolism to produce and secrete ammonia, which raises extracellular pH, triggers hyphal development, and escapes from immune killing by macrophages (Vylkova et al. 2011; Vylkova and Lorenz 2014). Interestingly, clinical studies showed that lactobacilli were frequently co-isolated with *C. albicans* in the vaginal epithelium of women with VVC and vaginal administration of different lactobacilli strains after conventional treatment can dramatically decrease recurrences of *Candida* vaginitis (Ehrstrom et al. 2010). This is presumably achieved by an improvement in the vaginal pH value, as displayed with inhibition of hyphal growth and repression of the genes involved in biofilm formation in *C. albicans* (De Seta et al. 2014).

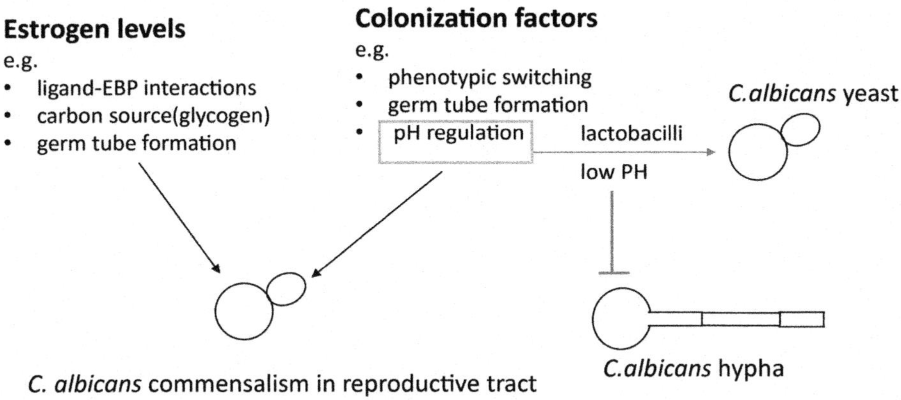

Fig. 10.2 *Candida albicans* colonization in the vaginal tract is determined by levels of estrogen and a list of characterized colonization factors derived from the fungus and its microbiota counterparts. *C. albicans* harbors a specific estrogen-binding protein (EBP) and evolves a cytosol receptor or binding system for estrogen (Skowronski and Feldman 1989). A functional ligand–EBP interaction significantly promotes estrogen-sensitive vaginal colonization of *C. albicans* (Tarry et al. 2005). Moreover, the production and accumulation of glycogen from human vaginal epithelial cells depend on activation of an estrogen pathway that significantly contributes to fungal colonization by providing a carbon source for *Candida* growth (Dennerstein and Ellis 2001; McCourtie and Douglas 1981). Finally, estrogen induces expression of soluble heat-shock proteins in *C. albicans* and significantly promotes germ tube formation, a strategy that also enhances its colonization (White and Larsen 1997; Burt et al. 2003). In addition to hormonal regulation, vaginal colonization with *C. albicans* was also affected by factors involved in high-frequency phenotypic switching (Soll 1988; Soll et al. 1989), germ-tube formation (Sobel et al. 1984), and pH regulation (Ehrstrom et al. 2010; Han et al. 2011; Vylkova et al. 2011; De Seta et al. 2014; Vylkova and Lorenz 2014)

Understanding the mechanisms of *C. albicans* commensalism and pathogenicity in the vaginal tract is essential for developing effective therapies against *Candida* vaginitis (Fig. 10.2). Animal models have validated the effectiveness of using anti-*Candida* vaccines and systemically administered antibodies in preventing vaginal candidiasis, but the efficacy of these treatments in human beings remains elusive. Improvement of prevention strategies requires a better knowledge of host anti-*Candida* defense mechanisms in the vagina, host genetic susceptibility, and fungal genetic factors that induce host immune responses and facilitate vaginal persistence.

10.4 *C. albicans* Commensalism in the GI Tract

Candida albicans colonization in the GI tract may cause disseminated infection and severe disease symptoms upon disruption of the host defense (Odds 1987; Nucci and Anaissie 2001). Even though commensalism was thought to be the default lifestyle of *C. albicans* in the GI tract, obvious disease symptoms told us that its pathogenic state often overshadows the commensal tendencies of this fungus. A clinical survey from patients who were diagnosed as positive in a blood culture for *Candida* noted that *C. albicans* isolates from blood were genetically indistinguishable from those isolated from the GI tract, suggesting that GI colonization might be a source of invasive *C. albicans* candidemia (Miranda et al. 2009). Moreover, studies from patients with familial Crohn's disease (Standaert-Vitse et al. 2009) and ulcerative colitis (Ksiadzyna et al. 2009; Zwolinska-Wcislo et al. 2009) further indicate a tight correlation between *C. albicans* colonization in the GI tract and inflammatory bowel disease. Patients with ulcerative colitis showed a reduction in clinical symptoms and a decrease in the size of inflammatory lesions after treatment with the anti-fungal drug fluconazole (Zwolinska-Wcislo et al. 2009). *C. albicans* can be frequently cultured from gastric biopsies and brush mucosal samples of patients with gastric ulcers, and the clinical severity of the ulcer was found to be dependent on the degree of *Candida* colonization (Zwolinska-Wcislo et al. 2006), implying that *Candida* GI colonization might be a determinant in triggering host inflammation and promoting a pathological condition in this niche. Therefore, understanding mechanisms responsible for the adaptation of *C. albicans* to GI commensalism should shed light on the multilevel design of new strategies for combating fungal diseases (Fig. 10.3).

10.4.1 C. albicans *Polymorphism*

Candida albicans is a polymorphic fungus that undergoes morphological transition among the yeast, pseudohyphal, and hyphal forms (Brown 2002). The ability of *C. albicans* to switch from yeast to filament, and vice versa, is a major virulence determinant of this organism (Berman and Sudbery 2002). Interestingly, recent studies have proposed that morphological alterations of *C. albicans* in response to environmental stimuli have been associated with both commensalism and pathogenesis (Neville et al. 2015). *C. albicans* colonization seems to favor a yeast form that tolerates host immunity and could be maintained at low numbers on the epithelial surface through various mechanisms that inhibit transition to a filamentous form (Neville et al. 2015). However, even though a form of yeast is the dominant morphology of this fungus in the GI tract, genetic analyses indicate that *C. albicans* GI colonization was found to be associated with elevated expression levels of hyphae-specific genes (HSGs), such as *EFH1*, *ECE1*, *RBT4*, and *RBT1*; even the cells are in yeast forms (Doedt et al. 2004; White et al. 2007; d'Enfert 2009; Rosenbach et al. 2010). These seemingly contradictory results suggest that upregulation of HSGs might facilitate the maintenance of *C. albicans* in yeast forms in the GI tract in a morphogenesis-independent manner. This hypothesis was further supported by an assay monitoring expression of Efg1, which was previously known to be a major regulator of filamentation. A substantial body of literature based on diverse lines of research has indicated a dual role of Efg1 in regulating the

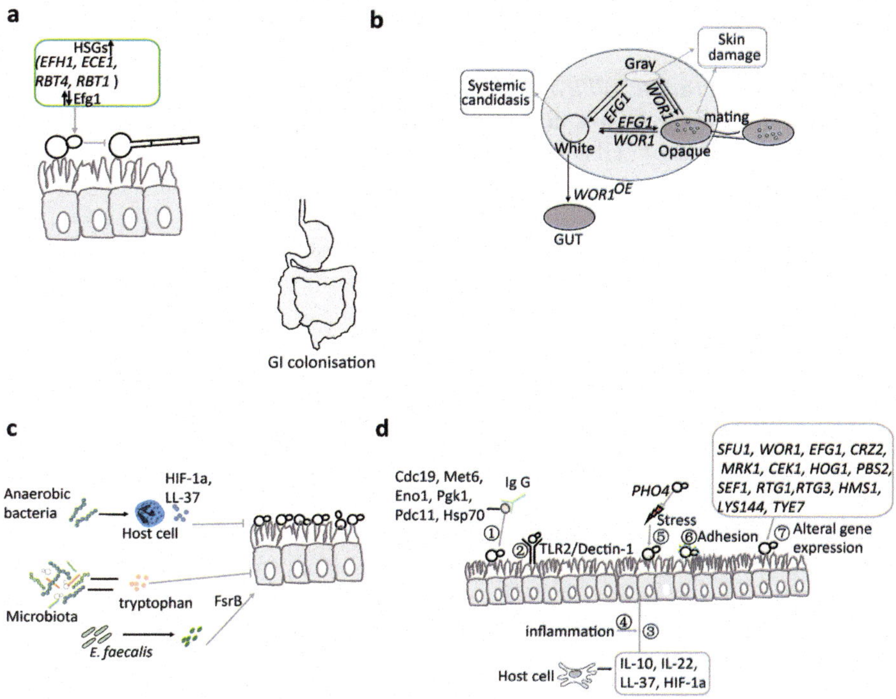

Fig. 10.3 Mechanisms responsible for the adaptation of *C. albicans* to commensalism in the gastrointestinal (GI) tract. Recently, the related mechanisms have been appreciated, in which the successful colonization of *C. albicans* in the GI tract depends on multiple factors, including (**a**) polymorphism (Doedt et al. 2004; White et al. 2007; d'Enfert 2009; Rosenbach et al. 2010), (**b**) phenotypic switch (Pande et al. 2013; Tao et al. 2014; Noble et al. 2016; Scaduto and Bennett 2015), (**c**) microbiota (Peleg et al. 2010; Zelante et al. 2013; Fan et al. 2015), (**d**) and interactions between fungus and host immune responses (Del Sero et al. 1999; Nucci and Anaissie 2001; Brown 2006; Jawhara et al. 2008; Mochon et al. 2010; Chen et al. 2011; Tsai et al. 2011; Gladiator et al. 2013; Pande et al. 2013; Perez et al. 2013; Fan et al. 2015; Trevijano-Contador et al. 2016; Urrialde et al. 2016). Note that *C. albicans* GI commensalism is associated with expression of tremendous numbers of genes from the host, microbiota and the fungus itself

pathogenic virulence of *C. albicans* and its commensal colonization in the GI tract (Kumamoto and Vinces 2005; d'Enfert 2009; Pierce et al. 2013; Pierce and Kumamoto 2012; Stoldt et al. 1997). Cell-to-cell variation in the levels of Efg1 expression in the GI tract may divide *C. albicans* cells into different subpopulations with multiple characteristics, enabling host-dependent shaping and diversity of the colonizing population (Pierce et al. 2013). Thus, a precise understanding of the attributes of the *C. albicans* polymorphism and their role in either commensalism or pathogenesis is very important.

10.4.2 C. albicans *Phenotypic Switch*

In its unicellular form, *C. albicans* may adopt distinct, nongenetically determined phenotypes to colonize in host niches, depending on environmental conditions. So far, at least four types of cells have been identified and characterized, named based on their colony appearance: white, opaque, gray, and gastrointestinally induced transition (GUT) (Tao et al. 2014; Pande et al. 2013; Noble et al. 2016). Cells of each type have distinct morphological phenotypes that are stable and inheritable. White cells exhibit a standard round-to-oval yeast morphology that can be easily distinguished from the other three more elongated yeast-like cell types. In-vitro assays have shown that these cell types display differences in a variety of cellular events such as transcriptional profiling, activities of secreted aspartyl proteases and mating competencies (Pande et al. 2013; Tao et al. 2014). In-vivo studies using mouse models also indicate that they have varying features in both pathogenesis and commensalism. Compared with the other three cell types, white cells were more virulent in a mouse model of systemic candidiasis, whereas gray and opaque cells appear to be important for skin damage (Tao et al. 2014), and GUT cells represent a unique phenotypic form specifically isolated from the GI tract (Pande et al. 2013).

Different cell types evolve to adapt to a variety of host niches and phenotypic switching in response to various environmental triggers have been verified to be associated with *C. albicans* commensalism (Gow et al. 2011). The white-to-opaque switch, firstly described by Soll in 1987 (cited in Slutsky et al. 1987) significantly affects *C. albicans* survival in the host by regulating mating competence, cell diversity, and host immune recognition. Opaque cells are essential for mating, rendering *C. albicans* competent in mating (Miller and Johnson 2002). Moreover, opaque cells have lost the ability to secrete important chemoattractants that can be specifically recognized by polymorphonuclear neutrophils and therefore successfully escape from recognition by immune cells under certain conditions (Geiger et al. 2004; Sasse et al. 2013). An in-vivo murine skin model suggests that compared with white cells, opaque cells are better colonizers at the mucosal surfaces of the skin (Kvaal et al. 1997, 1999; Geiger et al. 2004; Sasse et al. 2013), but appear to have a negative impact on mucosal commensalism because of its ability to damage skin tissues (Pande et al. 2013). However, white cells outcompete opaque cells in the GI tract of antibiotics-treated mice (Pande et al. 2013). In addition to the well-characterized white-to-opaque switch, recent studies add a novel "white–gray–opaque" tristable phenotypic switching system. Gray cells had a faster growth rate than either the white or the opaque cells in an ex-vivo murine tongue infection assay, which can be explained by a better adapted nutrient acquisition of this type of cells from host tissues (Tao et al. 2014). This tristable system seems to be a general feature of *Candida* species, as an in-vitro study in *Candida dubliniensis* showed that switching to a gray phenotype can be fostered by a combined treatment with *N*-acetylglucosamine (GlcNAc) and CO_2 (Yue et al. 2016; Pande et al. 2013). Taking into consideration the release of GlcNAc and CO_2

by commensal bacteria present in the GI tract, this study suggests that the "white–gray–opaque" tristable phenotypic switching system might facilitate *Candida* to compete with its bacterial counterparts and colonize the preferred biological niches. Furthermore, our unpublished in-vivo results demonstrate that gray cells are extremely unstable in the GI tract and rapidly switch to either white or opaque form, thus preventing the host from a necrotic cell death.

Passage through the murine GI tract triggers formation of another new *C. albicans* cell type, GUT, which primes this fungus for commensalism (Pande et al. 2013). A major difference between GUT and gray-type cells is the stability under various environmental conditions. GUT cells are identified only during in vivo GI colonization, whereas gray cells can be obtained in vitro and confirmed to be stable under a variety of culturing conditions (Pande et al. 2013; Tao et al. 2014). GUT is defined as the first specialized commensal cell type, adding more complexities in host–*Candida* interactions. Although colonizing *C. albicans* cells in the GI tract are the yeast form, "white" cell type (White et al. 2007; Pierce and Kumamoto 2012), the presence of the GUT cell type in this niche suggests that yeast-form cells may undergo phenotypic switching by altering their morphologies, transcriptomes, and metabolism, and unknown host factors may contribute to this transition. For example, GUT cells promote *C. albicans* commensalism in the GI tract through metabolic adaptation, as genes related to glucose catabolism and iron uptake were repressed whereas transcripts related to the catabolism of fatty acids and *N*-acetylglucosamine were activated. Importantly, transcription patterns operated by the GUT cells are consistent with its apparent optimization for GI colonization.

Thus, *C. albicans* is capable of undergoing multiple-stable phenotypic transitions under certain environmental conditions. The phenotypic switching increases fitness flexibility and genetic heterogeneity that greatly promotes the evolution of different phenotypic variants, and this strategy may be important for *C. albicans* to be better adapted to unfavorable conditions and support its survival in the host. A certain cell type expressed by *C. albicans* may reflect different functional specializations and developmental programs that have been optimized for commensalism in different niches (Soll 2002). In addition to those characterized cell types, we believe that more unidentified phenotypic variants of *C. albicans* may exist in certain host niches. A certain type of phenotypic variant represents a specialized growth form under environmental pressures and it is reasonable to hypothesize that phenotypic switching may be a general feature of natural *C. albicans* strains and the highly frequent and diverse phenotypic and morphological switching more likely contributes to survival of *C. albicans* within the host.

10.4.3 Other Factors

Candida albicans has no environmental reservoir and always resides in human individuals or other mammals (Hube 2004), meaning that its commensal state within the GI tract is controlled by a tripartite interaction involving the fungus, host immunity, and the microbiota (Neville et al. 2015). Importantly, the population size of *C. albicans* in this niche is maintained by multiple factors including normal microbiota, host physical barriers and immune system with which the yeast continuously or transiently interacts (Hoffmann et al. 1999; Mochon et al. 2010), in addition to changes in fungal gene expression when *C. albicans* colonizes as a commensal in the GI tract (Neville et al. 2015). Therefore, each of these factors, including microbiota, host immune defenses, host stresses, adhesion, and fungal gene expression are evaluated in terms of their contributions to *C. albicans* colonization in the GI tract.

10.4.3.1 Microbiota in the GI Tract

The gut microbial community (microbiota), which is composed of members of bacteria, fungi, archaea, viruses, and protozoans, plays a fundamental role in the well-being of its host. Microbiota constituents have been shown to exert specific functions by interacting with one another and with the host immune system and greatly influence disease development (Clemente et al. 2012). For example, gut microbiota is required for development and for homeostasis in human life because it actively participates in the regulation of a variety of key cellular events, including nutrient and drug metabolism, maintenance of mucosal integrity, immunomodulation and protection against pathogens. If the homeostasis is disrupted, inappropriate inflammation can result in host cell damage and/or autoimmunity. For example, dysbiosis of the intestinal microbiota has been largely related to chronic disease of the intestinal tract, including ulcerative colitis, Crohn's disease, celiac disease, and irritable bowel syndrome, in addition to more systemic diseases such as obesity and types 1 and 2 diabetes (Jandhyala et al. 2015).

The microbiota also has a major influence on the success of *C. albicans* as a commensal in the GI tract, where *C. albicans* co-exists and interacts with numerous bacteria (Lindsay et al. 2012). It has been appreciated that adult mice without any treatments normally showed resistance to GI colonization of *C. albicans*, given that the mouse is not a natural host of this fungus.

The associated mechanisms have been demonstrated recently. Anaerobic bacteria, e.g., *Bacteroides thetaiotaomicron* and *Bacteroides fragilis*, are highly enriched in the murine GI tract and mediate resistance to *C. albicans* colonization by activating the innate immune effector hypoxia-inducible factor (HIF)-1a and the AMP, LL-37 (Fan et al. 2015). Studies using a mouse commensal model revealed that endogenous tryptophan metabolites derived from the gut microbiota specifically trigger an IL-22-dependent immune response to tip the balance back toward homeostasis and thereby suppress *C. albicans* colonization in the GI tract (Zelante et al. 2013). Moreover, *C. albicans* is able to persistently colonize the GI tract in neonatal mice (Pope et al. 1979; Field et al. 1981), antibiotics-treated adult mice (White et al. 2007; Koh et al. 2008), and germ-free mice (Schofield et al. 2005), supporting the notion that the composition of resident bacteria in the gut microbiota is an important determinant of *C. albicans* colonization.

Resident probiotic bacteria directly inhibit fungal growth by blocking epithelial binding sites and competing for nutrients, and thus providing a balanced and protective immunity. *C. albicans* colonization was successfully established in the alimentary tract of germ-free mice, but superinfection of *Escherichia coli* imposed an antagonistic effect on *C. albicans* colonization (Nishikawa et al. 1969). The nonpathogenic yeast *Saccharomyces boulardii* was also found to antagonize colonization of *C. albicans* in the GI tract, as both in-vitro and in-vivo evidence has shown that large amounts of this nonpathogenic yeast significantly reduce the population of *C. albicans* in feces (Algin et al. 2005).

It needs to be clarified that the relationship between bacteria and *C. albicans* is not always deleterious. For example, commensal bacterium *E. faecalis* interacts with *C. albicans* and promotes a mutually beneficial association with the host, and the resulting homeostasis ensures a commensal lifestyle over a pathogenic one, as this commensal bacterium significantly inhibits the transition of *C. albicans* from being a commensal to a pathogen by secreting molecule(s) in an FsrB-dependent manner (Peleg et al. 2010). However, this beneficial commensal relationship could be disrupted by changing microbes or host factors that have been known to promote invasive infection (Mason et al. 2012a, b).

In summary, although we have obtained some knowledge of mechanisms by which commensal bacteria and other nonpathogenic yeasts diminish or prevent *C. albicans* colonization in the GI tract, there is still a long way to go before we obtain a comprehensive picture of how bacteria or host-derived factors function to promote or decrease fungal colonization. Moreover, prevention of invasive fungal disease in humans may be approached by interfering with *C. albicans* GI colonization, for example, by reducing the ability of *C. albicans* to compete for resources, supplementing probiotic bacteria that inhibit fungal growth, and/or boosting GI mucosal immune responses to reduce the fungal burden.

10.4.3.2 Host Immune Defenses

In the GI lumen of immunocompetent hosts, *C. albicans* is predominantly present as yeast-like cells in small numbers that do not cause epithelial damage, implying that host immunity must play a major role in preventing invasive infection of the yeast. Indeed, alterations of host defenses have been shifting *C. albicans* from commensal colonization to systemic dissemination and disease. It has been clear that the host immune system could selectively induce IgG antibodies to recognize those immunogenic proteins abundant in *C. albicans*, e.g., Eno1, Pgk1, Cdc19, Met6, Pdc11, and Hsp70, and to restrict fungal replication in the GI tract (Mochon et al. 2010). *C. albicans* GI colonization is modulated by a series of key host receptors, such as TLR2 and Dectin-1. Dectin-1 is a C-type lectin involved in recognition of beta-glucans enriched in the cell walls of fungi (Brown 2006; Brown et al. 2006; Graham et al. 2006; Heinsbroek et al. 2006; Ozment-Skelton et al. 2006; Pyz et al. 2006; Taylor et al. 2007). Oral inoculation of *C. albicans* in the conditional macrophage-specific Dectin-1-deficient mice results in greater

C. albicans loads in stomachs and pyloric ceca compared with wild-type mice, suggesting that activation of Dectin-1 signaling might negatively regulate *C. albicans* colonization of the GI tract (Carvalho et al. 2012). However, the role of Dectin-1 in *C. albicans* immune recognition remains controversial. Studies from Vautier et al. (2012) showed that Dectin-1 does not play a role in host immune responses to *C. albicans* carriage in the GI tract, which contradicts an earlier study in which Dectin-1 was found to play a positive role (Gales et al. 2010). It became somewhat clear that the opposite conclusions drawn from different groups are due to two facts. One is that genetic backgrounds of the mouse line of each group are different and have been shown to yield altered phenotypes in Dectin-1-deficient mice (Carvalho et al. 2012), and apparently may affect experimental results and reproducibility. Another possibility is that different mouse lines may trigger different innate and adaptive immune responses, and indeed, Dectin-1 deficiency produces different cytokine profiles upon *C. albicans* challenge in each of the different mouse lines (Carvalho et al. 2012). The implications of other specific cytokines for *C. albicans* GI colonization were also assessed. For example, IL-10-deficient mice are actually more resistant to *C. albicans* GI colonization (Del Sero et al. 1999). *Candida* colonization of the GI tract is associated with elevated levels of IL-17, a pro-inflammatory cytokine secreted by Th17 cells (Kumamoto 2011). Similar to IL-17, IL-22, which is produced by T cells and innate lymphoid cells, has been shown to be crucial for controlling colonizing numbers of *C. albicans* in the mouse GI tract. Consistently, altered IL-22 levels in humans correlate significantly with the progression of chronic mucocutaneous candidiasis (De Luca et al. 2010; Puel et al. 2010; Gladiator et al. 2013). In addition to secreted cytokines, host immune effectors activated by commensal bacteria were found to be crucial for the modulation of GI colonization by *C. albicans*. For example, the AMP, LL-37, produced by epithelial cells and by some phagocytes, belongs to the family of AMPs, which are small molecules previously known to mediate protection against infections (Lopez-Garcia et al. 2005; Tsai et al. 2011). The HIF-1a functions as an essential transcriptional regulator in controlling the expression of antimicrobial cathelicidin peptides such as LL-37 and therefore promoting mammalian innate defense (Peyssonnaux et al. 2005; Nizet and Johnson 2009). Commensal anaerobic bacteria, specifically the Bacteroides and clostridial Firmicutes clusters IV and XIVa, were found to confer resistance to *C. albicans* gut colonization by markedly promoting the expression of HIF-1a, a key transcriptional regulator of mammalian innate immunity that induces expression of the AMP, LL-37, with anti-*Candida* activity (Fan et al. 2015). These findings are highly suggestive that an intact gut microbiota, which helps to boost GI mucosal immune defense reduces GI colonization by *C. albicans*, may represent a feasible strategy for preventing *Candida* infection. Strikingly, *C. albicans* is able to hijack other inflammation-related mechanisms to enhance its ability to colonize the GI tract, as a recent finding revealed that mice pretreated with dextran-sulfate sodium, a chemical that damages the epithelial cells and causes inflammation, could establish sustained *C. albicans* colonization in the GI tract, whereas untreated mice still show colonization resistance to *C. albicans* (Jawhara et al. 2008).

Collectively, a complex network of host immune defenses is crucial for GI colonization of *C. albicans* in its commensal state, but this is dependent on the site and stage of infection, as well as host immune status. Modulating host immune effectors may also be a therapeutic strategy to prevent invasive candidiasis.

10.4.3.3 Environmental Stresses in the GI Tract

During its colonization as a commensal in the GI tract, *C. albicans* needs to cope with different environmental stresses including temperature, pH, nutrients, reactive oxygen species (ROS), microbiota counterparts, and host immunity. Most studies about host environmental stresses have been focusing on their contributions to *C. albicans* pathogenicity. In comparison, studies on their role in *C. albicans* GI commensalism remain in their infancy. In a search for uncharacterized transcription factors involved in stress responses, a genetic screening was conducted and the transcription factor Pho4 was identified as a key fungal mediator responding to host stresses (Urrialde et al. 2016). Mutant cells lacking *PHO4* were hypersensitive to osmotic and oxidative stresses and oral inoculation of this mutant exhibited significant attenuated GI colonization in a competitive commensalism mouse model. In this model, an equilibrated mix of fluorescently labeled strains (wt-GFP/pho4-dTOM2) was inoculated intragastrically in C57BL/6 mice and *C. albicans* gut colonization was determined by plating and counting CFUs from stools. Moreover, compared with wild-type cells, equally mixed *pho4* mutant cells displayed a reduced adherence to the intestinal mucosa surfaces in an ex-vivo competition assay. Successful colonization of commensal microbes such as *C. albicans* in the GI tract also indicate that these microbial organisms have evolved multiple evasion strategies to avoid removal from the niche that may trigger different kinds of cellular stresses upon immune recognition and/or host inflammatory responses.

10.4.3.4 Adhesion

Similar to its role in *C. albicans* colonization of the oral cavity, adhesion to host tissues is also important for commensal colonization of this fungus in the GI tract. Regardless of the site at which *C. albicans* colonization occurs, e.g., oral cavity, vagina, or GI tract, the process of adhesion depends on a number of well-characterized fungal and host factors, as discussed in numerous excellent reviews (Douglas 1985; Tronchin et al. 1991; Kennedy et al. 1992; Pendrak and Klotz 1995; Hostetter 1996; Fukazawa and Kagaya 1997; Sundstrom 1999, 2002; Calderone et al. 2000; Cotter and Kavanagh 2000; Verstrepen and Klis 2006; Chaffin 2008; Hiller et al. 2011; Liu and Filler 2011; Silva et al. 2011; de Groot et al. 2013; Fan et al. 2013; Cho et al. 2014; Modrzewska and Kurnatowski 2015; Moyes et al. 2015; Nobile and Johnson 2015; Polke et al. 2015; Hofs et al. 2016; Trevijano-Contador et al. 2016). For example, most *C. albicans* adhesins belong to fungal cell wall proteins such as the agglutinin-like sequence (Als) family, hyphal wall protein

1 (Hwp1), Sun41, cell surface hydrophobicity protein 1 (Csh1), Eap1, and hyphally regulated protein 1 (Hyr1), in addition to the family of secreted aspartyl protease (Saps). Among them, the *ALS* gene family encodes large cell surface glycoproteins that are important for *C. albicans* adhesion to host surfaces (Hoyer 2001; Hoyer and Cota 2016). Hwp1, a major *C. albicans* hypha-specific cell wall protein, functions as a substrate of mammalian transglutaminases and promotes the cross-link of the fungus to epithelial cells (Staab et al. 1999). The SUN family protein, Sun41, was reported to be a putative glycosidase and involved in *C. albicans* adhesion to host tissue (Hiller et al. 2007). Changes in Csh1 expression could influence a multitude of interactions with the host, including adhesion to various host cells (Hazen 1989). The gene *EAP1* was isolated as a putative cell wall adhesin and its expression significantly promoted *C. albicans* attachment to human epithelial cells (Li and Palecek 2003). The role of Hyr1 in adhesion is still under debate. Some studies noted that hyr1 is important in adherence (Sundstrom 1999); however, others questioned its irrelevance to the interaction with epithelial cells (Wächtler et al. 2011). Production of the secreted aspartyl proteinases (Saps) is associated with a number of putative colonization attributes of *C. albicans*, including adhesion (Naglik et al. 2003, 2004; Zhu and Filler 2010). In addition to fungal cell wall proteins, a number of transcription factors, kinases, heat shock proteins, and mediators, such as Czf1, Efg1, Tup1, Tpk1, Tpk2, Hgc1, Ras1, Rim101, Vps11, Ecm1, Cka2, Bcr1, Bud2, Rsr1, Irs4, Chs2, Scs7, Ubi4, Ume6, Tec1, and Gat2, have been characterized to contribute to adhesion as well (Modrzewska and Kurnatowski 2015). As for host factors related to *C. albicans* commensal colonization, it has become clear that the fungal adhesion receptors such as fibronectins and integrins, and features of epithelial cells, including morphology, cell type and differentiation phase, were found to influence adhesion. However, whether the presence or absence of specific adhesion molecules is more conducive to commensalism than pathogenesis requires further investigation.

10.4.3.5 Fungal Gene Expression

Transcription regulators act as central elements in regulating the gene expression network of any organisms. It is reasonable to speculate that *C. albicans* GI commensalism involves dynamic, but tightly regulated gene transcription programs. Studies have shown that the most important commensalism-associated transcriptional outputs include activation of processes involved in nutrition assimilation, such as carbon and nitrogen metabolism, morphogenesis, and stress tolerance. An iron-responsive transcription factor, Sfu1, acts as one of the main commensal factors, possibly through regulation of a unique tripartite iron utilization system of *C. albicans* (Chen et al. 2011). Deletion of Sfu1 confers susceptibility to toxic levels of iron in the gut and therefore significantly diminishes *C. albicans* colonization in the GI tract (Chen et al. 2011). In addition, activation of another transcription factor Wor1 was found to be essential for gut commensalism (Pande et al. 2013). Wor1 is the master regulator of the white-to-opaque switch (Morschhauser

2010). Cells lacking Wor1 were rapidly cleared from the mouse gut, whereas overexpression of *WOR1* enhanced GI colonization by *C. albicans*, and strikingly, promoted a phenotypic switch from the white cell type to a GI-specific new cell type, GUT (Pande et al. 2013; Noble et al. 2016). Efg1, another central regulator for *C. albicans* yeast-to-hypha transition, has also been identified as a major regulator of GI colonization in mice (Pierce and Kumamoto 2012). Even more importantly, mutants depleting Efg1 completely outcompete the wild type (White et al. 2007; Pande et al. 2013; Pierce et al. 2013) and expression levels of *EFG1* are greatly influenced by both host immune status and the time course of colonization (Pierce and Kumamoto 2012). In addition, a *Candida*-specific zinc finger transcription factor *CRZ2*, when overexpressed, significantly increases intestinal colonization by *C. albicans*; the inside mechanism may involve a substantial reprogramming of amino acid and metal/ion homeostasis (Znaidi et al. 2015). Given the complex nature of the GI tract, various signals triggered by the fungus, host immunity, microbiota, and/or microenvironments of different niches, should also be considered. Although the classical MAPK signaling pathways have been characterized to influence multiple aspects of fungal physiology, such as dimorphism, cell wall, oxidative stress and virulence (Monge et al. 2006; Roman et al. 2007), their role in regulating *C. albicans* GI fitness has recently been addressed. A recent in-vivo study using a murine commensal model indicated that all three classic MAPK pathways, mediated by the MAPK Mkc1, Cek1, and Hog1, are required to maintain *C. albicans* colonization in the GI tract (Nucci and Anaissie 2001). Interestingly, the Hog1 pathway appears to play a more critical role in establishing GI colonization of the yeast. Mutants lacking either Hog1 or Pbs2 showed defects in adhesion to the mucosa surfaces of the gut, susceptibility to bile salts, and rapid clearance from the gut in a competitive gut colonization assay. In comparison, colonization defects in mutants affecting the other two MAPKs can only be observed after a long course of inoculation, indicating that the contribution of these two kinases is much smaller than Hog1.

Transcriptional programs regulating the disparate behaviors of *C. albicans* as commensal versus pathogen are much more complicated than expected. Some regulators may act to execute individually, whereas others may operate by integrating their roles in both behaviors. For example, transcription regulators, including Efg1, Sef1, Rtg1, Rtg3, and Hms1, were found to be functional in both pathogenesis and commensalism (Chen et al. 2011; Pierce and Kumamoto 2012; Perez et al. 2013; Pierce et al. 2013). However, transcription factors such as Lys144 and Tye7 were characterized to regulate *C. albicans* colonization of the GI tract only (Perez et al. 2013).

10.4.3.6 Nutritional Immunity

For successful colonization in a specialized host niche, commensal microbes must acquire nutrients efficiently to be able to compete effectively. Regulation of nutrient acquisition is pivotal for *C. albicans* colonization in the mammalian gut, given changes in nutrient availability in the gut due to dietary intake. In-vivo screening for *C. albicans* mutants defective in GI colonization was carried out and 6 out of 77 transcription regulators were identified (Perez et al. 2013). Surprisingly, 4 of these 6 candidates, i.e., Rtg1, Rtg3, Tye7, and Lys144, were found to regulate expression of genes responsible for the acquisition and metabolism of nutrients, particularly carbon and nitrogen sources (Perez et al. 2013), implying the importance of locally available nutrients in GI colonization of the yeast.

Candida albicans is able to assimilate sugars and alternative carbon sources simultaneously (Niimi et al. 1988). It was shown that this type of metabolic flexibility enhances the ability of yeast to colonize a diverse range of complex niches in its mammalian host. Interestingly, most fermentative carbon sources such as glucose, fructose, and galactose, although routinely used in laboratory cell culture medium, are actually present at extremely low levels in many host niches. For example, glucose concentrations in the colon are thought to be vanishingly low (Childers et al. 2016), mainly because glucose derived either from hydrolysis of starch or from sucrose is rapidly absorbed in the small intestine by glucose transporters and any remaining amounts are easily taken up by microbes in the ileum/proximal colon (Barelle et al. 2006; Wright et al. 2011). *C. albicans* is exposed to complex mixtures of alternative carbon sources in the GI tract that include amino acids, fatty acids, and carboxylic acids, suggesting that alternative, nonfermentative carbon sources might contribute to *C. albicans* colonization in these glucose-poor niches. Indeed, our recent work indicates that mitochondrial complex I somehow regulates assimilation of alternative carbon sources by integrating its role in triggering an ROS-dependent, alternative carbon-specific signaling pathway and regulating hyphal growth and biofilm formation, and thereby influencing *C. albicans* colonization in the GI tract (Huang et al. 2017).

In addition, a heterogeneous range of complex plant-derived polysaccharides constitutes sources of great diversity in dietary fiber types and fermentation of fiber carbohydrates by intestinal bacteria greatly contributes to the maintenance of a favorable intestinal microbial balance (Fuller 1989; Saarela et al. 2002). As a commensal in the GI tract, *C. albicans* is certainly exposed to the fiber carbohydrates; however, contribution of dietary fiber to *C. albicans* gut colonization remains elusive. Dietary coconut oil was found to reduce GI colonization with *C. albicans*, as the number of *C. albicans* cells was much lower in mice fed with a coconut oil-rich diet than those fed with diets rich in beef tallow or soybean oil (Gunsalus et al. 2016). Moreover, consumption of a high-glucose diet leads to increased GI colonization of *C. albicans* in a neutropenic mouse model (Vargas et al. 1993). Intriguingly, clinical studies in healthy individuals did not find a positive correlation between subjects' normal dietary carbohydrate intake and GI colonization with *C. albicans* (Weig et al. 1999). Even doubling the daily carbohydrate intake still has no impact on GI colonization with *C. albicans*, suggesting that unlike animal models, dietary carbohydrates might have a minimal effect on GI colonization with *C. albicans* in humans.

Catabolite inactivation of the glyoxylate cycle was reported to reduce the fitness of *C. albicans* in the GI tract (Childers et al. 2016). Moreover, the relevance of metabolic adaptation to *C. albicans* colonization of the GI tract can be extended to other nutrients, including oxygen, phosphorus, or micronutrients. Iron is an essential nutrient for microorganisms, acting as a cofactor for many enzymatic reactions, but becoming extremely toxic at high levels because of the production of ROS. As expected, regulation of iron metabolism is vital for *C. albicans* gut commensalism (Chen et al. 2011; Chen and Noble 2012; Noble 2013). Moreover, further trace metals, such as zinc, manganese, and copper, are also essential for growth and survival of *C. albicans* in a commensal state. However, related regulators remain unidentified.

10.5 Conclusions

Here, we summarize recent advances in understanding *C. albicans* commensalism in host niches including the oral cavity, vagina, and GI tract. Apparently, successful colonization of *C. albicans* in these niches depends on both intrinsic factors (fungal gene regulation, cell morphology, adaptation, fungal burden) and extrinsic factors (competitive microbiome, diet, and host immune status), and requires a tripartite interaction involving the fungus, microbiota counterparts, and host immunity. In this chapter, we reviewed the influence of each of these factors on *C. albicans* commensalism and highlighted key events related to this process.

Nevertheless, some questions remain unsolved. Animal models have been most commonly used to study *C. albicans* commensalism; however, it should be noted that *C. albicans* does not normally colonize mice and all studies involving animal models require extra treatments such as antibiotics. Undoubtedly, this may cause a problem in that results obtained from animal studies may not truly reflect *C. albicans* behaviors in humans owing to altered host immune response and microbiota after treatment. Recent identification of the GUT cell type, which favors fitness in the mouse gut, also raises several biological questions. GUT cells appear to differ from the standard opaque cells based on in-vitro transcriptomic analyses, and this should be further validated by profiling an in-vivo expression pattern after inoculation of GUT cells in the GI tract. Moreover, the GUT cell type is only observed by overexpressing the transcription factor *WOR1* in vitro; the existence of this type of cell in vivo still needs to be explored further.

Genetic factors and potential environmental stimuli triggering transitions of different cell types should also be studied in more depth. Moreover, other morphological switches have been discovered, but it is not clear whether they are involved in adaptation to commensalism.

Understanding of the gene networks involved in *C. albicans* commensalism could provide therapeutic targets to prevent the transition from commensalism to pathogenesis. As such, investigations of expression and regulation of these signaling networks during in-vivo commensalism are worthwhile. So far, some forward genetics screens have been performed and have yielded a considerable insight into the gene regulons responsible for *C. albicans* commensalism in the GI tract, but transcript profiling in other niches, such as the vagina and oral cavity, should also be considered. In addition, most identified commensal factors are transcription factors. Downstream targets of these regulators need to be identified and the potential mechanisms by which they function have yet to be understood. Furthermore, other nontranscription factors should be assayed to obtain a more comprehensive picture of *C. albicans* commensalism.

It is worth noting that studying genetic programs contributing to *C. albicans* commensalism could potentially allow for development of novel therapeutic strategies that would reduce the risk of *C. albicans* infections. Although much progress has already been made toward this goal, more is needed, and research in this area is likely to continue into the future.

Commensalism behaves as a precursor to invasive disease. Studying host and microbial factors governing *C. albicans* colonization should shed light on our understanding of the pathogenesis of fungal disease and even the development of more effective therapies, as those will be highly dependent on how well we uncover mechanisms by which this microbe colonizes the human host. However, thus far, current studies in *C. albicans* have been heavily biased toward its pathogenesis. Further attention will be required to focus on the persistence, colonization, and commensalism of this fungus, as those behaviors may have been evolved over many hundreds of thousands of years of co-evolution within the human host.

References

Aballéa S, Guelfucci F, Wagner J, Khemiri A, Dietz JP, Sobel J, Toumi M (2013) Subjective health status and health-related quality of life among women with recurrent vulvovaginal candidosis (RVVC) in Europe and the USA. Health Qual Life Outcomes 11:169

Algin C, Sahin A, Kiraz N, Sahinturk V, Ihtiyar E (2005) Effectiveness of bombesin and *Saccharomyces boulardii* against the translocation of *Candida albicans* in the digestive tract in immunosuppressed rats. Surg Today 35(10):869–873. https://doi.org/10.1007/s00595-005-3049-9

Arendorf TM, Walker DM (1979) Oral candidal populations in health and disease. Br Dent J 147 (10):267–272

Arendorf TM, Walker DM (1980) The prevalence and intra-oral distribution of *Candida albicans* in man. Arch Oral Biol 25(1):1–10

Babu JP, Dabbous MK (1986) Interaction of salivary fibronectin with oral streptococci. J Dent Res 65(8):1094–1100

Barelle CJ, Priest CL, Maccallum DM, Gow NA, Odds FC, Brown AJ (2006) Niche-specific regulation of central metabolic pathways in a fungal pathogen. Cell Microbiol 8(6):961–971

Berman J, Sudbery PE (2002) *Candida albicans*: a molecular revolution built on lessons from budding yeast. Nat Rev Genet 3(12):918–930. https://doi.org/10.1038/nrg948

Borromeo GL, McCullough MJ, Reade PC (1992) Quantitation and morphotyping of *Candida albicans* from healthy mouths and from mouths affected by erythematous candidosis. J Med Vet Mycol 30(6):477–480

Brown AJ (2002) Morphogenetic signaling pathways in *Candida albicans*. In: Calderone RA (ed) Candida and candidiasis. ASM Press, Washington, DC, pp 95–106

Brown GD (2006) Dectin-1: a signalling non-TLR pattern-recognition receptor. Nat Rev Immunol 6(1):33–43. https://doi.org/10.1038/nri1745

Brown GD, Morin C, Lewandowsky S (2006) Evidence for time-based models of free recall. Psychon Bull Rev 13(4):717–723

Burt ET, Daly R, Hoganson D, Tsirulnikov Y, Essmann M, Larsen B (2003) Isolation and partial characterization of Hsp90 from *Candida albicans*. Ann Clin Lab Sci 33(1):86–93

Buurman ET, Westwater C, Hube B, Brown AJ, Odds FC, Gow NA (1998) Molecular analysis of CaMnt1p, a mannosyl transferase important for adhesion and virulence of *Candida albicans*. Proc Natl Acad Sci USA 95(13):7670–7675

Calderone R, Suzuki S, Cannon R, Cho T, Boyd D, Calera J, Chibana H, Herman D, Holmes A, Jeng HW, Kaminishi H, Matsumoto T, Mikami T, O'Sullivan JM, Sudoh M, Suzuki M, Nakashima Y, Tanaka T, Tompkins GR, Watanabe T (2000) *Candida albicans*: adherence, signaling and virulence. Med Mycol 38(Suppl 1):125–137

Calderone RA (2002) Candida and candidiasis. ASM Press, Washington, DC

Cannon RD, Holmes AR, Mason AB, Monk BC (1995) Oral candida: clearance, colonization, or candidiasis? J Dent Res 74(5):1152–1161

Carvalho A, Giovannini G, De Luca A, D'Angelo C, Casagrande A, Iannitti RG, Ricci G, Cunha C, Romani L (2012) Dectin-1 isoforms contribute to distinct Th1/Th17 cell activation in mucosal candidiasis. Cell Mol Immunol 9(3):276–286. https://doi.org/10.1038/cmi.2012.1

Cassone A (2015) Vulvovaginal *Candida albicans* infections: pathogenesis, immunity and vaccine prospects. BJOG 122(6):785–794. https://doi.org/10.1111/1471-0528.12994

Cassone A, Sobel JD (2016) Experimental models of vaginal candidiasis and their relevance to human candidiasis. Infect Immun 84(5):1255–1261. https://doi.org/10.1128/IAI.01544-15

Chaffin WL (2008) *Candida albicans* cell wall proteins. Microbiol Mol Biol Rev 72(3):495–544. https://doi.org/10.1128/MMBR.00032-07

Chen C, Noble SM (2012) Post-transcriptional regulation of the Sef1 transcription factor controls the virulence of *Candida albicans* in its mammalian host. PLoS Pathog 8(11):e1002956. https://doi.org/10.1371/journal.ppat.1002956

Chen C, Pande K, French SD, Tuch BB, Noble SM (2011) An iron homeostasis regulatory circuit with reciprocal roles in *Candida albicans* commensalism and pathogenesis. Cell Host Microbe 10(2):118–135. https://doi.org/10.1016/j.chom.2011.07.005

Childers DS, Raziunaite I, Mol Avelar G, Mackie J, Budge S, Stead D, Gow NA, Lenardon MD, Ballou ER, MacCallum DM, Brown AJ (2016) The rewiring of ubiquitination targets in a pathogenic yeast promotes metabolic flexibility, host colonization and virulence. PLoS Pathog 12(4):e1005566. https://doi.org/10.1371/journal.ppat.1005566

Cho M, Choi E, Kim JH, Kim H, Kim HM, Lee JI, Hwang KC, Kim HJ, Han G (2014) Lactam-based HDAC inhibitors for anticancer chemotherapy: restoration of RUNX3 by posttranslational modification and epigenetic control. ChemMedChem 9(3):649–656. https://doi.org/10.1002/cmdc.201300393

Clemente JC, Ursell LK, Parfrey LW, Knight R (2012) The impact of the gut microbiota on human health: an integrative view. Cell 148(6):1258–1270. https://doi.org/10.1016/j.cell.2012.01.035

Cleveland WW, Fogel BJ, Brown WT, Kay HE (1968) Foetal thymic transplant in a case of Digeorge's syndrome. Lancet 2(7580):1211–1214

Conti HR, Shen F, Nayyar N, Stocum E, Sun JN, Lindemann MJ, Ho AW, Hai JH, Yu JJ, Jung JW, Filler SG, Masso-Welch P, Edgerton M, Gaffen SL (2009) Th17 cells and IL-17 receptor signaling are essential for mucosal host defense against oral candidiasis. J Exp Med 206 (2):299–311. https://doi.org/10.1084/jem.20081463

Cotter G, Kavanagh K (2000) Adherence mechanisms of *Candida albicans*. Br J Biomed Sci 57 (3):241–249

D'Enfert C (2009) Hidden killers: persistence of opportunistic fungal pathogens in the human host. Curr Opin Microbiol 12(4):358–364. https://doi.org/10.1016/j.mib.2009.05.008

De Groot PW, Bader O, de Boer AD, Weig M, Chauhan N (2013) Adhesins in human fungal pathogens: glue with plenty of stick. Eukaryot Cell 12(4):470–481. https://doi.org/10.1128/EC.00364-12

De Luca A, Zelante T, D'Angelo C, Zagarella S, Fallarino F, Spreca A, Iannitti RG, Bonifazi P, Renauld JC, Bistoni F, Puccetti P, Romani L (2010) IL-22 defines a novel immune pathway of antifungal resistance. Mucosal Immunol 3(4):361–373. https://doi.org/10.1038/mi.2010.22

De Seta F, Parazzini F, De Leo R, Banco R, Maso GP, De Santo D, Sartore A, Stabile G, Inglese S, Tonon M, Restaino S (2014) Lactobacillus plantarum P17630 for preventing Candida vaginitis recurrence: a retrospective comparative study. Eur J Obstet Gynecol Reprod Biol 182:136–139. https://doi.org/10.1016/j.ejogrb.2014.09.018

Del Sero G, Mencacci A, Cenci E, d'Ostiani CF, Montagnoli C, Bacci A, Mosci P, Kopf M, Romani L (1999) Antifungal type 1 responses are upregulated in IL-10-deficient mice. Microbes Infect 1(14):1169–1180

Dennerstein GJ, Ellis DH (2001) Oestrogen, glycogen and vaginal candidiasis. Aust NZ J Obstet Gynaecol 41(3):326–328

Doedt T, Krishnamurthy S, Bockmuhl DP, Tebarth B, Stempel C, Russell CL, Brown AJ, Ernst JF (2004) APSES proteins regulate morphogenesis and metabolism in *Candida albicans*. Mol Biol Cell 15(7):3167–3180. https://doi.org/10.1091/10.1091/mbc.E03-11-0782

Douglas LJ (1985) Adhesion of pathogenic Candida species to host surfaces. Microbiol Sci 2 (8):243–247

Ehrstrom S, Daroczy K, Rylander E, Samuelsson C, Johannesson U, Anzen B, Pahlson C (2010) Lactic acid bacteria colonization and clinical outcome after probiotic supplementation in conventionally treated bacterial vaginosis and vulvovaginal candidiasis. Microbes Infect 12 (10):691–699. https://doi.org/10.1016/j.micinf.2010.04.010

Fan D, Coughlin LA, Neubauer MM, Kim J, Kim MS, Zhan X, Simms-Waldrip TR, Xie Y, Hooper LV, Koh AY (2015) Activation of HIF-1alpha and LL-37 by commensal bacteria inhibits *Candida albicans* colonization. Nat Med 21(7):808–814. https://doi.org/10.1038/nm.3871

Fan Y, He H, Dong Y, Pan H (2013) Hyphae-specific genes HGC1, ALS3, HWP1, and ECE1 and relevant signaling pathways in *Candida albicans*. Mycopathologia 176(5–6):329–335. https://doi.org/10.1007/s11046-013-9684-6

Fidel PL Jr (2011) Candida-host interactions in HIV disease: implications for oropharyngeal candidiasis. Adv Dent Res 23(1):45–49. https://doi.org/10.1177/0022034511399284

Fidel PL Jr, Cutright J, Steele C (2000) Effects of reproductive hormones on experimental vaginal candidiasis. Infect Immun 68(2):651–657

Field LH, Pope LM, Cole GT, Guentzel MN, Berry LJ (1981) Persistence and spread of *Candida albicans* after intragastric inoculation of infant mice. Infect Immun 31(2):783–791

Foxman B, Marsh JV, Gillespie B, Sobel JD (1998) Frequency and response to vaginal symptoms among white and African American women: results of a random digit dialing survey. J Womens Health 7(9):1167–1174

Foxman B, Barlow R, D'Arcy H, Gillespie B, Sobel JD (2000) Candida vaginitis: self-reported incidence and associated costs. Sex Transm Dis 27(4):230–235

Fukazawa Y, Kagaya K (1997) Molecular bases of adhesion of *Candida albicans*. J Med Vet Mycol 35(2):87–99

Fuller R (1989) Probiotics in man and animals. J Appl Bacteriol 66(5):365–378

Gales A, Conduche A, Bernad J, Lefevre L, Olagnier D, Beraud M, Martin-Blondel G, Linas MD, Auwerx J, Coste A, Pipy B (2010) PPARgamma controls Dectin-1 expression required for host antifungal defense against *Candida albicans*. PLoS Pathog 6(1):e1000714. https://doi.org/10.1371/journal.ppat.1000714

Geiger J, Wessels D, Lockhart SR, Soll DR (2004) Release of a potent polymorphonuclear leukocyte chemoattractant is regulated by white-opaque switching in *Candida albicans*. Infect Immun 72(2):667–677

Ghannoum M, Abu Elteen K (1986) Correlative relationship between proteinase production, adherence and pathogenicity of various strains of *Candida albicans*. J Med Vet Mycol 24 (5):407–413

Gladiator A, Wangler N, Trautwein-Weidner K, LeibundGut-Landmann S (2013) Cutting edge: IL-17-secreting innate lymphoid cells are essential for host defense against fungal infection. J Immunol 190(2):521–525. https://doi.org/10.4049/jimmunol.1202924

Glocker E, Grimbacher B (2010) Chronic mucocutaneous candidiasis and congenital susceptibility to Candida. Curr Opin Allergy Clin Immunol 10(6):542–550. https://doi.org/10.1097/ACI.0b013e32833fd74f

Goplerud CP, Ohm MJ, Galask RP (1976) Aerobic and anaerobic microbiota of the cervix during pregnancy and the puerperium. Am J Obstet Gynecol 126(7):858–868

Gow NA, Hube B (2012) Importance of the *Candida albicans* cell wall during commensalism and infection. Curr Opin Microbiol 15(4):406–412. https://doi.org/10.1016/j.mib.2012.04.005

Gow NA, van de Veerdonk FL, Brown AJ, Netea MG (2011) *Candida albicans* morphogenesis and host defence: discriminating invasion from colonization. Nat Rev Microbiol 10 (2):112–122. https://doi.org/10.1038/nrmicro2711

Graham LM, Tsoni SV, Willment JA, Williams DL, Taylor PR, Gordon S, Dennehy K, Brown GD (2006) Soluble Dectin-1 as a tool to detect beta-glucans. J Immunol Methods 314 (1–2):164–169. https://doi.org/10.1016/j.jim.2006.05.013

Gunsalus KT, Tornberg-Belanger SN, Matthan NR, Lichtenstein AH, Kumamoto CA (2016) Manipulation of host diet to reduce gastrointestinal colonization by the opportunistic pathogen *Candida albicans*. mSphere 1(1). https://doi.org/10.1128/mSphere.00020-15

Han TL, Cannon RD, Villas-Boas SG (2011) The metabolic basis of *Candida albicans* morphogenesis and quorum sensing. Fungal Genet Biol 48(8):747–763. https://doi.org/10.1016/j.fgb.2011.04.002

Hazen KC (1989) Participation of yeast cell surface hydrophobicity in adherence of *Candida albicans* to human epithelial cells. Infect Immun 57(7):1894–1900

Heinsbroek SE, Taylor PR, Rosas M, Willment JA, Williams DL, Gordon S, Brown GD (2006) Expression of functionally different dectin-1 isoforms by murine macrophages. J Immunol 176 (9):5513–5518

Hiller E, Heine S, Brunner H, Rupp S (2007) *Candida albicans* Sun41p, a putative glycosidase, is involved in morphogenesis, cell wall biogenesis, and biofilm formation. Eukaryot Cell 6 (11):2056–2065

Hiller E, Zavrel M, Hauser N, Sohn K, Burger-Kentischer A, Lemuth K, Rupp S (2011) Adaptation, adhesion and invasion during interaction of *Candida albicans* with the host—focus on the function of cell wall proteins. Int J Med Microbiol 301(5):384–389. https://doi.org/10.1016/j.ijmm.2011.04.004

Hoffmann JA, Kafatos FC, Janeway CA, Ezekowitz RA (1999) Phylogenetic perspectives in innate immunity. Science 284(5418):1313–1318

Hofs S, Mogavero S, Hube B (2016) Interaction of *Candida albicans* with host cells: virulence factors, host defense, escape strategies, and the microbiota. J Microbiol 54(3):149–169. https://doi.org/10.1007/s12275-016-5514-0

Holmes AR, Gopal PK, Jenkinson HF (1995) Adherence of *Candida albicans* to a cell surface polysaccharide receptor on *Streptococcus gordonii*. Infect Immun 63(5):1827–1834

Holmes AR, McNab R, Jenkinson HF (1996) *Candida albicans* binding to the oral bacterium *Streptococcus gordonii* involves multiple adhesin-receptor interactions. Infect Immun 64 (11):4680–4685

Hostetter MK (1996) Adhesion and morphogenesis in *Candida albicans*. Pediatr Res 39(4 Pt 1):569–573. https://doi.org/10.1203/00006450-199604000-00001

Hoyer LL (2001) The ALS gene family of *Candida albicans*. Trends Microbiol 9(4):176–180

Hoyer LL, Cota E (2016) *Candida albicans* Agglutinin-like sequence (Als) family vignettes: a review of Als protein structure and function. Front Microbiol 7:280. https://doi.org/10.3389/fmicb.2016.00280. eCollection 02016

Huang X, Chen X, He Y, Yu X, Li S, Gao N, Niu L, Mao Y, Wang Y, Wu X, Wu W, Wu J, Zhou D, Zhan X, Chen C (2017) Mitochondrial complex I bridges a connection between regulation of carbon flexibility and gastrointestinal commensalism in the human fungal pathogen *Candida albicans*. PLoS Pathog 13(6):e1006414

Hube B (2004) From commensal to pathogen: stage- and tissue-specific gene expression of *Candida albicans*. Curr Opin Microbiol 7(4):336–341. https://doi.org/10.1016/j.mib.2004.06.003

Imbert-Bernard C, Valentin A, Mallie M, Bastide JM (1995) Involvement of *Candida albicans* cell wall proteins in the adherence of blastospores to human buccal epithelial cells. Exp Mycol 19 (4):247–253

Jacobsen MD, Duncan AD, Bain J, Johnson EM, Naglik JR, Shaw DJ, Gow NA, Odds FC (2008) Mixed *Candida albicans* strain populations in colonized and infected mucosal tissues. FEMS Yeast Res 8(8):1334–1338. https://doi.org/10.1111/j.1567-1364.2008.00438.x

Jandhyala SM, Talukdar R, Subramanyam C, Vuyyuru H, Sasikala M, Nageshwar Reddy D (2015) Role of the normal gut microbiota. World J Gastroenterol 21(29):8787–8803. https://doi.org/10.3748/wjg.v21.i29.8787

Jawhara S, Thuru X, Standaert-Vitse A, Jouault T, Mordon S, Sendid B, Desreumaux P, Poulain D (2008) Colonization of mice by *Candida albicans* is promoted by chemically induced colitis and augments inflammatory responses through galectin-3. J Infect Dis 197(7):972–980. https://doi.org/10.1086/528990

Kalo-Klein A, Witkin SS (1989) *Candida albicans*: cellular immune system interactions during different stages of the menstrual cycle. Am J Obstet Gynecol 161(5):1132–1136

Kennedy MJ, Calderone RA, Cutler JE, Kanabe T, Riesselman MH, Robert R, Senet JM, Annaix V, Bouali A, Mahaza C et al (1992) Molecular basis of *Candida albicans* adhesion. J Med Vet Mycol 30(Suppl 1):95–122

Kett DH, Azoulay E, Echeverria PM, Vincent JL, Extended prevalence of infection in ICUSGoI (2011) Candida bloodstream infections in intensive care units: analysis of the extended prevalence of infection in intensive care unit study. Crit Care Med 39(4):665–670. https://doi.org/10.1097/CCM.0b013e318206c1ca

Koh AY, Kohler JR, Coggshall KT, Van Rooijen N, Pier GB (2008) Mucosal damage and neutropenia are required for *Candida albicans* dissemination. PLoS Pathog 4(2):e35. https://doi.org/10.1371/journal.ppat.0040035

Ksiadzyna D, Semianow-Wejchert J, Nawrot U, Wlodarczyk K, Paradowski L (2009) Serum concentration of interleukin 10, anti-mannan Candida antibodies and the fungal colonization of the gastrointestinal tract in patients with ulcerative colitis. Adv Med Sci 54(2):170–176. https://doi.org/10.2478/v10039-009-0023-6

Kumamoto CA (2011) Inflammation and gastrointestinal Candida colonization. Curr Opin Microbiol 14(4):386–391

Kumamoto CA, Vinces MD (2005) Contributions of hyphae and hypha-co-regulated genes to *Candida albicans* virulence. Cell Microbiol 7(11):1546–1554. https://doi.org/10.1111/j.1462-5822.2005.00616.x

Kvaal C, Lachke SA, Srikantha T, Daniels K, McCoy J, Soll DR (1999) Misexpression of the opaque-phase-specific gene PEP1 (SAP1) in the white phase of *Candida albicans* confers increased virulence in a mouse model of cutaneous infection. Infect Immun 67(12):6652–6662

Kvaal CA, Srikantha T, Soll DR (1997) Misexpression of the white-phase-specific gene WH11 in the opaque phase of *Candida albicans* affects switching and virulence. Infect Immun 65 (11):4468–4475

Li F, Palecek S (2003) EAP1, a *Candida albicans* gene involved in binding human epithelial cells. Eukaryot Cell 2(6):1266–1273

Lindsay AK, Deveau A, Piispanen AE, Hogan DA (2012) Farnesol and cyclic AMP signaling effects on the hypha-to-yeast transition in *Candida albicans*. Eukaryot Cell 11(10):1219–1225. https://doi.org/10.1128/EC.00144-12

Liu Y, Filler SG (2011) *Candida albicans* Als3, a multifunctional adhesin and invasin. Eukaryot Cell 10(2):168–173. https://doi.org/10.1128/EC.00279-10

Lopez-Garcia B, Lee PH, Yamasaki K, Gallo RL (2005) Anti-fungal activity of cathelicidins and their potential role in *Candida albicans* skin infection. J Invest Dermatol 125(1):108–115. https://doi.org/10.1111/j.0022-202X.2005.23713.x

Macura AB, Tondyra E (1989) Influence of some carbohydrates and concanavalin A on the adherence of *Candida albicans* in vitro to buccal epithelial cells. Zentralbl Bakteriol 272 (2):196–201

Mardh PA, Rodrigues AG, Genc M, Novikova N, Martinez-de-Oliveira J, Guaschino S (2002) Facts and myths on recurrent vulvovaginal candidosis—a review on epidemiology, clinical manifestations, diagnosis, pathogenesis and therapy. Int J STD AIDS 13(8):522–539. https://doi.org/10.1258/095646202760159639

Mason KL, Erb Downward JR, Falkowski NR, Young VB, Kao JY, Huffnagle GB (2012a) Interplay between the gastric bacterial microbiota and *Candida albicans* during postantibiotic recolonization and gastritis. Infect Immun 80(1):150–158. https://doi.org/10.1128/IAI.05162-11

Mason KL, Erb Downward JR, Mason KD, Falkowski NR, Eaton KA, Kao JY, Young VB, Huffnagle GB (2012b) *Candida albicans* and bacterial microbiota interactions in the cecum during recolonization following broad-spectrum antibiotic therapy. Infect Immun 80 (10):3371–3380. https://doi.org/10.1128/IAI.00449-12

Mavor AL, Thewes S, Hube B (2005) Systemic fungal infections caused by Candida species: epidemiology, infection process and virulence attributes. Curr Drug Targets 6(8):863–874

Mayer FL, Wilson D, Hube B (2013) *Candida albicans* pathogenicity mechanisms. Virulence 4 (2):119–128. https://doi.org/10.4161/viru.22913

McCarthy GM, Mackie ID, Koval J, Sandhu HS, Daley TD (1991) Factors associated with increased frequency of HIV-related oral candidiasis. J Oral Pathol Med 20(7):332–336

McCourtie J, Douglas LJ (1981) Relationship between cell surface composition of *Candida albicans* and adherence to acrylic after growth on different carbon sources. Infect Immun 32 (3):1234–1241

Miller MG, Johnson AD (2002) White-opaque switching in *Candida albicans* is controlled by mating-type locus homeodomain proteins and allows efficient mating. Cell 110(3):293–302

Millsap KW, van der Mei HC, Bos R, Busscher HJ (1998) Adhesive interactions between medically important yeasts and bacteria. FEMS Microbiol Rev 21(4):321–336

Miranda LN, van der Heijden IM, Costa SF, Sousa AP, Sienra RA, Gobara S, Santos CR, Lobo RD, Pessoa VP Jr, Levin AS (2009) Candida colonisation as a source for candidaemia. J Hosp Infect 72(1):9–16. https://doi.org/10.1016/j.jhin.2009.02.009

Mochon AB, Jin Y, Kayala MA, Wingard JR, Clancy CJ, Nguyen MH, Felgner P, Baldi P, Liu H (2010) Serological profiling of a *Candida albicans* protein microarray reveals permanent host-pathogen interplay and stage-specific responses during candidemia. PLoS Pathog 6(3): e1000827. https://doi.org/10.1371/journal.ppat.1000827

Modrzewska B, Kurnatowski P (2015) Adherence of Candida sp. to host tissues and cells as one of its pathogenicity features. Ann Parasitol 61(1):3–9

Monge RA, Roman E, Nombela C, Pla J (2006) The MAP kinase signal transduction network in *Candida albicans*. Microbiology 152(Pt 4):905–912. https://doi.org/10.1099/mic.0.28616-0

Morschhauser J (2010) Regulation of white-opaque switching in *Candida albicans*. Med Microbiol Immunol 199(3):165–172. https://doi.org/10.1007/s00430-010-0147-0

Moyes DL, Runglall M, Murciano C, Shen C, Nayar D, Thavaraj S, Kohli A, Islam A, Mora-Montes H, Challacombe SJ, Naglik JR (2010) A biphasic innate immune MAPK response discriminates between the yeast and hyphal forms of *Candida albicans* in epithelial cells. Cell Host Microbe 8(3):225–235. https://doi.org/10.1016/j.chom.2010.08.002

Moyes DL, Richardson JP, Naglik JR (2015) *Candida albicans*-epithelial interactions and pathogenicity mechanisms: scratching the surface. Virulence 6(4):338–346. https://doi.org/10.1080/21505594.2015.1012981

Mukherjee PK, Chandra J, Retuerto M, Sikaroodi M, Brown RE, Jurevic R, Salata RA, Lederman MM, Gillevet PM, Ghannoum MA (2014) Oral mycobiome analysis of HIV-infected patients:

identification of Pichia as an antagonist of opportunistic fungi. PLoS Pathog 10(3):e1003996. https://doi.org/10.1371/journal.ppat.1003996
Naglik J, Albrecht A, Bader O, Hube B (2004) *Candida albicans* proteinases and host/pathogen interactions. Cell Microbiol 6(10):915–926
Naglik JR, Challacombe S, Hube B (2003) *Candida albicans* secreted aspartyl proteinases in virulence and pathogenesis. Microbiol Mol Biol Rev 67(3):400–428
Naglik JR, Moyes DL, Wachtler B, Hube B (2011) *Candida albicans* interactions with epithelial cells and mucosal immunity. Microbes Infect 13(12–13):963–976. https://doi.org/10.1016/j.micinf.2011.06.009
Neville BA, d'Enfert C, Bougnoux ME (2015) *Candida albicans* commensalism in the gastrointestinal tract. FEMS Yeast Res 15(7). https://doi.org/10.1093/femsyr/fov081
Niimi M, Kamiyama A, Tokunaga M (1988) Respiration of medically important Candida species and *Saccharomyces cerevisiae* in relation to glucose effect. J Med Vet Mycol 26(3):195–198
Nishikawa T, Hatano H, Ohnishi N, Sasaki S, Nomura T (1969) Establishment of *Candida albicans* in the alimentary tract of the germ-free mice and antagonism with *Escherichia coli* after oral inoculation. Jpn J Microbiol 13(3):263–276
Nizet V, Johnson RS (2009) Interdependence of hypoxic and innate immune responses. Nat Rev Immunol 9(9):609–617. https://doi.org/10.1038/nri2607
Nobile CJ, Johnson AD (2015) *Candida albicans* biofilms and human disease. Annu Rev Microbiol 69:71–92. https://doi.org/10.1146/annurev-micro-091014-104330
Noble SM (2013) *Candida albicans* specializations for iron homeostasis: from commensalism to virulence. Curr Opin Microbiol 16(6):708–715. https://doi.org/10.1016/j.mib.2013.09.006
Noble SM, Gianetti BA, Witchley JN (2016) *Candida albicans* cell-type switching and functional plasticity in the mammalian host. Nat Rev Microbiol. https://doi.org/10.1038/nrmicro.2016.157
Nucci M, Anaissie E (2001) Revisiting the source of candidemia: skin or gut? Clin Infect Dis 33 (12):1959–1967. https://doi.org/10.1086/323759
O'Sullivan JM, Cannon RD, Sullivan PA, Jenkinson HF (1997) Identification of salivary basic proline-rich proteins as receptors for *Candida albicans* adhesion. Microbiology 143 (Pt 2):341–348. https://doi.org/10.1099/00221287-143-2-341
Odds FC (1987) Candida infections: an overview. Crit Rev Microbiol 15(1):1–5. https://doi.org/10.3109/10408418709104444
Odds FC (1988) Candida and candidosis: a review and bibliography. Bailliere Tindall, London
Odds FC, Davidson AD, Jacobsen MD, Tavanti A, Whyte JA, Kibbler CC, Ellis DH, Maiden MC, Shaw DJ, Gow NA (2006) *Candida albicans* strain maintenance, replacement, and microvariation demonstrated by multilocus sequence typing. J Clin Microbiol 44 (10):3647–3658. https://doi.org/10.1128/JCM.00934-06
Oriel JD, Partridge BM, Denny MJ, Coleman JC (1972) Genital yeast infections. Br Med J 4 (5843):761–764
Ozment-Skelton TR, Goldman MP, Gordon S, Brown GD, Williams DL (2006) Prolonged reduction of leukocyte membrane-associated Dectin-1 levels following beta-glucan administration. J Pharmacol Exp Ther 318(2):540–546. https://doi.org/10.1124/jpet.106.102293
Pande K, Chen C, Noble SM (2013) Passage through the mammalian gut triggers a phenotypic switch that promotes *Candida albicans* commensalism. Nat Genet 45(9):1088–1091. https://doi.org/10.1038/ng.2710
Peleg AY, Hogan DA, Mylonakis E (2010) Medically important bacterial-fungal interactions. Nat Rev Microbiol 8(5):340–349. https://doi.org/10.1038/nrmicro2313
Pendrak ML, Klotz SA (1995) Adherence of *Candida albicans* to host cells. FEMS Microbiol Lett 129(2–3):103–113
Perez JC, Kumamoto CA, Johnson AD (2013) *Candida albicans* commensalism and pathogenicity are intertwined traits directed by a tightly knit transcriptional regulatory circuit. PLoS Biol 11 (3):e1001510. https://doi.org/10.1371/journal.pbio.1001510

Peyssonnaux C, Datta V, Cramer T, Doedens A, Theodorakis EA, Gallo RL, Hurtado-Ziola N, Nizet V, Johnson RS (2005) HIF-1alpha expression regulates the bactericidal capacity of phagocytes. J Clin Invest 115(7):1806–1815. https://doi.org/10.1172/JCI23865

Pierce JV, Kumamoto CA (2012) Variation in *Candida albicans* EFG1 expression enables host-dependent changes in colonizing fungal populations. MBio 3(4):e00117–e00112. https://doi.org/10.1128/mBio.00117-12

Pierce JV, Dignard D, Whiteway M, Kumamoto CA (2013) Normal adaptation of *Candida albicans* to the murine gastrointestinal tract requires Efg1p-dependent regulation of metabolic and host defense genes. Eukaryot Cell 12(1):37–49. https://doi.org/10.1128/EC.00236-12

Polke M, Hube B, Jacobsen ID (2015) Candida survival strategies. Adv Appl Microbiol 91:139–235. https://doi.org/10.1016/bs.aambs.2014.12.002

Pope LM, Cole GT, Guentzel MN, Berry LJ (1979) Systemic and gastrointestinal candidiasis of infant mice after intragastric challenge. Infect Immun 25(2):702–707

Puel A, Doffinger R, Natividad A, Chrabieh M, Barcenas-Morales G, Picard C, Cobat A, Ouachee-Chardin M, Toulon A, Bustamante J, Al-Muhsen S, Al-Owain M, Arkwright PD, Costigan C, McConnell V, Cant AJ, Abinun M, Polak M, Bougneres PF, Kumararatne D, Marodi L, Nahum A, Roifman C, Blanche S, Fischer A, Bodemer C, Abel L, Lilic D, Casanova JL (2010) Autoantibodies against IL-17A, IL-17F, and IL-22 in patients with chronic mucocutaneous candidiasis and autoimmune polyendocrine syndrome type I. J Exp Med 207 (2):291–297. https://doi.org/10.1084/jem.20091983

Pyz E, Marshall AS, Gordon S, Brown GD (2006) C-type lectin-like receptors on myeloid cells. Ann Med 38(4):242–251. https://doi.org/10.1080/07853890600608985

Ribeiro FC, de Barros PP, Rossoni RD, Junqueira JC, Jorge AO (2016) Lactobacillus rhamnosus inhibits *Candida albicans* virulence factors in vitro and modulates immune system in *Galleria mellonella*. J Appl Microbiol. https://doi.org/10.1111/jam.13324

Roman E, Arana DM, Nombela C, Alonso-Monge R, Pla J (2007) MAP kinase pathways as regulators of fungal virulence. Trends Microbiol 15(4):181–190. https://doi.org/10.1016/j.tim.2007.02.001

Rosenbach A, Dignard D, Pierce JV, Whiteway M, Kumamoto CA (2010) Adaptations of *Candida albicans* for growth in the mammalian intestinal tract. Eukaryot Cell 9(7):1075–1086. https://doi.org/10.1128/EC.00034-10

Saarela M, Lähteenmäki L, Crittenden R, Salminen S, Mattila-Sandholm T (2002) Gut bacteria and health foods—the European perspective. Int J Food Microbiol 78(1–2):99–117

Samaranayake LP, Hughes A, Weetman DA, MacFarlane TW (1986) Growth and acid production of Candida species in human saliva supplemented with glucose. J Oral Pathol 15(5):251–254

Sasse C, Hasenberg M, Weyler M, Gunzer M, Morschhauser J (2013) White-opaque switching of *Candida albicans* allows immune evasion in an environment-dependent fashion. Eukaryot Cell 12(1):50–58. https://doi.org/10.1128/EC.00266-12

Scaduto CM, Bennett R (2015) *Candida albicans* the chameleon: transitions and interactions between multiple phenotypic states confer phenotypic plasticity. Curr Opin Microbiol 26:102–108

Schofield DA, Westwater C, Balish E (2005) Divergent chemokine, cytokine and beta-defensin responses to gastric candidiasis in immunocompetent C57BL/6 and BALB/c mice. J Med Microbiol 54(Pt 1):87–92. https://doi.org/10.1099/jmm.0.45755-0

Scully C, el-Kabir M, Samaranayake LP (1994) Candida and oral candidosis: a review. Crit Rev Oral Biol Med 5(2):125–157

Silva S, Negri M, Henriques M, Oliveira R, Williams DW, Azeredo J (2011) Adherence and biofilm formation of non-*Candida albicans* Candida species. Trends Microbiol 19(5):241–247. https://doi.org/10.1016/j.tim.2011.02.003

Skowronski R, Feldman D (1989) Characterization of an estrogen-binding protein in the yeast *Candida albicans*. Endocrinology 124(4):1965–1972. https://doi.org/10.1210/endo-124-4-1965

Slutsky B, Staebell M, Anderson J, Risen L, Pfaller M, Soll DR (1987) "White-opaque transition": a second high-frequency switching system in *Candida albicans*. J Bacteriol 169(1):189–197
Sobel JD (1990) Vaginal infections in adult women. Med Clin North Am 74(6):1573–1602
Sobel JD, Muller G, Buckley HR (1984) Critical role of germ tube formation in the pathogenesis of candidal vaginitis. Infect Immun 44(3):576–580
Sobel JD, Faro S, Force RW, Foxman B, Ledger WJ, Nyirjesy PR, Reed BD, Summers PR (1998) Vulvovaginal candidiasis: epidemiologic, diagnostic, and therapeutic considerations. Am J Obstet Gynecol 178(2):203–211
Soll DR (1988) High-frequency switching in *Candida albicans* and its relations to vaginal candidiasis. Am J Obstet Gynecol 158(4):997–1001
Soll DR (2002) Candida commensalism and virulence: the evolution of phenotypic plasticity. Acta Trop 81(2):101–110
Soll DR, Galask R, Isley S, Rao TV, Stone D, Hicks J, Schmid J, Mac K, Hanna C (1989) Switching of *Candida albicans* during successive episodes of recurrent vaginitis. J Clin Microbiol 27(4):681–690
Staab JF, Bradway S, Fidel PL, Sundstrom P (1999) Adhesive and mammalian transglutaminase substrate properties of *Candida albicans* Hwp1. Science 283(5407):1535–1538
Standaert-Vitse A, Sendid B, Joossens M, Francois N, Vandewalle-El Khoury P, Branche J, Van Kruiningen H, Jouault T, Rutgeerts P, Gower-Rousseau C, Libersa C, Neut C, Broly F, Chamaillard M, Vermeire S, Poulain D, Colombel JF (2009) *Candida albicans* colonization and ASCA in familial Crohn's disease. Am J Gastroenterol 104(7):1745–1753. https://doi.org/10.1038/ajg.2009.225
Stoldt VR, Sonneborn A, Leuker CE, Ernst JF (1997) Efg1p, an essential regulator of morphogenesis of the human pathogen *Candida albicans*, is a member of a conserved class of bHLH proteins regulating morphogenetic processes in fungi. EMBO J 16(8):1982–1991. https://doi.org/10.1093/emboj/16.8.1982
Sundstrom P (1999) Adhesins in *Candida albicans*. Curr Opin Microbiol 2(4):353–357. https://doi.org/10.1016/S1369-5274(99)80062-9
Sundstrom P (2002) Adhesion in Candida spp. Cell Microbiol 4(8):461–469
Susic D (1988) The role of the renal medulla in blood pressure control. Am J Med Sci 295 (4):234–240
Tao L, Du H, Guan G, Dai Y, Nobile CJ, Liang W, Cao C, Zhang Q, Zhong J, Huang G (2014) Discovery of a "white-gray-opaque" tristable phenotypic switching system in *Candida albicans*: roles of non-genetic diversity in host adaptation. PLoS Biol 12(4):e1001830. https://doi.org/10.1371/journal.pbio.1001830
Tarry W, Fisher M, Shen S, Mawhinney M (2005) *Candida albicans*: the estrogen target for vaginal colonization. J Surg Res 129(2):278–282. https://doi.org/10.1016/j.jss.2005.05.019
Taylor PR, Tsoni SV, Willment JA, Dennehy KM, Rosas M, Findon H, Haynes K, Steele C, Botto M, Gordon S, Brown GD (2007) Dectin-1 is required for beta-glucan recognition and control of fungal infection. Nat Immunol 8(1):31–38. https://doi.org/10.1038/ni1408
Trevijano-Contador N, Rueda C, Zaragoza O (2016) Fungal morphogenetic changes inside the mammalian host. Semin Cell Dev Biol 57:100–109. https://doi.org/10.1016/j.semcdb.2016.04.008
Tronchin G, Bouchara JP, Annaix V, Robert R, Senet JM (1991) Fungal cell adhesion molecules in *Candida albicans*. Eur J Epidemiol 7(1):23–33
Tsai PW, Yang CY, Chang HT, Lan CY (2011) Human antimicrobial peptide LL-37 inhibits adhesion of *Candida albicans* by interacting with yeast cell-wall carbohydrates. PLoS One 6 (3):e17755. https://doi.org/10.1371/journal.pone.0017755
Urrialde V, Prieto D, Pla J, Alonso-Monge R (2016) The *Candida albicans* Pho4 transcription factor mediates susceptibility to stress and influences fitness in a mouse commensalism model. Front Microbiol 7:1062. https://doi.org/10.3389/fmicb.2016.01062

Vargas SL, Patrick CC, Ayers GD, Hughes WT (1993) Modulating effect of dietary carbohydrate supplementation on *Candida albicans* colonization and invasion in a neutropenic mouse model. Infect Immun 61(2):619–626

Vautier S, Drummond RA, Redelinghuys P, Murray GI, MacCallum DM, Brown GD (2012) Dectin-1 is not required for controlling *Candida albicans* colonization of the gastrointestinal tract. Infect Immun 80(12):4216–4222. https://doi.org/10.1128/IAI.00559-12

Verstrepen KJ, Klis FM (2006) Flocculation, adhesion and biofilm formation in yeasts. Mol Microbiol 60(1):5–15. https://doi.org/10.1111/j.1365-2958.2006.05072.x

Vylkova S, Lorenz MC (2014) Modulation of phagosomal pH by *Candida albicans* promotes hyphal morphogenesis and requires Stp2p, a regulator of amino acid transport. PLoS Pathog 10 (3):e1003995. https://doi.org/10.1371/journal.ppat.1003995

Vylkova S, Carman AJ, Danhof HA, Collette JR, Zhou H, Lorenz MC (2011) The fungal pathogen *Candida albicans* autoinduces hyphal morphogenesis by raising extracellular pH. MBio 2(3): e00055–e00011. https://doi.org/10.1128/mBio.00055-11

Wächtler B, Wilson D, Haedicke K, Dalle F, Hube B (2011) From attachment to damage: defined genes of *Candida albicans* mediate adhesion, invasion and damage during interaction with oral epithelial cells. PLoS One 6(2):e17046. 10.11371/journal.pone.0017046

Watson C, Calabretto H (2007) Comprehensive review of conventional and non-conventional methods of management of recurrent vulvovaginal candidiasis. Aust NZ J Obstet Gynaecol 47 (4):262–272

Watts HJ, Cheah FS, Hube B, Sanglard D, Gow NA (1998) Altered adherence in strains of *Candida albicans* harbouring null mutations in secreted aspartic proteinase genes. FEMS Microbiol Lett 159(1):129–135

Webb BC, Willcox MD, Thomas CJ, Harty DW, Knox KW (1995) The effect of sodium hypochlorite on potential pathogenic traits of *Candida albicans* and other Candida species. Oral Microbiol Immunol 10(6):334–341

Weig M, Werner E, Frosch M, Kasper H (1999) Limited effect of refined carbohydrate dietary supplementation on colonization of the gastrointestinal tract of healthy subjects by *Candida albicans*. Am J Clin Nutr 69(6):1170–1173

White S, Larsen B (1997) *Candida albicans* morphogenesis is influenced by estrogen. Cell Mol Life Sci 53(9):744–749

White SJ, Rosenbach A, Lephart P, Nguyen D, Benjamin A, Tzipori S, Whiteway M, Mecsas J, Kumamoto CA (2007) Self-regulation of *Candida albicans* population size during GI colonization. PLoS Pathog 3(12):e184. https://doi.org/10.1371/journal.ppat.0030184

Wright EM, Loo D, Hirayama BA (2011) Biology of human sodium glucose transporters. Physiol Rev 91(2):733–794

Yue H, Hu J, Guan G, Tao L, Du H, Li H, Huang G (2016) Discovery of the gray phenotype and white-gray-opaque tristable phenotypic transitions in *Candida dubliniensis*. Virulence 7 (3):230–242. https://doi.org/10.1080/21505594.2015.1135287

Zelante T, Iannitti RG, Cunha C, De Luca A, Giovannini G, Pieraccini G, Zecchi R, D'Angelo C, Massi-Benedetti C, Fallarino F, Carvalho A, Puccetti P, Romani L (2013) Tryptophan catabolites from microbiota engage aryl hydrocarbon receptor and balance mucosal reactivity via interleukin-22. Immunity 39(2):372–385. https://doi.org/10.1016/j.immuni.2013.08.003

Zhu W, Filler SG (2010) Interactions of *Candida albicans* with epithelial cells. Cell Microbiol 12 (3):273–282. https://doi.org/10.1111/j.1462-5822.2009.01412.x

Zipfel PF, Wurzner R, Skerka C (2007) Complement evasion of pathogens: common strategies are shared by diverse organisms. Mol Immunol 44(16):3850–3857. https://doi.org/10.1016/j.molimm.2007.06.149

Znaidi S, Cervantes AH, Desseyn J-L, Sertour N, Vincent F, Gouyer V, Dalle F, Jouault T, Bougnoux M-É, d'Enfert C (2015) A highly parallel conditional overexpression screen, in vivo, identifies a new regulator of *Candida albicans* commensalism. J Mycol Méd/J Med Mycol 25(2):e102

Zordan R, Cormack B (2012) Adhesins on opportunistic fungal pathogens. In: Calderone RA, Clancy CJ (eds) Candida and candidiasis. ASM Press, Washington, DC, pp 243–259

Zwolinska-Wcislo M, Brzozowski T, Mach T, Budak A, Trojanowska D, Konturek PC, Pajdo R, Drozdowicz D, Kwiecien S (2006) Are probiotics effective in the treatment of fungal colonization of the gastrointestinal tract? Experimental and clinical studies. J Physiol Pharmacol 57 (Suppl 9):35–49

Zwolinska-Wcislo M, Brzozowski T, Budak A, Kwiecien S, Sliwowski Z, Drozdowicz D, Trojanowska D, Rudnicka-Sosin L, Mach T, Konturek SJ, Pawlik WW (2009) Effect of Candida colonization on human ulcerative colitis and the healing of inflammatory changes of the colon in the experimental model of colitis ulcerosa. J Physiol Pharmacol 60(1):107–118

Chapter 11
Fecal Microbiota Transplants: Current Knowledge and Future Directions

Roxana Chis, Prameet M. Sheth, and Elaine O. Petrof

Abstract The human gastrointestinal tract (GI) is home to an exceptionally diverse bacterial community, collectively referred to as the gut microbiota. These microbes play a pivotal role in modulating host physiology and immune response, in both gastrointestinal health and disease. There is evidence that alterations leading to an imbalance of the constituents of the gut microbiota, or dysbiosis, may contribute to several intestinal and extra-intestinal pathological states. These discoveries have led to exciting new microbe-based therapeutic developments, including reconstitution of bacterial communities, to reverse and correct dysbiosis. For instance, fecal microbiota transplantation (FMT), the process of infusing fecal matter from a healthy donor into a sick patient to restore a healthy microbiome in the recipient, has been used to treat recurrent *Clostridium difficile* infection (rCDI) and may hold promise for other gastrointestinal conditions. This chapter focuses on the key aspects of FMT, including methodology, physician and patient attitudes, safety and regulation, and its therapeutic potential for the treatment of rCDI and other gastrointestinal conditions, including inflammatory bowel disease, obesity, irritable bowel syndrome, and colorectal cancer.

R. Chis
Department of Internal Medicine, Kingston General Hospital, Queen's University, Kingston, ON, Canada

P.M. Sheth
Department of Pathology and Molecular Medicine, Kingston General Hospital, Queen's University, Kingston, ON, Canada

E.O. Petrof (✉)
Division of Infectious Diseases/GI Diseases Research Unit, Department of Medicine, Kingston General Hospital, Queen's University, Kingston, ON, Canada
e-mail: eop@queensu.ca

J. Sun, P.K. Dudeja (eds.), *Mechanisms Underlying Host-Microbiome Interactions in Pathophysiology of Human Diseases*, Physiology in Health and Disease,
https://doi.org/10.1007/978-1-4939-7534-1_11

11.1 Introduction

11.1.1 Human Gut Microbiome

The human gut microbiota, a community of bacteria that inhabit the human gastrointestinal (GI) tract, is made up of over 1000 bacterial species (Qin et al. 2010). This microbial ecosystem harbours over 100-fold more unique genes than our own genome and is increasingly being recognized as an essential determinant of host health including vitamin production, energy extraction, immune modulation, and protection against infection (Lozupone et al. 2012). Gut microbial composition varies significantly between individuals, with variations attributed to numerous biotic and abiotic factors, including medication use, health status, lifestyle, genetics, diet, age, and sex (Falony et al. 2016). However, despite such variation, there seem to be general trends that are conserved among healthy adults, with Bacteroidetes and Firmicutes being the most prevalent bacterial phyla, followed by Actinobacteria and Proteobacteria (Eckburg et al. 2005). These trends toward conservation of keystone organisms in the gut imply that the microbiome might play a role in physiological gut function. As such, alterations of this "normal" gut microbiota composition have important implications for health, often being associated with negative functional outcomes.

11.1.2 Gut Dysbiosis

Gut dysbiosis is defined as a state of imbalance in the gut microbial populations resulting from alterations in the composition or function of the microbial community (Frank et al. 2011).

Perhaps the most dramatic alteration of the gut microbiota occurs following the use of antibiotic therapy, leading to decreased overall microbial diversity (Chang et al. 2008; Chow et al. 2011). Diet has also been shown to play a significant role in altering the gut microbiota. For example, depending on dietary intake, gut microbial metabolism can be induced toward proteolytic fermentation, which in turn leads to the production of compounds such as amines, ammonia, phenols and sulfides (Vipperla and O'Keefe 2016). Given their pro-inflammatory nature, these compounds have been implicated in gut dysbiosis, the development of inflammatory bowel disease (IBD), and even colorectal cancer (Vipperla and O'Keefe 2016).

Gut dysbiosis, characterized by decreased gut microbial diversity, in addition to being linked to IBD (Frank et al. 2007), has also been implicated in irritable bowel syndrome (IBS) (Codling et al. 2010), cancer (Sobhani et al. 2011), and obesity (Turnbaugh et al. 2006). For instance, dysbiosis in Crohn's Disease (CD) is characterized by a deficiency of *Faecalibacterium prausnitzii*, an anti-inflammatory bacterium (Sokol et al. 2008). Other studies have shown that patients with ulcerative colitis (UC) also have dysbiotic microbiota, with less bacterial diversity and a greater abundance of Actinobacteria and Proteobacteria than healthy controls (Lepage et al. 2011). The intestinal microbiota in IBS is also thought to be different than the microbiota of healthy controls marked by a reduction in *Lactobacillus* and *Bifidobacterium*, combined with increased numbers of Enterobacteriaceae, Bacteroides, and Firmicutes species in patients with IBS (Bolino and Bercik 2010; Jeffery et al. 2012). Given the pathological conditions associated with intestinal dysbiosis, restoration of a healthy gut microbiome has been attempted by various methods (Preidis and Versalovic 2009), including probiotics (organisms beneficial to the host), prebiotics (aimed at improving growth and/or activity of commensal microbes), and fecal microbiota transplantation (FMT), the latter of which is the focus of this review.

11.2 Fecal Microbiota Transplantation

Fecal microbiota transplantation refers to the process of infusing fecal matter from a healthy donor into a sick patient to restore a healthy microbiome in the recipient. The infused donor bacteria restore microbiome diversity by colonizing the gut and existing alongside or taking over the existing microbiota (Li et al. 2016; Seekatz et al. 2014). Although there has been a marked increase in FMT over the past decade, its use as a therapeutic approach was described as early as fourth-century China, when the oral intake of human feces was used to treat patients with food poisoning or severe diarrhea (Zhang et al. 2012). The first report in modern medicine of FMT use for the treatment of pseudomembranous colitis was in 1958 (Eiseman et al. 1958).

11.2.1 Attitudes to FMT

With the advent of antimicrobials, returning to an "archaic" stool-based therapy might be perceived negatively by the general population. In a survey of physicians on their experience with FMT for the treatment of recurrent *C. difficile* infection (rCDI), the most common reason for not offering patients FMT or referring them for FMT was a belief that they would find it unappealing (Zipursky et al. 2014). However, this has proven to not be the case. A survey of nearly 200 participants examining patients' attitudes toward FMT revealed that although most patients do find the treatment itself unappealing, most would nonetheless opt for FMT to treat rCDI over multiple courses of antibiotic therapy (Xu et al. 2016). Furthermore, most patients who had already undergone FMT for rCDI reported a preference for FMT as a first treatment over antibiotics in the event of future rCDI (Xu et al. 2016). In another survey of 95 patients with UC, the authors found that despite excellent or very good control of their disease, nearly half of the patients were willing to consider FMT as an alternative to their treatment, and patients who had been hospitalized were even more willing to undergo FMT (Kahn et al. 2013). This underscores the notion that currently there is an interest in and acceptance of FMT as a therapeutic alternative by both physicians and patients alike. Therefore, dispelling any inaccurate preconceptions of patient attitudes could lead to the widespread utilization of FMT for a multitude of conditions.

11.2.2 FMT Methodology

At present there are multiple FMT protocols available for donor stool preparation and administration. FMT broadly consists of suspending donor fecal matter in liquid, blending the mixture, straining it to remove large particles and then infusing it into a recipient for the purpose of restoring a healthy microbiome (Fig. 11.1). The administration routes for fecal suspension vary and include gastroscopy, nasogastric tube, nasojejunal tube, rectal tube/enema, colonoscopy or transendoscopic enteral tubing to the appendix (Peng et al. 2016; Kassam et al. 2013), Frozen encapsulated microbial preparations have also been used by some investigators and are available as standard care at certain hospitals (Youngster et al. 2014). No ideal method of instillation has been determined, however, each method has its own advantages and disadvantages, lower gastrointestinal FMT infusion by colonoscopy being marginally more effective than upper gastrointestinal infusion (Kassam et al. 2013). Generally, fecal donation is a very involved process. Donors are screened at many levels and can be disqualified at any level (Table 11.1). A medical questionnaire is taken to exclude people with high-risk sexual behaviors, use of antibiotics within the preceding 3 months, and a history of gastrointestinal comorbidities such as IBS, IBD, and chronic diarrhea (Bakken et al. 2011). Another level of screening involves testing for blood-borne infections of both the donor and the recipient to exclude transmissible infections, including hepatitis A, B, C viruses, human immunodeficiency virus, human T-cell lymphotrophic virus, *Cytomegalovirus*, Epstein–Barr virus, and syphilis. Stool testing of the donor includes testing for *C. difficile*, ova and parasites, *Salmonella, Shigella, Escherichia coli* O157:H7, and *Yersinia* (Merenstein et al. 2014). All of these tests are undertaken to ensure the safety of the transplant.

11.2.3 Safety of FMT

The safety of FMT has been studied mostly in the context of the treatment of rCDI, adverse effects and complications being fairly uncommon.

In a systematic review of 11 studies with follow-up ranging from weeks to months, Kassam et al. reported that 90% of patients achieved clinical resolution in the absence of any adverse events (Kassam et al. 2013). Other studies have shown that FMT is generally a safe treatment, even in immunocompromised patients

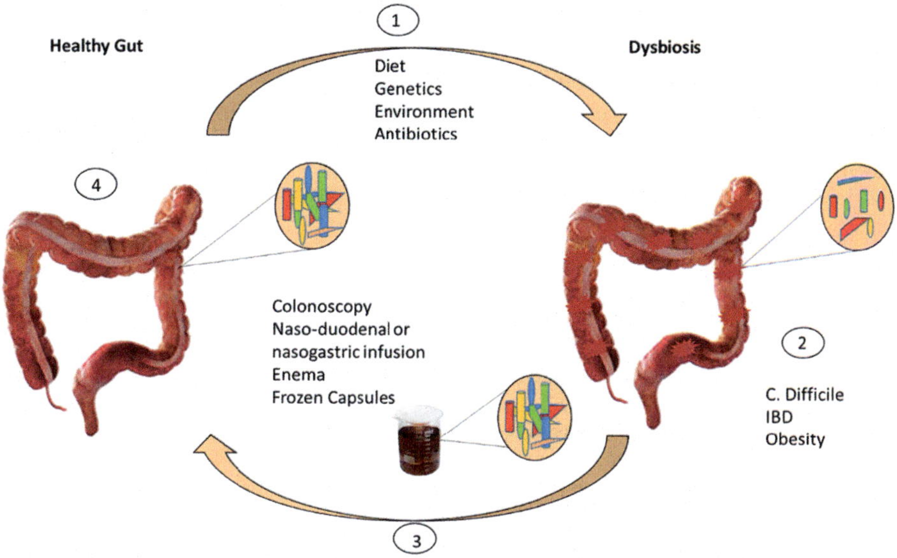

Fig. 11.1 Fecal microbiota transplantation (FMT) methodology. The normal human intestinal microbiome is made up of a wide variety of bacteria that interact with the host genetics, the environment, and the host dietary intake to modulate host physiology, and contributes to energy production and immune modulation. Certain alterations in genetics, environment, diet or use of antibiotics (*1*) can disrupt the normal intestinal microbiome and lead to a state of dysbiosis, an imbalance in the gut microbiota that is associated with the development of several pathological states including *Clostridium difficile* infection, inflammatory bowel disease, or obesity (*2*). FMT involves instilling stool-based material from a healthy donor to a recipient with the end goal of restoring the healthy microbiome in the sick recipient. The stool is resuspended, the suspension is mixed to a homogeneous consistency, and filtered to remove large particles. The mixture is then delivered to the recipient via various methods, including colonoscopy, enema, naso-gastric or naso-duodenal route; frozen pills have also proven effective (*3*). The healthy microbiome of the donor then colonizes the recipient's gastrointestinal tract and restores the healthy microbiome by existing alongside or overtaking the pre-existing microbiota (*4*)

(Kelly et al. 2014) and in the elderly (Agrawal et al. 2016). In a study of the efficacy of FMT in 168 elderly patients with recurrent, severe, or complicated CDI FMT, primary, and secondary cure rates were 82.9% and 95.9% respectively, and serious adverse events occurred in only six patients, with one culminating in death (Agrawal et al. 2016). Nevertheless, adverse events have been reported in a number of encounters. For instance, at the Virginia Mason Medical Center in Seattle, despite rigorous testing of donors and appropriate handling by staff of fecal material, 2 out of 13 FMT recipients developed norovirus gastroenteritis after receiving FMT (Schwartz et al. 2013). Bacteremia caused by *Escherichia coli* have also been reported post-FMT (Quera et al. 2014). More seriously, a limited number of studies have also reported deaths as adverse events of FMT administration. For instance, a multicenter retrospective series on the use of FMT in immunocompromised patients with recurrent, refractory, or severe CDI reported a death

Table 11.1 Summarized donor screening recommendations

Regulatory body	Stool donor characteristics	Donor laboratory screening	
Consensus of IDSA, AGA, ACG	Initial screening questionnaire similar to current protocols for screening blood donors. *Donor exclusion criteria* – History of antibiotics treatment during the 3 months before donation – History of gastrointestinal illnesses, including inflammatory bowel disease, irritable bowel syndrome, gastrointestinal malignancies or major gastrointestinal surgical procedures. – History of autoimmune or atopic illnesses or ongoing immune modulating therapy – History of chronic pain syndromes (fibromyalgia, chronic fatigue) or neurological, neurodevelopmental disorders – Metabolic syndrome, obesity (BMI of >30), or moderate-to-severe undernutrition – History of malignant illnesses or ongoing oncological therapy	Serological testing • HAV IgM • HBV surface antigen • HCV antibody • HIV 1/2 immunoassay • Syphilis antibody	Stool testing • *Clostridium difficile* (PCR or EIA test for toxin A and B) • Routine culture for enteric bacterial pathogens • Ova and parasite studies, if pertinent travel history
Health Canada Guidelines	Donor Health History/Lifestyle Questionnaire for risk behaviors, and physical examination *Donor exclusion criteria* – Donors who may transfer undesirable agents (i.e., *C. difficile* toxins, antibiotics, systemic immunosuppressive or biological agents, systemic antineoplastic agents and exogenous glucocorticoids, anti-diarrheal drugs, mineral oil, bismuth, magnesium, or kaolin)	Serological testing • HBV surface antigen • HCV antibody • HIV 1/2 immunoassay • HTLV-I/II • Syphilis antibody • Chagas disease[a] • *Strongyloides*[a] Other screen tests[a] • *Helicobacter pylori* • Malaria • *Neisseria gonorrhoeae* • *Chlamydia trachomatis*	Stool testing • Enteric pathogens (*Shigella*, *Salmonella* • *Yersinia*, *Campylobacter*, *E. coli* 0157:H7, *Plesiomonas* spp., *Aeromonas* spp.) • Shiga-toxin producing *E. coli* • Norovirus • Rotavirus • Adenovirus • *Vibrio* spp. • *Listeria monocytogenes* • Ova and parasites • Vancomycin-resistant enterococci • Methicillin-resistant *Staphylococcus aureus*

(continued)

Table 11.1 (continued)

Regulatory body	Stool donor characteristics	Donor laboratory screening	
		• Babesiosis • Creuztfeldt–Jakob disease • Prion-related diseases	

Note: The guidelines from Health Canada list a number of pathogens and permit the physician to risk-stratify which ones are appropriate. Non-infectious diseases (e.g., cancer) are also mentioned, but not listed here. See also: http://www.hc-sc.gc.ca/dhp-mps/consultation/biolog/fecal_microbiota-bacterio_fecale-eng.php

Source: Bousvaros, A., Relman, D., Rustgi, A., Vender, R., Wang, K. (2013). "Letter to FDA re. Current Consensus Guidance on Donor Screening and Stool Testing for FMT. Accessed February 25, 2017."; HealthCanada (2016). "Guidance Document: Fecal Microbiota Therapy Used in the Treatment of *Clostridium difficile* Infection Not Responsive to Conventional Therapies."

IDSA Infectious Diseases Society of America, *AGA* American Gastroenterological Association, *ACG* American College of Gastroenterology

[a]If clinically indicated

that was the result of aspiration during sedation for FMT (Kelly et al. 2014). Another death due to aspiration pneumonia following stool infusion by endoscopy to the distal duodenum has been described in the literature (Baxter et al. 2015). Non-infectious adverse events have also been reported. Interestingly, weight gain post-FMT was described in a case report of a woman successfully treated with FMT for rCDI, who subsequently became obese after receiving the transplant (Alang and Kelly 2015). Autoimmune diseases such as rheumatoid arthritis, Sjögren syndrome, idiopathic thrombocytopenic purpura, and peripheral neuropathy have been reported in other patients post-FMT, but without clear causality between FMT and the autoimmune disease (Brandt et al. 2012).

Finally, the safety of FMT in other disease states warrants further study. In UC patients, transient flares of IBD have been reported post-FMT treatment (De Leon et al. 2013). Such findings underscore the need for further research into the subtleties of FMT as a therapeutic option for different gastrointestinal diseases.

11.2.4 Regulation of FMT

Given the composition of fecal matter, it should come as no surprise that regulations surrounding it are complex and still evolving. A lack of consensus exists with respect to the designation of FMT as a drug by regulatory bodies across different countries. Although certain countries do not categorize FMT as a drug, Health Canada and the Federal Drug Administration (FDA 2013; HealthCanada 2016) do label FMT a drug. In practice, this requires an Investigational New Drug application for approval of the use of stool-based therapeutics. The exception to this is FMT used for rCDI, provided that several requirements are met, including adequate screening of the donor and donor stool, and adequate consent having been obtained by the health care provider from the patient after discussion of the risks (FDA 2013; HealthCanada 2016). Most recently, an additional stipulation has been added—that the donor be "known" to the treating physician (IDSA 2016). Even with such restrictions, in Ontario, the provincial advisor on the quality of health has released a recommendation in favor of publicly funding FMT for patients with rCDI, in light of the "large improvement of outcomes that are important to patients" in addition to the significant cost savings to the healthcare system (HealthQualityOntario 2016).

11.3 FMT in Recurrent *C. Difficile* Infection

11.3.1 Clostridium Difficile *Infection*

Clostridium difficile is an anaerobic, toxin-producing, Gram-positive, spore-forming bacterium which is transmitted via the fecal–oral route. It is also the etiological agent of *C. difficile* colitis, a gastrointestinal disease caused by *C. difficile* toxin production (Abt et al. 2016). In CDI, toxin production by *C. difficile* causes increased intestinal permeability and fluid secretion in addition to intense colonic inflammation. If left untreated, CDI can have devastating consequences, including toxic megacolon and even death (Bartlett and Gerding 2008). CDI has achieved notoriety as a widespread nosocomial infection, accounting for 15–25% of cases of antibiotics-induced diarrhea (Bartlett and Gerding 2008) and affecting up to 1.2% of hospitalized patients in the USA, costing the health care system $433–$797 million per year (Ghantoji et al. 2010). Furthermore, in recent years, CDI has been found to affect people previously thought to be at a low risk of infection. For instance, there are increasing concerns regarding community-associated CDI in a healthy young population, in antibiotic-naïve patients, and in people with no recent health care exposure (Gupta and Khanna 2014; Wilcox et al. 2008). Such changes in the epidemiology of CDI amplify the magnitude of the threat that CDI poses and have broad implications for the health care system.

It has been known for a long time that exposure to antibiotics, most commonly broad-spectrum cephalosporins, fluoroquinolones, and clindamycin (Aldeyab et al. 2012), is a major pre-disposing factor for CDI, as antibiotics use creates an altered microbial environment in the GI tract that is conducive to the overgrowth of *C. difficile* (Hensgens et al. 2013). Under normal circumstances, the resident gut microbiota is thought to create an environment that is resistant to *C. difficile* colonization, suppressing its pathogenic activity in the colon (Theriot et al. 2014). However, broad-spectrum antibiotics use disrupts the host microbiota, decreasing its abundance and diversity and altering its metabolic activity. This results in decreases in metabolites such as secondary bile acids, glucose, free fatty acids and dipeptides, and increases in primary bile acids and sugar alcohols, all changes that promote *C. difficile* propagation (Theriot et al. 2014).

11.3.2 Recurrent C. difficile *Infection*

With regard to therapeutic intervention, antibiotics (vancomycin, metronidazole, and fidaxomicin) remain the treatment for an initial CDI presentation. However, despite appropriate antibiotics treatment, CDI recurs in 20% of patients (Surawicz et al. 2000). Recurrent CDI is defined as the return of symptoms within 8 weeks of the initial episode, after successful resolution of symptoms following appropriate antibiotics therapy (Bakken 2009). Studies have shown that rCDI can be due to either re-infection with a different strain (Barbut, Richard et al. 2000) or relapse of the infection with the original strain (Figueroa et al. 2012) against a background of an impaired immune response and altered gut microbiota, which allows for the proliferation of *C. difficile* (Kyne et al. 2001).

After a first recurrence, the risk of a second recurrence increases to 40–65% (McFarland et al. 2002), and standard therapy with antimicrobials such as vancomycin is only around 30% effective in these cases (Leong and Zelenitsky 2013). Given such high failure rates with appropriate antibiotics treatment, attention has shifted toward other viable therapeutic alternatives, such as FMT.

Other than previous CDI infection, risk factors that predispose to recurrence include hospitalization, further antibiotics use, chemotherapy, use of proton pump inhibitors, and advanced age (Abdelfatah et al. 2015). A previous appendectomy has also been shown to be a risk factor for rCDI. Patients without an appendix have been shown to have a 2.5-fold increase in the risk of recurrence of CDI compared with patients who still have an appendix (Im et al. 2011). The postulated reason for this is that the appendix serves as a natural reservoir of normal commensal bacteria, which subsequently restores and recolonizes the colon with normal microbiota to protect against the development of rCDI (Im et al. 2011).

11.3.3 Efficacy of FMT in rCDI

Fecal microbiota transplantation has been extensively used in the treatment of rCDI with moderate success. In a 2013 clinical trial, van Nood et al. (2013) compared three treatment arms: FMT preceded by bowel lavage and vancomycin, vancomycin treatment alone, and vancomycin treatment with bowel lavage. The authors found that 94% of the FMT group achieved clinical resolution, 81% after the first FMT and the rest following a second infusion. This outcome was significantly better than the other treatment arms, with 23% of patients achieving clinical resolution after vancomycin treatment alone, and 31% after treatment with vancomycin and bowel lavage. The study was discontinued early as most patients in the control groups (the groups not receiving FMT) had suffered a relapse and witholding FMT was deemed unethical. Furthermore, in that same study increased microbiota diversity, a positive predictor of gut health, was noted in FMT recipients with increases in Bacteroidetes, *Clostridium* clusters IV and XIVa, and decreases in Proteobacteria (van Nood et al. 2013).

A more recent randomized double-blind clinical trial investigated the efficacy of FMT for the treatment of rCDI. The patients were treated with either FMT from a donor or received an autologous FMT. The authors found that 91% of the patients who received donor FMT compared with 63% of those who received autologous stool achieved clinical cure, thus rendering donor FMT more effective than autologous stool at preventing rCDI (Kelly et al. 2016). A double-blind randomized clinical trial conducted over 2 years at six medical academic centers in Canada sought to determine whether frozen FMT is non-inferior to fresh FMT for patients with recurrent or refractory CDI. Patients in this study were randomized to receive either fresh FMT within 24 h of collection or frozen FMT, within 24 h of thawing. The authors found that frozen FMT was non-inferior to fresh FMT, with 83.5% of patients in the frozen FMT groups achieving clinical resolution compared with 85.1% in the fresh FMT group (Lee et al. 2016).

Similarly, Youngster et al. demonstrated the effectiveness of oral frozen fecal capsules from unrelated donors in the treatment of rCDI, with significant clinical resolution of diarrhea in 90% of participants (Youngster et al. 2014). In a follow-up to this small feasibility study, the authors treated 180 patients with rCDI with frozen, encapsulated FMT and found that 82% experienced clinical resolution after one treatment and 91% after two treatments at 8 weeks (Youngster et al. 2016).

Regardless of fresh or frozen, or the method of infusion, the success of FMT is likely due to the ability of the donor microbial ecosystem to replace the microbiota that has been lost or altered in the recipient and thus suppress *C. difficile* overgrowth, promoting patient recovery (Seekatz et al. 2014). Although failure of a first FMT is uncommon, it occurs most frequently in patients with a previous history of CDI-related hospitalization events, pre-existing IBD, or severe/complicated rCDI status (Fischer et al. 2016; Khoruts et al. 2016). Given the therapeutic success of FMT in rCDI, current recommendations state that if there are three or more recurrences of CDI following vancomycin therapy, FMT should be considered as the next therapeutic option (Surawicz et al. 2013).

11.3.4 Other Stool-Based Therapies for rCDI

The appeal of FMT is closely linked to its effectiveness. However, theoretical concerns about its safety profile have led some to investigate other alternatives in the form of stool-substitute therapies. Stool-substitute therapies take a similar approach to FMT, but use a more defined community of microbes, consisting of specifically chosen bacterial strains to compete with *C. difficile*. To date, this approach has been used only for the treatment of rCDI. The first report of using stool-derived isolated microbial strains to cure CDI occurred in the late 1980s before the emergence of hypervirulent *C. difficile* strains such as NAP1/ribotype 027 (Tvede and Rask-Madsen 1989). Currently, more efforts are targeting progress in this area. For instance, Khanna et al. investigated the efficacy of a spore mixture derived from healthy donor stool treated with ethanol to prevent rCDI in a phase Ib clinical trial (Khanna et al. 2016). Overall, the mixture leads to resolution of diarrhea in 87% of patients within 8 weeks and was associated with increased intestinal microbiota diversity, including beneficial bacterial species not present in the spore preparation. Unfortunately, despite the promising results, the study failed to show diarrhea resolution in phase II (SeresTherapeutics 2016).

In a separate phase II randomized control trial, a spore preparation of nontoxigenic *C. difficile* has been tested for the prevention of rCDI and has shown some success. CDI recurrence was decreased in patients receiving the spore treatment compared with the placebo group, with the authors reporting recurrence in 11% of patients in the treatment group and in 30% of placebo controls (Gerding et al. 2015). Another success story is the proof of concept "RePOOPulate" study, by Petrof et al. (2013) in which 33 representative bacterial species were isolated from a donor stool sample and used as a "stool substitute" in an attempt to repopulate the gut microbiota of two patients with rCDI. Both patients formed normal stool within 2–3 days of treatment and remained free of rCDI at 6 months' follow-up. However, despite its success and the potential advantages of using a more defined microbial composition, several hurdles remain. The complex regulatory requirements that apply to the production of microbial ecosystems present challenges that must be addressed before this approach is available for widespread use.

11.4 FMT in Inflammatory Bowel Disease

11.4.1 Inflammatory Bowel Disease

Inflammatory bowel disease is a chronic and relapsing inflammatory disorder of the intestine that includes both UC and CD. IBD is a complex disease thought to arise from interactions among environmental factors, dysregulated immune responses and modifications of the microbiota in genetically susceptible individuals (Zhang and Li 2014). There is evidence that the intestinal microbiota in IBD is different than the normal gut microbiota and that the gut microbiota plays an important role in driving inflammation in IBD, by enhancing immune stimulation, causing epithelial dysfunction, or increasing mucosal permeability (Sartor 2008). For instance, the intestinal microbiota of IBD patients shows decreased bacterial diversity compared with healthy individuals. A study of both CD and UC patients with active disease showed that microbial diversity in CD patients was reduced to 50% and that of UC patients to 30% compared with healthy controls; this reduction was due to decreases in normal anaerobic bacteria including depletion of *Bacteroides, Eubacterium* species and *Lactobacillus* species (Ott et al. 2004). In addition to reduced microbiota diversity, other studies have detected a reduced complexity of the bacterial phylum Firmicutes in the microbiota of CD patients (Manichanh et al.

2006). *Faecalibacterium prausnitzii* is a major member of the Firmicutes phylum and has been found to be associated with anti-inflammatory effects in patients with CD, partly as a result of metabolites (such as butyrate) being able to block pro-inflammatory NF-κB activation and IL-8 production (Sokol et al. 2008). It seems that bacterial diversity in IBD can also change within the same individual at different time points. For instance, a study examining microbiota diversity of UC patients found a further loss of bacterial species such as *Bacteroides*, *Escherichia*, *Eubacterium*, *Lactobacillus*, and *Ruminococcus* during a UC relapse compared with remission in the same patients (Ott et al. 2008). Similar to the CD microbiota, *F. prausnitzii* has been found to be depleted in patients with UC, and its recovery after relapse was associated with maintenance of clinical remission (Varela et al. 2013).

Differences in bacterial communities between CD and UC suggest that the microbiome might play a different role in the pathogenesis of these two separate diseases. Overall though, the IBD microbiome has been shown to be pro-inflammatory and profoundly altered in function, with changes characterized by increased oxidative stress and perturbed micronutrient availability, in addition to increases in virulence and secretion pathways (Morgan et al. 2012). As a result of its pro-inflammatory nature, current medical treatments for IBD are aimed mainly at reducing the immune inflammatory response, via immune suppressive therapies (Zenlea and Peppercorn 2014). Given that the intestinal microbiota can modulate the host inflammatory response as well, there is significant interest in using FMT therapy in IBD to modulate or alleviate the existing pathological inflammatory state.

11.4.2 Efficacy of FMT in IBD

Fecal microbiota transplantation use in the IBD population to treat rCDI has been shown to have beneficial effects without serious adverse effects. A systematic review of FMT in IBD, which included 18 studies and one randomized controlled trial, showed positive results, finding a remission rate of 22% in UC patients in a subgroup analysis and 61% in CD patients treated with FMT (Colman and Rubin 2014). A multicenter retrospective study evaluating the use of FMT for rCDI in immunocompromised patients that included patients with IBD found that adverse effects occurred in 14% of the patients with IBD and consisted of disease exacerbation (Kelly et al. 2014). In another study evaluating the impact of IBD on FMT outcomes for rCDI, IBD patients were found to have a lower efficacy of clearing *C. difficile* infections with only 74% clearing the infection compared with 91% of non-IBD patients following a single FMT (Khoruts et al. 2016). However, Khoruts et al. (2016) also reported that one in four patients in their study experienced an IBD flare after FMT. Studies on treating CD exclusively with FMT, however, are scarce. An early case report of a patient with refractory CD demonstrated short-term clinical improvement, relapsing 18 months later (Borody et al. 1989). A more recent case report demonstrated complete clinical, endoscopic, and histological remission following FMT in a patient with Crohn's colitis who had failed immunosuppressive therapy (Kao et al. 2014). However, despite changes in the fecal microbiota following FMT in a patient with Crohn's colitis, these changes did not persist once FMT was discontinued (suggesting that maintenance FMT might be required in patients with CD). Other positive results have been observed in a pediatric population of CD. In a 2015 study by Suskind et al., seven of the nine patients enrolled were in remission after FMT and a 12-week follow-up demonstrated that more than half of the patients remained in remission (Suskind et al. 2015). Furthermore, in a study of 30 adult patients with refractory CD, FMT led to clinical improvement in 87% and remission in 77% after a month of follow-up (Cui et al. 2015) These studies suggest that FMT might be a therapeutic option in the management of CD, but larger clinical studies are required to assess the efficacy and safety of FMT in CD.

With regard to UC, the data are more abundant. The first case report of the use of FMT for the treatment of UC dates back to 1989 and included one of the authors who suffered from UC as a subject. Following self-transplantation of healthy donor stool by enema, he achieved symptomatic resolution for 6 months and resolution of acute inflammation on colonic biopsies at 3 months (Bennet and Brinkman 1989). A more recent report in 2003 describes a retrospective case series of six patients with endoscopically and histologically confirmed UC in whom traditional therapy had failed and disease remission lasting 1–13 years had been achieved following FMT administered as retention enemas daily for 5 days (Borody et al. 2003). More studies reflect a positive outcome in UC patients treated with FMT. In the largest randomized control trial to date of FMT in UC, it was found that 24% of the treated patients achieved remission compared with only 5% in the placebo arm. These patients also had a statistically significant increase in microbial diversity similar to the fecal microbial profile of their donors compared with the placebo group at week 6 (Moayyedi et al. 2015), supporting the theory that FMT can be used to restore a healthy gut microbiota. Another study of six patients with UC refractory to conventional therapy who were treated with FMT demonstrated transient clinical improvement in all patients (Kump et al. 2013). A recent systematic review assessing the efficacy and safety of FMT in UC, which included 2 randomized controlled trials, 15 cohort studies, 8 case studies, and 234 patients, also had positive results, finding that 42% of patients achieved clinical remission and 65% achieved a clinical response (Shi et al. 2016).

Given that UC is a relapsing disease, the role of FMT maintenance therapy is of interest to the management of the disease. In fact, a 2016 Australian study evaluated the effectiveness of maintenance therapy. In this study, patients received either FMT or placebo infusion delivered via colonoscopy, after which they were assigned to receive either an intensive treatment of five FMTs or placebo enemas per week, for 8 weeks. The study found a trend toward significance in endoscopic remission rates between the groups, with the treated group achieving a better remission rate (Paramsothy 2016).

However, despite the encouraging results, not all studies have been positive. A recent randomized control trial by Rossen et al. (2015) was stopped early owing to unexpected inferior results in the treatment arm compared with the control arm. In this study, patients with UC were randomized to receive FMT from healthy donors or to receive autologous fecal material. Only 30% of patients receiving FMT from donors achieved symptomatic relief versus 32% in the autologous FMT group. In another prospective study using FMT in five adult patients with moderate to severe UC not responding to immunosuppressive therapies, the authors found that none of the five patients achieved clinical remission and only one showed clinical improvement during the 12-week follow-up (Angelberger et al. 2013).

Overall, it seems that FMT is not as effective in treating IBD as it is in the treatment of CDI. A potential explanation for this might be that IBD is not caused solely by microbiota alterations in the absence of environmental and genetic inputs (Willing et al. 2010; Hansen et al. 2010). Therefore, FMT restoration of the microbiota may be only one of the therapeutic components of a multifactorial cocktail for IBD. Given the mixed results in the literature regarding the role and efficacy of FMT in IBD, more trials are needed to further our understanding of how FMT influences the microbiota and inflammatory response in IBD.

11.5 FMT in Obesity

11.5.1 Obesity

Obesity, defined as a body mass index (BMI) >30 in adults due to the accumulation of excess fat, is associated with negative health effects and is a serious worldwide public health epidemic (Villanueva-Millan et al. 2015). Recent evidence suggests that gut microbiota can modulate nutrient uptake and energy regulation, potentially playing a role in metabolic syndrome and obesity-related disorders (Boulange et al. 2016). For instance, studies have shown that the ratio of gut Firmicutes to Bacteroidetes was different in obese versus lean animal and human models and this was also associated with different energy harvesting capacity (Turnbaugh et al. 2009; Ley et al. 2006). However, other studies have shown no link between the gut Firmicutes to Bacteroidetes ratio and obesity (Fernandes et al. 2014), and the role that the intestinal microbiota plays in modulating nutrient absorption has not been fully defined.

11.5.2 Efficacy of FMT in Obesity

To date, studies evaluating the effect of FMT on the treatment of obesity are scarce, but promising. Studies in animal models have shown a potential link among the gut microbiota, obesity or metabolic syndrome, and associated insulin resistance. With regard to microbiome diversity, FMT from twin mice discordant for obesity into germ-free mice led to differences in the microbiome in addition to the phenotype of the recipient mice (Ridaura et al. 2013). The mice that had received the microbiome of the obese twin had decreased gut bacterial diversity and increased adiposity, whereas mice populated with the microbiota of the lean twin had increased gut bacterial diversity and a leaner phenotype. These results suggest an association between the microbiome and the metabolic phenotype of the host. In another study by Di Luccia et al. the authors found that insulin resistance in adult rats on a high-fructose diet could be reversed with orally administered FMT from control rats (Di Luccia et al. 2015).

Despite the encouraging results and the ongoing interest in FMT for the treatment of obesity, there is to our knowledge only one published study of FMT for the treatment of metabolic disorders in humans. In the study by Vrieze et al., obese individuals diagnosed with type 2 diabetes received stool transplants from either healthy, lean donors or an autologous stool transplant. Although there were no significant differences in BMI, the authors did report a significant increase in gut microbiota diversity and insulin sensitivity in the FMT recipients of healthy stool compared with the recipients who received an FMT comprising their own, autologous stool (Vrieze et al. 2012). It stands to reason that the increased insulin sensitivity could be due to increased gut microbial diversity (including butyrate-producing bacteria), as butyrate has been shown to play an important role in promoting insulin sensitivity in mice (Lin et al. 2012). FMT as a potential therapy for obesity and metabolic syndrome is an area of intense interest; with 11 clinical trials at various phases currently in progress (ClinicalTrials.gov), more data on this application should soon be available.

11.6 FMT and Irritable Bowel Syndrome

11.6.1 Irritable Bowel Syndrome

Irritable bowel syndrome is a functional bowel disorder characterized by abdominal pain, bloating, and stool irregularities; it is often associated with extra-intestinal symptoms such as pain syndromes or psychiatric conditions such as depression and anxiety (Enck et al. 2016). IBS is a commonly diagnosed gastrointestinal disease in the USA, with a significant impact on health-care costs (Sandler et al. 2002). The pathogenesis of IBS cannot be explained by a single mechanism and traditional treatments of IBS (including dietary change, psychological therapies, probiotics, and antibiotics) have had only limited success, highlighting the need for additional therapeutic options. As a result, IBS-associated alterations in the intestinal microbiota are increasingly a focus of interest. Studies have shown that alterations in the gut microbiota exist in patients with IBS with a dysbiosis characterized by a reduction in species of *Bifidobacterium* (Tojo et al. 2014).

11.6.2 FMT in IBS and Slow-Transit Constipation

While there is much interest in this area, the data on FMT efficacy in IBS is limited. In a single center study, Pinn et al. report that out of a total of 13 patients with IBS who received FMT treatment, 70% experienced resolution or improvement of symptoms, including improvement of abdominal pain, dyspepsia, bloating, and flatus and nearly half achieved improvement of overall well-being (Pinn et al. 2014). The role of FMT in treating IBS is now increasingly being explored, with seven clinical trials actively examining the impact of FMT on IBS (ClinicalTrials.gov).

Focus has also been placed on the role of FMT in slow transit constipation. In a study by Ge et al., the authors investigated the effect of FMT on constipation symptoms. They found that 67% of patients showed clinical improvement and remission of constipation, increased stool frequency, and improved consistency (Ge et al. 2016). The authors concluded that FMT combined with fiber may improve constipation symptoms by modulating the intestinal microbiota.

11.7 Other Areas of Research

11.7.1 Colorectal Cancer

Colorectal cancer has become the third most common cancer in the world (Song and Li 2016). Although genetic predisposition, (Hahn et al. 2016), environmental factors (Lichtenstein et al. 2000), and diet (Bingham 2000) have been linked to the development of colorectal cancer, attention has also been given to the role of the microbiota leading to colorectal cancer (Hope et al. 2005; Tjalsma et al. 2012). Several studies have linked gut dysbiosis with colorectal carcinoma. For instance, a study that examined the microbial makeup of colon tumors and normal tissue reported an increased abundance of *Fusobacterium* in tumors compared with normal tissue (Castellarin et al. 2012), with *Fusobacterium* having been shown to modulate the tumor immune microenvironment to promote inflammation and tumorigenesis (Kostic et al. 2013). Other studies that have investigated the dysbiosis in colorectal cancer have found increases in potential pathogenic bacteria *Fusobacterium* and *Campylobacter* and decreases in beneficial butyrate producers (Wu et al. 2013). At the level of the metabolome, studies in humans have shown that metabolites related to cancer, inflammatory response, carbohydrate metabolism, and GI disease pathways are significantly increased in colorectal adenomas compared with normal tissue, suggesting that alterations in the function of the microbiota might be linked to the development of colorectal carcinoma (Nugent et al. 2014).

11.7.2 FMT in Colorectal Cancer

Given that differences in the microbiota of cancer-free patients and patients with colorectal cancer have been identified, it is of interest to determine whether manipulation of the microbiota can induce tumorigenesis. A study in a murine model showed that changes in the gut microbiota through FMT can lead to the development of colorectal cancer. In the study, the authors inoculated and colonized germ-free mice with stool from either tumor-bearing mice (tumours that had been induced by the authors with intraperitoneal injections of a chemical carcinogen, azoxymethane) or healthy mice. The study found that FMT from tumor-bearing mice promoted tumor formation in the recipient mice, which developed twice as many colon tumors as mice that received FMT with a healthy microbiota (Zackular et al. 2013).

In addition, these same mice showed significant microbiota enrichment in *Bacteroides* and Erysipelotrichaceae and decreases in *Prevotella* and Porphyromonadaceae, suggesting that the alterations of the gut microbiome that were associated with chronic inflammation and tumor formation in the original mice were effectively transmitted to germ-free mice and promoted colon tumorigenesis (Zackular et al. 2013).

Similar studies have evidently not been performed in humans and although a link between dysbiosis and colorectal cancer has been established, whether the dysbiosis precedes the cancer or is caused by it has not been fully elucidated. It seems that manipulation of the gut microbiota to restore normal physiological balance may be beneficial in preventing colorectal carcinoma, but further research is needed to understand the mechanism of interaction of the gut microbiota and the development of colorectal carcinoma.

11.8 Future Directions

The potential therapeutic applications of FMT are significant and growing interest in the field promises newer and broader developments and applications of stool-based therapies. However, the exact science behind the manipulation of the intestinal microbiota remains a challenge. To date, FMT has been successfully used in rCDI, but its success in IBD and obesity remain equivocal, likely because of the more complex dysbiosis in these conditions. Ongoing research and clinical trials are being undertaken in areas such as IBS, fatty liver disease, sepsis, multi-organ dysfunction, and stem cell transplantation, attesting to the therapeutic potential that FMT is believed to have. Whole stool transplantation in its various forms has shown promise, but its use has also been associated with rigorous government regulations in the United States and Canada. Currently, the FDA and Health Canada classify stool samples as a drug and biologic. However, as FMT becomes more widely accepted and adopted, there will be a need to streamline the procedure. As such, specific questions that will need to be addressed in future research include protocol standardization (including donor selection and screening, stool mixture preparation, delivery method), safety and regulation, long-term follow-up, and the role of maintenance therapy.

Fecal microbiota transplantation is a rapidly evolving area of study, and as a result, standard delivery methods via enema, colonoscopy, or naso-duodenal tube are being evolved into microbiota-based pills. Attention has also turned to stool-based therapies that have focused on isolating specific bacterial strains from donor stool. These can potentially replicate the efficiency of FMT while bypassing the safety concerns that accompany FMT. In a time of patient-centered, personalized medicine, rather than relying on a one-microbiome-for-all approach, future research into the specific disturbances that characterize each gastrointestinal condition could lead to the development of targeted microbiome therapies to treat each individual dysbiosis. Regardless of the area of research, the future of FMT seems to hold great promise.

References

Abdelfatah M, Nayfe R, Nijim A, Enriquez K, Ali E, Watkins RR, Kandil H (2015) Factors predicting recurrence of *Clostridium difficile* infection (CDI) in hospitalized patients: retrospective study of more than 2000 patients. J Investig Med 63:747–751

Abt MC, McKenney PT, Pamer EG (2016) *Clostridium difficile* colitis: pathogenesis and host defence. Nat Rev Microbiol 14:609–620

Agrawal M, Aroniadis OC, Brandt LJ, Kelly C, Freeman S, Surawicz C, Broussard E, Stollman N, Giovanelli A, Smith B, Yen E, Trivedi A, Hubble L, Kao D, Borody T, Finlayson S, Ray A, Smith R (2016) The long-term efficacy and safety of fecal microbiota transplant for recurrent, severe, and complicated *Clostridium difficile* infection in 146 elderly individuals. J Clin Gastroenterol 50:403–407

Alang N, Kelly CR (2015) Weight gain after fecal microbiota transplantation. Open Forum Infect Dis 2:ofv004

Aldeyab MA, Kearney MP, Scott MG, Aldiab MA, Alahmadi YM, Darwish Elhajji FW, Magee FA, McElnay JC (2012) An evaluation of the impact of antibiotic stewardship on reducing the use of high-risk antibiotics and its effect on the incidence of *Clostridium difficile* infection in hospital settings. J Antimicrob Chemother 67(12):2988–2996

Angelberger S, Reinisch W, Makristathis A, Lichtenberger C, Dejaco C, Papay P, Novacek G, Trauner M, Loy A, Berry D (2013) Temporal bacterial community dynamics vary among ulcerative colitis patients after fecal microbiota transplantation. Am J Gastroenterol 108:1620–1630

Barbut F, Richard A, Hamadi K, Chomette V, Burghoffer B, Petit JC (2000) Epidemiology of recurrences or reinfections of *Clostridium difficile*-associated diarrhea. J Clin Microbiol 38(6):2386–2388

Bakken JS (2009) Fecal bacteriotherapy for recurrent *Clostridium difficile* infection. Anaerobe 15:285–289

Bakken JS, Borody T, Brandt LJ, Brill JV, Demarco DC, Franzos MA, Kelly C, Khoruts A, Louie T, Martinelli LP, Moore TA, Russell G, Surawicz C (2011) Treating *Clostridium difficile* infection with fecal microbiota transplantation. Clin Gastroenterol Hepatol 9:1044–1049

Bartlett JG, Gerding DN (2008) Clinical recognition and diagnosis of *Clostridium difficile* infection. Clin Infect Dis 46(Suppl 1):S12–S18

Baxter M, Ahmad T, Colville A, Sheridan R (2015) Fatal aspiration pneumonia as a complication of fecal microbiota transplant. Clin Infect Dis 61:136–137
Bennet JD, Brinkman M (1989) Treatment of ulcerative colitis by implantation of normal colonic flora. Lancet 1:164
Bingham SA (2000) Diet and colorectal cancer prevention. Biochem Soc Trans 28:12–16
Bolino CM, Bercik P (2010) Pathogenic factors involved in the development of irritable bowel syndrome: focus on a microbial role. Infect Dis Clin North Am 24:961–975. ix
Borody TJ, George L, Andrews P, Brandl S, Noonan S, Cole P, Hyland L, Morgan A, Maysey J, Moore-Jones D (1989) Bowel-flora alteration: a potential cure for inflammatory bowel disease and irritable bowel syndrome? Med J Aust 150:604
Borody TJ, Warren EF, Leis S, Surace R, Ashman O (2003) Treatment of ulcerative colitis using fecal bacteriotherapy. J Clin Gastroenterol 37:42–47
Boulange CL, Neves AL, Chilloux J, Nicholson JK, Dumas ME (2016) Impact of the gut microbiota on inflammation, obesity, and metabolic disease. Genome Med 8:42
Brandt LJ, Aroniadis OC, Mellow M, Kanatzar A, Kelly C, Park T, Stollman N, Rohlke F, Surawicz C (2012) Long-term follow-up of colonoscopic fecal microbiota transplant for recurrent *Clostridium difficile* infection. Am J Gastroenterol 107:1079–1087
Castellarin M, Warren RL, Freeman JD, Dreolini L, Krzywinski M, Strauss J, Barnes R, Watson P, Allen-Vercoe E, Moore RA, Holt RA (2012) Fusobacterium nucleatum infection is prevalent in human colorectal carcinoma. Genome Res 22:299–306
Chang JY, Antonopoulos DA, Kalra A, Tonelli A, Khalife WT, Schmidt TM, Young VB (2008) Decreased diversity of the fecal microbiome in recurrent *Clostridium difficile*-associated diarrhea. J Infect Dis 197:435–438
Chow J, Tang H, Mazmanian SK (2011) Pathobionts of the gastrointestinal microbiota and inflammatory disease. Curr Opin Immunol 23:473–480
Clinicaltrials.gov
Codling C, O'Mahony L, Shanahan F, Quigley EM, Marchesi JR (2010) A molecular analysis of fecal and mucosal bacterial communities in irritable bowel syndrome. Dig Dis Sci 55:392–397
Colman RJ, Rubin DT (2014) Fecal microbiota transplantation as therapy for inflammatory bowel disease: a systematic review and meta-analysis. J Crohns Colitis 8:1569–1581
Cui B, Feng Q, Wang H, Wang M, Peng Z, Li P, Huang G, Liu Z, Wu P, Fan Z, Ji G, Wang X, Wu K, Fan D, Zhang F (2015) Fecal microbiota transplantation through mid-gut for refractory Crohn's disease: safety, feasibility, and efficacy trial results. J Gastroenterol Hepatol 30:51–58
De Leon LM, Watson JB, Kelly CR (2013) Transient flare of ulcerative colitis after fecal microbiota transplantation for recurrent *Clostridium difficile* infection. Clin Gastroenterol Hepatol 11(1038):1036
Di Luccia B, Crescenzo R, Mazzoli A, Cigliano L, Venditti P, Walser JC, Widmer A, Baccigalupi L, Ricca E, Iossa S (2015) Rescue of fructose-induced metabolic syndrome by antibiotics or faecal transplantation in a rat model of obesity. PLoS One 10:e0134893
Eckburg PB, Bik EM, Bernstein CN, Purdom E, Dethlefsen L, Sargent M, Gill SR, Nelson KE, Relman DA (2005) Diversity of the human intestinal microbial flora. Science 308:1635–1638
Eiseman B, Silen W, Bascom GS, Kauvar AJ (1958) Fecal enema as an adjunct in the treatment of pseudomembranous enterocolitis. Surgery 44:854–859
Enck P, Aziz Q, Barbara G, Farmer AD, Fukudo S, Mayer EA, Niesler B, Quigley EMM, Rajilić-Stojanović M, Schemann M, Schwille-Kiuntke J, Simren M, Zipfel S, Spiller RC (2016) Irritable bowel syndrome. Nat Rev Dis Primers 2:16014
FDA (2013). Available: http://www.fda.gov/downloads/BiologicsBloodVaccines/GuidanceComplianceRegulatoryInformation/Guidances/Vaccines/UCM488223.pdf?source=govdelivery&utm_medium=email&utm_source=govdelivery [Accessed]
Falony G, Joossens M, Vieira-Silva S, Wang J, Darzi Y, Faust K, Kurilshikov A, Bonder MJ, Valles-Colomer M, Vandeputte D, Tito RY, Chaffron S, Rymenans L, Verspecht C, de Sutter L, Lima-Mendez G, D'Hoe K, Jonckheere K, Homola D, Garcia R, Tigchelaar EF, Eeckhaudt L, Fu J, Henckaerts L, Zhernakova A, Wijmenga C, Raes J (2016) Population-level analysis of gut microbiome variation. Science 352:560–564

Fernandes J, Su W, Rahat-Rozenbloom S, Wolever TM, Comelli EM (2014) Adiposity, gut microbiota and faecal short chain fatty acids are linked in adult humans. Nutr Diabetes 4:e121

Figueroa I, Johnson S, Sambol SP, Goldstein EJ, Citron DM, Gerding DN (2012) Relapse versus reinfection: recurrent *Clostridium difficile* infection following treatment with fidaxomicin or vancomycin. Clin Infect Dis 55(Suppl 2):S104–S109

Fischer M, Kao D, Mehta SR, Martin T, Dimitry J, Keshteli AH, Cook GK, Phelps E, Sipe BW, Xu H, Kelly CR (2016) Predictors of early failure after fecal microbiota transplantation for the therapy of *Clostridium difficile* infection: a multicenter study. Am J Gastroenterol 111:1024–1031

Frank DN, St Amand AL, Feldman RA, Boedeker EC, Harpaz N, Pace NR (2007) Molecular-phylogenetic characterization of microbial community imbalances in human inflammatory bowel diseases. Proc Natl Acad Sci USA 104:13780–13785

Frank DN, Zhu W, Sartor RB, Li E (2011) Investigating the biological and clinical significance of human dysbioses. Trends Microbiol 19:427–434

Ge X, Tian H, Ding C, Gu L, Wei Y, Gong J, Zhu W, Li N, Li J (2016) Fecal microbiota transplantation in combination with soluble dietary fiber for treatment of slow transit constipation: a pilot study. Arch Med Res 47:236–242

Gerding DN, Meyer T, Lee C, Cohen SH, Murthy UK, Poirier A, Van Schooneveld TC, Pardi DS, Ramos A, Barron MA, Chen H, Villano S (2015) Administration of spores of nontoxigenic *Clostridium difficile* strain M3 for prevention of recurrent *C. difficile* infection: a randomized clinical trial. JAMA 313:1719–1727

Ghantoji SS, Sail K, Lairson DR, Dupont HL, Garey KW (2010) Economic healthcare costs of *Clostridium difficile* infection: a systematic review. J Hosp Infect 74:309–318

Gupta A, Khanna S (2014) Community-acquired *Clostridium difficile* infection: an increasing public health threat. Infect Drug Resist 7:63–72

Hahn MM, De Voer RM, Hoogerbrugge N, Ligtenberg MJ, Kuiper RP, Van Kessel AG (2016) The genetic heterogeneity of colorectal cancer predisposition – guidelines for gene discovery. Cell Oncol (Dordr) 39:491–510

Hansen J, Gulati A, Sartor RB (2010) The role of mucosal immunity and host genetics in defining intestinal commensal bacteria. Curr Opin Gastroenterol 26:564–571

HealthCanada (2016) Fecal microbiota therapy for *Clostridium difficile* infection: OHTAC recommendation [Online]. Available: http://www.hqontario.ca/Evidence-to-Improve-Care/Health-Technology-Assessment/Reviews-And-Recommendations/Fecal-Microbiota-Therapy [Accessed]

HealthqualityOntario (2016) Fecal microbiota therapy for *Clostridium difficile* infection: OHTAC recommendation [Online]. Available: http://www.hqontario.ca/Portals/0/documents/evidence/open-comment/recommendation-fecal-transplantation-1604.pdf [Accessed]

Hensgens MP, Goorhuis A, Dekkers OM, Van Benthem BH, Kuijper EJ (2013) All-cause and disease-specific mortality in hospitalized patients with *Clostridium difficile* infection: a multicenter cohort study. Clin Infect Dis 56:1108–1116

Hope ME, Hold GL, Kain R, El-Omar EM (2005) Sporadic colorectal cancer—role of the commensal microbiota. FEMS Microbiol Lett 244:1–7

IDSA (2016) Fecal microbiota transplantation: investigational new drug protocol [Online]. Available: https://www.idsociety.org/FMT/ [Accessed]

Im GY, Modayil RJ, Lin CT, Geier SJ, Katz DS, Feuerman M, Grendell JH (2011) The appendix may protect against *Clostridium difficile* recurrence. Clin Gastroenterol Hepatol 9:1072–1077

Jeffery IB, O'Toole PW, Ohman L, Claesson MJ, Deane J, Quigley EM, Simren M (2012) An irritable bowel syndrome subtype defined by species-specific alterations in faecal microbiota. Gut 61:997–1006

Kahn SA, Vachon A, Rodriquez D, Goeppinger S, Surma B, Marks J, Rubin DT (2013) Patient perceptions of fecal microbiota transplantation for ulcerative colitis. Inflamm Bowel Dis 19:1506–1513

Kao D, Hotte N, Gillevet P, Madsen K (2014) Fecal microbiota transplantation inducing remission in Crohn's colitis and the associated changes in fecal microbial profile. J Clin Gastroenterol 48:625–628

Kassam Z, Lee CH, Yuan Y, Hunt RH (2013) Fecal microbiota transplantation for *Clostridium difficile* infection: systematic review and meta-analysis. Am J Gastroenterol 108:500–508

Kelly CR, Ihunnah C, Fischer M, Khoruts A, Surawicz C, Afzali A, Aroniadis O, Barto A, Borody T, Giovanelli A, Gordon S, Gluck M, Hohmann EL, Kao D, Kao JY, McQuillen DP, Mellow M, Rank KM, Rao K, Ray A, Schwartz MA, Singh N, Stollman N, Suskind DL, Vindigni SM, Youngster I, Brandt L (2014) Fecal microbiota transplant for treatment of *Clostridium difficile* infection in immunocompromised patients. Am J Gastroenterol 109:1065–1071

Kelly CR, Khoruts A, Staley C, Sadowsky MJ, Abd M, Alani M, Bakow B, Curran P, McKenney J, Tisch A, Reinert SE, Machan JT, Brandt LJ (2016) Effect of fecal microbiota transplantation on recurrence in multiply recurrent *Clostridium difficile* infection: a randomized trial. Ann Intern Med 165:609–616

Khanna S, Pardi DS, Kelly CR, Kraft CS, Dhere T, Henn MR, Lombardo MJ, Vulic M, Ohsumi T, Winkler J, Pindar C, McGovern BH, Pomerantz RJ, Aunins JG, Cook DN, Hohmann EL (2016) A novel microbiome therapeutic increases gut microbial diversity and prevents recurrent *Clostridium difficile* infection. J Infect Dis 214:173–181

Khoruts A, Rank KM, Newman KM, Viskocil K, Vaughn BP, Hamilton MJ, Sadowsky MJ (2016) Inflammatory bowel disease affects the outcome of fecal microbiota transplantation for recurrent *Clostridium difficile* infection. Clin Gastroenterol Hepatol 14:1433–1438

Kostic AD, Chun E, Robertson L, Glickman JN, Gallini CA, Michaud M, Clancy TE, Chung DC, Lochhead P, Hold GL, El-Omar EM, Brenner D, Fuchs CS, Meyerson M, Garrett WS (2013) Fusobacterium nucleatum potentiates intestinal tumorigenesis and modulates the tumor-immune microenvironment. Cell Host Microbe 14:207–215

Kump PK, Grochenig HP, Lackner S, Trajanoski S, Reicht G, Hoffmann KM, Deutschmann A, Wenzl HH, Petritsch W, Krejs GJ, Gorkiewicz G, Hogenauer C (2013) Alteration of intestinal dysbiosis by fecal microbiota transplantation does not induce remission in patients with chronic active ulcerative colitis. Inflamm Bowel Dis 19:2155–2165

Kyne L, Warny M, Qamar A, Kelly CP (2001) Association between antibody response to toxin A and protection against recurrent *Clostridium difficile* diarrhoea. Lancet 357:189–193

Lee CH, Steiner T, Petrof EO, Smieja M, Roscoe D, Nematallah A, Weese JS, Collins S, Moayyedi P, Crowther M, Ropeleski MJ, Jayaratne P, Higgins D, Li Y, Rau NV, Kim PT (2016) Frozen vs fresh fecal microbiota transplantation and clinical resolution of diarrhea in patients with recurrent *Clostridium difficile* infection: a randomized clinical trial. JAMA 315:142–149

Leong C, Zelenitsky S (2013) Treatment strategies for recurrent *Clostridium difficile* Infection. Can J Hosp Pharm 66:361–368

Lepage P, Hasler R, Spehlmann ME, Rehman A, Zvirbliene A, Begun A, Ott S, Kupcinskas L, Dore J, Raedler A, Schreiber S (2011) Twin study indicates loss of interaction between microbiota and mucosa of patients with ulcerative colitis. Gastroenterology 141:227–236

Ley RE, Turnbaugh PJ, Klein S, Gordon JI (2006) Microbial ecology: human gut microbes associated with obesity. Nature 444:1022–1023

Li SS, Zhu A, Benes V, Costea PI, Hercog R, Hildebrand F, Huerta-Cepas J, Nieuwdorp M, Salojarvi J, Voigt AY, Zeller G, Sunagawa S, De Vos WM, Bork P (2016) Durable coexistence of donor and recipient strains after fecal microbiota transplantation. Science 352:586–589

Lichtenstein P, Holm NV, Verkasalo PK, Iliadou A, Kaprio J, Koskenvuo M, Pukkala E, Skytthe A, Hemminki K (2000) Environmental and heritable factors in the causation of cancer—analyses of cohorts of twins from Sweden, Denmark, and Finland. N Engl J Med 343:78–85

Lin HV, Frassetto A, Kowalik EJ Jr, Nawrocki AR, Lu MM, Kosinski JR, Hubert JA, Szeto D, Yao X, Forrest G, Marsh DJ (2012) Butyrate and propionate protect against diet-induced obesity and regulate gut hormones via free fatty acid receptor 3-independent mechanisms. PLoS One:7, e35240

Lozupone CA, Stombaugh JI, Gordon JI, Jansson JK, Knight R (2012) Diversity, stability and resilience of the human gut microbiota. Nature 489:220–230

Manichanh C, Rigottier-Gois L, Bonnaud E, Gloux K, Pelletier E, Frangeul L, Nalin R, Jarrin C, Chardon P, Marteau P, Roca J, Dore J (2006) Reduced diversity of faecal microbiota in Crohn's disease revealed by a metagenomic approach. Gut 55:205–211

McFarland LV, Elmer GW, Surawicz CM (2002) Breaking the cycle: treatment strategies for 163 cases of recurrent *Clostridium difficile* disease. Am J Gastroenterol 97:1769–1775

Merenstein D, El-Nachef N, Lynch SV (2014) Fecal microbial therapy: promises and pitfalls. J Pediatr Gastroenterol Nutr 59:157–161

Moayyedi P, Surette MG, Kim PT, Libertucci J, Wolfe M, Onischi C, Armstrong D, Marshall JK, Kassam Z, Reinisch W, Lee CH (2015) Fecal microbiota transplantation induces remission in patients with active ulcerative colitis in a randomized controlled trial. Gastroenterology 149:102–109.e6

Morgan XC, Tickle TL, Sokol H, Gevers D, Devaney KL, Ward DV, Reyes JA, Shah SA, Leleiko N, Snapper SB, Bousvaros A, Korzenik J, Sands BE, Xavier RJ, Huttenhower C (2012) Dysfunction of the intestinal microbiome in inflammatory bowel disease and treatment. Genome Biol 13:R79

Nugent JL, McCoy AN, Addamo CJ, Jia W, Sandler RS, Keku TO (2014) Altered tissue metabolites correlate with microbial dysbiosis in colorectal adenomas. J Proteome Res 13:1921–1929

Ott SJ, Musfeldt M, Wenderoth DF, Hampe J, Brant O, Folsch UR, Timmis KN, Schreiber S (2004) Reduction in diversity of the colonic mucosa associated bacterial microflora in patients with active inflammatory bowel disease. Gut 53:685–693

Ott SJ, Plamondon S, Hart A, Begun A, Rehman A, Kamm MA, Schreiber S (2008) Dynamics of the mucosa-associated flora in ulcerative colitis patients during remission and clinical relapse. J Clin Microbiol 46:3510–3513

Paramsothy SEA (2016) Multi donor intense faecal microbiota transplantation is an effective treatment for resistant ulcerative colitis: a randomised placebo-controlled trial. Gastroenterology 150:s122–s123

Peng Z, Xiang J, He Z, Zhang T, Xu L, Cui B, Li P, Huang G, Ji G, Nie Y, Wu K, Fan D, Zhang F (2016) Colonic transendoscopic enteral tubing: a novel way of transplanting fecal microbiota. Endosc Int Open 4:E610–E613

Petrof EO, Gloor GB, Vanner SJ, Weese SJ, Carter D, Daigneault MC, Brown EM, Schroeter K, Allen-Vercoe E (2013) Stool substitute transplant therapy for the eradication of *Clostridium difficile* infection: 'RePOOPulating' the gut. Microbiome 1:3

Pinn DM, Aroniadis OC, Brandt LJ (2014) Is fecal microbiota transplantation the answer for irritable bowel syndrome? A single-center experience. Am J Gastroenterol 109:1831–1832

Preidis GA, Versalovic J (2009) Targeting the human microbiome with antibiotics, probiotics, and prebiotics: gastroenterology enters the metagenomics era. Gastroenterology 136:2015–2031

Qin J, Li R, Raes J, Arumugam M, Burgdorf KS, Manichanh C, Nielsen T, Pons N, Levenez F, Yamada T, Mende DR, Li J, Xu J, Li S, Li D, Cao J, Wang B, Liang H, Zheng H, Xie Y, Tap J, Lepage P, Bertalan M, Batto JM, Hansen T, Le Paslier D, Linneberg A, Nielsen HB, Pelletier E, Renault P, Sicheritz-Ponten T, Turner K, Zhu H, Yu C, Li S, Jian M, Zhou Y, Li Y, Zhang X, Li S, Qin N, Yang H, Wang J, Brunak S, Dore J, Guarner F, Kristiansen K, Pedersen O, Parkhill J, Weissenbach J, Meta HITC, Bork P, Ehrlich SD, Wang J (2010) A human gut microbial gene catalogue established by metagenomic sequencing. Nature 464:59–65

Quera R, Espinoza R, Estay C, Rivera D (2014) Bacteremia as an adverse event of fecal microbiota transplantation in a patient with Crohn's disease and recurrent *Clostridium difficile* infection. J Crohns Colitis 8:252–253

Ridaura VK, Faith JJ, Rey FE, Cheng J, Duncan AE, Kau AL, Griffin NW, Lombard V, Henrissat B, Bain JR, Muehlbauer MJ, Ilkayeva O, Semenkovich CF, Funai K, Hayashi DK, Lyle BJ, Martini MC, Ursell LK, Clemente JC, Van Treuren W, Walters WA, Knight R, Newgard CB, Heath AC, Gordon JI (2013) Gut microbiota from twins discordant for obesity modulate metabolism in mice. Science 341:1241214

Rossen NG, Fuentes S, Van der Spek MJ, Tijssen JG, Hartman JH, Duflou A, Lowenberg M, Van den Brink GR, Mathus-Vliegen EM, De Vos WM, Zoetendal EG, D'Haens GR, Ponsioen CY (2015) Findings from a randomized controlled trial of fecal transplantation for patients with ulcerative colitis. Gastroenterology 149:110–118.e4

Sandler RS, Everhart JE, Donowitz M, Adams E, Cronin K, Goodman C, Gemmen E, Shah S, Avdic A, Rubin R (2002) The burden of selected digestive diseases in the United States. Gastroenterology 122:1500–1511

Sartor RB (2008) Microbial influences in inflammatory bowel diseases. Gastroenterology 134:577–594

Schwartz M, Gluck M, Koon S (2013) Norovirus gastroenteritis after fecal microbiota transplantation for treatment of *Clostridium difficile* infection despite asymptomatic donors and lack of sick contacts. Am J Gastroenterol 108:1367

Seekatz AM, Aas J, Gessert CE, Rubin TA, Saman DM, Bakken JS, Young VB (2014) Recovery of the gut microbiome following fecal microbiota transplantation. MBio 5:e00893–e00814

Serestherapeutics (2016). Available: http://ir.serestherapeutics.com/phoenix.zhtml?c=254006&p=RssLanding&cat=news&id=2190006 [Accessed 2016]

Shi Y, Dong Y, Huang W, Zhu D, Mao H, Su P (2016) Fecal microbiota transplantation for ulcerative colitis: a systematic review and meta-analysis. PLoS One 11:e0157259

Sobhani I, Tap J, Roudot-Thoraval F, Roperch JP, Letulle S, Langella P, Corthier G, Tran Van Nhieu J, Furet JP (2011) Microbial dysbiosis in colorectal cancer (CRC) patients. PLoS One 6:e16393

Sokol H, Pigneur B, Watterlot L, Lakhdari O, Bermudez-Humaran LG, Gratadoux JJ, Blugeon S, Bridonneau C, Furet JP, Corthier G, Grangette C, Vasquez N, Pochart P, Trugnan G, Thomas G, Blottiere HM, Dore J, Marteau P, Seksik P, Langella P (2008) *Faecalibacterium prausnitzii* is an anti-inflammatory commensal bacterium identified by gut microbiota analysis of Crohn disease patients. Proc Natl Acad Sci USA 105:16731–16736

Song LL, Li YM (2016) Current noninvasive tests for colorectal cancer screening: an overview of colorectal cancer screening tests. World J Gastrointest Oncol 8:793–800

Surawicz CM, McFarland LV, Greenberg RN, Rubin M, Fekety R, Mulligan ME, Garcia RJ, Brandmarker S, Bowen K, Borjal D, Elmer GW (2000) The search for a better treatment for recurrent *Clostridium difficile* disease: use of high-dose vancomycin combined with Saccharomyces boulardii. Clin Infect Dis 31:1012–1017

Surawicz CM, Brandt LJ, Binion DG, Ananthakrishnan AN, Curry SR, Gilligan PH, McFarland LV, Mellow M, Zuckerbraun BS (2013) Guidelines for diagnosis, treatment, and prevention of *Clostridium difficile* infections. Am J Gastroenterol 108:478–498. quiz 499

Suskind DL, Brittnacher MJ, Wahbeh G, Shaffer ML, Hayden HS, Qin X, Singh N, Damman CJ, Hager KR, Nielson H, Miller SI (2015) Fecal microbial transplant effect on clinical outcomes and fecal microbiome in active Crohn's disease. Inflamm Bowel Dis 21:556–563

Theriot CM, Koenigsknecht MJ, Carlson PE Jr, Hatton GE, Nelson AM, Li B, Huffnagle GB, Li JZ, Young VB (2014) Antibiotic-induced shifts in the mouse gut microbiome and metabolome increase susceptibility to *Clostridium difficile* infection. Nat Commun 5:3114

Tjalsma H, Boleij A, Marchesi JR, Dutilh BE (2012) A bacterial driver-passenger model for colorectal cancer: beyond the usual suspects. Nat Rev Microbiol 10:575–582

Tojo R, Suarez A, Clemente MG, De Los Reyes-Gavilan CG, Margolles A, Gueimonde M, Ruas-Madiedo P (2014) Intestinal microbiota in health and disease: role of bifidobacteria in gut homeostasis. World J Gastroenterol 20:15163–15176

Turnbaugh PJ, Ley RE, Mahowald MA, Magrini V, Mardis ER, Gordon JI (2006) An obesity-associated gut microbiome with increased capacity for energy harvest. Nature 444:1027–1031

Turnbaugh PJ, Hamady M, Yatsunenko T, Cantarel BL, Duncan A, Ley RE, Sogin ML, Jones WJ, Roe BA, Affourtit JP, Egholm M, Henrissat B, Heath AC, Knight R, Gordon JI (2009) A core gut microbiome in obese and lean twins. Nature 457:480–484

Tvede M, Rask-Madsen J (1989) Bacteriotherapy for chronic relapsing *Clostridium difficile* diarrhoea in six patients. Lancet 1:1156–1160

Van Nood E, Vrieze A, Nieuwdorp M, Fuentes S, Zoetendal EG, De Vos WM, Visser CE, Kuijper EJ, Bartelsman JF, Tijssen JG, Speelman P, Dijkgraaf MG, Keller JJ (2013) Duodenal infusion of donor feces for recurrent *Clostridium difficile*. N Engl J Med 368:407–415

Varela E, Manichanh C, Gallart M, Torrejon A, Borruel N, Casellas F, Guarner F, Antolin M (2013) Colonisation by *Faecalibacterium prausnitzii* and maintenance of clinical remission in patients with ulcerative colitis. Aliment Pharmacol Ther 38:151–161

Villanueva-Millan MJ, Perez-Matute P, Oteo JA (2015) Gut microbiota: a key player in health and disease. A review focused on obesity. J Physiol Biochem 71:509–525

Vipperla K, O'Keefe SJ (2016) Diet, microbiota, and dysbiosis: a 'recipe' for colorectal cancer. Food Funct 7:1731–1740

Vrieze A, Van Nood E, Holleman F, Salojarvi J, Kootte RS, Bartelsman JF, Dallinga-Thie GM, Ackermans MT, Serlie MJ, Oozeer R, Derrien M, Druesne A, Van Hylckama Vlieg JE, Bloks VW, Groen AK, Heilig HG, Zoetendal EG, Stroes ES, De Vos WM, Hoekstra JB, Nieuwdorp M (2012) Transfer of intestinal microbiota from lean donors increases insulin sensitivity in individuals with metabolic syndrome. Gastroenterology 143:913–916.e7

Wilcox MH, Mooney L, Bendall R, Settle CD, Fawley WN (2008) A case-control study of community-associated *Clostridium difficile* infection. J Antimicrob Chemother 62:388–396

Willing BP, Dicksved J, Halfvarson J, Andersson AF, Lucio M, Zheng Z, Jarnerot G, Tysk C, Jansson JK, Engstrand L (2010) A pyrosequencing study in twins shows that gastrointestinal microbial profiles vary with inflammatory bowel disease phenotypes. Gastroenterology 139:1844–1854.e1

Wu N, Yang X, Zhang R, Li J, Xiao X, Hu Y, Chen Y, Yang F, Lu N, Wang Z, Luan C, Liu Y, Wang B, Xiang C, Wang Y, Zhao F, Gao GF, Wang S, Li L, Zhang H, Zhu B (2013) Dysbiosis signature of fecal microbiota in colorectal cancer patients. Microb Ecol 66:462–470

Xu L, Zhang T, Cui B, He Z, Xiang J, Long C, Peng Z, Li P, Huang G, Ji G, Zhang F (2016) Clinical efficacy maintains patients' positive attitudes toward fecal microbiota transplantation. Medicine (Baltimore) 95:e4055

Youngster I, Sauk J, Pindar C, Wilson RG, Kaplan JL, Smith MB, Alm EJ, Gevers D, Russell GH, Hohmann EL (2014) Fecal microbiota transplant for relapsing *Clostridium difficile* infection using a frozen inoculum from unrelated donors: a randomized, open-label, controlled pilot study. Clin Infect Dis 58:1515–1522

Youngster I, Mahabamunuge J, Systrom HK, Sauk J, Khalili H, Levin J, Kaplan JL, Hohmann EL (2016) Oral, frozen fecal microbiota transplant (FMT) capsules for recurrent *Clostridium difficile* infection. BMC Med 14:134

Zackular JP, Baxter NT, Iverson KD, Sadler WD, Petrosino JF, Chen GY, Schloss PD (2013) The gut microbiome modulates colon tumorigenesis. MBio 4:e00692-13

Zenlea T, Peppercorn MA (2014) Immunosuppressive therapies for inflammatory bowel disease. World J Gastroenterol 20:3146–3152

Zhang F, Luo W, Shi Y, Fan Z, Ji G (2012) Should we standardize the 1,700-year-old fecal microbiota transplantation? Am J Gastroenterol 107:1755–1756

Zhang YZ, Li YY (2014) Inflammatory bowel disease: pathogenesis. World J Gastroenterol 20:91–99

Zipursky JS, Sidorsky TI, Freedman CA, Sidorsky MN, Kirkland KB (2014) Physician attitudes toward the use of fecal microbiota transplantation for the treatment of recurrent *Clostridium difficile* infection. Can J Gastroenterol Hepatol 28:319–324

Chapter 12
Statistical Models and Analysis of Microbiome Data from Mice and Humans

Yinglin Xia and Jun Sun

Abstract After the initiation of the Human Microbiome Project in 2007, numerous statistical and bioinformatic tools for data analysis and computational methods were developed and applied to meet the needs of microbiome studies. One of the popular platforms is to implement the newly developed statistical and bioinformatic methods and models using R packages.

In this chapter, we introduce the widely used and newly developed statistical methods and models in the ecology and microbiome fields. We show readers how to use the current available statistical tools based on the R programming language to analyze microbiome data. Our purpose is to provide the analytical steps and tools to be implemented by microbiome researchers, who may not have advanced knowledge of statistical models and R programming language. Specifically, this chapter covers frequently used univariate and multivariate statistical models and visualization tools, in addition to alpha and beta metrics and R programming skills, using real data from mouse and human microbiome studies.

Keywords Gut microbiome • Statistical methods • Statistical analysis • R package

Y. Xia (✉)
Division of Academic Internal Medicine and Geriatrics, Department of Medicine, University of Illinois at Chicago, Chicago, IL, USA

Division of Gastroenterology and Hepatology, Department of Medicine, University of Illinois at Chicago, Chicago, IL, USA
e-mail: yxia@uic.edu

J. Sun (✉)
Division of Gastroenterology and Hepatology, Department of Medicine, University of Illinois at Chicago, Chicago, IL, USA
e-mail: junsun7@uic.edu

J. Sun, P.K. Dudeja (eds.), *Mechanisms Underlying Host-Microbiome Interactions in Pathophysiology of Human Diseases*, Physiology in Health and Disease,
https://doi.org/10.1007/978-1-4939-7534-1_12

12.1 Introduction

RStudio (RStudio Team 2016) is a free and open-source integrated development environment (IDE) for R (R Core Team 2016), a widely used open-source programming language and environment for statistical computing and graphics. Both R and RStudio are under active development by a dedicated team of researchers with a commitment to documentation and software design. The purpose of this chapter is to provide an understanding of the statistical models and a step-by-step introduction to microbiome data analysis using RStudio. The analysis steps and tools described in this chapter could be implemented by microbiome researchers without advanced knowledge of statistical models and the R programming language.

12.2 Materials

12.2.1 Software

R can be downloaded from http://www.r-project.org and installed on all three mainstream operating systems (Windows, Mac, Unix/Linux). RStudio can be downloaded from https://www.rstudio.com/products/rstudio/download3/ and installed on four supported platforms (Windows, Mac, Ubuntu, and Fedora). The general installation manual and introductory tutorials can be obtained from the same website. Similar to other statistical software packages, R provides a statistical framework and terminal-based interface for users to input commands for data manipulation. As an IDE, all statistical analyses and graphics can be implemented through RStudio. Additional packages (Table 12.1) from R (http://www.r-project.org) are required before starting the analysis. Details of the package installation can be found in Sect. 12.3.1. The R terminal output is highlighted in `Lucida Console` font throughout the chapter.

12.2.2 Datasets

12.2.2.1 Vdr$^{-/-}$ Mice

The overall purpose of this study is to explore if vitamin D receptor (VDR) status regulates the composition and functions of the intestinal bacterial community. The murine microbiome data sets include samples from both the fecal and cecal locations of five VDR knock-out mice (Vdr$^{-/-}$) and three wild-type (WT) mice (Jin et al. 2015; Wang et al. 2016). Bacterial DNA was extracted and sequenced with 454 pyrosequencing. The post-sequencing data have six taxonomic ranks,

Table 12.1 List of add-on R packages required for analysis

Package	Description
Vegan; Oksanen et al. (2016)	R package originally developed for the analysis of ecological communities. It has tools for analyzing ecological diversity, and for the multivariate analysis of communities. It also has been widely used for analyzing microbiome data
GUniFrac; Chen (2012)	Generalized UniFrac distance for comparing microbial communities. Permutational multivariate analysis of variance using multiple distance matrices
dplyr; Wickham and Francois (2016)	dplyr is a powerful R-package for transforming and summarizing tabular data with rows and columns
BiodiversityR; Kindt and Coe (2005)	A graphical user interface package (via the R-commander) and utility functions (often based on the vegan package) for statistical analysis of biodiversity and ecological communities

Table 12.2 Example sample information file in comma-separated values format

Fecal sample	Group	Cecal sample	Group
1_11_drySt-28F	$Vdr^{-/-}$	19_11_CeSt-28F	$Vdr^{-/-}$
2_12_drySt-28F	$Vdr^{-/-}$	20_12_CeSt-28F	$Vdr^{-/-}$
3_13_drySt-28F	$Vdr^{-/-}$	21_13_CeSt-28F	$Vdr^{-/-}$
4_14_drySt-28F	$Vdr^{-/-}$	22_14_CeSt-28F	$Vdr^{-/-}$
5_15_drySt-28F	$Vdr^{-/-}$	23_15_CeSt-28F	$Vdr^{-/-}$
7_22_drySt-28F	WT	25_22_CeSt-28F	WT
8_23_drySt-28F	WT	26_23_CeSt-28F	WT
9_24_drySt-28F	WT	27_24_CeSt-28F	WT

including phylum, class, family, order, genus, and species. The null hypothesis is that VDR status and intestinal location are not associated with taxonomic alterations of the bacterial community in the gut. For better differentiation from sample to sample, we analyzed the intestinal microbiota at the genus level from the cecum (Table 12.2).

The original data set is a matrix or table with rows for bacteria and columns for samples.

12.2.2.2 Cigarette Smokers

The cigarette smokers' data set (Charlson et al. 2010; Chen 2012), used here to illustrate compositional data analysis, is part of a microbiome data set for studying the effect of smoking on the upper respiratory tract microbiome. The original data set contains samples of microbiomes from the throat and nose. It contains 60 subjects consisting of 32 nonsmokers and 28 smokers. As an illustration, only the first 20 subjects' data from the throat microbiome were used (Table 12.3).

Table 12.3 Partial operational taxonomic units and meta tables of the cigarette smokers dataset

```
> head(throat.otu.tab)
            4695 2983 2554 3315 879 1313 5661 4125 2115 3309 3225 514 3427 484
ESC_1.1_OPL    1    0    0    0   0    0    0    0    0    0    0   0    0   0
ESC_1.3_OPL    0    0    0    0   0    0    0    0    0    0    0   0    0   0
ESC_1.4_OPL    0    0    0    0   0    0    0    0    0    0    0   0    0   0
ESC_1.5_OPL    1    0    0    0   0    0    0    0    0    0    0   0    0   0
ESC_1.6_OPL    0    0    0    0   0    0    0    0    0    0    0   0    0   0
ESC_1.10_OPL   0    0    0    0   0    0    0    0    0    0    0   0    0   0

> head(throat.meta)
             BarcodeSequence LinkerPrimerSequence SmokingStatus PatientID
ESC_1.1_OPL         ACGTCATG      CTGCTGCCTYCCGTA     NonSmoker         1
ESC_1.3_OPL         ACTCGTGA      CTGCTGCCTYCCGTA        Smoker         3
ESC_1.4_OPL         ACTGCTGA      CTGCTGCCTYCCGTA        Smoker         4
ESC_1.5_OPL         AGACTGTC      CTGCTGCCTYCCGTA        Smoker         5
ESC_1.6_OPL         AGCTGATC      CTGCTGCCTYCCGTA        Smoker         6
ESC_1.10_OPL        ATGCGCTA      CTGCTGCCTYCCGTA        Smoker        10
```

12.3 Methods

12.3.1 An Introduction to R

The computer language R consists of many user-written packages aimed at solving different statistical problems. We begin with the base version of R, which can be downloaded from the official R website. The capabilities of R can be expanded by installing additional packages, e.g., via using the **install.packages()** command. To see what packages are installed, we can use the **installed.packages()** command. This returns a matrix with a row for each installed package. Below are the first five R packages:

```
> installed.packages()[1:5,]
        Package   LibPath                               Version    Priority
ade4    "ade4"    "C:/Users/Yinglin/R/win-library/3.3"  "1.7-4"    NA
ALDEx2  "ALDEx2"  "C:/Users/Yinglin/R/win-library/3.3"  "1.4.0"    NA
bayesm  "bayesm"  "C:/Users/Yinglin/R/win-library/3.3"  "3.0-2"    NA
BH      "BH"      "C:/Users/Yinglin/R/win-library/3.3"  "1.60.0-2" NA
cluster "cluster" "C:/Users/Yinglin/R/win-library/3.3"  "2.0.5"    "recommended"
```

We can also check if we have already installed a specific package (e.g., ALDEx2) by typing the command:

```
> a<-installed.packages()
> packages<-a[,1]
> is.element("ALDEx2", packages)
[1] TRUE
```

After downloading and installing R and RStudio software, an R or RStudio terminal can be started to install the required additional packages. Any package that does not appear in the installed package matrix must be installed and loaded before its functions can be used. A package can be installed using **install.packages** ("*<package name>*"):

```
> install.packages("vegan")
```

In R, additional packages can also be installed from the R terminal menu "Packages" → "select the CRAN mirror" → "select repositories."

In RStudio, you can also click "Packages" → "Install" → type package name (e.g., vegan in the column "Packages") and choose to install from "Repository (CRAN, CRANextra)" or "Package Archive File(.zip;,tar.gz)" if you downloaded the R package to your computer → click "Install" to install additional packages.

After installing, the packages can be loaded in either R or RStudio with the following command:

```
> library(vegan)
```

Or check this package from the User Library in RStudio.

We can check the version information about R and the attached packages by using function sessionInfo():

```
> sessionInfo()
R version 3.2.2 (2015-08-14)
Platform: x86_64-w64-mingw32/x64 (64-bit)
Running under: Windows 7 x64 (build 7601) Service Pack 1

locale:
[1] LC_COLLATE=English_United States.1252
[2] LC_CTYPE=English_United States.1252
[3] LC_MONETARY=English_United States.1252
[4] LC_NUMERIC=C
[5] LC_TIME=English_United States.1252

attached base packages:
[1] stats     graphics  grDevices utils     datasets  methods   base

loaded via a namespace (and not attached):
[1] tools_3.2.2
```

We can also check the current working directory using the getwd() command:

```
> getwd()
 [1] "C:/Users/Yinglin"
```

This shows the current working directory or folder location. However, if we want to work in a specific folder for our R program, data, and to save results in specific folders, we can set the working directory to this folder. Organizing data and results is very helpful for the flexible use of the scripts. For this project, we created a directory "SpringerBookChapter" to store the raw and intermediate data files and the analysis results:

```
> setwd("E:/Home/SpringerBookChapter/")
```

In RStudio, we can choose "Session" → "Set Working Directory" → "To Source File Location" to set the working directory. Use the getwd() function again, and you will find that your directory has been changed.

```
> getwd()
[1] "E:/Home/SpringerBookChapter"
```

By typing the command below, we can set the working directory back to the file pane folder:

```
> setwd("~/")
```

In RStudio, we can also choose "Session"→ "Set Working Directory"→ "To Files Pane Location" to do the same thing.

```
> getwd()
[1] "C:/Users/Yinglin"
```

The first step is always to import data into R. Community microbiome data are usually in the form of spreadsheet tables, with rows usually denoting the various taxa (bacteria) and columns denoting samples. However, some R packages need to transpose the data into rows of samples and columns of taxa (bacteria). R has functions such as read.table(), read.delim(), read.csv(), and read.csv2() to import the data from files into the R working space. In this book chapter, we used function read.csv() to import microbiome sample data into a data frame object in R and named the dataset "abund_table":

```
> abund_table=read.csv("VdrMice.csv",row.names=1,check.names=FALSE)
```

To check the first six rows to see what the data look like, apply the following command:

```
> head(abund_table)
                        20_12_CeSt-28F 19_11_CeSt-28F 21_13_CeSt-28F 22_14_CeSt-28F 23_15_CeSt-28F
Tannerella                          67              0              6             20             37
Lactococcus                        737            422            173            580           4867
Lactobacillus                      597            330            639            633           1819
Lactobacillus::Lactococcus          12              7              0              3             25
Parasutterella                       0              0              0              0              2
Helicobacter                         0              0              0              0              0
                        25_22_CeSt-28F 26_23_CeSt-28F 27_24_CeSt-28F
Tannerella                          38             81            235
Lactococcus                        707           1404           1913
Lactobacillus                      625           1361            365
Lactobacillus::Lactococcus           9             10             13
Parasutterella                       0              2              0
Helicobacter                         0              0              0
```

The package "vegan" that we use later requires rows to be samples and columns to be taxa (bacteria). We use the t() function to transpose the dataset "abund_table" and name the transposed version "abund_table_t." Following this, we check the number of rows and columns:

```
> abund_table_t<-t(abund_table)
> ncol(abund_table_t)  # for the number of genera
[1] 248
>
> nrow(abund_table_t)  # for the number of samples
[1] 8
```

We can use the write.table() function to output a data frame object like this:

```
> write.table(abund_table_t, "Genus_by_row.csv", quote=F, sep="\t")
```

"Quote" means there are quotation marks in the table. As we do not want these, we use the command quote = F to suppress them.

12.3.2 Data Exploration Using R

We can obtain group information from the sample names. For easier understanding and model fitting, the groups are transformed into factors from characters.

The function strsplit() is used to split the row names of "abund_table" (strings) into substrings according to the presence of the substring "_" within them. We name the object "grouping."

```
> grouping<-data.frame(row.names=rownames(abund_table_t),t(as.data.
frame(strsplit(rownames(abund_table_t),"_"))))
> grouping
                  X1 X2       X3
20_12_CeSt-28F 20 12 CeSt-28F
19_11_CeSt-28F 19 11 CeSt-28F
21_13_CeSt-28F 21 13 CeSt-28F
22_14_CeSt-28F 22 14 CeSt-28F
23_15_CeSt-28F 23 15 CeSt-28F
25_22_CeSt-28F 25 22 CeSt-28F
26_23_CeSt-28F 26 23 CeSt-28F
27_24_CeSt-28F 27 24 CeSt-28F
```

Then, the function ifelse() is used to set comparison groups:

```
> grouping$Group <- with(grouping,ifelse(as.factor(X2)%in% c(11,
12,13,14,15),c("Vdr-/-"), c("WT")))
> grouping
               X1 X2       X3  Group
20_12_CeSt-28F 20 12 CeSt-28F Vdr-/-
19_11_CeSt-28F 19 11 CeSt-28F Vdr-/-
21_13_CeSt-28F 21 13 CeSt-28F Vdr-/-
22_14_CeSt-28F 22 14 CeSt-28F Vdr-/-
23_15_CeSt-28F 23 15 CeSt-28F Vdr-/-
25_22_CeSt-28F 25 22 CeSt-28F     WT
26_23_CeSt-28F 26 23 CeSt-28F     WT
27_24_CeSt-28F 27 24 CeSt-28F     WT
> names(grouping)
[1] "X1"    "X2"    "X3"    "Group"
```

Finally, the function c() is used to extract group information:

```
> grouping_inf<- grouping[,c(4)]
> grouping_inf
[1] "Vdr-/-" "Vdr-/-" "Vdr-/-" "Vdr-/-" "Vdr-/-" "WT"    "WT"    "WT"
```

In the data set, the samples are ordered based on the genetic conditions Vdr−/− and WT, so that we can simply set comparison groups using the rep() function. The vector of conditions is consistent with the order of the samples in the input counts table.

```
> conds <- c(rep("Vdr-/-", 5), rep("WT", 3))
> conds
[1] "Vdr-/-" "Vdr-/-" "Vdr-/-" "Vdr-/-" "Vdr-/-" "WT"    "WT"    "WT"
```

We can examine the basic characteristics of the sequenced mice microbiome by looking at the distribution of all genera and find the most represented bacterial genus members of the communities. As many statistical methods in ecology and microbiome studies are sensitive to the total abundance of taxa in a sample, we should convert these absolute abundance estimates to a relative abundance estimate. We can do this with the function decostand() from the vegan package.

First, total abundance in each sample is checked:

```
> apply(abund_table_t, 1, sum)
20_12_CeSt-28F 19_11_CeSt-28F 21_13_CeSt-28F 22_14_CeSt-28F 23_15_CeSt-28F
          3239           5067           2397           3788           9264
25_22_CeSt-28F 26_23_CeSt-28F 27_24_CeSt-28F
          2072           6903           6327
```

Then, we can use the function decostand() to convert the genus abundances of community data to relative abundances (that is, in all samples, all genera add up to 1). The function decostand() provides some popular and effective standardization methods for community ecologists and microbiome researchers.

If the vegan package is not loaded yet, we can load it with the following command:

```
> library(vegan)
```

In the arguments, the standardization method = “total” indicates that the abundance accounts will be divided by the margin total:

```
> abund_table_stand<- decostand(abund_table, method = "total")
```

First, we check the total abundance in each sample to see if the sum of all of them is 1:

```
> apply(abund_table_stand, 1, sum)
20_12_CeSt-28F 19_11_CeSt-28F 21_13_CeSt-28F 22_14_CeSt-28F 23_15_CeSt-28F
             1              1              1              1              1
25_22_CeSt-28F 26_23_CeSt-28F 27_24_CeSt-28F
             1              1              1
```

Then, we can look at the transformed data to see if we have obtained relative abundances:

```
> abund_table_stand [1:5, 1:5]
                Tannerella Lactococcus Lactobacillus Lactobacillus::Lactococcus
20_12_CeSt-28F 0.020685397  0.22753936    0.18431615                0.0037048472
19_11_CeSt-28F 0.000000000  0.08328399    0.06512729                0.0013814881
21_13_CeSt-28F 0.002503129  0.07217355    0.26658323                0.0000000000
22_14_CeSt-28F 0.005279831  0.15311510    0.16710665                0.0007919747
23_15_CeSt-28F 0.003993955  0.52536701    0.19635147                0.0026986183
                Parasutterella
20_12_CeSt-28F    0.0000000000
19_11_CeSt-28F    0.0000000000
21_13_CeSt-28F    0.0000000000
22_14_CeSt-28F    0.0000000000
23_15_CeSt-28F    0.0002158895
```

As we standardized the abundance count into the relative abundances, we can now find the most abundant genera by calculating per-column mean values:

```
> Genus_Mean <- colMeans(abund_table_stand)
> head(Genus_Mean)
  Tannerella                       Lactococcus              Lactobacillus
  1.245981e-02                      2.385550e-01               1.794970e-01
Lactobacillus::Lactococcus          Parasutterella            Helicobacter
  2.052986e-03                      6.320232e-05               0.000000e+00
```

We can assess the distribution by plotting a histogram (Fig. 12.1):

```
> hist(colMeans(abund_table_stand))
```

Now, we can sort the mean genus of relative abundances from lowest to highest to identify the most abundant genera:

```
> sort(colMeans(abund_table_stand))
```

The order is from lowest to highest now; thus, the most abundant genera are at the bottom of this list. We can pick the five most abundant genera and bind them to a new data frame. The matrix object can only be allowed by using the column number or row number, not by a name. Thus, we need to define "abund_table_stand" as a data.frame object by using abund_table_stand_f

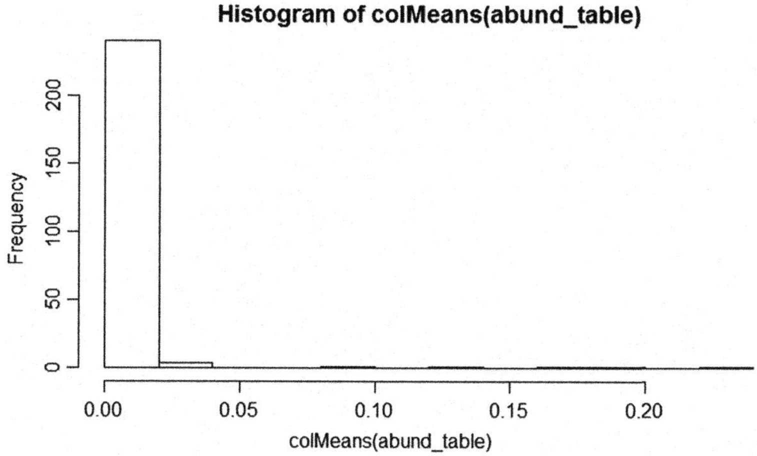

Fig. 12.1 Histogram of the distribution of genus means

<- data.frame (abund_table_stand) to make it easy to extract a certain column by its name.

```
> abund_table_stand_f <- data.frame(abund_table_stand)
```

From the genus list, we can see that the top five most abundant genera are: *Lactococcus*, *Butyrivibrio*, *Lactobacillus*, *Clostridium*, and *Turicibacter*.

The following cbind() function is used to bind them to a new data frame called "genera_major5":

```
> genera_major5 <- cbind(abund_table_stand_f$Lactococcus, abund_
table_stand_f$Butyrivibrio,    abund_table_stand_f$Lactobacillus,
abund_table_stand_f$Clostridium,             abund_table_stand_f
$Turicibacter)

> genera_major5
           [,1]       [,2]       [,3]       [,4]         [,5]
[1,] 0.22753936 0.13893177 0.18431615 0.12380364 0.0546464958
[2,] 0.08328399 0.63410302 0.06512729 0.08387606 0.0868363923
[3,] 0.07217355 0.24280350 0.26658323 0.09762203 0.1989987484
[4,] 0.15311510 0.02402323 0.16710665 0.19878564 0.3210137276
[5,] 0.52536701 0.01047064 0.19635147 0.10924007 0.0003238342
[6,] 0.34121622 0.03281853 0.30164093 0.15781853 0.0038610039
[7,] 0.20338983 0.04447342 0.19716065 0.15022454 0.0024626974
[8,] 0.30235499 0.31942469 0.05768927 0.11348190 0.0015805279
```

How many percentages of these five genera make up each sample? We can check using the rowSums function:

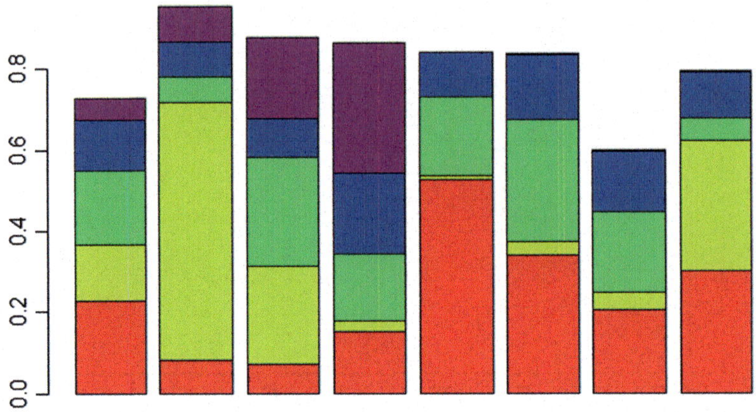

Fig. 12.2 Bar plot of the five most abundant genera in each sample

```
> rowSums(genera_major5)
[1] 0.7292374 0.9532268 0.8781811 0.8640444 0.8417530 0.8373552
0.5977111 0.7945314
```

We now can visualize these five most abundant genera by using the barplot() function. The function barplot() takes transposed matrix to plot bars. The default color for plotting is gray, but here we set five different colors from the rainbow to present five major abundant genera by assigning col = rainbow(5) (Fig. 12.2):

```
> barplot(t(genera_major5), col=rainbow(5))
```

We can also order the samples based on the value of the first column (*Lactococcus*) by using the order() function (Fig. 12.3):

```
> barplot(t(genera_major5[order(genera_major5[,1]), ]), col=rainbow(5))
```

12.3.3 Univariate Community Analysis of Alpha Diversity

12.3.3.1 Alpha Diversity and Measures

Alpha diversity is an essential concept in both ecology and microbiome study. Two fundamental questions encountered by researchers in these fields are how many

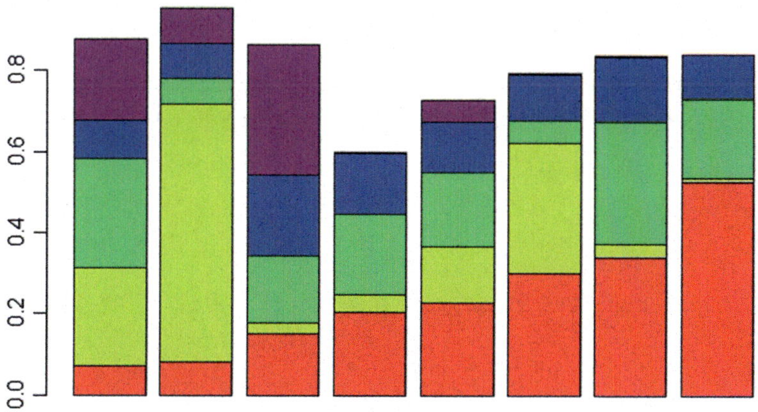

Fig. 12.3 Bar plot of the ordered five most abundant genera

species are present in the community (richness) and how evenly each species makes up the community (evenness).

In the next section, we illustrate how to calculate the most widely used alpha diversities in microbiome study including a qualitative taxa-based measure, the Chao 1 index (Chao 1984), and quantitative taxa-based measures, the Shannon/Shannon–Wiener index (Shannon 1948; Shannon and Weaver 1949) and Simpson's index (Simpson 1949).

12.3.3.2 Calculating Alpha Diversities

Chao 1 Index

We use the estimateR() function to calculate the Chao 1 index.

```
> index=estimateR(abund_table_t*1000)
> index
         20_12_CeSt-28F 19_11_CeSt-28F 21_13_CeSt-28F 22_14_CeSt-28F
S.obs                52             37             42             46
S.chao1              52             37             42             46
se.chao1              0              0              0              0
S.ACE               NaN            NaN            NaN            NaN
se.ACE              NaN            NaN            NaN            NaN
         23_15_CeSt-28F 25_22_CeSt-28F 26_23_CeSt-28F 27_24_CeSt-28F
S.obs                54             48             70             68
S.chao1              54             48             70             68
se.chao1              0              0              0              0
```

```
S.ACE                   NaN              NaN              NaN              NaN
se.ACE                  NaN              NaN              NaN              NaN
> head(index)
        20_12_CeSt-28F 19_11_CeSt-28F 21_13_CeSt-28F 22_14_CeSt-28F
S.obs                   52               37               42               46
S.chao1                 52               37               42               46
se.chao1                 0                0                0                0
S.ACE                  NaN              NaN              NaN              NaN
se.ACE                 NaN              NaN              NaN              NaN
        23_15_CeSt-28F 25_22_CeSt-28F 26_23_CeSt-28F 27_24_CeSt-28F
S.obs                   54               48               70               68
S.chao1                 54               48               70               68
se.chao1                 0                0                0                0
S.ACE                  NaN              NaN              NaN              NaN
se.ACE                 NaN              NaN              NaN              NaN
```

The function estimateR() generates five indices, in which chao1 is listed as the second row. Thus, we can extract the Chao 1 index using the following codes:

```
> chao1_genus=estimateR(abund_table_t*1000)[2,]
> chao1_genus
20_12_CeSt-28F 19_11_CeSt-28F 21_13_CeSt-28F 22_14_CeSt-28F 23_15_CeSt-28F
            52             37             42             46             54
25_22_CeSt-28F 26_23_CeSt-28F 27_24_CeSt-28F
            48             70             68
```

Then, we can make a data frame of Chao 1 richness:

```
> CH=estimateR(abund_table_t*1000)[2,]
>     df_CH     <-data.frame(sample=names(CH),value=CH,measure=rep
("Chao1",length(CH)))
> df_CH
                       sample value measure
20_12_CeSt-28F 20_12_CeSt-28F    52   Chao1
19_11_CeSt-28F 19_11_CeSt-28F    37   Chao1
21_13_CeSt-28F 21_13_CeSt-28F    42   Chao1
22_14_CeSt-28F 22_14_CeSt-28F    46   Chao1
23_15_CeSt-28F 23_15_CeSt-28F    54   Chao1
25_22_CeSt-28F 25_22_CeSt-28F    48   Chao1
26_23_CeSt-28F 26_23_CeSt-28F    70   Chao1
27_24_CeSt-28F 27_24_CeSt-28F    68   Chao1
```

Richness Based on Rarefaction

We can also calculate richness based on rarefaction. As various species (or taxa) are sampled in the community, an issue arises: the higher the number of individuals sampled, the larger amounts of that species are found. Rarefaction was developed to allow the calculation of species richness for a given number of individual samples based on rarefaction curves. This curve is a plot of the number of species as a function of the number of samples. It generally grows rapidly at first, as the most common species are found, but the curves plateau when only the rarest species remain to be sampled. Thus, rarefaction techniques are used to quantify species diversity of newly studied ecosystems, including microbiomes and community ecology. We use rarefaction to assess genera richness from the results of genera abundance sampling here:

```
> R<-rarefy(abund_table,min(rowSums(abund_table_t)))
> R
20_12_CeSt-28F 19_11_CeSt-28F 21_13_CeSt-28F 22_14_CeSt-28F 23_15_CeSt-28F
      45.78968       25.93238       40.13438       38.37177       36.69034
25_22_CeSt-28F 26_23_CeSt-28F 27_24_CeSt-28F
      48.00000       45.97843       46.13922
attr(,"Subsample")
[1] 2072
```

After rarefying, we store the results as a data frame for later use.

```
> df_R<-data.frame(sample=names(R),value=R,measure=rep("Richness",
length(R)))
> df_R
                       sample    value  measure
20_12_CeSt-28F 20_12_CeSt-28F 45.78968 Richness
19_11_CeSt-28F 19_11_CeSt-28F 25.93238 Richness
21_13_CeSt-28F 21_13_CeSt-28F 40.13438 Richness
22_14_CeSt-28F 22_14_CeSt-28F 38.37177 Richness
23_15_CeSt-28F 23_15_CeSt-28F 36.69034 Richness
25_22_CeSt-28F 25_22_CeSt-28F 48.00000 Richness
26_23_CeSt-28F 26_23_CeSt-28F 45.97843 Richness
27_24_CeSt-28F 27_24_CeSt-28F 46.13922 Richness
```

Shannon Index

The Shannon index can be calculated using either plain R functions or the diversity() function in the vegan R package. First, we illustrate how to use plain R functions to calculate Shannon diversity. Initially, we need the function decostand() to convert the data into proportions. We name the dataset "abund_table_total" to indicate the "total" method used for this conversion.

```
> abund_table_total<-decostand(abund_table_t, MARGIN=1, method="total")
```

With default settings, MARGIN = 1(1 = rows, and 2 = columns).
Then we multiply that matrix by a natural log-transformed matrix-- p*ln(p):

```
> abund_table_p_lnp<-abund_table_total*log(abund_table_total)
```

Finally, we sum the values by sample and multiply by −1 to obtain the Shannon index.

```
> rowSums(abund_table_p_lnp,na.rm=TRUE)*-1
20_12_CeSt-28F 19_11_CeSt-28F 21_13_CeSt-28F 22_14_CeSt-28F 23_15_CeSt-28F
      2.339725       1.344813       2.016113       1.955433       1.614456
25_22_CeSt-28F 26_23_CeSt-28F 27_24_CeSt-28F
      1.958839       2.270818       2.002195
```

The Shannon index can also be calculated using the diversity() function in the vegan package.

We can either explicitly specify index = "shannon," MARGIN = 1, or use the default argument. Either argument used gives the same estimated results.

```
> H <- diversity(abund_table_t,index="shannon",MARGIN=1)
> H
20_12_CeSt-28F 19_11_CeSt-28F 21_13_CeSt-28F 22_14_CeSt-28F 23_15_CeSt-28F
      2.339725       1.344813       2.016113       1.955433       1.614456
25_22_CeSt-28F 26_23_CeSt-28F 27_24_CeSt-28F
      1.958839       2.270818       2.002195
```

```
> H <- diversity(abund_table_t,MARGIN=1)
> H
20_12_CeSt-28F 19_11_CeSt-28F 21_13_CeSt-28F 22_14_CeSt-28F 23_15_CeSt-28F
      2.339725       1.344813       2.016113       1.955433       1.614456
25_22_CeSt-28F 26_23_CeSt-28F 27_24_CeSt-28F
      1.958839       2.270818       2.002195

> H <- diversity(abund_table_t)
> H
20_12_CeSt-28F 19_11_CeSt-28F 21_13_CeSt-28F 22_14_CeSt-28F 23_15_CeSt-28F
      2.339725       1.344813       2.016113       1.955433       1.614456
25_22_CeSt-28F 26_23_CeSt-28F 27_24_CeSt-28F
      1.958839       2.270818       2.002195
```

We can see from above that the two approaches give the same results. We can make a data frame of Shannon evenness:

```
> df_H<-data.frame(sample=names(H),value=H,measure=rep("Shannon",
length(H)))
> df_H
                        sample    value measure
20_12_CeSt-28F 20_12_CeSt-28F 2.339725 Shannon
19_11_CeSt-28F 19_11_CeSt-28F 1.344813 Shannon
21_13_CeSt-28F 21_13_CeSt-28F 2.016113 Shannon
22_14_CeSt-28F 22_14_CeSt-28F 1.955433 Shannon
23_15_CeSt-28F 23_15_CeSt-28F 1.614456 Shannon
25_22_CeSt-28F 25_22_CeSt-28F 1.958839 Shannon
26_23_CeSt-28F 26_23_CeSt-28F 2.270818 Shannon
27_24_CeSt-28F 27_24_CeSt-28F 2.002195 Shannon
```

Simpson Index

Similarly, the Simpson index can be calculated using either plain R functions or the diversity() function in the vegan R package. We use the plain R function first. The following R codes convert data into proportions by using the function decostand():

```
> abund_table_total<-decostand(abund_table_t, MARGIN=1, method="total")
```

Then, we square the proportions:

```
> abund_table_total_p2<-abund_table_total^2
```

Finally, we obtain the Simpson index by subtracting the row sums from one:

```
> 1-rowSums(abund_table_total_p2, na.rm=TRUE)
20_12_CeSt-28F 19_11_CeSt-28F 21_13_CeSt-28F 22_14_CeSt-28F 23_15_CeSt-28F
     0.8648902      0.5717474      0.8138716      0.8020054      0.6687232
25_22_CeSt-28F 26_23_CeSt-28F 27_24_CeSt-28F
     0.7649460      0.8545929      0.7844728
```

The follow R codes use the function diversity() to achieve the same result:

```
> simp <- diversity(abund_table_t, "simpson")
> simp
20_12_CeSt-28F 19_11_CeSt-28F 21_13_CeSt-28F 22_14_CeSt-28F 23_15_CeSt-28F
     0.8648902      0.5717474      0.8138716      0.8020054      0.6687232
25_22_CeSt-28F 26_23_CeSt-28F 27_24_CeSt-28F
     0.7649460      0.8545929      0.7844728
```

We can store the Simpson index as a new data frame:

```
> df_simp<-data.frame(sample=names(simp),value=simp,measure=rep
("Simpson",length(simp)))
> df_simp
                       sample     value measure
20_12_CeSt-28F 20_12_CeSt-28F 0.8648902 Simpson
19_11_CeSt-28F 19_11_CeSt-28F 0.5717474 Simpson
21_13_CeSt-28F 21_13_CeSt-28F 0.8138716 Simpson
22_14_CeSt-28F 22_14_CeSt-28F 0.8020054 Simpson
23_15_CeSt-28F 23_15_CeSt-28F 0.6687232 Simpson
25_22_CeSt-28F 25_22_CeSt-28F 0.7649460 Simpson
26_23_CeSt-28F 26_23_CeSt-28F 0.8545929 Simpson
27_24_CeSt-28F 27_24_CeSt-28F 0.7844728 Simpson
```

Combine All the Calculated Indices Together to Make a Data Frame

We can use the function rbind() to combine all the indices together to make a data frame for future use.

```
> df<-rbind(df_CH,df_R,df_H,df_simp)
> rownames(df)<-NULL
> df
            sample      value  measure
1  20_12_CeSt-28F 52.0000000    Chao1
2  19_11_CeSt-28F 37.0000000    Chao1
3  21_13_CeSt-28F 42.0000000    Chao1
4  22_14_CeSt-28F 46.0000000    Chao1
5  23_15_CeSt-28F 54.0000000    Chao1
6  25_22_CeSt-28F 48.0000000    Chao1
7  26_23_CeSt-28F 70.0000000    Chao1
8  27_24_CeSt-28F 68.0000000    Chao1
9  20_12_CeSt-28F 45.7896767 Richness
10 19_11_CeSt-28F 25.9323806 Richness
11 21_13_CeSt-28F 40.1343835 Richness
12 22_14_CeSt-28F 38.3717737 Richness
13 23_15_CeSt-28F 36.6903357 Richness
14 25_22_CeSt-28F 48.0000000 Richness
15 26_23_CeSt-28F 45.9784344 Richness
16 27_24_CeSt-28F 46.1392236 Richness
17 20_12_CeSt-28F  2.3397253  Shannon
18 19_11_CeSt-28F  1.3448129  Shannon
19 21_13_CeSt-28F  2.0161125  Shannon
20 22_14_CeSt-28F  1.9554325  Shannon
21 23_15_CeSt-28F  1.6144563  Shannon
22 25_22_CeSt-28F  1.9588390  Shannon
23 26_23_CeSt-28F  2.2708178  Shannon
24 27_24_CeSt-28F  2.0021951  Shannon
25 20_12_CeSt-28F  0.8648902  Simpson
26 19_11_CeSt-28F  0.5717474  Simpson
27 21_13_CeSt-28F  0.8138716  Simpson
28 22_14_CeSt-28F  0.8020054  Simpson
29 23_15_CeSt-28F  0.6687232  Simpson
30 25_22_CeSt-28F  0.7649460  Simpson
31 26_23_CeSt-28F  0.8545929  Simpson
32 27_24_CeSt-28F  0.7844728  Simpson
```

12.3.3.3 Comparisons of Diversities in Two Groups

We already defined the "grouping_inf" data frame in Sect. 12.3.2 and calculated Shannon diversity in section "Shannon Index". Here, we combine these two data frames to create a new data frame called "df_H_G".

```
> # combine df_H and grouping data frames to a new data frame for
later use
> df_H_Grouping <-cbind(df_H, grouping)
> df_H_Grouping
                           sample    value measure X1 X2       X3  Group
20_12_CeSt-28F 20_12_CeSt-28F 2.339725 Shannon 20 12 CeSt-28F Vdr-/-
19_11_CeSt-28F 19_11_CeSt-28F 1.344813 Shannon 19 11 CeSt-28F Vdr-/-
21_13_CeSt-28F 21_13_CeSt-28F 2.016113 Shannon 21 13 CeSt-28F Vdr-/-
22_14_CeSt-28F 22_14_CeSt-28F 1.955433 Shannon 22 14 CeSt-28F Vdr-/-
23_15_CeSt-28F 23_15_CeSt-28F 1.614456 Shannon 23 15 CeSt-28F Vdr-/-
25_22_CeSt-28F 25_22_CeSt-28F 1.958839 Shannon 25 22 CeSt-28F     WT
26_23_CeSt-28F 26_23_CeSt-28F 2.270818 Shannon 26 23 CeSt-28F     WT
27_24_CeSt-28F 27_24_CeSt-28F 2.002195 Shannon 27 24 CeSt-28F     WT
> rownames(df_H_Grouping)<-NULL
> df_H_G<-df_H_Grouping[,c(1,2,3,7)]
> df_H_G
          sample    value measure  Group
1 20_12_CeSt-28F 2.339725 Shannon Vdr-/-
2 19_11_CeSt-28F 1.344813 Shannon Vdr-/-
3 21_13_CeSt-28F 2.016113 Shannon Vdr-/-
4 22_14_CeSt-28F 1.955433 Shannon Vdr-/-
5 23_15_CeSt-28F 1.614456 Shannon Vdr-/-
6 25_22_CeSt-28F 1.958839 Shannon     WT
7 26_23_CeSt-28F 2.270818 Shannon     WT
8 27_24_CeSt-28F 2.002195 Shannon     WT
```

Now we can use the boxplot() function to plot Shannon diversities with Vdr$^{-/-}$ and WT groups (Fig. 12.4).

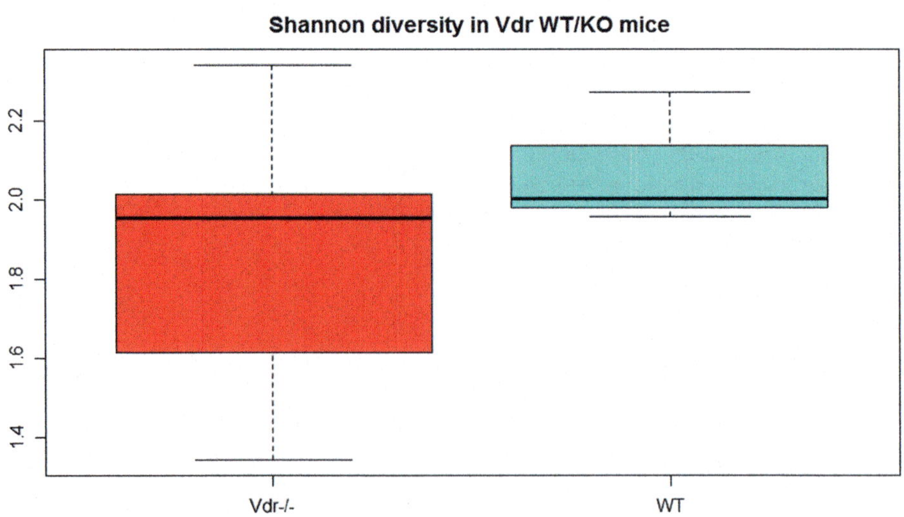

Fig. 12.4 Box plot of the Shannon index with Vdr$^{-/-}$ and wild-type groups

```
> boxplot(value ~ Group,data=df_H_G, col=rainbow(2),main="Shannon
diversity
in Vdr WT/KO mice")
```

Welch's Two Sample *t*-Test

Welch's *t*-test or the unequal variances *t*-test is adapted from the *t*-test. It is considered to be more reliable when the two samples have unequal variances and unequal sample sizes. Thus, here we use Welch's *t*-test to test the differences in Shannon diversity between the $Vdr^{-/-}$ and WT mice data we calculated above. The test value of the function *t*-test() should be numeric, and "Group" should be a binary factor.

```
> fit_t <- t.test(value ~ Group, data=df_H_G)
> fit_t

Welch Two Sample t-test

data:  value by Group
t = -1.1304, df = 5.7931, p-value = 0.3029
alternative hypothesis: true difference in means is not equal to 0
95 percent confidence interval:
 -0.7104788  0.2641267
sample estimates:
mean in group Vdr-/-    mean in group WT
            1.854108            2.077284
```

Wilcoxon Rank Sum Test

The Wilcoxon rank-sum test is equivalent to the Mann–Whitney *U* test. It is a nonparametric alternative to the two-sample *t*-test. It uses ranks of data from two independent samples to test the null hypothesis that the two populations are identical. Unlike the *t*-test, the Wilcoxon rank-sum test does not require the assumption of normal distributions. However, it is nearly as efficient as the *t*-test and therefore widely used in microbiome study. The Wilcoxon rank-sum test is fitted by using the function wilcox.test() below:

```
> fit_w <- wilcox.test(value ~ Group, data=df_H_G)
> fit_w
```

```
Wilcoxon rank sum test

data:  value by Group
W = 5, p-value = 0.5714
alternative hypothesis: true location shift is not equal to 0
```

12.3.3.4 Comparisons Among More Than Two Groups

One-Way ANOVA

Analysis of variance (ANOVA) generalizes the two-sample *t*-test to more than two groups. The null hypothesis of ANOVA is that the means of the groups compared are equal. ANOVA analysis relies on the assumption of normality of the underlying data. However, most of the micriobiome community composition data are not normally distributed; thus, in this book chapter, we only illustrate the capabilities of ANOVA for comparing diversity measures. For multivariate community composition data, either a nonparametric version of ANOVA or another suitable statistical method should be applied.

First, we load the abundance table, which includes both fecal and cecal samples:

```
> abund_table=read.csv("VdrMiceBacteria.csv",row.names=1,check.
names=FALSE)
```

Then, we will transpose the data into samples by taxa format:

```
> abund_table_t<-t(abund_table)
```

Next, we obtain grouping information:

```
> grouping<-data.frame(row.names=rownames(abund_table_t),t(as.data.
frame(strsplit(rownames(abund_table_t),"_"))))
```

We use the with() function to assign the variable “location” to identify fecal and cecal sites:

```
> grouping$Location <- with(grouping, ifelse(X3%in%"drySt-28F",
"Fecal", "Cecal"))
```

We then assign the group variable to the two genetic conditions: $Vdr^{-/-}$ and WT:

```
> grouping$Group <- with(grouping,ifelse(as.factor(X2)%in% c(11,
12,13,14,15),c("Vdr-/-"), c("WT")))
```

We then use the diversity() function in the vegan package to calculate Shannon diversity. If the vegan package has not been loaded, it should be loaded at this point.

```
> library(vegan)
```

We can now calculate Shannon diversity and make a data frame of it, as previously in Sect. 12.3.3.2:

```
> H<-diversity(abund_table_t, "shannon")
> df_H<-data.frame(sample=names(H),value=H,measure=rep("Shannon",
length(H)))
```

We can then combine Shannon diversity and group data frames together for analysis later:

```
> df_H_G <-cbind(df_H, grouping_inf)
```

We also use the with() and interaction() functions to create four levels of groups to identify Fecal.Vdr$^{-/-}$, Fecal.WT, Cecal.Vdr$^{-/-}$, and Cecal.WT. Note the default separator of '.'

```
> df_H_G$Group4<- with(df_H_G, interaction(Location,Group))
```

The created data frame looks like this:

```
> df_H_G
          sample    value measure Location  Group      Group4
1  5_15_drySt-28F 2.460729 Shannon    Fecal Vdr-/- Fecal.Vdr-/-
2  20_12_CeSt-28F 2.339725 Shannon    Cecal Vdr-/- Cecal.Vdr-/-
3  1_11_drySt-28F 2.228023 Shannon    Fecal Vdr-/- Fecal.Vdr-/-
4  2_12_drySt-28F 2.734405 Shannon    Fecal Vdr-/- Fecal.Vdr-/-
5  3_13_drySt-28F 2.077282 Shannon    Fecal Vdr-/- Fecal.Vdr-/-
6  4_14_drySt-28F 2.466830 Shannon    Fecal Vdr-/- Fecal.Vdr-/-
7  7_22_drySt-28F 1.777171 Shannon    Fecal     WT     Fecal.WT
8  8_23_drySt-28F 1.999559 Shannon    Fecal     WT     Fecal.WT
9  9_24_drySt-28F 1.971996 Shannon    Fecal     WT     Fecal.WT
10 19_11_CeSt-28F 1.344813 Shannon    Cecal Vdr-/- Cecal.Vdr-/-
11 21_13_CeSt-28F 2.016113 Shannon    Cecal Vdr-/- Cecal.Vdr-/-
12 22_14_CeSt-28F 1.955433 Shannon    Cecal Vdr-/- Cecal.Vdr-/-
13 23_15_CeSt-28F 1.614456 Shannon    Cecal Vdr-/- Cecal.Vdr-/-
14 25_22_CeSt-28F 1.958839 Shannon    Cecal     WT     Cecal.WT
```

```
15 26_23_CeSt-28F 2.270818 Shannon    Cecal     WT     Cecal.WT
16 27_24_CeSt-28F 2.002195 Shannon    Cecal     WT     Cecal.WT
```

To simplify the data, we only select value (Shannon diversity) and groups using the dplyr package:

```
> library(dplyr)
> df_H_G4 <- select(df_H_G, Group4,value)
> df_H_G4
        Group4    value
1  Fecal.Vdr-/- 2.460729
2  Cecal.Vdr-/- 2.339725
3  Fecal.Vdr-/- 2.228023
4  Fecal.Vdr-/- 2.734405
5  Fecal.Vdr-/- 2.077282
6  Fecal.Vdr-/- 2.466830
7      Fecal.WT 1.777171
8      Fecal.WT 1.999559
9      Fecal.WT 1.971996
10 Cecal.Vdr-/- 1.344813
11 Cecal.Vdr-/- 2.016113
12 Cecal.Vdr-/- 1.955433
13 Cecal.Vdr-/- 1.614456
14     Cecal.WT 1.958839
15     Cecal.WT 2.270818
16     Cecal.WT 2.002195
```

The boxplot() function is used to explore the four groups (Fig. 12.5).

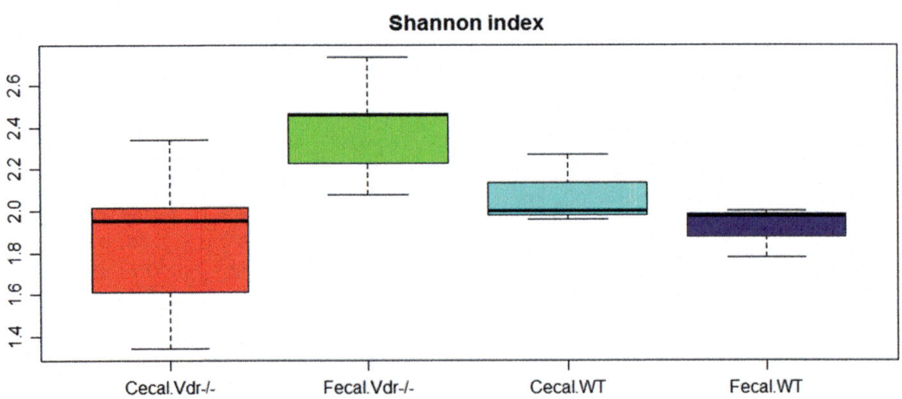

Fig. 12.5 Box plot of the Shannon index with $Vdr^{-/-}$ and wild-type groups in both fecal and cecal samples

```
> boxplot(value~Group4, data=df_H_G, col=rainbow(4), main="Shannon index")
```

The ANOVA is fitted by using the lm() and aov() functions respectively. Both functions give the same results:

```
> fit = lm(formula = value~Group4,data=df_H_G)
> anova (fit)
Analysis of Variance Table
Response: value
          Df  Sum Sq Mean Sq F value  Pr(>F)
Group4     3 0.82664 0.27555  3.5587 0.04753 *
Residuals 12 0.92915 0.07743
---
Signif. Codes:  0 '***' 0.001 '**' 0.01 '*' 0.05 '.' 0.1 ' ' 1

> summary(aov(value~Group4, data=df_H_G))
            Df Sum Sq Mean Sq F value Pr(>F)
Group4       3 0.8266 0.27555   3.559 0.0475 *
Residuals   12 0.9292 0.07743
---
Signif. Codes:  0 '***' 0.001 '**' 0.01 '*' 0.05 '.' 0.1 ' ' 1

> aov_fit <- aov(value~Group4,data=df_H_G) #calculate the anova
> summary(aov_fit, intercept=T)
            Df Sum Sq Mean Sq F value    Pr(>F)
(Intercept)  1  68.97   68.97 890.698 1.25e-12 ***
Group4       3   0.83    0.28   3.559    0.0475 *
Residuals   12   0.93    0.08
---
Signif. Codes:  0 '***' 0.001 '**' 0.01 '*' 0.05 '.' 0.1 ' ' 1
```

The following codes are used to do the pair-wise comparisons of mean differences among the four groups using the function pairwise.t.test():

```
> pairwise.t.test(df_H_G$value, df_H_G$Group4, p.adjust="none",
pool.sd = T)

    Pairwise comparisons using t tests with pooled SD

data:  df_H_G$value and df_H_G$Group4

               Cecal.Vdr-/- Fecal.Vdr-/- Cecal.WT
Fecal.Vdr-/-   0.0098       -            -
Cecal.WT       0.2936       0.1457       -
Fecal.WT       0.7650       0.0368       0.4920

P value adjustment method: none
```

The results show that the means of Shannon diversities between Fecal Vdr$^{-/-}$ and Cecal Vdr$^{-/-}$ groups are statistically significant with a p value of 0.0098. The Shannon diversities from the Fecal Vdr$^{-/-}$ group are also different than the Fecal WT group, with a p value of 0.0368. However, the p values were not adjusted by multiple group comparisons.

Several methods can be used to adjust the p values. Among them, we illustrate the Bonferroni, Holm, Benjamini and Hochberg (BH), Benjamini and Yekutieli (BY), and Tukey's honestly significant difference (Tukey's HSD) test methods.

(i) Bonferroni method

```
> pairwise.t.test(df_H_G$value, df_H_G$Group4, p.adjust="bonferroni", pool.
sd = T)

    Pairwise comparisons using t tests with pooled SD

data:  df_H_G$value and df_H_G$Group4

             Cecal.Vdr-/- Fecal.Vdr-/- Cecal.WT
Fecal.Vdr-/- 0.059        -            -
Cecal.WT     1.000        0.874        -
Fecal.WT     1.000        0.221        1.000

P value adjustment method: bonferroni
```

After adjustment with the Bonferroni method, only the Fecal.Vdr$^{-/-}$ versus Cecal.Vdr$^{-/-}$ comparison remains marginally significantly different, with a p value of 0.059.

(ii) Holm method

```
> pairwise.t.test(df_H_G$value, df_H_G$Group4, p.adjust="holm",
pool.sd = T)

Pairwise comparisons using t tests with pooled SD

data:  df_H_G$value and df_H_G$Group4

             Cecal.Vdr-/- Fecal.Vdr-/- Cecal.WT
Fecal.Vdr-/- 0.059        -            -
Cecal.WT     0.881        0.583        -
Fecal.WT     0.984        0.184        0.984

P value adjustment method: holm
```

When adjusted using the Holm method, the pair Fecal.Vdr$^{-/-}$ vs Cecal.Vdr$^{-/-}$ is at the same significant level as that found by using the Bonferroni method; all other pairs have slightly lower *p* values than using Bonferroni method. However, none of them are statistically significant.

(iii) Benjamini and Hochberg (BH) method

```
> pairwise.t.test(df_H_G$value, df_H_G$Group4, p.adjust="BH",
pool.sd = T)

    Pairwise comparisons using t tests with pooled SD

data:  df_H_G$value and df_H_G$Group4

             Cecal.Vdr-/- Fecal.Vdr-/- Cecal.WT
Fecal.Vdr-/- 0.059        -            -
Cecal.WT     0.440        0.291        -
Fecal.WT     0.765        0.110        0.590

P value adjustment method: BH
```

The BH method gives the same *p* value for the pair of Fecal.Vdr$^{-/-}$ versus Cecal.Vdr$^{-/-}$ and slightly lower *p* values for all other pairs compared with the Bonferroni and Holm methods.

(iv) BY method

```
> pairwise.t.test(df_H_G$value, df_H_G$Group4, p.adjust="BY",
pool.sd = T)

    Pairwise comparisons using t tests with pooled SD

data: df_H_G$value and df_H_G$Group4

             Cecal.Vdr-/- Fecal.Vdr-/- Cecal.WT
Fecal.Vdr-/- 0.14         -            -
Cecal.WT     1.00         0.71         -
Fecal.WT     1.00         0.27         1.00

P value adjustment method: BY
```

After adjustment using the BY method, the only significant pair of Fecal.Vdr$^{-/-}$ versus Cecal.Vdr$^{-/-}$ using the Bonferroni, Holm, or the BH methods is no longer significant.

(v) Tukey's HSD method

Finally, we use Tukey's HSD to conduct multiple comparisons of means to create confidence intervals for all pairwise differences between factor level means, while controlling the family-wise error rate to a 5% significant level. The 95% family-wise confidence level is also plotted:

```
> TukeyHSD(aov_fit, conf.level=.95)
  Tukey multiple comparisons of means
    95% family-wise confidence level

Fit: aov(formula = value ~ Group4, data = df_H_G)

$Group4
                                diff         lwr       upr     p adj
Fecal.Vdr-/--Cecal.Vdr-/-  0.53934581  0.01685443 1.0618372 0.0423608
Cecal.WT-Cecal.Vdr-/-      0.22317604 -0.38014504 0.8264971 0.6972467
Fecal.WT-Cecal.Vdr-/-      0.06213415 -0.54118693 0.6654552 0.9895699
Cecal.WT-Fecal.Vdr-/-     -0.31616977 -0.91949085 0.2871513 0.4373640
Fecal.WT-Fecal.Vdr-/-     -0.47721166 -1.08053274 0.1261094 0.1412485
Fecal.WT-Cecal.WT         -0.16104189 -0.83557536 0.5134916 0.8916878
```

The adjustment for comparison of Fecal.Vdr$^{-/-}$ vs Cecal.Vdr$^{-/-}$ using Tukey's HSD is statistically significant (p value $= 0.04$).

The Bonferroni method is considered more conservative. However, we actually found that the BY method is more conservative in this case, given that it results in the largest p value for pair comparison of Fecal.Vdr$^{-/-}$ versus Cecal.Vdr$^{-/-}$ among all the adjustment methods (Fig. 12.6):

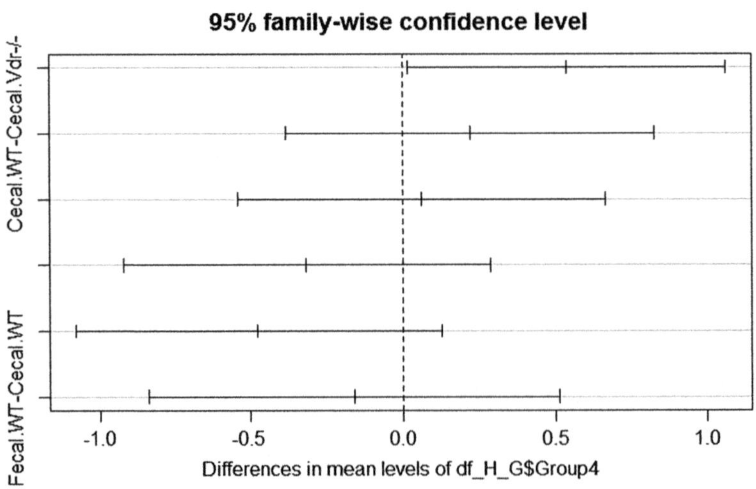

Fig. 12.6 Differences in means of groups with 95% family-wise confidence interval

```
> plot(TukeyHSD(aov(df_H_G$value~df_H_G$Group4), conf.level=.95))
```

Now, we test the homogeneity of variances, using two widely used functions: bartlett.test() and fligner.test().

```
> bartlett.test(df_H_G4, Group4)

    Bartlett test of homogeneity of variances

data: df_H_G4
Bartlett's K-squared = 17.013, df = 1, p-value = 3.713e-05

> fligner.test(df_H_G4, Group4)

    Fligner-Killeen test of homogeneity of variances

data:  df_H_G4
Fligner-Killeen:med chi-squared = 5.4968, df = 1, p-value =
0.01905
```

Both test methods show that the four groups do not have the same or similar variances (p value < 0.05), which indicates a violation of the assumption of homogeneity of variance using ANOVA. Thus, it is likely that conducting the nonparametric equivalent of the analysis (Kruskal–Wallis rank sum test) is more appropriate.

Kruskal–Wallis Rank Sum Test

The Kruskal–Wallis rank sum test or one-way ANOVA on ranks is a nonparametric method for testing whether samples originate from the same distribution. It extends the Mann–Whitney U test for use with more than two groups. The null hypothesis of the Kruskal–Wallis test is that the mean ranks of the groups are the same. Unlike the analogous one-way ANOVA, the nonparametric Kruskal–Wallis test does not assume a normal distribution of the underlying data. Thus, it has been widely used in microbiome research literature. The following codes use the function kruskal.test() to perform the Kruskal–Wallis rank sum test of Shannon indices among four genetic conditions:

```
> kruskal.test(value ~ Group4, data = df_H_G)

    Kruskal-Wallis rank sum test
```

```
data:  value by Group4
Kruskal-Wallis chi-squared = 7.8118, df = 3, p-value = 0.05007
```

The results show that the mean ranks of the groups are not same with a p value of 0.05007.

12.3.4 Multivariate Community Analysis of Beta Diversity

One important purpose of microbiome study is to determine whether the microbiome communities can be classified together or need to be separated by their bacteria to differentiate treatment from control, or healthy from disease, or genetic deficiency from WT groups. The questions of community classification lead us to measure the similarity or dissimilarity between two such community samples (beta-diversity). In this section, we show how to calculate beta diversities and then present the most frequently used exploratory techniques including classification and ordination. Finally, we focus on tests of significance of beta diversity.

12.3.4.1 Beta Diversity and Measures

In general, beta diversity evaluates differences between two or more local assemblages or between local and regional assemblages. This allows us to elucidate how much diversity is unique to a local assemblage and describe how many taxa are shared between communities.

There are more than two dozen measures of similarity or beta diversity indices available in the literature. These can be grouped into two broad classes of similarity measures: *binary* similarity coefficients, which only measure the presence or absence data, and *quantitative* similarity coefficients, which require some measures of relative abundance to be available for each species. In this chapter, we focus on the three very popular measures in microbiome studies: Bray–Curtis distance, binary Jaccard, and Sørensen similarity.

12.3.4.2 Calculate Beta Diversities

We use the vegdist() function from the vegan package to calculate beta diversities. If vegan has not been loaded, it should be loaded now:

```
> library(vegan)
```

Bray–Curtis Index

The function vegdist () returns a distance object (matrix), which is suitable for certain ordination methods and distance-based analyses. We name it "bc_dist" for later use. The method = "bray" is a character input to specify that the Bray–Curtis method is to be used:

```
> bc_dist <- vegdist(abund_table_t, method = "bray")
> bc_dist
                 20_12_CeSt-28F 19_11_CeSt-28F 21_13_CeSt-28F 22_14_CeSt-28F
23_15_CeSt-28F
19_11_CeSt-28F      0.5290152
21_13_CeSt-28F      0.3403123      0.5045552
22_14_CeSt-28F      0.3704283      0.5814794      0.4098626
23_15_CeSt-28F      0.6253699      0.7958272      0.7705171      0.6210542
25_22_CeSt-28F      0.2818678      0.6467292      0.4522264      0.4034130
0.6557869
26_23_CeSt-28F      0.4612502      0.7161236      0.6552688      0.5347489
0.4517226
27_24_CeSt-28F      0.4664437      0.4028436      0.6329665      0.5685615
0.5127958
                  25_22_CeSt-28F 26_23_CeSt-28F
19_11_CeSt-28F
21_13_CeSt-28F
22_14_CeSt-28F
23_15_CeSt-28F
25_22_CeSt-28F
26_23_CeSt-28F      0.5610028
27_24_CeSt-28F      0.5911418      0.4631897
```

Jaccard Index

We can calculate the Jaccard index with the following function:

```
> j_dist <-vegdist(abund_table_t, "jaccard")
> j_dist
                 20_12_CeSt-28F 19_11_CeSt-28F 21_13_CeSt-28F 22_14_CeSt-28F
23_15_CeSt-28F
19_11_CeSt-28F      0.6919685
21_13_CeSt-28F      0.5078104      0.6707035
22_14_CeSt-28F      0.5406023      0.7353613      0.5814220
23_15_CeSt-28F      0.7695109      0.8863071      0.8703865      0.7662350
25_22_CeSt-28F      0.4397767      0.7854712      0.6228043      0.5749027
```

```
0.7921151
26_23_CeSt-28F      0.6313090      0.8345828      0.7917370      0.6968552
0.6223264
27_24_CeSt-28F      0.6361563      0.5743243      0.7752352      0.7249464
0.6779445
                 25_22_CeSt-28F 26_23_CeSt-28F
19_11_CeSt-28F
21_13_CeSt-28F
22_14_CeSt-28F
23_15_CeSt-28F
25_22_CeSt-28F
26_23_CeSt-28F      0.7187723
27_24_CeSt-28F      0.7430410      0.6331233
```

Sørensen Index

The Sørensen index is calculated in the same way:

```
> S_dist <-vegdist(abund_table,binary=TRUE)
> S_dist
                 20_12_CeSt-28F 19_11_CeSt-28F 21_13_CeSt-28F 22_14_CeSt-28F
23_15_CeSt-28F
19_11_CeSt-28F      0.3932584
21_13_CeSt-28F      0.4042553      0.3670886
22_14_CeSt-28F      0.3877551      0.4939759      0.3181818
23_15_CeSt-28F      0.4528302      0.4505495      0.3958333      0.4000000
25_22_CeSt-28F      0.3400000      0.4823529      0.4000000      0.4255319
0.3529412
26_23_CeSt-28F      0.4098361      0.4205607      0.4107143      0.4655172
0.4193548
27_24_CeSt-28F      0.4166667      0.4666667      0.4727273      0.4035088
0.4262295
                 25_22_CeSt-28F 26_23_CeSt-28F
19_11_CeSt-28F
21_13_CeSt-28F
22_14_CeSt-28F
23_15_CeSt-28F
25_22_CeSt-28F
26_23_CeSt-28F      0.3898305
27_24_CeSt-28F      0.3620690      0.3913043
```

12.3.4.3 Exploratory Analysis of Microbiome Data

Clustering (or classification) and ordination are the two main classes of multivariate methods that microbiome researchers and community ecologists employ. To some degree, these two approaches are complementary. Clustering reduces complexity (dimensionalities) of data via ordering samples into hierarchical classes. However, community data are continuous; thus, reducing complexity to two or three dimensions is usually desired. Therefore, many multivariate methods have been developed in the study of microbiome and ecology based on ordination.

Clustering

Several different clustering methods including average-link, complete-link, and single-link, are available. In this chapter, we use the Bray–Curtis distance method to illustrate sample classification. Other distances can also be applied.

As we calculated the Bray–Curtis dissimilarities in section "Bray–Curtis Index", we can now apply the function hclust() with the three different clustering algorithms—"average," "complete," and "single" linkage methods, respectively—to perform hierarchical clustering. Then, we plotted a cluster dendrogram using the calculated values:

```
> cluster_average <- hclust (bc_dist, method = 'average')
> cluster_complete <- hclust (bc_dist, method = 'complete')
> cluster_single <- hclust (bc_dist, method = 'single')
```

Finally, we can draw the results together into one diagram using par (mfrow = c (1,3)) to create one graph with one row of three panels (Fig. 12.7).

```
> par (mfrow = c(1,3))
> plot (cluster_single)
> plot (cluster_complete)
> plot (cluster_average)
```

The above two steps can be combined into one step as shown below, which generates the same plots:

```
> plot(hclust(bc_dist, "average"))
> plot(hclust(bc_dist, "complete"))
> plot(hclust(bc_dist, "single"))
```

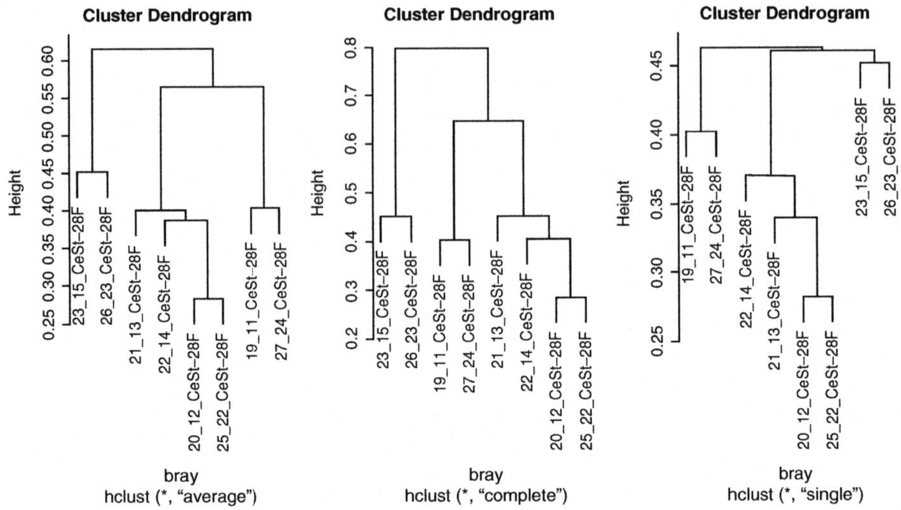

Fig. 12.7 Cluster dendrogram of the Bray–Curtis dissimilarity

Ideally, clustering can provide a distinction between Vdr$^{-/-}$ and WT samples. However, although the three clustering methods provide different results, not all of them can make a clear distinction between the Vdr$^{-/-}$ and WT samples. We attempt ordination in the next section.

Ordination

The primary aim of ordination is to represent sample and species (or operational taxonomic units [OTUs]/taxa) relationships as faithfully as possible in a low-dimensional space. This objective is desirable because although community data consist of multiple dimensions mixed with noise, low dimensions most ideally represent important and intuitive interpretations of species (or OTUs/taxa)–environment relationships. Ordination endeavors to represent multiple objects in a reduced number of orthogonal (i.e., independent) axes. The first axis of an ordination explains the most variation in the data set, followed by the second axis, the third, and so on, where the total number of axes is less than or equal to the number of objects.

The ordination plots are particularly useful for visualizing similarity among objects. For example, in the context of beta diversity, samples that are closer in ordination space have species assemblages that are more alike one another than samples that are further apart in ordination space.

Depending on whether the ordination axes are to be constrained by environmental factors (variables), ordination methods can be divided into two types: unconstrained and constrained ordinations. As the names suggest, in unconstrained ordination, ordination axes are not constrained by environmental factors; in contrast, in constrained ordination, ordination axes are constrained by environmental factors. Unconstrained ordination is primarily a descriptive method and does not really involve hypothesis testing in multivariate data. On the other hand, constrained ordination is a hypothesis-testing method that directly tests hypotheses on the influence of environmental factors on species (or OTUs/taxa) composition. In this chapter, we cover the most common unconstrained ordinations in the microbiome literature: principle component analysis (PCA), principal coordinate analysis (PcoA), nonmetric multidimensional scaling (NMDS), and constrained ordinations (CAPs).

Principal Component Analysis

In terms of the vegan package, the variable denoting genetic conditions (Vdr$^{-/-}$ and WT) in our case is an environmental variable. We want to know if the Vdr deficiency can interpret beta diversity in genus composition in mice samples. We can conduct principal component analysis to explore whether the changes in genus composition of communities (beta diversities) are caused by the genetic conditions.

Principle component analysis plots samples based on abundances of genus A on axis 1, genus B on axis 2, genus C on axis 3, and so on, until N samples are plotted in a very high dimensional space. The first straight line going through the space created by all these samples is called PC1. It is the most important PC and explains the most variations among all samples. The second line is called PC2, perpendicular to PC1; this explains the second most variations, and so the third most PC3, and so on untill the less important PC(N-1).

Several R functions can be used to conduct PCA, including prcomp() in the preinstalled stats package, rda() in the vegan package, and pca() in the labdsv package. Two extensional functions are evplot() (Borcard et al. 2011) and PCAsignificance() in the BiodiversityR package. The evplot() provides visual methods to decide the importance of ordination axes by using the Keiser–Guttman criterion and the broken stick model. PCAsignificance() calculates the broken-stick model for PCA axes.

We use the function rda() from the vegan package to conduct PCA here. The function performs unconstrained ordination PCA by not specifying the environmental data matrix (in this case, group variable). PCA can be performed via the following steps:

(i) Standardize abundance counts

In microbiome data analysis, the absolute abundance counts are not appropriate owing to sampling. The largest values have too great an influence on the analysis. Thus, we need to standardize the abundance read data before analysis. There are two ways of obtaining the relative abundance: we can either call the decostand() function or write a simple R function. Here, we use the decostand() function to standardize read with total method:

```
> stand_abund_table <- decostand(abund_table_t, method = "total")
> PCA <-rda(stand_abund_table)
> PCA
Call: rda(X = stand_abund_table)

                Inertia Rank
Total           0.09728
Unconstrained 0.09728     7
Inertia is variance

Eigenvalues for unconstrained axes:
    PC1     PC2     PC3     PC4     PC5     PC6     PC7
0.05884 0.02445 0.00799 0.00458 0.00074 0.00044 0.00023
```

The total variation of the whole dataset is 0.09728 in this case, and the first axis explains 60.4% of the total variation (0.05884/0.09728 = 0.604). Total variation is a sum of variations of each genus in the analyzed matrix. First, we check the total variance:

```
> sum (apply (stand_abund_table, 2, var))
[1] 0.09727763
```

(ii) Draw biplot and ordiplot diagrams

Then, we draw the diagrams using the function biplot() (Fig. 12.8):

```
> biplot(PCA, display = 'species')
```

The display option "species" is the vegan package label for OTUs/taxa. The default is "sites" (label for samples).

The above diagrams plotted by biplot() simply draw arrows for the genus and are not informative. The more informative plot is to use the function ordiplot() to draw both genus and sample scores as centroids:

```
> ordiplot(PCA, display = "sites", type = "text")
```

In the above augments, type = "text" or "t" added text labels to the figure (the default setting adds only points) (Fig. 12.9).

(iii) Draw PCA scalings

As an alternative, we can use the function cleanplot.pca() to intend to draw PCA results to two diagrams with different scalings. The cleanplot.pca() function draws two biplots (scaling 1 and scaling 2) from an object of class "rda" in the PCA or redundancy analysis (RDA) result from vegan's rda() function. It was written by Francois Gillet and Daniel Borcard and is provided in Borcard et al. (2011):

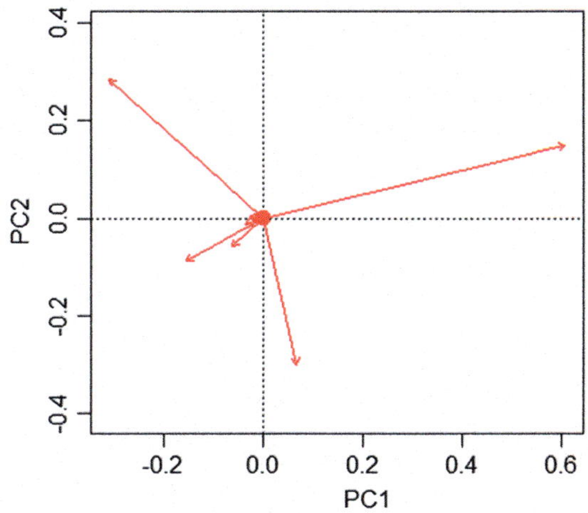

Fig. 12.8 Biplot of two principal components

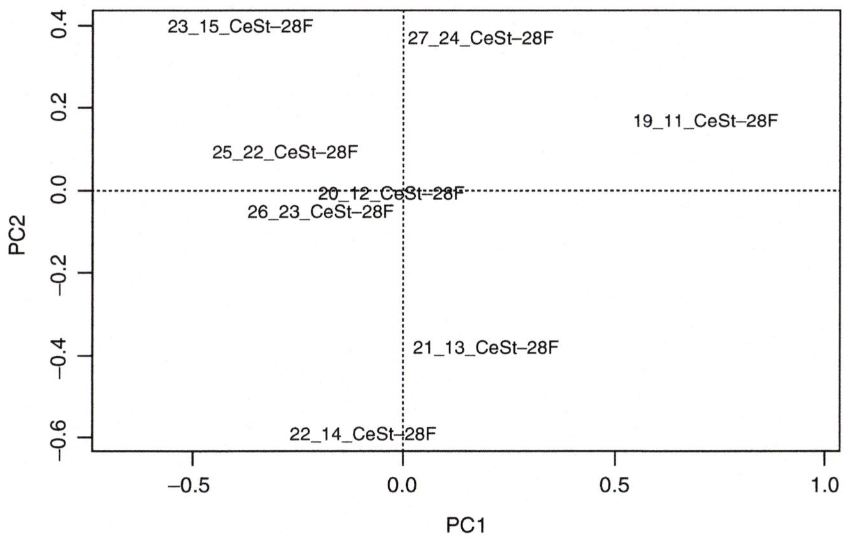

Fig. 12.9 Ordiplot of two principal components with samples labeled

```
> "cleanplot.pca" <- function(res.pca, ax1=1, ax2=2, point=FALSE,
+                               ahead=0.07, cex=0.7)
+ {
+   # A function to draw two biplots (scaling 1 and scaling 2) from
an object
+   # of class "rda" (PCA or RDA result from vegan's rda() function)
```

```
+   #
+   # License: GPL-2
+   # Authors: Francois Gillet & Daniel Borcard, 24 August 2012
+
+   require("vegan")
+
+   par(mfrow=c(1,2))
+   p <- length(res.pca$CA$eig)
+
+   # Scaling 1: "species" scores scaled to relative eigenvalues
+    sit.sc1 <- scores(res.pca, display="wa", scaling=1, choices=c
(1:p))
+    spe.sc1 <- scores(res.pca, display="sp", scaling=1, choices=c
(1:p))
+     plot(res.pca, choices=c(ax1, ax2), display=c("wa", "sp"),
type="n",
+        main="PCA - scaling 1", scaling=1)
+   if (point)
+   {
+     points(sit.sc1[,ax1], sit.sc1[,ax2], pch=20)
+     text(res.pca, display="wa", choices=c(ax1, ax2), cex=cex,
pos=3, scaling=1)
+   }
+   else
+   {
+   text(res.pca, display="wa", choices=c(ax1, ax2), cex=cex, scal-
ing=1)
+   }
+   text(res.pca, display="sp", choices=c(ax1, ax2), cex=cex,
pos=4,
+        col="red", scaling=1)
+  arrows(0, 0, spe.sc1[,ax1], spe.sc1[,ax2], length=ahead,
angle=20, col="red")
+   pcacircle(res.pca)
+
+   # Scaling 2: site scores scaled to relative eigenvalues
+   sit.sc2 <- scores(res.pca, display="wa", choices=c(1:p))
+   spe.sc2 <- scores(res.pca, display="sp", choices=c(1:p))
+   plot(res.pca, choices=c(ax1,ax2), display=c("wa","sp"),
type="n",
+        main="PCA - scaling 2")
+   if (point) {
+     points(sit.sc2[,ax1], sit.sc2[,ax2], pch=20)
+       text(res.pca, display="wa", choices=c(ax1 ,ax2), cex=cex,
pos=3)
```

```
+   }
+   else
+   {
+     text(res.pca, display="wa", choices=c(ax1, ax2), cex=cex)
+   }
+     text(res.pca,  display="sp",  choices=c(ax1,  ax2),  cex=cex,
pos=4, col="red")
+     arrows(0,  0,  spe.sc2[,ax1],  spe.sc2[,ax2],  length=ahead,
angle=20, col="red")
+ }

> "pcacircle" <- function (pca)
+ {
+   # Draws a circle of equilibrium contribution on a PCA plot
+   # generated from a vegan analysis.
+   # vegan uses special constants for its outputs, hence
+   # the 'const' value below.
+
+   eigenv <- pca$CA$eig
+   p <- length(eigenv)
+   n <- nrow(pca$CA$u)
+   tot <- sum(eigenv)
+   const <- ((n - 1) * tot)^0.25
+   radius <- (2/p)^0.5
+   radius <- radius * const
+   symbols(0, 0, circles=radius, inches=FALSE, add=TRUE, fg=2)
+ }
>
> cleanplot.pca (PCA)
```

The left panel of Fig. 12.10 for scaling 1 focuses on distances among samples (distance biplot). The circle is called the circle of equilibrium contribution, representing the equilibrium contribution of the genera. For the given combination of axes, the genera with vectors longer than the radius of the circle could be interpreted with confidence as the most important genera, whereas the genera with vectors shorter than the radius of the equilibrium contribution circle contribute little to a given reduced space. The most abundant genera we identified here are the same as those we found in Sect. 12.3.5. The right panel of Fig. 12.10 for scaling 2 describes the correlation among genera (correlation biplot), which is reflected in the angle of particular vectors. The cosine of the angle approximates correlation between genera, and the length of the vector approximates the standard deviations of the genera.

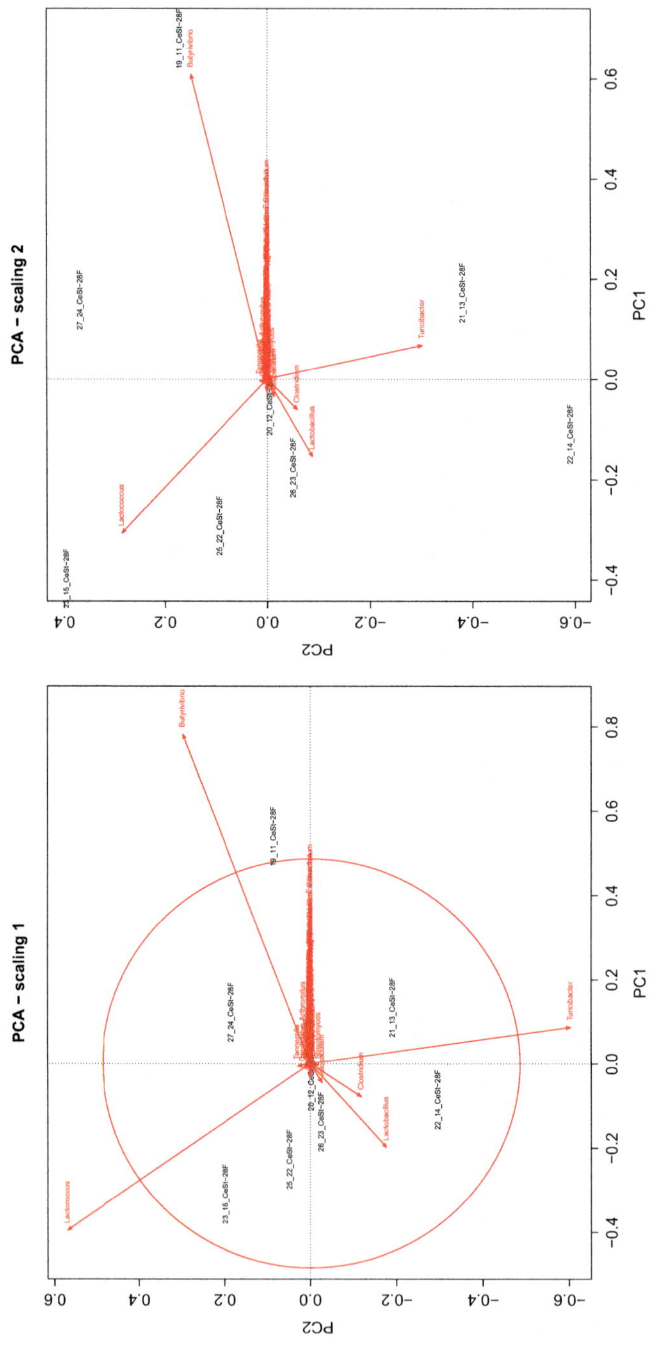

Fig. 12.10 Plots of two principal components with different scalings

Principal Coordinate Analysis

Principal coordinate analysis is also referred to as metric multidimensional scaling. It is calculated on Euclidean distances among samples. Like PCA, PCoA uses eigenvalues to measure the importance of a set of returned orthogonal axes. This means that PCoA yields the same results as PCA calculated on the covariance matrix of the same dataset (if scaling 1 is used). The major benefit of PCoA is its flexibility: it allows the user to choose virtually any distance metric (e.g., Jaccard, Bray–Curtis, Euclidean, etc.). With PCoA, the dimensionality of the matrix is reduced by determining each eigenvector and eigenvalue. Then, each eigenvector is scaled to obtain the principal coordinates.

Principal coordinate analysis can be performed using the R functions cmdscale() in the vegan package and pcoa() in the ape package. With the vegan package, the input data can be calculated by using the function vegdist() (the default is Bray–Curtis dissimilarity), and the ordination diagram can be drawn with the function ordiplot(). The ordination diagram could also be drawn with the function biplot. pcoa() from the ape package. Here, we illustrate the cmdscale() function to conduct a PCoA using the same Vdr$^{-/-}$ cecal data from mice.

(i) Call function cmdscale()

The function needs a resemblance matrix as the input data. In the previous section, the Bray–Curtis dissimilarity has already been calculated using the function vegdist() and is named "bc_dist." Here, we directly use this resemblance matrix as the input data of the cmdscale() function. We are going to explicitly set $k = 2$ (the default values for the number of dimensions we want to return) and *eig* = TRUE (which saves the eigenvalues).

```
> PCoA <- cmdscale (bc_dist, eig = TRUE, k = 2)
```

```
Or
> PCoA <- cmdscale (bc_dist, eig = TRUE)
> PCoA
$points
                       [,1]        [,2]
20_12_CeSt-28F  0.10195331 -0.10211223
19_11_CeSt-28F  0.23837868  0.36188333
21_13_CeSt-28F  0.30283902 -0.08951510
22_14_CeSt-28F  0.09299027 -0.15021372
23_15_CeSt-28F -0.43515987 -0.01709577
25_22_CeSt-28F  0.10127521 -0.23036598
26_23_CeSt-28F -0.27932562 -0.05803318
27_24_CeSt-28F -0.12295101  0.28545266
$eig
[1]   4.603381e-01   3.101754e-01   1.073265e-01   8.969401e-02
```

```
6.668375e-02
[6]  2.152519e-02  4.610549e-03 -2.431400e-17

$x
NULL

$ac
[1] 0

$GOF
[1] 0.7266572 0.7266572
```

The function cmdscale() produces a list of outputs. The first output *points* contain the coordinates for each sample in each reduced dimension. The second output *eig* contains the eigenvalues. The last three outputs pertain to other options of the analysis that we do not cover here.

(ii) Assess the explained variation

The following chunk of R code is used to examine the percentage variation in the data set that is explained by the first two axes of the PCoA.

```
> explainedvar1 <- round(PcoA$eig[1] / sum(PcoA$eig), 2) * 100
> explainedvar1
[1] 43

> explainedvar2 <- round(PcoA$eig[2] / sum(PcoA$eig), 2) * 100
> explainedvar2
[1] 29

> sum_eig <- sum(explainedvar1, explainedvar2)
> sum_eig
[1] 72
```

We can see that the first axis explains 43% variations in the data, and the second axis explains 29%. Thus, a large amount of variation in the data (72%) can be explained by these two axes alone.

There are two criteria to assess whether or not the first few PCoA axes capture a disproportionately large amount of the total explained variation. First, there is the Kaiser–Guttman criterion, which states that the eigenvalues associated with the first few axes should be larger than the average of all the eigenvalues; second, we can compare the eigenvalues associated with the first few axes with the expectations of the broken-stick model. The broken stick model assumes that the total sum of eigenvalues decreases sequentially with the ordered PCoA axes. We evaluate these two criteria with the following plots:

```
> # Define Plot Parameters
> par(mar = c(5, 5, 1, 2) + 0.1)
>
> # Plot Eigenvalues
> plot(PCoA$eig, xlab = "PcoA", ylab = "Eigenvalue",
+      las = 1, cex.lab = 1.5, pch = 16)
>
> # Add Expectation based on Kaiser-Guttman criterion and
Broken Stick Model
> abline(h = mean(PcoA$eig), lty = 2, lwd = 2, col = "blue")
> b_stick <- bstick(8, sum(PcoA$eig))
> lines(1:8, b_stick, type = "l", lty = 4, lwd = 2, col =
"red")
> # Add Legend
> legend("topright", legend = c("Avg Eigenvalue", "Broken-
Stick"),
+        lty = c(2, 4), bty = "n", col = c("blue", "red"))
```

The Fig. 12.11 shows that the eigenvalues associated with the first two axes are larger than the average of all the eigenvalues and the expectations of the broken-stick model. Thus, based on the two criteria described above, the PcoA is efficient at explaining variations in the Vdr data set of mice (Fig. 12.11).

(iii) Create an ordination plot

After evaluating the PcoA output, we create an ordination plot for the two PcoA axes (Fig. 12.12):

```
> # Initiate Plot
> plot(PCoA$points[ ,1], PCoA$points[ ,2], ylim = c(-0.5,
0.5),
+      xlab = paste("PCoA 1 (", explainedvar1, "%)", sep = ""),
+      ylab = paste("PCoA 2 (", explainedvar2, "%)", sep = ""),
+      pch = 5, cex = 1.0, type = "n", cex.lab = 1.0, cex.axis
```

(continued)

```
= 1.2, axes = FALSE)
>
> # Add Axes
> axis(side = 1, labels = T, lwd.ticks = 2, cex.axis = 1.2,
las = 1)
> axis(side = 2, labels = T, lwd.ticks = 2, cex.axis = 1.2,
las = 1)
> abline(h = 0, v = 0, lty = 3)
> box(lwd = 2)
>
> # Add Points & Labels
> points(PCoA$points[ ,1], PCoA$points[ ,2],
+         pch = 19, cex = 3, bg = "blue", col = "blue")
> text(PCoA$points[ ,1], PCoA$points[ ,2],
+       labels = row.names(PCoA$points))
```

(iv) Identify and visualize influential genera

Basic ordination plots allow us to see how samples separate from one another. In our example, samples are separated along the PCoA axes owing to the variation in the abundance of different mice genera. A logical follow-up question is to ask what genera of the data set are driving the observed divergence among points. Can we

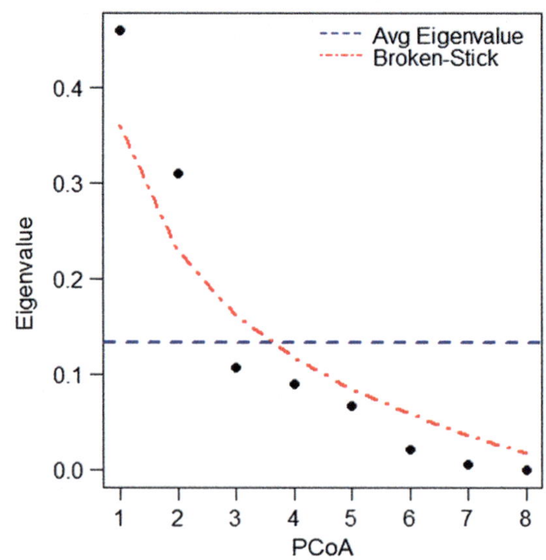

Fig. 12.11 Principal coordinate analysis (PCoA) with the Kaiser–Guttman criterion and the broken-stick model

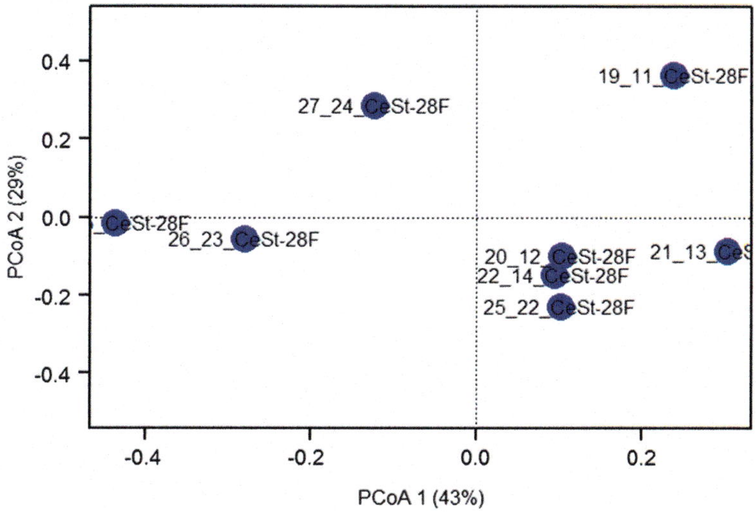

Fig. 12.12 Ordination plot for the two PCoA axes

identify and visualize these influential genera in PCoA? We can obtain this information using the add.spec.scores() function in the BiodiversityR package.

First, the relative abundance is calculated:

```
> cecalREL <- abund_table
> for(i in 1:nrow(abund_table)){
+   cecalREL[i, ] = abund_table[i, ] / sum(abund_table[i, ])
+ }
```

Then, the genera scores are calculated and added to the figure (Fig. 12.13):

```
> require("BiodiversityR")

> PCoA <- add.spec.scores(PCoA,cecalREL,method = "pcoa.scores")
> text(PCoA$cproj[ ,1], PCoA $cproj[ ,2],
+      labels = row.names(PCoA$cproj), col = "black")
```

(v) Determine the correlation between the genera and PCoA axes

A more quantitative way of identifying influential genera is to determine the correlation of each genus along the PCoA axes. To do this, we use the function add.spec.scores() again:

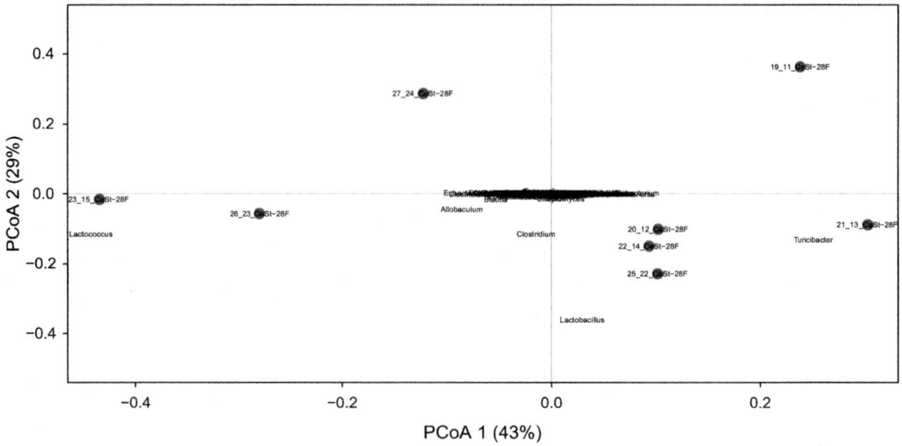

Fig. 12.13 Ordination plot for the two PCoA axes with influential genera

```
> Genus_corr <- add.spec.scores(PCoA, cecalREL, method = "cor.
scores")$cproj
```

Then, we can define a correlation-coefficient cutoff (e.g., $r = 0.70$) to identify and pull out the important genera:

```
> corrcut <- 0.7
> genus_corr <- add.spec.scores(PCoA, cecalREL, method = "cor.
scores")$cproj
> import_genus <- genus_corr[abs(genus_corr[, 1]) >= corrcut | abs
(genus_corr[, 2]) >= corrcut, ]
```

The 11 important genera with correlation greater than or equal to 0.7 along the PCoA Axes are printed:

```
> import_genus
                                  Dim1        Dim2
Lactococcus                -0.77626735 -0.16132862
Lactobacillus               0.08811740 -0.88909362
Parasutterella             -0.82084741 -0.11725257
Butyrivibrio                0.48692185  0.85374107
Coprococcus                -0.73449719  0.22969525
Streptomyces                0.52193360 -0.75463293
Candidatus Arthromitus     -0.70896069 -0.11611886
Roseburia::Clostridium     -0.08179291  0.83835843
Clostridium::Butyrivibrio  -0.82084741 -0.11725257
Atopobium                  -0.87531442  0.02819833
Butyrivibrio::Ruminococcus  0.27176540  0.76961869
```

(vi) Test abundances across PCoA axes on correlations

Finally, we use the envfit() function from the vegan package to conduct a permutation test for general abundances across axes on these correlations:

```
> fit <- envfit(PCoA, cecalREL, perm = 999)
> fit
```

The partial output is given below:

```
                                                Pr(>r)
Lactococcus                                      0.081 .
Lactobacillus                                    0.018 *
Parasutterella                                   0.073 .
Butyrivibrio                                     0.001 ***
Coprococcus                                      0.088 .
Streptomyces                                     0.008 **
Adlercreutzia                                    0.068 .
Roseburia::Clostridium                           0.049 *
Clostridium::Butyrivibrio                        0.073 .
Atopobium                                        0.038 *
Butyrivibrio::Ruminococcus                       0.059 .

Signif. codes:  0 '***' 0.001 '**' 0.01 '*' 0.05 '.' 0.1 ' ' 1
Permutation: free
Number of permutations: 999
```

Nonmetric Multidimensional Scaling

Nonmetric multidimensional scaling is the nonmetric alternative to PCoA analysis. The primary advantages of NMDS analysis are that it can use ecologically meaningful ways of measuring community dissimilarities and any distance (dissimilarities) measure among samples as the input. Thus, it is the recommended method for community ordination. The main focus of NMDS analysis is to project the relative position of sample points into low-dimensional ordination space (typically, two or three axes).

The function metaMDS() in the vegan package performs NMDS analysis. To simplify, the algorithm of NMDS analysis can be summarized by the following procedures:

First, we use the function vegdist() to obtain adequate dissimilarity measures; then, we run NMDS several times with random starting configurations, compare the results via the function procrustes(), and stop after finding a similar minimum stress solution twice. Finally, we scale and rotate the solution and add species (or OTUs/taxa) scores to the configuration as weighted averages using the function wascores

(). After the algorithm is finished, the final solution is rotated using PCA to ease its interpretation.

In our example, NMDS analysis is illustrated using the same Vdr$^{-/-}$ mice cecal data.

(i) Call the function metaMDS()

First, we call the metaMDS() function. Here, the Bray–Curtis dissimilarity measure is used (the default setting of metaMDS()). The function automatically transforms data and checks the robustness of the solution.

```
> bc_nmds <- metaMDS(abund_table_t, dist = "bray")
Square root transformation
Wisconsin double standardization
Run 0 stress 0.09904295
Run 1 stress 0.09904295
... New best solution
... Procrustes: rmse 8.550326e-06  max resid 1.599522e-05
... Similar to previous best
Run 2 stress 0.09904299
... Procrustes: rmse 4.754853e-05  max resid 9.35321e-05
... Similar to previous best
Run 3 stress 0.2300661
Run 4 stress 0.1064542
Run 5 stress 0.155098
Run 6 stress 0.09904295
... Procrustes: rmse 6.263572e-05  max resid 0.0001215291
... Similar to previous best
Run 7 stress 0.3083099
Run 8 stress 0.09904295
... Procrustes: rmse 4.229725e-05  max resid 8.391946e-05
... Similar to previous best
Run 9 stress 0.09904294
... New best solution
... Procrustes: rmse 3.22047e-05  max resid 6.339528e-05
... Similar to previous best
Run 10 stress 0.09904303
... Procrustes: rmse 0.0001707007  max resid 0.0003355989
... Similar to previous best
Run 11 stress 0.09904302
... Procrustes: rmse 5.571893e-05  max resid 0.0001150672
... Similar to previous best
Run 12 stress 0.09904299
... Procrustes: rmse 0.0001162634  max resid 0.0002282454
... Similar to previous best
Run 13 stress 0.09904294
```

```
... New best solution
... Procrustes: rmse 1.498919e-06  max resid 2.460273e-06
... Similar to previous best
Run 14 stress 0.1576271
Run 15 stress 0.09904295
... Procrustes: rmse 1.978028e-05  max resid 4.043441e-05
... Similar to previous best
Run 16 stress 0.1064542
Run 17 stress 0.1850305
Run 18 stress 0.1677913
Run 19 stress 0.09904294
... Procrustes: rmse 1.015472e-05  max resid 1.9617e-05
... Similar to previous best
Run 20 stress 0.09904294
... Procrustes: rmse 1.10496e-05  max resid 2.043547e-05
... Similar to previous best
*** Solution reached
```

In this case, a combination of Wisconsin double standardization and square-root transformation was used. This stress value is 0.09.

(ii) Draw the results of NMDS

Second, we use the function ordiplot() to draw the results of NMDS. The default setting adds only points to the figure, in this case, the type = 't' or type = 'text' adds text labels (Fig. 12.14).

```
> ordiplot (bc_nmds, type = 't')
```

To plot site (sample) scores as text (Fig. 12.15):

```
> ordiplot(bc_nmds, display = "sites", type = "text")
```

(iii) Draw a Shepard stress plot

Finally, we use the function stressplot() to draw the Shepard stress plot (where ordination distances are plotted against the chosen community dissimilarities and the fit is shown as a monotone step line) and to assess the goodness of ordination using the function goodness() to return the goodness-of-fit values of particular samples.

We want to divide the plotting window into two panels using this function:

```
> par (mfrow = c(1,2))
```

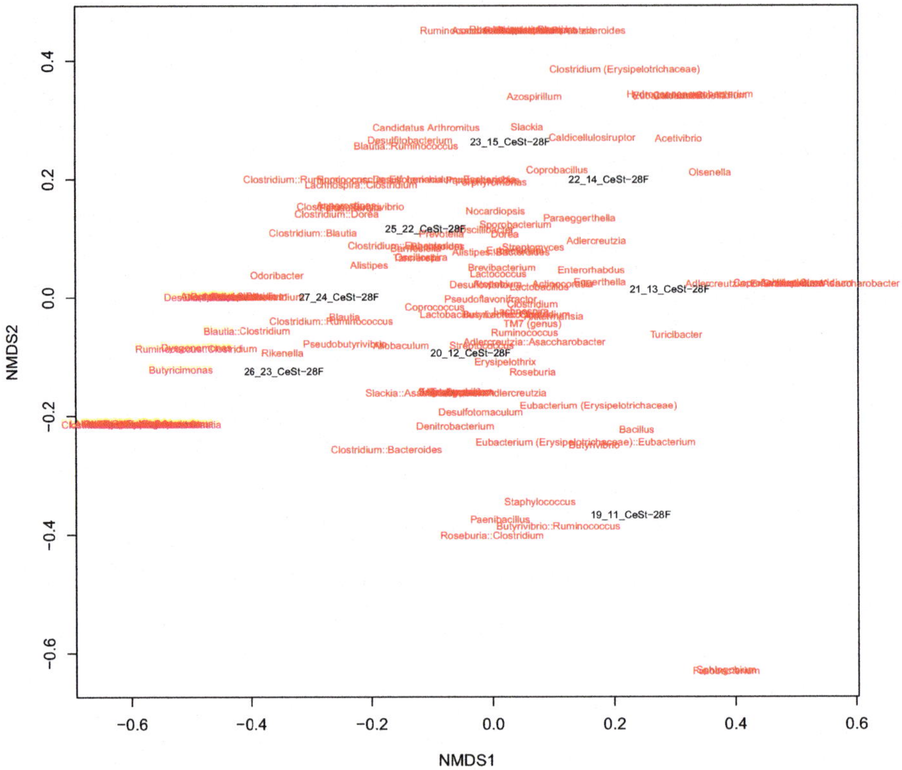

Fig. 12.14 Ordiplot of two nonmetric multidimensional scaling (NMDS) axes with text labels

The following plot() function draws an NMDS ordination diagram with sites (samples) (Fig. 12.16):

```
> stressplot (bc_nmds)
> plot (bc_nmds, display = 'sites', type = 't', main = 'Goodness
of fit')
```

The following points() function adds the points, with size reflecting the goodness of fit (bigger = worse fit):

```
> points (bc_nmds, display = 'sites', cex = goodness (bc_nmds)
*300))
```

The stress plot shows the relationship between the real distances between samples in the resulting *m* dimensional ordination solution and their particular

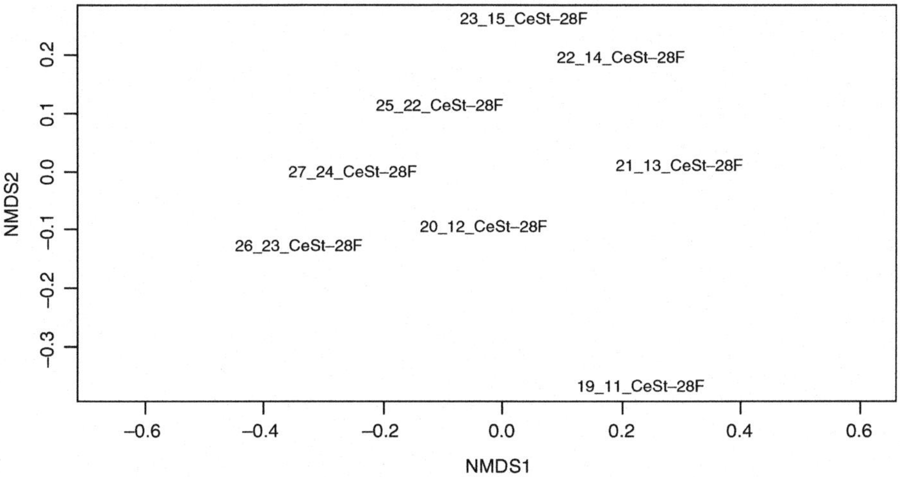

Fig. 12.15 Ordiplot of two NMDS axes with sample labels

compositional dissimilarities expressed by the selected Bray–Curtis dissimilarity measure. There are two correlation-like statistics of goodness of fit: the correlation based on stress, $R^2 = 1-S^2$ (non-metric fit = 0.99), and the correlation between the fitted values and the ordination distances or between the step line and the points: "fit-based R^2" (linear fit = 0.927).

Constrained Analysis of Proximities

Constrained ordination is a "hypothesis-driven" ordination; the factors tested are based on our hypothesis. Thus, constrained ordination is related to multivariate linear models with "dependent" variables or the community in the left side as responses, and "independent" variables or constraints in the right side as explained factors. The vegan package has three versions of constrained ordination: constrained analysis of proximities (CAP), RDA, and constrained correspondence analysis (CCA). Here, we illustrate the CAP with the function capscale() using the same Vdr$^{-/-}$ data set we used in section "Non-metric Multidimensional Scaling".

(i) Call function capscale()

The function capscale() is used to conduct CAP. In the function capscale(), CAP is related to metric scaling (cmdscale). It can handle any dissimilarity measures. Here, we use Bray–Curtis dissimilarity to perform linear mapping to genetic groups. The Bray–Curtis dissimilarity (bc_dist) was estimated in section "Bray–Curtis Index".

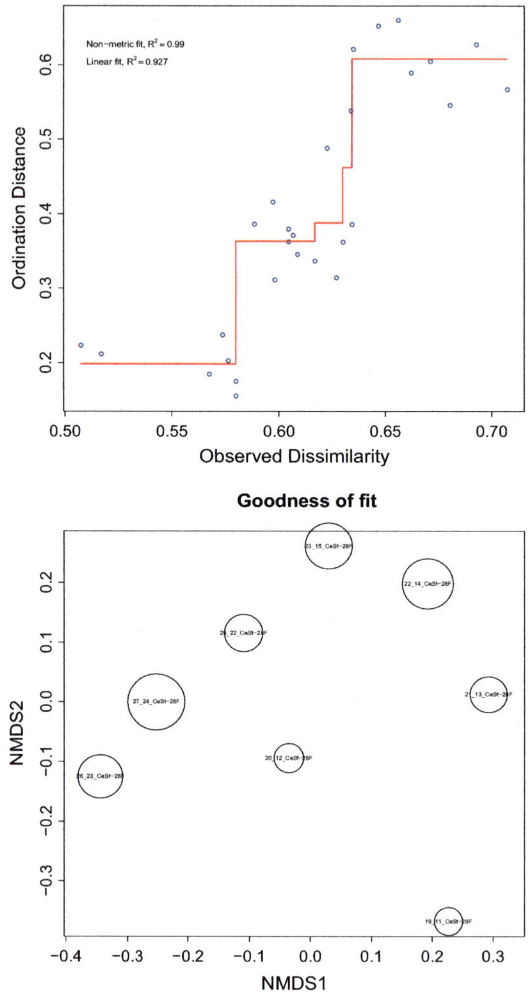

Fig. 12.16 Shepard stress plot and goodness of fit

```
> cap_constrained=capscale(bc_dist ~ grouping_inf)
> summary(cap_constrained)

Call:
capscale(formula = bc_dist ~ grouping_inf)
```

```
Partitioning of squared Bray distance:
              Inertia Proportion
Total          1.0604     1.0000
Constrained    0.1080     0.1018
Unconstrained  0.9524     0.8982

Eigenvalues, and their contribution to the squared Bray distance

Importance of components:
                         CAP1   MDS1   MDS2    MDS3    MDS4    MDS5    MDS6
Eigenvalue             0.1080 0.4196 0.3102 0.09081 0.07241 0.04395 0.01543
Proportion Explained   0.1018 0.3957 0.2925 0.08564 0.06829 0.04144 0.01455
Cumulative Proportion  0.1018 0.4976 0.7901 0.87572 0.94401 0.98545 1.00000

Centroids for factor constraints

                      CAP1 MDS1 MDS2 MDS3 MDS4 MDS5
grouping_infVdr-/- -0.4520    0    0    0    0    0
grouping_infWT      0.7534    0    0    0    0    0
```

The first axis is called "CAP1" and is then followed by the original MDS. Based on the factor levels, there can be multiple CAPs; however, only the first axis is extracted to show the most important separation by the factors. The most important separation is Vdr genetic deficiency, which explains 10.18% of the total variation of the whole data set.

(ii) Plot the results of CAP

The results of CAP can be plotted by using the following string of R codes:

First, the function plot() generates an empty CAP ordination diagram.

Then, the function points() (low-level plotting function) adds points to the ordination diagram created by plot().

Third, the function ordispider() creates a spider plot by connecting individual members of the group with the group centroid.

Last, the function ordiellipse() encircles the clouds of points within the group by ellipse-like envelopes (Fig. 12.17).

```
> plot(cap_constrained, type="n")
> points(cap_constrained, col=as.numeric(as.factor(grouping_inf)),
+          pch=as.numeric(as.factor(grouping_inf)))
> ordispider(cap_constrained, grouping_inf, lty=2, col="grey", label=T)
> ordiellipse(cap_constrained, grouping_inf, lty=2, col="grey", label=F)
```

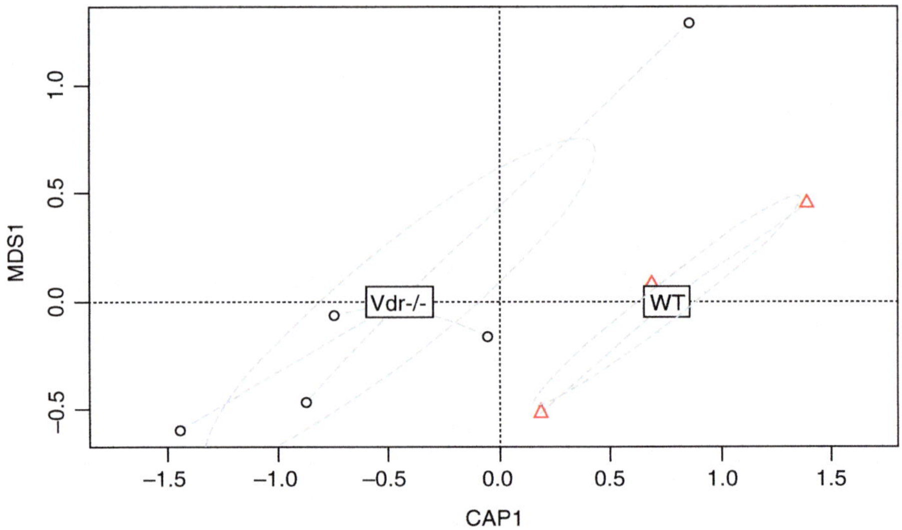

Fig. 12.17 Ordispider and ordiellipse plots of constrained analysis of proximities

12.3.4.4 Statistical Testing for Differences Between Groups

Permutational MANOVA

Permutational multivariate analysis of variance (PERMANOVA) is a multivariate analog to univariate ANOVA and has less restriction than the parametric MANOVA. PERMANOVA partitions distance matrices among sources of variation and fits linear models to distance matrices, testing differences according to a specified model by randomly permuting the data.

We can implement PERMANOVA using the function adonis() (adonis = "analysis of dissimilarity") in the vegan package; adonis allows the use of any semimetric (e.g., Bray–Curtis, and Sørensen) or metric (e.g., Euclidean) distance matrix. Typically, the function adonis() is used to analyze ecological and microbiome community data (samples by species/OTUs/taxa matrices) or genetic data (samples by gene expression).

To run adonis, we first need a factor vector or matrix that specifies the treatments and replicates. The following is one syntax: adonis (formula, data, permutations = 999, method = "bray")

where, formula is a typical model formula such as $Y \sim A + B \times C$: Y is either a dissimilarity object (inheriting from class "dist"), data frame, or matrix; A, B, and C may be factors or continuous variables. If a dissimilarity object is supplied, no species coefficients can be calculated.

The argument data are the data frame including variables A, B, and C. The number of replicate permutations is needed to be specified for the hypothesis tests (*F* tests).

If the left-hand side of the formula is a data frame or a matrix, then the vegdist() function is required to calculate pairwise distances by specifying a method before running adonis().

As we already calculated the Bray–Curtis dissimilarity and named it "bc_dist" in section "Bray–Curtis Index", the adonis() function can now be called to implement PERMANOVA.

```
> adonis(bc_dist ~ grouping_inf)

Call:
adonis(formula = bc_dist ~ grouping_inf)

Permutation: free
Number of permutations: 999

Terms added sequentially (first to last)

             Df SumsOfSqs MeanSqs F.Model      R2 Pr(>F)
grouping_inf  1   0.10797 0.10797 0.68019 0.10182  0.638
Residuals     6   0.95239 0.15873         0.89818
Total         7   1.06035                 1.00000
```

If Bray–Curtis dissimilarity is not given, we can use the abundance table to specify the formula. However, we also need to calculate the dissimilarity object by specifying a method such as "bray."

```
> adonis(abund_table_t ~ grouping_inf,permutations = 999, method = "bray")

Call:
adonis(formula = abund_table_t ~ grouping_inf, permutations = 999, method
= "bray")

Permutation: free
Number of permutations: 999

Terms added sequentially (first to last)

             Df SumsOfSqs MeanSqs F.Model      R2 Pr(>F)
grouping_inf  1   0.10797 0.10797 0.68019 0.10182  0.665
Residuals     6   0.95239 0.15873         0.89818
Total         7   1.06035                 1.00000
```

Similarly, we specify method = "jaccard" to conduct PERMANOVA using the Jaccard method.

```
> adonis(abund_table_t ~ grouping_inf,permutations = 999, method
= "jaccard")

Call:
adonis(formula = abund_table_t ~ grouping_inf, permutations =
999,    method = "jaccard")

Permutation: free
Number of permutations: 999

Terms added sequentially (first to last)

             Df SumsOfSqs MeanSqs F.Model      R2 Pr(>F)
grouping_inf  1    0.1968 0.19680 0.78415 0.11559  0.715
Residuals     6    1.5059 0.25098         0.88441
Total         7    1.7027                 1.00000
```

The following R codes call the adonis() function to conduct PERMANOVA using the Sørensen method:

```
> adonis(S_dist  ~ grouping_inf,permutations = 999)

Call:
adonis(formula = S_dist ~ grouping_inf, permutations = 999)

Permutation: free
Number of permutations: 999

Terms added sequentially (first to last)

             Df SumsOfSqs  MeanSqs F.Model      R2 Pr(>F)
grouping_inf  1   0.11856 0.118560  1.4819 0.19807  0.077 .
Residuals     6   0.48003 0.080005         0.80193
Total         7   0.59859                  1.00000
---
Signif. codes:  0 '***' 0.001 '**' 0.01 '*' 0.05 '.' 0.1 ' ' 1
```

Analysis of Similarity

Analysis of similarity (ANOSIM) is a nonparametric procedure based on a permutation test of among-and-within group similarities for two or more groups of sampling units. The algorithm behind the hypothesis testing is as follows: if two groups of sampling units are very different in their species (or other taxon) composition, then the compositional dissimilarities between the groups should be greater than those within the groups.

The statistic R ranges from 0 to 1; an R value close to 1 means that there is dissimilarity between the groups, whereas an R value close to 0 indicates no significant dissimilarity between the groups. Theoretically, it is also possible that $R < 0$, but in practice, such case is unlikely in ecological and microbiome studies. The extreme case, $R = -1$, indicates that the most similar samples are not all in the groups.

The function anosim() in the vegan package performs ANOSIM. The input data are a dissimilarity matrix, which can be produced using the functions dist() or vegdist(), and the p value is obtained by permutation. The function has summary() and plot() methods to perform post-modeling analysis.

One of the syntax examples is given below:

$$\text{anosim}\left(\text{data, grouping, permutations} = 999, \text{distance} = \text{``bray''}\right)$$

where,

data	data matrix or data frame in which rows are samples and columns are response variable(s), a dissimilarity object, or a symmetric square matrix of dissimilarities.
grouping	grouping variable (a factor).
permutations	number of permutations to assess the significance of the ANOSIM statistic.
distance	distance or dissimilarity measure.

If the input data comprise a dissimilarity structure or a symmetric square matrix, then the distance also needs to be specified.

The following codes run ANOSIM using the Bray–Curtis dissimilarity matrix as input data:

```
> anosim(bc_dist, grouping_inf,permutations = 999)
```

```
Call:
anosim(dat = bc_dist, grouping = grouping_inf, permutations = 999)
Dissimilarity: bray

ANOSIM statistic R: -0.07692
      Significance: 0.561

Permutation: free
Number of permutations: 999
```

The following codes run ANOSIM using the abundance data frame as input data:

```
> anosim(abund_table_t, grouping_inf, permutations = 999, distance
= "bray")
Call:
anosim(dat = abund_table_t, grouping = grouping_inf, permutations
= 999,      distance = "bray")
Dissimilarity: bray

ANOSIM statistic R: -0.07692
      Significance: 0.563

Permutation: free
Number of permutations: 999
```

We can use the function summary() to summarize the results:

```
> fit <- anosim(bc_dist, grouping_inf,permutations = 999)
> summary(fit)

Call:
anosim(dat = bc_dist, grouping = grouping_inf, permutations = 999)
Dissimilarity: bray

ANOSIM statistic R: -0.07692
      Significance: 0.547

Permutation: free
Number of permutations: 999

Upper quantiles of permutations (null model):
  90%   95% 97.5%   99%
     0.313 0.325 0.559 0.682
```

```
Dissimilarity ranks between and within classes:
        0%  25% 50%   75% 100%  N
Between  1  7.5  13 22.50   26 15
Vdr-/-   2  7.5  16 20.75   28 10
WT      10 13.0  16 17.50   19  3
```

The ANOSIM statistic R is -0.07692 and it is not statistically significant with the permutation test.

Similarly, we fit anosim() using the Jaccard method:

```
> anosim(abund_table_t, grouping_inf, permutations = 999, distance
= "jaccard")

Call:
anosim(dat = abund_table_t, grouping = grouping_inf, permutations
= 999,   distance = "jaccard")
Dissimilarity: jaccard

ANOSIM statistic R: -0.07692
      Significance: 0.573

Permutation: free
Number of permutations: 999
```

We can also fit anosim() using the Sørensen method:

```
> fit_S <- anosim(s_dist, grouping_inf, permutations = 999)
> summary(fit_S)

Call:
anosim(dat = s_dist, grouping = grouping_inf, permutations = 999)
Dissimilarity: binary bray

ANOSIM statistic R: 0.4103
      Significance: 0.016

Permutation: free
Number of permutations: 999

Upper quantiles of permutations (null model):
  90%   95% 97.5%   99%
   0.287 0.364 0.374 0.410

Dissimilarity ranks between and within classes:
        0%   25%   50%  75% 100%  N
```

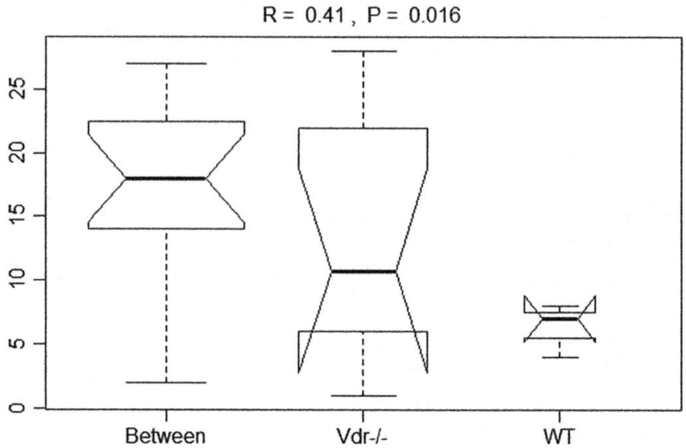

Fig. 12.18 Plots of between and within means of Sørensen dissimilarity

```
Between   2 14.00 18.00 22.5   27 15
Vdr-/-    1  6.75 10.75 20.0   28 10
WT        4  5.50  7.00  7.5    8  3
```

Finally, we can plot the results (Fig. 12.18):

```
> plot(fit_S)
```

We conclude that the Vdr$^{-/-}$ condition significantly differs from the WT condition in terms of the Sørensen dissimilarity measure.

12.3.5 Identifying Major Contributors to Community Differences Using SIMPER

The similarity percentage (SIMPER) is a statistical test method developed by Clarke (1993) for assessing which species (variables) are primarily responsible for an observed dissimilarity between groups of ecological objects (sampling units). Applied to microbiome study, variables can be any taxa, and sampling units can be either human subjects or experimental samples. For example, in our Vdr$^{-/-}$ mice case, the Bray–Curtis dissimilarity for a pair of samples is basically the differences between the samples for each genus, summed over all the genera. SIMPER computes the percentage contribution of each genus to the dissimilarities between all pairs of samples in Vdr$^{-/-}$ and wild-type mice, in addition to the percentage contribution of each genus to the similarities between all pairs of samples within each group. It then calculates the average of these percentage contributions and the standard deviation. Genera with a large ratio of average vs standard deviation percentages contribution to dissimilarity between samples in the various groups are those genera that best discriminate between the groups.

The similarity percentage assumes that the samples are independent. It was originally implemented in the commercial software, PRIMER-E. It is now available in the vegan package as a paired-test. The author is Eduard Szöcs. There are no formal hypothesis tests with SIMPER; the output only lists the taxa in order of their percentage contributions to dissimilarities between groups or similarities within groups.

The input data are two or more groups of multivariate abundance samples (taxa in columns). If there are more than two groups, two options are available: either pairwise comparisons of two groups of samples can be performed or all samples can be pooled to perform one overall multi-group SIMPER.

One syntax is as follows:

$$\text{simper}(\text{comm}, \text{group}, \text{permutations} = 0, \text{trace} = \text{FALSE})$$

where comm is a community data matrix, group is a factor describing the group structure, which requires at least two levels, permutations are the number of permutations required, and trace specifies whether the trace permutations are true or false.

The results can be summarized using the function summary():

$$\text{summary}\left(\text{object}, \text{ordered} = \text{TRUE}, \text{digits} = \max\left(3, \text{getOption}\left(''\text{digits}''\right) - 3\right)\right)$$

where object is an object returned by SIMPER; ordered is a logical option to specify whether the genera can be ordered by their average contributions; and digits is the number of digits in the output.

We use the function simper() to find the most influential genera in our Vdr$^{-/-}$ cecal dataset:

```
> library(vegan)
>  simper(abund_table_t,grouping_inf,permutations  =  0,  trace  =
FALSE)
cumulative contributions of most influential species:
$'Vdr-/-_WT'
  Lactococcus  Butyrivibrio  Turicibacter Lactobacillus   Allobaculum
    0.2422170     0.4591818     0.5670344     0.6525181     0.7335440
```

The five most influential genera are identified.

With the argument ordered = TRUE, the data frames also include the cumulative contributions and are ordered by genera contribution.

```
> sim <- simper(abund_table_t, grouping_inf)
> summary(sim, ordered = TRUE, digits = max(3,getOption("digits")
- 3))

Contrast: Vdr-/-_WT

                          average    sd         ratio   ava     avb       cumsum
Lactococcus               1.266e-01  9.108e-02 1.3904  1355.8  1341.3333 0.2422
Butyrivibrio              1.134e-01  1.186e-01 0.9568   886.6   798.6667 0.4592
Turicibacter              5.639e-02  5.721e-02 0.9857   462.6    11.6667 0.5670
Lactobacillus             4.470e-02  3.627e-02 1.2322   803.6   783.6667 0.6525
Allobaculum               4.236e-02  4.721e-02 0.8974    62.6   462.3333 0.7335
Clostridium               3.644e-02  2.639e-02 1.3804   565.0   694.0000 0.8032
Eubacterium               1.808e-02  1.211e-02 1.4925   185.6   190.6667 0.8378
Blautia                   1.729e-02  2.330e-02 0.7421    14.4   194.0000 0.8709
Akkermansia               1.147e-02  1.034e-02 1.1094   146.0    52.0000 0.8928
Tannerella                9.609e-03  8.117e-03 1.1837    26.0   118.0000 0.9112
Blautia::Clostridium      6.552e-03  9.068e-03 0.7225     1.0    72.6667 0.9237
Bacteroides               4.792e-03  4.130e-03 1.1602    18.0    62.3333 0.9329
```

Only the first 12 genera are reproduced.

In the output, the average is the average contribution to overall dissimilarity, sd is the standard deviation of contribution, ratio is the average-to-standard deviation ratio, ava and avb are the average abundances per group, and cumsum is the ordered cumulative contribution.

The permutation p value (the probability of obtaining a larger or equal average contribution in random permutation of the group factor) can be obtained using the permutation test as follows:

```
> mod <- simper(abund_table_t,grouping_inf,permu=999)
> summary(mod)

Contrast: Vdr-/-_WT

                           average        sd  ratio   ava      avb cumsum     p
Tannerella               9.609e-03 8.117e-03 1.1837  26.0 118.0000 0.9112 0.056 .
Ruminococcus             2.788e-03 2.054e-03 1.3577  27.0  53.0000 0.9382 0.026 *
Dorea                    2.336e-03 2.679e-03 0.8719   3.4  18.0000 0.9478 0.021 *
Butyrivibrio::Clostridium 1.186e-03 1.434e-03 0.8273  2.8   9.0000 0.9598 0.071 .
Desulfotomaculum         6.522e-04 4.784e-04 1.3633   1.6   8.0000 0.9714 0.094 .
Denitrobacterium         5.943e-04 3.930e-04 1.5123   1.8   7.0000 0.9737 0.054 .
Alistipes                4.395e-04 3.180e-04 1.3821   0.8   5.3333 0.9806 0.008 **
Erysipelothrix           3.656e-04 3.679e-04 0.9938   0.8   3.0000 0.9836 0.069 .
Butyricimonas            2.720e-04 2.347e-04 1.1589   0.0   3.0000 0.9872 0.094 .
Anaerostipes             2.476e-04 3.161e-04 0.7833   0.0   1.6667 0.9887 0.021 *
```

```
Clostridium::Ruminococcus:      2.163e-04 3.350e-04 0.6456   0.0   1.3333 0.9891 0.051 .
Lachnospira::Clostridium        1.420e-04 1.270e-04 1.1180   0.2   1.3333 0.9917 0.021 *
Clostridium::Dorea              1.395e-04 1.517e-04 0.9194   0.0   1.0000 0.9922 0.021 *
Clostridium::Blautia            1.167e-04 9.270e-05 1.2591   0.0   1.0000 0.9929 0.021 *
Odoribacter                     1.151e-04 4.625e-05 2.4881   0.0   1.0000 0.9934 0.008 **
Clostridium::Ruminococcus       1.013e-04 8.364e-05 1.2106   0.2   1.0000 0.9946 0.092 .
Dysgonomonas                    9.233e-05 8.264e-05 1.1173   0.0   1.0000 0.9953 0.094 .
Ruminococcus::Clostridium       9.233e-05 8.264e-05 1.1173   0.0   1.0000 0.9955 0.094 .
Ruminococcus::Escherichia       5.407e-05 8.375e-05 0.6456   0.0   0.3333 0.9982 0.051 .
Signif. Codes:  0 '***' 0.001 '**' 0.01 '*' 0.05 '.' 0.1 ' ' 1
Permutation: free
Number of permutations: 999
```

The average dissimilarity between the two compared genetic communities, Vdr $^{-/-}$ and WT, can be found by using the lapply() function:

```
> lapply(sim, function(x){x$overall})
$'Vdr-/-_WT'
[1] 0.5228499
```

These results show that the genetic conditions Vdr$^{-/-}$ and WT are 52.28% different from each other.

12.3.6 Compositional Data Analysis

It is widely recognized in the field of microbiome study that it is inappropriate to draw inferences regarding the total abundance in the ecosystem from the abundance of OTUs in the sample. Instead, researchers prefer to use the relative abundance or the logarithm of the ratio of counts to analyze microbiome composition. The reason behind using the log-ratios approach is that there is a compositional constraint: all relative microbial abundances within a specimen add up to one, which results in the compositional data residing in a simplex (Aitchison 1982), rather than the Euclidean space. Thus, the compositional constraint results in violating assumptions of most statistical models for use in the analysis of micriobiome data.

The existing tools for compositional data analysis in geology, ecology, and other fields have been shown to be readily adapted and valid in analyzing microbiome high-throughput sequencing data (Gloor et al. 2016; Gloor and Reid 2016). In the microbiome literature, analysis of composition of microbiomes (ANCOM) (Mandal et al. 2015) and ANOVA-like differential express (ALDEx and ALDEx2) (Gloor et al. 2016; Fernandes et al. 2013) have been recently developed. In this section, we illustrate the capabilities of ALDEx2 by analyzing data on cigarette smokers.

To use the cigarette smoker data, we first need to install and load the GUniFrac package:

```
>install.packages(GuniFrac)
>library(GuniFrac)
```

Then, the data can be loaded using the data() function:

```
> data(throat.otu.tab)
> head(throat.otu.tab)
              4695 2983 2554 3315 879 1313 5661 4125 2115 3309 3225 514 3427 484
ESC_1.1_OPL      1    0    0    0   0    0    0    0    0    0    0   0    0   0
ESC_1.3_OPL      0    0    0    0   0    0    0    0    0    0    0   0    0   0
ESC_1.4_OPL      0    0    0    0   0    0    0    0    0    0    0   0    0   0
ESC_1.5_OPL      1    0    0    0   0    0    0    0    0    0    0   0    0   0
ESC_1.6_OPL      0    0    0    0   0    0    0    0    0    0    0   0    0   0
ESC_1.10_OPL     0    0    0    0   0    0    0    0    0    0    0   0    0   0
```

The original format of the OTU table consisted of samples by taxa, but we need to transpose it to the format with taxa by samples for ALDEx2() use:

```
> otu_table <-t(throat.otu.tab)
> head(otu_table)
ESC_1.1_OPL ESC_1.3_OPL ESC_1.4_OPL ESC_1.5_OPL ESC_1.6_OPL ESC_1.10_OPL
4695           1           0           0           1           0            0
2983           0           0           0           0           0            0
2554           0           0           0           0           0            0
3315           0           0           0           0           0            0
879            0           0           0           0           0            0
1313           0           0           0           0           0            0
```

In the analysis, we subset the first 20 samples for use and name the subset “throat”.

```
> throat<- data.frame(otu_table[,1:20])
```

Similarly, we load and subset the first 20 metadata.

```
> data(throat.meta)
> head(throat.meta)
```

```
             BarcodeSequence LinkerPrimerSequence SmokingStatus PatientID
ESC_1.1_OPL         ACGTCATG      CTGCTGCCTYCCGTA     NonSmoker         1
ESC_1.3_OPL         ACTCGTGA      CTGCTGCCTYCCGTA        Smoker         3
ESC_1.4_OPL         ACTGCTGA      CTGCTGCCTYCCGTA        Smoker         4
ESC_1.5_OPL         AGACTGTC      CTGCTGCCTYCCGTA        Smoker         5
ESC_1.6_OPL         AGCTGATC      CTGCTGCCTYCCGTA        Smoker         6
ESC_1.10_OPL        ATGCGCTA      CTGCTGCCTYCCGTA        Smoker        10
             SampleIndex AirwaySite SideOfBody    SampleType
ESC_1.1_OPL            1     Throat       Left patientsample
ESC_1.3_OPL            1     Throat       Left patientsample
ESC_1.4_OPL            1     Throat       Left patientsample
ESC_1.5_OPL            1     Throat       Left patientsample
ESC_1.6_OPL            1     Throat       Left patientsample
ESC_1.10_OPL           1     Throat       Left patientsample
> throat_meta <- data.frame(throat.meta[1:20,])
```

The following R codes extract group information from the meta-table and assign a group variable:

```
> groups <- throat_meta$SmokingStatus
> groups
 [1] NonSmoker Smoker    Smoker    Smoker    Smoker    Smoker    NonSmoker
 [8] NonSmoker NonSmoker NonSmoker Smoker    NonSmoker NonSmoker Smoker
[15] NonSmoker Smoker    Smoker    NonSmoker NonSmoker NonSmoker
Levels: NonSmoker Smoker
```

ALDEx2 has two approaches, aldex wrapper and modular, ALDEx2 can be run using either approach. To illustrate the capabilities of ALDEx2, we run both of them. To use the aldex() function, we need to install the "ALDEx2" package and call the ALDEx2 library.

First, we run the aldex wrapper. Currently, the aldex wrapper is limited to two sample tests and a one-way ANOVA design. When you run the aldex wrapper, it links the modular elements together to emulate ALDEx2 before the modular approach.

```
> throat <- aldex(throat, groups, mc.samples=128, test="t",
effect=TRUE,include.sample.summary=FALSE,
+           verbose=FALSE)
[1] "aldex.clr: generating Monte-Carlo instances and clr values"
[1] "operating in serial mode"
[1] "aldex.ttest: doing t-test"
```

As there are two test groups, smokers versus nonsmokers, this is two-sample *t*-test. We specify test = "t," and then the effect should be set to TRUE. The "t" option evaluates the data as a two-factor experiment using both the Welch's *t*-test and the Wilcoxon rank test. If the test is "glm," then the effect should be specified as FALSE. The "glm" option evaluates the data as a one-way ANOVA using the glm and Kruskal–Wallis tests. All tests include a BH correction of the raw *p* values.

Now, we run the aldex modular step-by-step. The aldex modular offers the user the ability to build a data analysis pipeline for their experimental designs and tests. To simplify, the procedure of this approach is just to call aldex.clr, aldex.ttest, and aldex.effect in turn and then merge the data into one object. Readers can check the manual of aldex software for more details.

Step 1: Generate instances of the centered log-ratio transformed values using the function aldex.clr().
The function has three inputs: counts table, number of Monte-Carlo instances, and level of verbosity (TRUE or FALSE). The authors of this software recommend 128 or more mc.samples for the *t*-test, 1,000 for a rigorous effect size calculation, and at least 16 for ANOVA.

```
> throat <- aldex.clr(throat, mc.samples=128, verbose=TRUE)
[1] "operating in serial mode"
[1] "removed rows with sums equal to zero"
[1] "data format is OK"
[1] "dirichlet samples complete"
[1] "clr transformation complete"
```

Step 2: Perform Welch's *t*-test and Wilcoxon rank test using aldex.ttest().
As in other statistical testing using only two conditions, Welch's *t*-test and the Wilcoxon rank test can both be used. The function aldex.ttest() has three inputs: the aldex object from aldex.clr(), the vector of conditions, and whether or not a paired test should be conducted (TRUE or FALSE). The aldex.ttest() function returns the values of we.ep (Expected *p* value of Welch's t test), we.eBH (expected Benjamini–Hochberg corrected *p* value of Welch's *t*-test), wi.ep (expected *p* value of the Wilcoxon rank test), and wi.eBH (expected Benjamini–Hochberg corrected *p* value of the Wilcoxon test).

```
> throat_tt <- aldex.ttest(throat, groups, paired.test=FALSE)
```

As an alternative to step 2, we can perform the glm and Kruskal–Wallis tests for one-way ANOVA using the function aldex.glm() here; however, this is slow. The aldex.glm() function returns the values of kw.ep (expected *p* value of Kruskal–Wallis test), kw.eBH (expected Benjamini–Hochberg corrected *p* value of Kruskal–Wallis test), glm.ep (expected *p* value of glm test), and glm.eBH (expected Benjamini–Hochberg corrected *p* value of the glm test).

```
> throat_glm <- aldex.glm(throat, groups)
```

Step 3: Estimate the effect size using the function aldex.effect().
The aldex.effect() estimates effect size and the within and between condition values in the case of two conditions. It has four inputs: the aldex object from aldex.clr(), the vector of conditions, a flag to indicate whether or not to include values for all samples, and the level of verbosity. The aldex.effect function returns all the values, including:

rab.all (median clr value for all samples in the feature)
rab.win.NS (median clr value for the NS group of samples)
rab.win.S (median clr value for the S group of samples)
dif.btw (median difference in clr values between the S and NS groups)
dif.win (median of the largest difference in clr values within the S and NS groups)
effect (median effect size: diff.btw/max(diff.win) for all instances
and overlap (proportion of effect size that overlaps 0 (i.e. no effect).

```
> throat_tt <- aldex.ttest(throat, groups, paired.test=FALSE)
> throat_effect <- aldex.effect(throat, groups, include.sample.
summary=FALSE, verbose=TRUE)
[1] "operating in serial mode"
[1] "sanity check complete"
[1] "rab.all  complete"
[1] "rab.win  complete"
[1] "rab of samples complete"
[1] "within sample difference calculated"
[1] "between group difference calculated"
[1] "group summaries calculated"
[1] "effect size calculated"
[1] "summarizing output"
```

Step 4: Merge all data into one object and make a data frame for result viewing.

```
> throat_all <- data.frame(throat_tt, throat_glm, throat_effect)
> head(x.all)
          we.ep     we.eBH     wi.ep     wi.eBH     kw.ep     kw.eBH    glm.ep    glm.eBH
4695 0.5243678 0.9261159 0.5433570 0.9351110 0.5161343 0.8841405 0.5022318 0.8653634
2983 0.4901260 0.9234518 0.5364041 0.9299947 0.5091792 0.8816335 0.4665601 0.8553843
2554 0.5097563 0.9213185 0.5135721 0.9290968 0.4871505 0.8766474 0.4890758 0.8534886
3309 0.4786177 0.9191998 0.5167427 0.9401704 0.4900083 0.8857985 0.4509357 0.8477759
484  0.4609612 0.9165449 0.4492607 0.9195321 0.4246162 0.8652998 0.4355394 0.8314702
4194 0.4100868 0.9043515 0.4091460 0.9015420 0.3867599 0.8482407 0.3931106 0.8101037
rab.all rab.win.NonSmoker rab.win.Smoker    diff.btw diff.win      effect    overlap
4695 0.178503316         0.15839031     0.23155921 -0.01133157 4.002614 -0.00233237 0.4965278
2983 0.093689745        -0.03740126     0.18874937  0.27905207 3.870483  0.05842798 0.4618056
```

```
2554 0.004862142     -0.10256651     0.12723399  0.29658974 3.894425  0.05191272 0.4644714
3309 0.280715492      0.44082557     0.01110872 -0.56235299 4.025415 -0.09632831 0.4496528
484  0.019478837      0.16268272    -0.18217474 -0.18299580 4.189606 -0.04429169 0.4826389
4194 0.578551098      1.18629088     0.05764254 -1.34695288 4.236469 -0.25317094 0.3726170
```

12.4 Summary

In this chapter, we demonstrated a general workflow of the statistical analysis of microbiome data from mice and human samples using R. We described frequently used univariate and multivariate statistical models and visualization tools, in addition to alpha and beta metrics (Xia and Sun 2017). R and its modules comprise a widely used open-source software project for analyzing and comprehending the high-throughput data sets from different platforms. The main advantage of R is access to a wide range of powerful statistical and graphical methods for the analysis of ecology and microbiome data, coupled with the rapid development of extensible software based on the most advanced and updated analysis algorithms.

We focused our analysis on commonly used statistical techniques for studying microbiome data, including data management, calculating metrics, hypothesis testing of univariate and multivariate data, and exploratory data analysis with clustering and ordination. However, alternative statistical methods and approaches can also be used for modeling and analyzing microbiome data. For each individual project, not all of the skills may be required. The reader may choose the appropriate ones to apply to his or her own study. Also, although the R packages and statistical techniques used in this chapter are frequently used in ecology and microbiome fields, it does not mean that other packages or statistical tools are inferior. Additional models and analyses (such as zero-inflated and over-dispersion models) can allow for a more in-depth understanding of the microbiome data.

Acknowledgments We would like to acknowledge the NIDDK/National Institutes of Health grant R01 DK105118 and DOD BC160450P1 to Jun Sun. We thank the two anonymous reviewers whose comments/suggestions helped to improve and clarify this manuscript.

References

Aitchison J (1982) The statistical analysis of compositional data. J R Stat Soc Ser B (Methodological) 44(2):139–177

Borcard D, Gillet F et al (2011) Numerical ecology with R. Springer, New York

Chao A (1984) Nonparametric estimation of the number of classes in a population. Scand J Stat 11:265–270

Charlson ES, Chen J et al (2010) Disordered microbial communities in the upper respiratory tract of cigarette smokers. PLoS One 5(12):0015216

Chen J (2012) GUniFrac: generalized UniFrac distances. R package version 1.0. http://CRAN.R-project.org/package=GUniFrac
Clarke KR (1993) Non-parametric multivariate analysis of changes in community structure. Aust J Ecol 18:117–143
Fernandes AD, Macklaim JM et al (2013) ANOVA-like differential expression (ALDEx) analysis for mixed population RNA-Seq. PLoS One 8(7):e67019
Gloor GB, Reid G (2016) Compositional analysis: a valid approach to analyze microbiome high-throughput sequencing data. Can J Microbiol 62(8):692–703
Gloor GB, Wu JR et al (2016) It's all relative: analyzing microbiome data as compositions. Ann Epidemiol 26(5):322–329
Jin D, Wu S et al (2015) Lack of vitamin D receptor causes dysbiosis and changes the functions of the murine intestinal microbiome. Clin Ther 37(5):996–1009.e1007
Kindt R, Coe R (2005) Tree diversity analysis. A manual and software for common statistical methods for ecological and biodiversity studies. World Agroforestry Centre (ICRAF), Nairobi. ISBN: 92-9059-179-X
Mandal S, Van Treuren W et al (2015) Analysis of composition of microbiomes: a novel method for studying microbial composition. Microb Ecol Health Dis 26:27663
Oksanen J, Guillaume Blanchet F et al (2016) Vegan: community ecology package. R package version 2.4-1. http://CRAN.R-project.org/package=vegan
R Core Team (2016) R: a language and environment for statistical computing. R Foundation for Statistical Computing, Vienna. https://www.R-project.org/
RStudio Team (2016) RStudio: integrated development for R. RStudio, Boston. http://www.rstudio.com/
Shannon CE (1948) A mathematical theory of communication. Bell Syst Tech J 27:379–423
Shannon CE, Weaver W (1949) The mathematical theory of communication. University of Illinois Press, Urbana
Simpson EH (1949) Measurement of diversity. Nature 163:688
Wang J, Thingholm LB et al (2016) Genome-wide association analysis identifies variation in vitamin D receptor and other host factors influencing the gut microbiota. Nat Genet 48 (11):1396–1406
Wickham H, Francois R (2016). dplyr: a grammar of data manipulation. R package version 0.5.0. http://CRAN.R-project.org/package=dplyr
Xia Y, Sun J (2017) Hypothesis testing and statistical analysis of microbiome. Genes Dis 4 (3):138–148. https://doi.org/10.1016/j.gendis.2017.06.001

Index

A
Adaptive immunity, 147, 148
  deficiencies, functional maturation, 11
  intestinal commensals, 16
  lamina propria and GALTs, 10
  lymphoid follicles, 16
  MAMP–TLR interactions, 16
  microbial colonization, 10
  non-pathogenic strains, 16
  Peyer's patches, 10
Adenomatous polyposis coli gene (APC), 168
Adenosine monophosphate-activated protein kinase (AMPK), 134
Adhesion
  ALS gene family, 264
  EAP1, 264
  SUN family protein, 264
Aggregatibacter actinomycetemcomitans, 221–223
Airway mucosal dendritic cells (AMDCs), 44
Akkermansia muciniphila, 151
aldex.effect(), 369
Aldex modular, 368
Allergic diseases of childhood, 42, 44
  AD (*see* Atopic dermatitis (AD))
  asthma (*see* Asthma; Food allergy)
  "atopic march", 36
  inflammatory process, 37
  prevalence of, 36
  "signaling molecules", 36
  Tregs cells, 36
  types of, 36
Allergic rhinitis (AR), 47
Alveolar bone loss (ABL), 207
Alzheimer's disease (AD), 204–206
Amyotrophic lateral sclerosis (ALS), 120
Analysis of similarity (ANOSIM), 359–362
Animal models, 212
Anti-host clearance mechanisms, 251, 252
Antimicrobial proteins, 143
Asthma
  aero-allergen-specific Th2 cells, 44
  airway inflammation, 44
  airway microbiota analysis, 45
  AMDCs, 44
  AR, 47
  description, 44
  early-life colonization, upper respiratory tract, 46
  environmental and genetic influences, 45
  gut microbiome, 46
  host interactions, 47
  low diversity and low load, microorganisms, 45
  NPA samples, 46
  pathogen-free experimental mice models, 46
  sampling methods, 45
  Th2-related cytokines, 44
Atopic dermatitis (AD)
  atopy at early ages, 43
  birth cohort studies, 42
  cutaneous inflammation, 42
  description, 42
  family history, 42
  FLG mutation, 42
  fungi, 43
  healthy skin microbiome, 42
  lesion and nonlesion sites, 43
  mast cells (MCs), 43

J. Sun, P.K. Dudeja (eds.), *Mechanisms Underlying Host-Microbiome Interactions in Pathophysiology of Human Diseases*, Physiology in Health and Disease,
https://doi.org/10.1007/978-1-4939-7534-1

Atopic dermatitis (AD) (*cont.*)
  recurrent eczematous skin lesions, 42
  Th2-mediated inflammatory mechanisms, 43
"Atopic March", 36, 44
Azoxymethane (AOM), 181

B
Bacteremia, 199
Bacterial biofilm formation, 175
Bacterial toxin and virulence factors
  Aggregatibacter actinomycetemcomitans, 221–223
  Fusobacterium nucleatum, 220
  Gingipains, 217, 218
  GroEL, 215
  LPS, 214
  PADs, 218
  Porphyromonas gingivalis, 216
  Tannerella forsythia, 220
  Treponema denticola, 219
Benjamini and Hochberg (BH) method, 329
B. fragilis toxin (BFT), 170
Bile acids, 141, 142, 175, 176
Biofilms, 176–178
Bonferroni method, 328
Bray–Curtis index, 333, 335
Broken stick model, 345
Bronchoalveolar lavage (BAL) fluid, 45
Brush border (BB) membrane, 2
Buccal epithelial cells (BECs), 250
Butyrate, 176

C
Cancer
  ABL, 207
  CAL, 207
  dysbiosis, 211
  late complications, 211
  poly-microbial studies, 209, 210
  Porphyromonas gingivalis, 208
  qPCR, 211
  quality of life, 208
  treatments outcomes, 208
  xerostomia, 208
Candida albicans commensalism
  attributes, 248
  candidemia-associated mortalities, 248
  fungal gene expression, 264, 265
  GI tract, 248
    clinical survey, 256
    dysbiosis, 260
    Efg1, 256
    environmental stress, 263
    fungus and host immune responses, 257
    gastric ulcers, 256
    GUT, 259
    homeostasis, 260
    HSGs, 256
    microbiota, 257
    mouse model, 260
    nonpathogenic yeast, 261
    phenotypic switch, 257–259
    polymorphism, 257
    ulcerative colitis patients, 256
  host immune defenses, 261–263
  infectious syndromes, 248
  nutritional immunity, 265, 266
  oral cavity
    adhesion, 249, 250, 252, 263, 264
    antibiotics treatment, 251
    anti-host clearance, 251, 252
    BECs, 250
    OPC, 249
    saliva, 251
  reproductive tract
    estrogen levels, 253–255
    germ-tube formation, 255
    lactobacilli, 254
    pH regulation, 254, 255
    phenotypic switching, 255
    RVVC, 253
    symptomatic vaginitis, 254
    vaginal infection, 253
    yeast adherence, 254
Carbachol-/forskolin (FSK)-induced Cl^- secretion, 101
Carbohydrate digestion, 6
Cardiovascular diseases (CVDs), 200
Casein kinase 1α (CK1α), 81
Chao 1 index, 315, 316
Childhood Asthma Study (CAS), 46
Chronic intestinal pseudo-obstruction (CIPO), 115
  chronic alcohol abuse, 121
  description, 121
  inflammatory and neurodegenerative process, 121
  recurrent episodes before diagnosis, 121
Chronic kidney disease (CKD), 204
Cigarette smokers' data set, 305
Clinical attachment loss (CAL), 207
Clostridium difficile infection (CDI), 149, 286
Colibactin, 171
Colorectal cancer (CRC), 294
  chronic inflammatory diseases, 168

disruption/imbalances, 168
E. coli, 171, 172
E. faecalis, 170
ETBF, 170, 171
F. nucleatum, 172, 173
heritable, sporadic, and inflammation-associated, 168
immunopathophysiology, 173 (*see also* Innate immune system)
intestinal responses, 168
intrinsic and extrinsic factors, 169
metabolites
bile acids, 175, 176
NOCs, 174
SCFAs, 176
sulfides, 174
microbiota
anti-inflammatory cytokines, 183
biofilm formation, 176–178
cell cycle progression, 183
FMT, 185
mucosal immune system, 183
probiotics, 183–185
TLR agonists and antagonists, 185, 186
S. gallolyticus, 169, 170
tumorigenesis, 168
"Competitive exclusion" mechanism, 92
Compositional data analysis
aldex.clr(), 368
aldex.effect(), 369
ALDEx2, 367
cigarette smoker data, 366, 367
log-ratios approach, 365
Welch's *t*-test and Wilcoxon rank test, 368
Constrained analysis of proximities (CAP), 353–355
Constrained correspondence analysis (CCA), 353
CRC xenograft model, 186
Crohn's Disease (CD), 280
Cystic fibrosis transmembrane conductance regulator (CFTR), 101
Cytolethal distending toxin (CDT-V), 171

D
Dectin-1, 261, 262
Dextran sulfate sodium (DSS), 181
Diarrheal diseases
defined, 90
etiologies, 90
global disease burden and cause of death, 90 (*see also* Probiotics)
therapeutic modalities, 90
Disseminated candidiasis, 248
Dual-specificity phosphatase 6 (DUSP6), 132
Dysbiosis, 120–122, 296
CIPO (*see* Chronic intestinal pseudo-obstruction (CIPO))
definition, 130
dietary emulsifiers, 132
IBS (*see* Irritable Bowel Syndrome (IBS))
metabolic disorders, 132, 134
obese and lean microbiota, 131
saccharin, 132
SIBO (*see* Small intestinal bacterial overgrowth (SIBO))
Dysbiotic microbiome
description, 16
digestion/absorption, 19
immunity, 19, 20
motility, 19
NEC, 16, 17
necrotizing enterocolitis, 17
parenteral nutrition, 17, 18
preterm intestinal development, 18
Dysregulated proinflammatory reaction, 225

E
E-cadherin, 170
Emulsifiers, 132
Endocarditis, 199
Enteral nutrition (EN), 7
Enteric nervous system (ENS), 3
distal colon, 114
early development, 115
gastrointestinal disorders, 114
gram-negative and anaerobic bacteria, 114
gut microbiota, 114
host factors, 118
human GI tract, 114
metabolic, immunological and trophic functions, 114
metagenomic sequencing, fecal samples, 114
microbial metabolism, 116–118
neurohormones, 115, 116
Enterococcus faecalis, 170, 174
Enteroendocrine cells (EECs), 2
Enteroinvasive *E. coli* (EIEC), 104
Enterotoxigenic *Bacteroides fragilis* (ETBF), 170, 171, 174
Escherichia coli, 171, 172, 174
Estrogen-binding protein (EBP), 253–255
Estrogen-sensitive vaginal colonization, 254

F
Farnesoid X receptor (FXR), 141, 142
Fecal microbiota, 118–120, 122, 123

Fecal microbiota transplantation (FMT), 149, 185
attitudes, 281–282
Clostridium difficile infection, 286
colorectal cancer, 295
definition, 281
donor screening recommendations, 284, 285
dysbiosis, 295
IBD, 290–292
IBS and slow-transit constipation, 294
methodology, 282, 283
obesity, 292, 293
pseudomembranous colitis, treatment, 281
rCDI, 281, 287, 288
regulation, 285, 286
safety, 282, 285
stool-based therapies, 295
Firmicutes and bacteroidetes, 129–131
Focal infection theory, 196
Food allergy
breastfed infants, 39
defined, 37
food antigen (Ag), 37
groups, 37
gut colonization, early infancy, 39
human breast milk, 41
IgE-mediated food allergies, 41
inflammatory response, 41
intestinal microbiota, 39
oral tolerance, 37, 39
OVA challenge, 40
RORγt$^-$ Tregs, 39
SCFAs, 40
spore-forming Clostridia species, 39
Th2-specific cytokines, 40
TLR, 40
Fragilysin, 170
Functional maturation
during gestation, 4
motility, 3–5
normal digestion and absorption
carbohydrates, 6
intestinal size, 6, 7
lipids, 6
proteins, 5
Fungal gene expression, 264, 265
Fusobacterium nucleatum, 172–174, 220

G

Gastrointestinal (GI) tract, 260, 280
Gastrointestinally induced transition (GUT), 258, 259
Genomic mechanisms, 101
Germ-free (GF) mice, 168
Gingipains, 217, 218
Glucose-stimulated insulin secretion (GSIS), 138
Glycogen synthase kinase-3 (GSK-3), 81
Goblet cells (GCs), 2
GroEL, 215
Gut-associated lymphoid tissues (GALTs), 10
Gut bacteria, 175
Gut–brain axis
intestinal hormones and peptides, 7, 8
neural and hormonal signals, 7
Gut dysbiosis, 280, 281
Gut epithelial barrier function
culture supernatants, 95
intestinal epithelium, 94
Lactobacillus rhamnosus GG-produced soluble proteins, 95
pro-inflammatory cytokines, 95
TJs proteins, 95
ZO-1 and occludin, 95
Gut microbiome, 198
Gut microbiota, 130
adipocytes, 134
Akkermansia muciniphila, 151
antibiotics, 131
bacteroidetes and firmicutes, 129, 130
FMT, 149
metabolic disorders, 132
microbiota dysbiosis (*see* Dysbiosis)
NLRP3/NLRP6, 147
probiotics and prebiotics, 150
proteobacteria, 129, 130, 150
*Tlr5*KO mice, 137
TMA, 139

H

Heat shock protein family 60 (HSP60), 215
Hemagglutinin/adhesion (HA), 217
Hepatocellular carcinoma (HCC), 142
HGM-transplanted Gn pigs, 65, 66
Holm method, 328
Homeostasis, 260
Host immune defenses, 261–263
Host immunity, 259, 261, 263, 265, 267
Human gut microbiota (HGM), 65, 66, 280
Human intestinal development, 3–10
digestion/absorption, 7
during gestation, 2, 3
functional maturation (*see* Functional maturation)
gut–brain axis, 7, 8
immunity (*see* Immune system)

morphogenesis and differentiation, 2, 3
Human milk oligosaccharides (HMOs), 14
Human nanovirus (HuNoVs), 61
acute gastroenteritis and biopsy specimens, 58
animal models, 56, 57
anti-HuNoV drugs evaluation, 59
Balb/c mouse deficient, RAG, 59
γc/IL2RG, 59
description, 56
HBGA-expressing inactivated enteric bacteria, 58
IECs, 58
neonatal Gn pig model, 59, 60
NHPs, 58
robust cell culture system and small-animal model, 56
stem cell-derived and nontransformed human intestinal enteroids, 58
viral epidemic acute gastroenteritis, 56
Humans and animal models, 56–60, 62–64
HuNoVs (*see* Human noroviruses (HuNoVs))
rotavirus (*see* Rotaviruses)
Hydrogen sulfide, 174
4-Hydroxy-2-nonenal, 170
Hyperinsulinemia, 146
Hyphae-specific genes (HSGs), 256

I

Immune system
adaptive, 10, 11, 16
adherent inner non-adherent outer layer, 8
antigen detection and immune tolerance, 15
antigen recognition, 9, 10
distal ileum segments, 19
epithelial enterocytes, 8
genes and pathways, 19
GF rodents, 15
gut permeability, 9
inflammatory bowel diseases, 20
intestinal mucus, 8
intestinal tract, 8
long-term effects, 20
mucin-producing GCs and PCs, 8
passive immunity, 8
pathogen invasion, 9
preterm dysbiotic communities, 19
sIgA antibodies, 9, 15
TJs, 9
Immunological and inflammatory parameters
dysregulated proinflammatory reaction, 225
HSP60, 226
keystone pathogen, 225
MM, 226
platelets/thrombocytes, 225
Immunomodulation
adaptive immune response, 96
cytokine profiles modification, 97
innate and adaptive immune system, 96, 97
intestinal homeostasis, 96
NF-κB pathway, 96
pro-inflammatory mediators, 96
TLRs, 96
Infection, 196, 199, 202, 203
Inflammasomes, 147
Inflammation, 206, 211, 225
Inflammatory bowel disease (IBD), 280, 289, 290
Innate immune deficiency, 145, 146
Innate immune system, 179
chronic inflammation and tumorigenesis, 178
IBD patients, 178
infectious and non-infectious agents, 178
TLR (*see* Toll-like receptors (TLRs))
Integrated development environment (IDE), 304
Intestinal epithelial cell (IEC), 181
In-vitro and organotypic models, 212, 213
In-vivo murine skin model, 258
Irritable bowel syndrome (IBS), 280, 293
functional bowel disorders, 122
lack of consistency, 122
pathogenic species, 122
phylogenetic microarray analysis, 122
PI-IBS development, 122
symptom-based diagnosis, 122

J

Jaccard index, 333, 334

K

Kruskal–Wallis rank sum test, 331–332

L

Lactobacillus rhamnosus GG (LGG), 62, 68, 183
Leaky guts, 144
Ligature model, 212
Lipid digestion, 6
Lipopolysaccharide (LPS), 144, 145, 179, 205, 214
Lymphocytes, 251

M

Major outer sheath protein (MSP), 219
Mast cells (MCs), 43
Metabolic endotoxemia, 144, 145
Metabolic syndrome, 131, 137, 143
Metabolites, 226, 227
Microbes, 173, 176
Microbial-associated molecular patterns (MAMPs), 9
Microbial enzymatic activities, 226, 227
Microbial metabolism
  bile acids, 117
  gut bacteria, 117
  host innate immune system, 117
  low-grade inflammation, 116
  LPS, 117
  methane gas, 118
  SCFAs, 116
  TLRs, 117
Microbiomes
  antibiotic uses, 26
  breast milk feeding, 24, 25
  clinicians and researchers, 20
  community composition influences development, 21, 22
  conventional and germ-free mice, 64, 65
  HGM, 65, 66
  NEC patients, 21
  prebiotics, 24
  probiotics (*see* Probiotics)
  vaccination in humans, 66–67
  See also Allergic diseases in childhood
Microbiota
  adipocytes, 134
  AMPK, 134 (*see also* Colorectal cancer (CRC))
  energy harvest, 134
  FIAF, 134
  in silico techniques, 135
  neurological diseases
    ALS, 120
    PD, 119
Mitogen-activated protein kinase (MAPK), 219, 251
Molecular mimicry (MM), 226
Motility disorders, *see* Dysbiosis
Mucocutaneous candidiasis, 248
Multivariate community analysis, beta diversity calculation
    Bray–Curtis index, 333
    Jaccard index, 333, 334
    Sørensen index, 334
  exploratory analysis, microbiome data, 343–349
    CAP, 353–355
    clustering methods, 335, 336
    NMDS analysis, 349–351
    PCoA (*see* Principal coordinate analysis (PCoA))
    principal component analysis, 337–341
  measures, 332
  statistical testing
    ANOSIM, 359–362
    permutational MANOVA, 356–358
Murine norovirus (MNV), 61
Myelocytomatosis viral oncogene (MYC), 170

N

Nasopharyngeal aspirate (NPA) samples, 46
Necrotizing enterocolitis (NEC), 16, 17
Neonatal Gn pig model, 59, 60
Neonatal intensive care unit (NICU), 17
Neurohormones
  biosynthesis and release, 5-HT, 116
  enteropathogenic *Escherichia coli*, 116
  serotonin (5-HT) functions, 115
Neurological diseases
  ALS, 120
  neurological conditions, 119
  PD, 119
Neutrophil extracellular traps (NETs), 226
NIH Human Microbiome Project, 198
N-nitroso compounds (NOCs), 174
NOD-like receptors (NLRs), 9, 168
Nongenomic mechanisms, 102, 103
Nonhuman primates (NHPs), 58
Nonmetric multidimensional scaling (NMDS) analysis, 349–351
Nucleoside diphosphate kinase, 218

O

Obesity, 131, 200, 227, 292
Oligodeoxynucleotides (ODNs), 182
Oral and oropharyngeal candidiasis (OPC), 248
Oral candidiasis, 249, 251
Oral carcinogenesis models, 213
Oral microbiome
  alpha diversity, 198
  bacteremia, 199
  bacteria
    hypertension, 202
    PCR, 202

16S rRNA pyrosequencing, 202
breastfed newborns, 198
CVDs, 200, 201
diversity, 197
endocarditis, 199
experimental models, 203
habitats, 198
infections, 196
microorganisms, 196
P. gingivalis, 229
periodontal disease, 200, 201
16S rRNA sequencing, 196
serum antibodies, 201, 202
Streptococcus and *Staphylococcus*, 197
Organoid system, 213
bacterial infected mammalian intestines, 84
host–bacteria interactions, 84
intestinal stem cell markers, 84
Paneth-like cells and villus-like protrusions, 84
Osteoradionecrosis, 211

P
Paneth cells (PCs), 3
Parenteral nutrition (PN), 7, 17, 18
Parkinson's disease (PD), 119
Pathogen-associated molecular patterns, 179
Pattern recognition receptors (PRRs), 143
148, 179
Peptidylarginine deiminase (PAD), 206, 218
Periodontal disease, 200, 201
Periodontitis, 205
Permutational multivariate analysis of variance
(PERMANOVA), 356–358
Polyketide synthase (PKS), 171
Porphyromonas gingivalis, 206, 208, 216, 217
Post-infectious IBS (PI-IBS), 122
Postnatal development, 15–16
digestion/absorption, 14
immunity (*see* Immune system)
intestinal barrier function, 10
microbial colonization, 11–13
microbiome and functional maturation, 14
motility, 10, 13, 14
Prebiotics
bacterial community growth, 24
beneficial Bifidobacteria, 24
dysbiosis treatment, preterm infants, 24
gastrointestinal hormones, 24
nondigestible food ingredients, 24
Preterm infants, 10–16, 20–26
See also Postnatal development;
Microbiomes
Principal component analysis (PCA), 337–341
Principal coordinate analysis (PCoA)
benefit, 343
cmdscale() function
broken-stick model, 345
call function cmdscale(), 343, 344
influential genera identification
346, 347
ordination plot creation, 345, 347
percentage variation, 344
test abundances, 349
Probiotics, 183–185
anti-microbial and bioactive factors, 94
beneficial effects, 91
Bifidobacteria, 23
Bifidobacterium breve strains, 98
clinical evidence, 91, 92
Crohn's disease (CD) patients, 92
description, 22, 90
Firmicutes and Bacteroidetes, 98
gut epithelial barrier function, 94–96
human gastrointestinal system, 90
human gut microbiome, 97
immunomodulation, 96, 97
infection/IBD-associated diarrhea, 91
L. kefiri supplementation, 98
long-term effects, 23
NEC prevention, 23
pathogen exclusion, 92, 93
pathogenic organisms, 22
reduced microbial diversity, 23
regimens, 91
resident microbiota, gastrointestinal system, 97
single species supplementation, 23
therapeutic modality, 98
(*see also* Vectorial ion transport)
Probiotics/prebiotics mechanisms
NoV infection
animal models, features of, 57
gastroenteritis, 61
LGG, 62
vaccine adjuvants and therapeutic
agents, 61
vitamin A, 62
rotavirus infection
AttHRV-vaccinated Gn pigs, 68
Bifidobacterium spp., 68
Caco-2 cells and human intestinal organ
culture, 69
diarrhea prevention and treatment, 70, 71
dose effects, 69, 70
EcN and LGG, 69
IFN-inducible antiviral effectors, 68

Probiotics/prebiotics mechanisms (*cont.*)
LCG, 68
LGG, 67
Prokaryotic PAD (PPAD), 219
Protected specimen brushings (PSBs), 45
Protein digestion, 5
Proteobacteria, 129–131

Q
Quorum-sensing, 176

R
Reactive oxygen species (ROS), 169, 218
Recurrent Clostridium difficile infection (rCDI), 281, 287, 288
Recurrent VVC (RVVC), 253
Rgps, 217
Rheumatoid arthritis (RA), 206, 207
Rotaviruses
commercial vaccines, 62
description, 62
Gn pigs, 63
gnotobiotic calves, 64
HRV strains, 64
long-term protection, 63
mature, nondividing enterocytes, 63
murine rotavirus model, 64
pathogenesis and immunity, 63
R package
analysis, 305
data exploration, 309–315
functions, 308
getwd() function, 308
install.packages() command, 306
RStudio terminal, 307
sessionInfo(), 307
software, 304
RStudio, 304

S
Salmonella strains
AvrA protein, 81
axin-1 family and GSK-3β suppression, 83
description, 81
intestinal epithelial homeostasis, 82
in vivo, phosphorylated β-catenin (Ser552), 83
NF-κB pathway, 82
polymer chain reaction array, 82
ribonucleic acid analysis, 82
ubiquitinated forms, 82
Saturated fatty acids (SFAs), 227
Secretory immunoglobulin A (sIgA), 15
Serum antibodies, 201, 202
Shannon index, 318, 319
Short chain fatty acids (SCFAs), 13, 14, 40, 176, 177
acetate, 137, 138
anti-obesity properties, 136
butyrate, 139
^{13}C-labeled acetate, 137
GPR41 and GPR43, 136
propionate, 138
*Tlr5*KO mice, 137
vectorial ion transport, 104
Similarity percentage (SIMPER), 362–364
Simpson index, 319, 320
Small intestinal bacterial overgrowth (SIBO), 114, 144
antroduodenojejunal pressure tracings, 121
clinical manifestations, 120
defined, 120
delayed small intestinal transit time, 121
dysmotility, 121
gram-negative bacilli concentrations, 121
intestinal motor patterns, patients, 121
PPI users and non-users, 120
proton pump inhibitors, 120
reduced gastric acid production, 120
Sørensen index, 334
16S rRNA sequencing, 196
Stool-based therapies, 288, 289, 295
Streptococcus gallolyticus, 169, 170, 174
Systemic bacterial translocation, 223, 224

T
Tannerella forsythia, 220
Tauro-β-muricholic acid (T-βMCA), 141
Tight junctions (TJs), 8, 9
Toll-like receptors (TLRs), 9, 40, 168
adaptor molecules, 182
anti-tumor effect, 180
bacterial toxins, 179
CRC, 187, 188
dimerization, 179
intracellular compartment, 179
ligands, 179
signal transduction pathways, 179
sporadic and colitis-associated cancer models, 180
tissue microenvironment, 180
TLR1, 180, 181
TLR10, 182

TLR2, 181
TLR2/6, 182
TLR4, 181, 182
tumorigenesis, 180
type I interferons, 180
Treponema denticola, 219
Triggering receptor expressed on myeloid cells 1 (TREM1), 226
Trimethylamine-N-oxide (TMAO), 139, 140
Tukey's HSD method, 330, 331
Type three secretion system (TTSS), 80

U
Ulcerative colitis (UC), 280
Univariate community analysis
calculation
Chao 1 Index, 315, 316
function rbind(), 320
rarefaction richness, 317
Shannon index, 318, 319
Simpson index, 319, 320
Kruskal–Wallis rank sum test, 331
measures, 314–315
one-way ANOVA
abundance table, 324
BH method, 329
Bonferroni method, 328
boxplot() function, 326
data frame creation, 325
diversity() function, 325
dplyr package, 326
Holm method, 328
lm() and aov() functions, 327
pairwise.t.test(), 327
Tukey's HSD method, 330, 331
with() and interaction() functions, 324, 325
Welch's *t*-test, 323
Wilcoxon rank-sum test, 323
US Food and Drug Administration (FDA), 183

V
Vaginal candidiasis, 252, 255
Vaginal colonization, 254
Vectorial ion transport
CFTR and NKCCL1, 101–104
characterization, 105
fluid and electrolytes, 98
LGG, 105
NaCl absorption, 99–103
pro-absorptive effects, 99
SCFAs, 104
SGLT-1 and glucose transporter 2, 105
Vitamin D receptor (VDR), 304
Vulvovaginal candidiasis (VVC), 248

W
Welch's *t*-test, 323
White-to-opaque switch, 258, 259
Wilcoxon rank-sum test, 323
Wild-type (WT) mice, 304
Wnt/β-catenin signaling
colon chronically infection, 84, 85
destruction complex proteins, 81
effectors into host cells, 80
intestinal epithelial and mucosal homeostasis, 80
intestinal stem cells proliferation, 81 (*see also* Organoid system)
Salmonella infection, 80 (*see also Salmonella* strains)
TTSS/T3SS, 80

X
Xerostomia, 208

Y
Yersinia enterocolitica, 180

Printed by Printforce, the Netherlands